T0328915

Antibiotics and Antimicrobial Resistance Genes in the Environment

Advances in Environmental Pollution Research Series

Antibiotics and Antimicrobial Resistance Genes in the Environment

Volume I

Edited by

Muhammad Zaffar Hashmi

Elsevier
Radarweg 29, PO Box 211, 1000 AE Amsterdam, Netherlands
The Boulevard, Langford Lane, Kidlington, Oxford OX5 1GB, United Kingdom
50 Hampshire Street, 5th Floor, Cambridge, MA 02139, United States

Library of Congress Cataloging-in-Publication Data
A catalog record for this book is available from the Library of Congress

British Library Cataloguing-in-Publication Data
A catalogue record for this book is available from the British Library

ISBN: 978-0-12-818882-8

For information on all Elsevier publications visit our website at
https://www.elsevier.com/books-and-journals

Publisher: Candice Janco
Acquisition Editor: Marisa Lafleur
Editorial Project Manager: Laura Okidi
Production Project Manager: Vignesh Tamil
Cover Designer: Matthew Limbert

Typeset by TNQ Technologies

Contents

Contributors

Taha Arooj
Department of Botany, GC University, Lahore, Punjab, Pakistan

Asma Aftab
Research Centre for CO_2 Capture (RCCO2.C), Department of Chemical Engineering Universiti Teknologi PETRONAS Tronoh, Perak, Malaysia

Muhammad Afzaal
Sustainable Development Study Center (Env.Science) GC University, Lahore, Punjab, Pakistan

Ali Ahmad
Department of Basic Sciences, University of Veterinary and Animal Sciences, Narowal, Punjab, Pakistan

Fiaz Ahmad
Central Cotton Research Institute, Multan, Pakistan

Waqas Ahmad
Department of Clinical Sciences, University of Veterinary and Animal Sciences Lahore, Narowal Campus, Narowal, Pakistan

Sarfraz Ahmed
Department of Basic Siences, University of Veterinary and Animal Sciences, Narowal, Punjab, Pakistan

Iftikhar Ahmed
National Culture Collection of Pakistan (NCCP), Bioresource Conservation Institute (BCI), National Agricultural Research Centre (NARC), Islamabad, Pakistan

Noor Ul Ain
Department of Chemistry, University of Gujrat, Gujrat, Pakistan

Muhammad Sajid Hamid Akash
Department of Pharmaceutical Chemistry, Government College University Faisalabad, Faisalabad, Pakistan

Qaisar Akram
Department of Basic Sciences, University of Veterinary and Animal Sciences, Narowal, Punjab, Pakistan

Rizwan Ali
Centers for Biomedical Engineering, University of Science and Technology of China, Hefei, China

Jafar Ali
Key Laboratory of Environmental Nanotechnology and Health Effects, Research Center for Eco-Environmental Sciences, Chinese Academy of Sciences, Beijing, China

Muhammad Ishtiaq Ali
Environmental Microbiology Laboratory, Department of Microbiology, Quaid-i-Azam University, Islamabad, Pakistan

Zeshan Ali
Department of Biotechnology, University of Sialkot, Sialkot, Pakistan

Safdar Ali Mirza
Department of Botany, GC University, Lahore, Punjab, Pakistan

Muhammad Sulman Ali Taseer
Department of Basic Sciences, University of Veterinary and Animal Sciences, Narowal, Punjab, Pakistan

Muniza Almas
Department of Botany, GC University, Lahore, Punjab, Pakistan

Arshia Amin
Department of Bioinformatics and Biosciences, Capital University of Science and Technology, Islamabad, Pakistan

Mehroze Amin
Institute of Biochemistry and Biotechnology, University of Punjab, Lahore, Punjab, Pakistan

Saadia Andleeb
Atta-ur-Rahman School of Applied Biosciences (ASAB), National University of Sciences and Technology (NUST), Islamabad, Pakistan

Muzammil Anjum
Department of Environmental Sciences, Pir Mehr Ali Shah Arid Agriculture University, Rawalpindi, Pakistan; School of Materials Science and Engineering, Sun Yat-sen University, Guangzhou, China

Wajiha Anwar
Department of Biochemistry, Bahauddin Zakariya University, Multan, Pakistan

Muhammad Arshad
Institute of Environmental Sciences and Engineering (IESE), School of Civil and Environmental Engineering (SCEE), National University of Sciences and Technology (NUST), Islamabad, Pakistan

Muhammad Ashfaq
Department of Chemistry, University of Gujrat, Gujrat, Pakistan

Hajra Ashraf
Department of Biotechnology, Quaid-e-Azam University, Islamabad, Punjab, Pakistan

Basit Ateeq
Department of Pharmacy, University of Agriculture, Faisalabad, Pakistan

Muhammad Umer Farooq Awan
Department of Botany, GC University, Lahore, Punjab, Pakistan

B. Balabanova
Faculty of Agriculture, University “Goce Delčev”, Štip, Republic of North Macedonia

Isam Bashour
Department of Agriculture, Faculty of Agricultural and Food Sciences, AUB, Beirut, Lebanon

Tahira Batool
Institute of Biochemistry and Biotechnology, The University of Punjab, Pakistan

Sajida Begum
Department of Botany, GC University, Lahore, Punjab, Pakistan

Syeda Aniqa Bukhari
Department of Biochemistry and Biotechnology, University of Gujrat, Gujrat, Punjab, Pakistan

Zoma Chaudhry
Department of Biochemistry and Biotechnology, University of Gujrat, Gujrat, Punjab, Pakistan

Tahir Ali Chohan
Institute of Pharmaceutical Sciences, University of Veterinary and Animal Sciences, Lahore, Pakistan

Surojeet Das
Faculty of Biotechnology, Institute of Bio-Sciences and Technology, Shri Ramswaroop Memorial University, Barabanki, Uttar Pradesh, India

Erum Dilshad
Department of Bioinformatics and Biosciences, Capital University of Science and Technology, Islamabad, Pakistan

Lara El-Gemayel
Department of Agriculture, Faculty of Agricultural and Food Sciences, AUB, Beirut, Lebanon

Shahid Hussain Farooqi
Department of Clinical Sciences, University of Veterinary and Animal Sciences Lahore, Narowal Campus, Narowal, Pakistan

Fareeha Fiayyaz
Department of Pharmaceutical Chemistry, Government College University Faisalabad, Faisalabad, Pakistan; Department of Microbiology, Government College University Faisalabad, Faisalabad, Pakistan

Marium Fiaz
Institute of Environmental Sciences and Engineering (IESE), School of Civil and Environmental Engineering (SCEE), National University of Sciences and Technology (NUST), Islamabad, Pakistan

Sahrish Habib
Department of Biotechnology, University of Sialkot, Sialkot, Pakistan

Muhammad Zaffar Hashmi
Department of Chemistry, COMSATS University Islamabad, Pakistan

Jouman Hassan
Department of Nutrition and Food Sciences, Faculty of Agricultural and Food Sciences, American University of Beirut (AUB), Beirut, Lebanon

Munawar Hussain
Department of Chemistry, Khwaja Fareed University of Engineering & Information Technology, Rahim Yar Khan, Pakistan

Muhammad Ibrahim
Department of Biochemistry, Bahauddin Zakariya University, Multan, Pakistan

Gilberto Igrejas
MicroART- Microbiology and Antibiotic Resistance Team, University of Trás-os-Montes and Alto Douro, Quinta de Prados Vila Real, Portugal; Associated Laboratory for Green Chemistry (LAQV-REQUIMTE), University NOVA of Lisboa, Lisboa, Caparica, Portugal; Department of Genetics and Biotechnology, Functional Genomics and Proteomics' Unit, University of Trás-os-Montes and Alto Douro, Vila Real, Portugal; Functional Genomics and Proteomics Unit, University of Tras-os-Montes and Alto Douro (UTAD), Vila Real, Portugal

Ayesha Imran
Department of Biochemistry, Bahauddin Zakariya University, Multan, Pakistan

Muhammad Javed Iqbal
Department of Biotechnology, University of Sialkot, Sialkot, Pakistan; Department of Biochemistry and Biotechnology, University of Gujrat, Gujrat, Pakistan

Komal Jabeen
Department of Pharmacy, University of Agriculture, Faisalabad, Pakistan; Institute of Physiology and Pharmacology, University of Agriculture, Faisalabad, Pakistan

Deeba Javed
Department of Chemistry, University of Gujrat, Gujrat, Pakistan

Ayesha Kabeer
Department of Biotechnology, University of Sialkot, Sialkot, Pakistan

Ghulam Mustafa Kamal
Department of Chemistry, Khwaja Fareed University of Engineering & Information Technology, Rahim Yar Khan, Pakistan

Saira Hafeez Kamran
Institute of Pharmacy, Gulab Devi Educational Complex, Pakistan

Issmat I. Kassem
Department of Nutrition and Food Sciences, Faculty of Agricultural and Food Sciences, American University of Beirut (AUB), Beirut, Lebanon; Center for Food Safety, Department of Food Science and Technology, University of Georgia, GA, United States

Srujana Kathi
Guest Faculty, Department of Ecology and Environmental Sciences, Pondicherry University, Puducherry, India

Muhammad Khalid
Department of Chemistry, Khwaja Fareed University of Engineering & Information Technology, Rahim Yar Khan, Pakistan

Mohsin Khurshid
Department of Microbiology, Government College University Faisalabad, Faisalabad, Pakistan

Sunil Kumar
Faculty of Biotechnology, Institute of Bio-Sciences and Technology, Shri Ramswaroop Memorial University, Barabanki, Uttar Pradesh, India

Iram Liaqat
Department of Zoology, GC University, Lahore, Pakistan

Mahnoor Majid
Atta-ur-Rahman School of Applied Biosciences (ASAB), National University of Sciences and Technology (NUST), Islamabad, Pakistan

Bushra Manzoor
Institute of Biochemistry and Biotechnology, The University of Punjab, Pakistan

Iqra Mazhar
Sustainable development study center GC University Lahore, Punjab, Pakistan

Bisma Meer
Department of Biotechnology, Quaid-e-Azam University, Islamabad, Punjab, Pakistan

Sajid Mehmood
Department of Biochemistry and Biotechnology, University of Gujrat, Gujrat, Pakistan

Arooj Mumtaz
Department of Chemistry, University of Gujrat, Gujrat, Pakistan

Muhammad Naveed
Department of Biotechnology, University of Central Punjab, Lahore, Punjab, Pakistan

Sania Niaz
Department of Pharmacy, University of Agriculture, Faisalabad, Pakistan; Institute of Physiology and Pharmacology, University of Agriculture, Faisalabad, Pakistan

Waqar Pervaiz
Department of Pharmacy, University of Agriculture, Faisalabad, Pakistan

Patrícia Poeta
Veterinary Sciences Department, University of Trás-os-Montes and Alto Douro, Quinta de Prados Vila Real, Portugal; MicroART- Microbiology and Antibiotic Resistance Team, University of Trás-os-Montes and Alto Douro, Quinta de Prados Vila Real, Portugal; Associated Laboratory for Green Chemistry (LAQV-REQUIMTE), University NOVA of Lisboa, Lisboa, Caparica, Portugal

Hafsa Raja
Department of Bioinformatics and Biosciences, Capital University of Science and Technology, Islamabad, Pakistan

Ayesha Ramzan
Department of Biochemistry, Bahauddin Zakariya University, Multan, Pakistan

Hafsa Anwar Rana
Department of Biochemistry, Bahauddin Zakariya University, Multan, Pakistan

Tazeen Rao
Department of Biochemistry, Bahauddin Zakariya University, Multan, Pakistan

Umer Rashid
Department of Biochemistry and Biotechnology, University of Gujrat, Gujrat, Pakistan

Kanwal Rehman
Department of Pharmacy, University of Agriculture, Faisalabad, Pakistan

Muhammad Saif Ur Rehman
Department of Chemical Engineering, Khawaja Fareed University of Engineering & Information Technology, Rahim Yar Khan, Pakistan

Luqman Riaz
College of Life Sciences, Henan Normal University, Xinxiang, China; School of Materials Science and Engineering, Sun Yat-sen University, Guangzhou, China

Shakila Sabir
Department of Pharmaceutical Chemistry, Government College University Faisalabad, Faisalabad, Pakistan; Department of Pharmacology, Government College University Faisalabad, Faisalabad, Pakistan

Rabia Safeer
Department of Environmental Sciences, Pir Mehr Ali Shah Arid Agriculture University, Rawalpindi, Pakistan

Saima Saima
Institute of Environmental Sciences and Engineering (IESE), School of Civil and Environmental Engineering (SCEE), National University of Sciences and Technology (NUST), Islamabad, Pakistan

Hamza Saleem ur Rehman
Department of Biotechnology, University of Sialkot, Sialkot, Pakistan

Sumbal Sardar
Atta-ur-Rahman School of Applied Biosciences (ASAB), National University of Sciences and Technology (NUST), Islamabad, Pakistan

Asfandyar Shahab
College of Environmental Science and Engineering, Guilin University of Technology, Guilin, China

Sana Shifaqat
Department of Chemistry, University of Gujrat, Gujrat, Pakistan

Anila Sikandar
Department of Environmental Sciences, Pir Mehr Ali Shah Arid Agriculture University, Rawalpindi, Pakistan

Adriana Silva
Veterinary Sciences Department, University of Trás-os-Montes and Alto Douro, Quinta de Prados Vila Real, Portugal; MicroART- Microbiology and Antibiotic Resistance Team, University of Trás-os-Montes and Alto Douro, Quinta de Prados Vila Real, Portugal; Associated Laboratory for Green Chemistry (LAQV-REQUIMTE), University NOVA of Lisboa, Lisboa, Caparica, Portugal

Vanessa Silva
Veterinary Sciences Department, University of Trás-os-Montes and Alto Douro, Quinta de Prados Vila Real, Portugal; MicroART- Microbiology and Antibiotic Resistance Team, University of Trás-os-Montes and Alto Douro, Quinta de Prados Vila Real, Portugal; Associated Laboratory for Green Chemistry (LAQV-REQUIMTE), University NOVA of Lisboa, Lisboa, Caparica, Portugal

Aashna Srivastava
Faculty of Biotechnology, Institute of Bio-Sciences and Technology, Shri Ramswaroop Memorial University, Barabanki, Uttar Pradesh, India

Ayesha Tahir
Department of Pharmacy, University of Agriculture, Faisalabad, Pakistan; Institute of Physiology and Pharmacology, University of Agriculture, Faisalabad, Pakistan

Habib Ullah
CAS Key Laboratory of Crust Mantle Materials and the Environments, School of Earth and Space Sciences, University of Science and Technology of China, Hefei, China

Francis Victor
Department of Pharmacy, University of Agriculture, Faisalabad, Pakistan

Qianqian Wang
College of Life Sciences, Henan Normal University, Xinxiang, China; Henan International Joint Laboratory of Agricultural Microbial Ecology and Technology (Henan Normal University), Xinxiang, China

Hassan Waseem
Environmental Microbiology Laboratory, Department of Microbiology, Quaid-i-Azam University, Islamabad, Pakistan; Department of Biotechnology, University of Sialkot, Sialkot, Pakistan

Qingxiang Yang
College of Life Sciences, Henan Normal University, Xinxiang, China; Henan International Joint Laboratory of Agricultural Microbial Ecology and Technology (Henan Normal University), Xinxiang, China

Bushra Yaqub
Department of Biochemistry, Bahauddin Zakariya University, Multan, Pakistan

Muhammad Younus
Department of Basic Sciences, University of Veterinary and Animal Sciences, Narowal, Punjab, Pakistan

Wei Yuan
School of Environmental and Municipal Engineering, North China University of Water Resources and Electric Power, Zhengzhou, China

Rabeea Zafar
Institute of Environmental Sciences and Engineering (IESE), School of Civil and Environmental Engineering (SCEE), National University of Sciences and Technology (NUST), Islamabad, Pakistan; Department of Environmental Design, Health & Nutritional Sciences, Faculty of Sciences, Allama Iqbal Open University, Islamabad, Pakistan

Tehseen Zahra
Department of Bioinformatics and Biosciences, Capital University of Science and Technology, Islamabad, Pakistan

Acknowledgment

Special thanks to the Higher Education Commission of Pakistan NRPU projects 7954 and 7964. Further thanks to the Pakistan Science Foundation project PSF/Res/CP/C-CUI/Envr (151).

CHAPTER

1 Microorganisms and antibiotic production

Kanwal Rehman[1], Sania Niaz[1,2], Ayesha Tahir[1,2], Muhammad Sajid Hamid Akash[3]

[1]*Department of Pharmacy, University of Agriculture, Faisalabad, Pakistan;* [2]*Institute of Physiology and Pharmacology, University of Agriculture, Faisalabad, Pakistan;* [3]*Department of Pharmaceutical Chemistry, Government College University Faisalabad, Faisalabad, Pakistan*

1.1 Introduction

Microorganisms are organisms or infectious agents of microscopic or submicroscopic size, which include bacteria, fungi, protozoans, and viruses. For the treatment of infections, antimicrobial drugs are valuable due to selectivity of their toxicity, thereby having capability to kill the invading microorganisms without harming the host cells. Antimicrobial medicines can be classified according to their action against the microorganisms. For example, antibiotics are used against bacteria, whereas antifungals are specifically used against fungi. The term probiotic was introduced by Lilly and Stillwell (Lilly and Stillwell, 1965).

1.2 Probiotics

The use of probiotics for their health benefits is increasing worldwide (Agheyisi, 2005). The word *probiotic* is derived from the Greek word meaning for life and has had several different meanings over the years. Improving the host health by consumption of live microorganisms provides a basic concept of a probiotic. A probiotic can be defined as microorganism introduced into the body in sufficient quantity for its beneficial qualities into the host. Gut health or microflora can be improved by the utilization of typical microorganisms that are present in fermented products (Hill et al., 2014; Ndowa et al., 2012). According to the mechanistic approach, disorder or imbalance of important intestinal microflora leads to many gastrointestinal infirmity or infections. Probiotics are viable microbial cultures that maintain or balance the microflora of intestine, correct the microbial dysfunction, and enhance the host health and well-being (Fuller, 1989; Rokka and Rantamäki, 2010). Two of the most common microbes that are widely used as probiotics are *Lactobacillus* and *Bifidobacteria* strains. Growth of the concerned microorganism is stimulated by using the bacterial culture of probiotics, which improves the natural defensive mechanism of the body and also disrupts the harmful bacteria (Dunne, 2001).

Probiotics have shown a curative role against cancer, and they also have been shown to reduce cholesterol levels, modify lactose intolerance, and enhance immunity (Kailasapathy and Chin, 2000). As probiotics boost immunity, they provide beneficial health effects by the stimulation of cell-

Antibiotics and Antimicrobial Resistance Genes in the Environment. https://doi.org/10.1016/B978-0-12-818882-8.00001-2

mediated immune responses as well as enhance the antibody secretions. Probiotics are selected according to the protection point of view against microbial pathogens (Cross, 2002) and also play a vital role in maintaining the overweight of an obese adult (Kadooka et al., 2010).

1.3 Prebiotics

Prebiotic concepts were introduced in 1995 by Gibson and Roberfroid as a substitute approach to alter or modify the microbiota of the gut (Gibson and Roberfroid, 1995). A prebiotic is a nondigestible food ingredient, usually *bifidobacteria* and *lactobacilli*, that beneficially affects the host by enhancing the growth and/or activity of one or a limited number of specific species of bacteria in the gut, thus strengthening the host health. They are indigestible by human enzymes because they have short-chain carbohydrates (SCCs), so-called resistant SCCs (Quigley et al., 1999). To be considered as a prebiotic, a food ingredient must have specific properties. For example, (1) it should be resistant by passing the upper portion of gastrointestinal track for the absorption and hydrolysis; (2) it should provide a favorable environment by modifying the microflora of the colon and provide more healthy and favorable composition there; and (3) it should show specific property of selective substrate for one or a specific amount of colon bacteria (Park and Kroll, 1993). Hence there are numerous potential applications of prebiotics.

Prebiotics should be resistant to being hydrolyzed by intestinal enzymes of the human but should be fermented by specific bacteria and should have fruitful effects for the host. Upon administration, prebiotics should have beneficial outcomes including lowering the permeability of intestine, decreasing triglyceride levels, and improving glucose levels after eating (Cani et al., 2009; Gibson and Roberfroid, 1995). Prebiotics are widely used as a supplement and can be formulated in various ways such as syrups or powder and also into different food products, particularly in bread and yogurt, that provide beneficial health effects by enhancing the minerals' bioavailability (Roberfroid et al., 2010). They have also been recommended for improved bone and mineral metabolism.

1.4 Symbiotics

It has been suggested that symbiotics are the combination of probiotics and prebiotics, not only comprising the combined effects of these two probiotics and prebiotics but also purposed to have a synergistic effect (Rafter et al., 2007).

1.5 Antibiotics

Many of the antibiotics are the essential excretions of environmental bacteria and fungi. At present, these antibiotics are used as a major source of human medicines for the treatment of infections (Kieser et al., 2000).

1.5.1 Classification of antibiotics

The most important classification of antibiotics is based on their spectrum, mode of action, and molecular structure. There are certain ways to classify antibiotics (Calderón and Sabundayo, 2007), notably, one method is based on their route of administration such as topically, orally, or as an injectable. Other antibiotics that are related to the same structural class will show analogous patterning of efficiency, allergic side effects, and toxicity. Some common classes of antibiotics like macrolides, quinolones, tetracyclines, aminoglycosides, sulfonamides, oxazolidinones, glycopeptides, and beta-lactam are based on their molecular and chemical structures (Adzitey, 2015; Frank and Tacconelli, 2012; Van Hoek et al., 2011). For many years, antibiotics have proven efficacious in providing a curative response for many contagious diseases. Antibiotics include composites that hinder the growth of microorganisms, which are considered as "antimicrobial agents." Several natural antibiotics can also be used in the treatment of numerous diseases.

1.5.2 Mechanisms of antibiotic resistance

Antibiotic resistance came into existence between 1940 and 1970. There are several ways for the development of antibiotic resistance that are described in the following subsections.

1.5.3 Enzymatic inactivation

In enzymatic inactivation, the primitive enzyme undergoes modification by reacting with the antibiotic and then the antibiotic cannot kill the microorganism. The most common example is β-lactamase enzymes which causes hydrolysis of antibiotics and ultimately leads to antibiotic resistance against penicillins and cephalosporins.

1.5.4 Drug elimination

In *Pseudomonas aeruginosa* and *Acinetobacter* species, the most important resistance mechanism is drug elimination due to the excitation of efflux pump. Bacteria activate the proteins that cause the removal of compounds from periplasm to outside of the cell to remove the antibiotics.

1.5.5 Permeability changes

Due to the alterations in outer membrane portability, there is a decrease in uptake of administered antibiotics, due to which the adequate access to the antibiotics is blocked.

1.6 Modifications of antimicrobial targets

Three different types of antibiotic adjuvants have been invented that can be used to block the antibiotic resistance mechanisms. These may include the (1) inhibitors of β-lactamases, (2) inhibitors of efflux pump, and (3) permeabilization of outer membrane (Clatworthy et al., 2007; Rasko and Sperandio, 2010). The World Health Organization has recommended an antimicrobial resistance control policy that includes increased supervision, development of new molecules, and rational use of antibiotics.

1.7 Production of antibiotics

Most antibiotics are produced by staged fermentations in which strains of microorganisms producing high yields are grown under optimum conditions. It is important that the organism that is used for the production of antibiotic must be identified and isolated. The microorganism must be grown enough for the purification and chemical analysis of the isolated antibiotics. Sterile conditions must be followed during the purification and isolation of antibiotics because contamination by foreign microbes may ruin the fermentation of the antibiotics. Following are the most commonly used techniques for the production of antibiotics.

1.7.1 Natural production of antibiotics

In natural production, fermentation technique is used for the production of antibiotics. The most common example of an antibiotic produced by this method is penicillin.

1.7.2 Semisynthetic production of antibiotics

This method is used for the production of natural antibiotics, for example, ampicillin.

1.7.3 Synthetic production of antibiotics

This method is used for the production of antibiotics in a laboratory. For example, the production of quinoline is done by this method.

1.7.4 Industrial production of antibiotics

In this technique, the source microorganism is grown in large containers containing a liquid growth medium. In this technique, the oxygen concentration, temperature, pH, and nutrient levels must be optimum. As the antibiotics are secondary metabolites, their production must be controlled to ensure that the maximum yield of antibiotics is obtained before the cells die.

1.7.5 Methods for increased production of antibiotics

Species for the production of specific antibiotics are often genetically modified to yield the maximum amounts of antibiotics. Mutations and gene amplification techniques are used to increase the production of antibiotics.

1.8 Stability of antimicrobial agents

According to several research studies, many kinds of encapsulation procedures and materials are used for microencapsulation and coating of antibiotics (Hébrard et al., 2010; Nag et al., 2011; Papagianni and Anastasiadou, 2009). To preserve the antibiotics from the unpleasant conditions in the intestinal tract, microencapsulation technique is widely used (Anal and Singh, 2007; Kailasapathy, 2002).

Microencapsulation technique plays an important role by separating the core material from environmental conditions until it gets released, thereby modifying stability and viability and improving shelf life and helping to provide the controlled and sustained release of encapsulated products. The outer structure is formed by microencapsulation technique around the core. This property provides a core with characteristics of controlled-release product under favorable environmental conditions and also provides a way for small molecules to pass out of and into the membrane. At the time of release of encapsulated core material at the favorable site, it follows different mechanistic approaches including dissolution of the cell wall, melting of the cell wall, diffusion through the wall, and breakage of the cell wall (F. Gibbs, 1999; Franjione and Vasishtha, 1995).

1.9 Conclusion

For the better efficacy of antimicrobial agents against microorganisms, efficient methods should be chosen for the production and purification of antimicrobial agents. As the stability of antimicrobial agents is a major concern, it is mandatory that appropriate technique should be adopted for the encapsulation of antimicrobial agents.

References

Adzitey, F., 2015. Antibiotic Classes and Antibiotic Susceptibility of Bacterial Isolates From Selected Poultry; a Mini Review.

Agheyisi, R., 2005. Ga-121 Probiotics: Ingredients, Supplements, Foods. Publisher BCC Research Inc.

Anal, A.K., Singh, H., 2007. Recent advances in microencapsulation of probiotics for industrial applications and targeted delivery. Trends in Food Science & Technology 18, 240–251.

Calderón, C.B., Sabundayo, B.P., 2007. Antimicrobial classifications: drugs for bugs. In: Schwalbe, R., Steele-Moore, L., Goodwin, A.C. (Eds.), Antimicrobial Susceptibility Testing Protocols. CRC Press. Taylor & Frances Group.

Cani, P.D., Lecourt, E., Dewulf, E.M., Sohet, F.M., Pachikian, B.D., Naslain, D., De Backer, F., Neyrinck, A.M., Delzenne, N.M., 2009. Gut microbiota fermentation of prebiotics increases satietogenic and incretin gut peptide production with consequences for appetite sensation and glucose response after a meal. The American Journal of Clinical Nutrition 90, 1236–1243.

Clatworthy, A.E., Pierson, E., Hung, D.T., 2007. Targeting virulence: a new paradigm for antimicrobial therapy. Nature Chemical Biology 3, 541.

Cross, M.L., 2002. Microbes versus microbes: immune signals generated by probiotic lactobacilli and their role in protection against microbial pathogens. FEMS Immunology and Medical Microbiology 34, 245–253.

Dunne, C., 2001. Adaptation of bacteria to the intestinal niche: probiotics and gut disorder. Inflammatory Bowel Diseases 7, 136–145.

Franjione, J., Vasishtha, N., 1995. The art and science of microencapsulation. Technology Today 2, 1–6.

Frank, U., Tacconelli, E., 2012. The Daschner Guide to In-Hospital Antibiotic Therapy: European Standards. Springer Science & Business Media.

Fuller, R., 1989. Probiotics in man and animals. Journal of Applied Bacteriology 66, 365–378.

Gibbs, B.F., Kermasha, S., Alli, I., Mulligan, C.N., 1999. Encapsulation in the food industry: a review. International Journal of Food Sciences and Nutrition 50, 213–224.

Gibson, G.R., Roberfroid, M.B., 1995. Dietary modulation of the human colonic microbiota: introducing the concept of prebiotics. Journal of Nutrition 125, 1401–1412.

Hébrard, G., Hoffart, V., Beyssac, E., Cardot, J.-M., Alric, M., Subirade, M., 2010. Coated whey protein/alginate microparticles as oral controlled delivery systems for probiotic yeast. Journal of Microencapsulation 27, 292–302.

Hill, C., Guarner, F., Reid, G., Gibson, G.R., Merenstein, D.J., Pot, B., Morelli, L., Canani, R.B., Flint, H.J., Salminen, S., 2014. Expert consensus document: the International Scientific Association for Probiotics and Prebiotics consensus statement on the scope and appropriate use of the term probiotic. Nature Reviews Gastroenterology & Hepatology 11, 506.

Kadooka, Y., Sato, M., Imaizumi, K., Ogawa, A., Ikuyama, K., Akai, Y., Okano, M., Kagoshima, M., Tsuchida, T., 2010. Regulation of abdominal adiposity by probiotics (Lactobacillus gasseri SBT2055) in adults with obese tendencies in a randomized controlled trial. European Journal of Clinical Nutrition 64, 636.

Kailasapathy, K., 2002. Microencapsulation of probiotic bacteria: technology and potential applications. Current Issues in Intestinal Microbiology 3, 39−48.

Kailasapathy, K., Chin, J., 2000. Survival and therapeutic potential of probiotic organisms with reference to Lactobacillus acidophilus and Bifidobacterium spp. Immunology & Cell Biology 78, 80−88.

Kieser, T., Bibb, M.J., Buttner, M.J., Chater, K.F., Hopwood, D.A., 2000. Practical Streptomyces Genetics. John Innes Foundation, Norwich.

Lilly, D.M., Stillwell, R.H., 1965. Probiotics: growth-promoting factors produced by microorganisms. Science 147, 747−748.

Nag, A., Han, K.-S., Singh, H., 2011. Microencapsulation of probiotic bacteria using pH-induced gelation of sodium caseinate and gellan gum. International Dairy Journal 21, 247−253.

Ndowa, F., Lusti-Narasimhan, M., Unemo, M., 2012. The Serious Threat of Multidrug-Resistant and Untreatable Gonorrhoea: The Pressing Need for Global Action to Control the Spread of Antimicrobial Resistance, and Mitigate the Impact on Sexual and Reproductive Health. BMJ Publishing Group Ltd.

Papagianni, M., Anastasiadou, S., 2009. Encapsulation of Pediococcus acidilactici cells in corn and olive oil microcapsules emulsified by peptides and stabilized with xanthan in oil-in-water emulsions: studies on cell viability under gastro-intestinal simulating conditions. Enzyme and Microbial Technology 45, 514−522.

Park, S.F., Kroll, R.G., 1993. Expression of listeriolysin and phosphatidylinositol-specific phospholipase C is repressed by the plant-derived molecule cellobiose in Listeria monocytogenes. Molecular Microbiology 8, 653−661.

Quigley, M.E., Hudson, G.J., Englyst, H.N., 1999. Determination of resistant short-chain carbohydrates (non-digestible oligosaccharides) using gas−liquid chromatography. Food Chemistry 65, 381−390.

Rafter, J., Bennett, M., Caderni, G., Clune, Y., Hughes, R., Karlsson, P.C., Klinder, A., O'Riordan, M., O'Sullivan, G.C., Pool-Zobel, B., 2007. Dietary synbiotics reduce cancer risk factors in polypectomized and colon cancer patients. The American journal of clinical nutrition 85, 488−496.

Rasko, D.A., Sperandio, V., 2010. Anti-virulence strategies to combat bacteria-mediated disease. Nature Reviews Drug Discovery 9, 117.

Roberfroid, M., Gibson, G.R., Hoyles, L., McCartney, A.L., Rastall, R., Rowland, I., Wolvers, D., Watzl, B., Szajewska, H., Stahl, B., 2010. Prebiotic effects: metabolic and health benefits. British Journal of Nutrition 104, S1−S63.

Rokka, S., Rantamäki, P., 2010. Protecting probiotic bacteria by microencapsulation: challenges for industrial applications. European Food Research and Technology 231, 1−12.

Van Hoek, A.H., Mevius, D., Guerra, B., Mullany, P., Roberts, A.P., Aarts, H.J., 2011. Acquired antibiotic resistance genes: an overview. Frontiers in Microbiology 2, 203.

CHAPTER

2 Antibiotics and antimicrobial resistance: temporal and global trends in the environment

Kanwal Rehman[1], Fareeha Fiayyaz[2,3], Mohsin Khurshid[3], Shakila Sabir[2,4], Muhammad Sajid Hamid Akash[2]

[1]*Department of Pharmacy, University of Agriculture, Faisalabad, Pakistan;* [2]*Department of Pharmaceutical Chemistry, Government College University Faisalabad, Faisalabad, Pakistan;* [3]*Department of Microbiology, Government College University Faisalabad, Faisalabad, Pakistan;* [4]*Department of Pharmacology, Government College University Faisalabad, Faisalabad, Pakistan*

2.1 Introduction

Antibiotics are a backbone in the management and control of infectious diseases that are caused by microorganisms like bacteria, and at present, the major worldwide problem that we are facing is antibiotic resistance and increase in resistant bacterial species (Dunne et al., 2000; Kollef and Fraser, 2001; Reacher et al., 2000; Tenover and Hughes, 1996). One of the essential and important roles played by antimicrobial drugs is in decreasing infectious diseases and deaths. Selective pressure is one of the major driving forces behind the emergence and spread of drug-resistance traits among pathogenic and commensal bacteria that is exerted by the use of antimicrobial drugs (Aarestrup et al., 2008). A major risk factor for the emergence of resistance against antibiotics is the use of frequent and inappropriate antibiotics (De Man et al., 2000). Other factors that lead toward antimicrobial resistance are poor hygienic conditions, antibiotics used in food animals (Singer et al., 2003), and lifestyle such as overcrowded living conditions (Bartoloni et al., 2004; Lester et al., 1990).

2.2 Antimicrobial resistance

A worldwide challenge in pathogenic bacteria is antimicrobial resistance that leads to a high rate of morbidity and mortality (Akova, 2016). It is very difficult to treat infections or even those that are untreatable with conventional antimicrobials due to multidrug-resistant patterns that are present in both gram-positive and gram-negative bacteria. In many hospitals, antibiotics having broad spectrum are freely and mostly needlessly used, and this is due to lack of proper early identification of the causative agent and their antibiotic susceptibility patterns in patients with septicemia and the same kind of other severe infections (Akova, 2016). When combined with poor infection control practices, a dramatic increase in emerging resistance occurs, and bacteria that become resistant can be easily spread to the other individuals as well as the environment (Akova, 2016). Reorganized

Antibiotics and Antimicrobial Resistance Genes in the Environment. https://doi.org/10.1016/B978-0-12-818882-8.00002-4

epidemiological data on antimicrobial resistance about frequently encountered pathogenic bacteria is available. This data will be helpful not only for determining management policies but also for developing an operative stewardship program related to antimicrobial use in hospitals (Akova, 2016). Many important pathogenic bacteria are resistant to commonly used antimicrobial therapies. The emergence of bacteria that exhibit multidrug resistance is increasing at an alarming rate, and now antimicrobial resistance has become a global threat.

2.2.1 Escherichia coli

In humans and animals, *Escherichia coli* is usually considered as a commensal bacterium. Pathogenic strains are responsible for causing infections that are related to both intestinal and extraintestinal regions of the body, including peritonitis, gastroenteritis, meningitis, urinary tract infection (UTI), and septicemia (Control and Prevention, 2002; Von Baum and Marre, 2005). It is difficult to treat UTIs caused by *E. coli* due to the emergence of resistance of antibiotics against this organism (Giske et al., 2008). Resistance developed due to several mechanisms, such as inactivation of antibiotics due to enzymes, altered target sites, decreased permeability due to the presence of porins known in bacteria, especially gram-negative bacteria, and active efflux pump (Rao et al., 2014). Production of extended-spectrum beta-lactamase (ESBL) enzymes is one of the most common resistance mechanisms that can hydrolyze all penicillins, oximino-cephalosporins, cephalosporins, and monobactams. But ESBL cannot hydrolyze carbapenems or cephamycins (Bonnet, 2004; Eckert et al., 2005; Minond et al., 2011). In the Enterobacteriaceae family, approximately 400 types of enzymes are currently known (Feizabadi et al., 2010). The CTX-M beta-lactamase types have been increasing in various countries and now have become the most prevalent enzymes (Ruppe et al., 2009). These enzymes show susceptibility to inhibitors, such as sulbactam, tazobactam, and clavulanic acid (Cantón et al., 2012; Naseer and Sundsfjord, 2011). The occurrence of ESBL in pathogenic strains continues to be linked with higher health care costs and mortality (Paterson and Bonomo, 2005). UTIs caused by ESBL produce *E. coli* strains. The long-term and misuse of cephalosporin resulted in the incidence of UTIs and their prevalence is on the rise. Now a serious problem related to public health has become global is the prevalence of CTX-M beta-lactamase in commonly isolated organisms, such as *E. coli*. Currently, CTX-M-beta-lactamases are available as they are encoded in a plasmid and can hydrolyze both cefotaxime and ceftazidime. But there is a high-resistance level against cefotaxime and their level of activity is also low against ceftazidime (Edelstein et al., 2003; Tzouvelekis et al., 2000). These periplasmic enzymes were described for the first time in the late 1980s (Bonnet, 2004).

2.2.1.1 Temporal trends

For antimicrobial agents that have been in use in the human and veterinary medicine for the treatment of infectious diseases, according to surveillance data the highest level of resistance is consistently observed against *E. coli* (Walusansa, 2017). With the passage of time, microorganisms are becoming resistant to newer compounds such as certain cephalosporins and fluoroquinolones (Levy and Marshall, 2004). For example, during a 12-year period of study that was performed to check the susceptibility of *E. coli* isolates isolated from hospitals (1971–82), there was no major change in resistance presented by isolates of *E. coli* to any of the antimicrobial drugs that were tested (Atkinson and Lorian, 1984). But according to a retrospective analysis of *E. coli* isolates that was done during 1997–2007, this time period displayed the highest trends of resistance for amoxicillin/clavulanic acid, ciprofloxacin, and trimethoprim/sulfamethoxazole (Blaettler et al., 2009). Similarly, a 30-year (1979–2009) follow-up study done in Sweden on *E. coli* showed the highest rate of resistance for gentamicin, ampicillin, trimethoprim, and sulfonamide (Kronvall, 2010).

In the United States, a monitoring system was established in 1996 called the National Antimicrobial Resistance Monitoring System (NARMS); the purpose of this system is to monitor variations that come in susceptibilities of zoonotic foodborne bacteria against antimicrobial drug, including *E. coli* from trade meats (ground turkey, pork chops, chicken breast, ground beef), and chickens at butchery or slaughter. To determine the minimum inhibitory concentration to antimicrobial drugs essential in human and veterinary medicine, NARMS laboratories tested 13,521 isolates of *E. coli* from chicken during 2000–2008 (Tadesse et al., 2012). The isolates of *E. coli* collected from human presented an increasing rate of resistance only to a specific antimicrobial drug such as tetracycline (0.45% per year), ampicillin (0.59% per year), and sulfonamide (0.49% per year). This trend in resistance fluctuated during the period of study 0%–58% for tetracycline, from 0% to 66.7% for ampicillin, and 0%–50% for sulfonamide. There is no case of resistance reported against ciprofloxacin by the human *E. coli* isolates (Tadesse et al., 2012). The higher rates of ESBL-positive *E. coli* isolates showed that rates of *E. coli* rose during the last 2 years of the study period, reaching 12.3% in 2007 and 14.0% in 2008, after declining slightly in 2004, 2005, and 2006. From 4.0% in 2002 to 7.4% in 2007 and 6.5% in 2008, according to this data, it is noted that among community-associated (CA) IAIs the rates of ESBL-positive *E. coli* isolate remained relatively constant (Table 2.1; Hawser et al., 2010). From hospital-associated (HA) or CA infections, the percentages of *E. coli* isolates were reported as 57.9% and 38.5%. Another percentage of *E. coli* isolates taken from patients, about whom there is no information about the length of stay that they have during the phase of specimen collection, was 3.6%.

2.2.1.2 Global trends

At present *E. coli* is becoming a developing issue. Recently a surveillance study that was done in all areas of Japan tested 997 isolates of *E. coli* taken from the hospital. Their findings demonstrate that the

Table 2.1 Percentage of ESBL-positive *E. coli* from HA and CA IAIs from 2002 to 2008 in Europe.

	E. coli (HA only)			*E. coli* (CA only)		
Year of isolation	**% ESBL (+ve)**	**95% CI**	**No. of isolates**	**% ESBL (+ve)**	**95% CI**	**No. of isolates**
2002	4.8	2.2–9.7	146	4.00	1.9–7.8	200
2003	13.70	11.1–16.8	555	5.70	3.9–8.3	456
2004	11.80	9.7–14.3	754	7.00	5.1–9.5	530
2005	8.30	6.4–10.7	637	4.40	3.1–6.3	680
2006	7.60	5.8–9.9	660	5.50	4.1–7.4	776
2007	12.30	9.9–15.1	604	7.40	5.7–9.7	686
2008	14.00	11.9–16.5	863	6.50	4.7–8.8	574

Table 2.1 shows the number of isolates of E. coli *isolated from 2002 to 2008 from both hospital-associated (HA) and community-associated (CA) intra-abdominal isolates (IAIs). These isolates of* E. coli *are showing the positive percentage of extended-spectrum beta-lactamase (ESBL) with a 95% confidence interval.*

prevalence of resistance against cephalosporin, especially to third- and fourth-generation cephalosporins, was found to be less than 1.5% (Ishii et al., 2006). According to the study, about 1.3% isolates of *this* organism were established as major producers of ESBL. Out of 100,000 population, 5.5% cases per year is the overall reported infections caused by *E. coli* that is mostly responsible for the production of ESBL; this data is according to findings recorded by a Canadian surveillance study (Pitout et al., 2004). Seventy-one percent of subjects had disease onset that was associated with the community as reported by that study in those patients that are older than 65 years of age, and due to these organisms, considerably higher rates of infection were most commonly seen in women. A few current studies related to the epidemiology of *E. coli* also found that *E. coli* is resistant to cephalosporin, especially showing resistance against higher-generation cephalosporin in the community. Many infections are reported that are mostly acquired from the community due to *E. coli* isolates that are responsible for the production of ESBL (Paterson and Bonomo, 2005). To examine associated risk factors for acquisition of ESBL-producing *E. coli* or Klebsiella species, a study performed in Israel and compared about 311 patients who were not hospitalized with community-acquired UTIs (Colodner et al., 2004). The 128 carriers of bacteria specific to producing ESBL had antibiotic treatment with significantly higher rates in the last 3 months (with quinolones, with penicillin, and especially with second-and third-generation cephalosporins) and of hospitalization. This study also compared their findings with the 183 patients who did not carry bacteria producing ESBL. They were also more likely to be over 60 years of age, especially male gender and diabetics. A case-control study reported in Spain found *E. coli* producing extended-spectrum beta-lactamase affected 49 patients with community-onset infections (Rodríguez-Bano et al., 2004). There is a country-wise resistance distribution pattern of *E. coli* in Europe and percentage of hospital-associated and community-associated *E. coli* isolates and identification of *E. coli* as ESBL positive in Europe was done in 2008 (Hawser et al., 2010; Table 2.2).

2.2.1.3 Asymptomatic populations

Many studies have been performed in asymptomatic populations, and most of the studies used stool samples as sample material. The studies that examined healthy children included studies of Garau et al. (1999), Calva et al. (1996), and Dominguez et al. (2002). And the studies that examined adults included studies of Stürmer et al. (Erb et al., 2007), London et al. (1994), Gulay et al. (Briñas et al., 2002), Garau et al. (1999), and Bonten et al. (1992). In these populations, there is a variation in the pattern of resistance against ampicillin, varying between 13% and 100%. Lower resistance rates were seen against ampicillin in patients of general practitioners in Germany who are not selected and in healthy Dutch volunteers. The incidence of resistance of *E. coli* against antibiotics, especially trimethoprim and cotrimoxazole, fluctuated from 7.5% to 100% (Table 2.3; Erb et al., 2007).

2.2.1.4 Symptomatic populations

Symptomatic patients consisted of patients that were hospitalized and associated with infections caused by *E. coli*, or patients in outpatient centers with this organism that caused UTIs, or the patients of general practitioners (Barrett et al., 1999; Brumfitt et al., 1971; Huovinen and Toivanen, 1980; Karlowsky et al., 2002b; Vorland et al., 1985; Zhanel et al., 2006; Zhanel et al., 2000) in the study populations (Table 2.4). In these studies, urine or blood samples that were taken from patients were used as sample materials.

Table 2.2 Percentage of HA and CA *E. coli* isolates identified as ESBL positive in Europe in 2008 by country.

		E. coli		
Country	**Infection source**	**Number of isolates**	**% ESBL pos**	**95% CI**
Estonia	HA	0		
	CA	20	5.0	0–25.4
France	HA	73	9.6	4.4–18.8
	CA	122	6.6	3.2–12.6
Germany	HA	103	18.5	12.1–27.1
	CA	46	6.5	1.6–18.2
Greece	HA	20	30.0	14.3–52.1
	CA	7	14.3	0.5–53.4
Italy	HA	138	10.9	6.6–17.3
	CA	25	24.0	11.2–43.8
Latvia	HA	22	27.3	12.9–48.4
	CA	27	0.0	0–14.8
Lithuania	HA	14	0.0	0–25.2
	CA	37	0.0	0–11.2
Portugal	HA	77	20.8	13.1–31.2
	CA	31	9.7	2.6–25.7
Spain	HA	332	9.3	6.6–13.0
	CA	227	5.7	3.3–9.6
Switzerland	HA	4	0.0	0–54.6
	CA	0		
Turkey	HA	51	25.5	15.4–39.0
	CA	3	0.0	0–61.8
United Kingdom	HA	29	27.6	14.5–44.3
	CA	29	6.9	0.1–23.0

Table 2.2 shows the global trends of antimicrobial resistance developing in E. coli. *Resistance pattern of antimicrobial resistance fluctuates from country to country. Source of* E. coli *infection taken from both hospital-associated isolates and community-associated isolates.*

2.2.2 Klebsiella pneumoniae

Klebsiella pneumoniae is a major hospital-associated pathogen that poses significant risks like bacteremia, UTIs, and pneumonia (Tsay et al., 2002). Strains of *K. pneumoniae* responsible for the production of ESBL have become endemic in hospitals worldwide due to the widespread use of broad-spectrum cephalosporins over the last several decades (Kang et al., 2004; Paterson and Bonomo, 2005).

2.2.2.1 Temporal trends

The percentages of *K. pneumoniae* isolates that are taken from HA or CA infections were reported as 70.8% and 23.9% for *K. pneumoniae* (P 0.001), respectively. Another 5.3% of *K. pneumoniae* isolates

Table 2.3 Studies on the prevalence of *E. coli* resistance in asymptomatic patients.

Country (year)	First Author	Study population	Sample size	Sample material	Prevalence of resistance
Germany (2004)	Sturmor (18)	Unselected patients of general practitioners	n = 406	Stool	Ampicillin: 16.7% Ciprofloxacin: 0.7% Cotrimoxazole: 8.6%
Spain (2002)	Dominguez (14)	Healthy children <2 years	n = 41	Stool	Ampicillin: 58.5%; Ciprofloxacin: 4.9% Cotrimoxazole: 24.4%
Turkey (2000)	Gulay (16)	Healthy volunteers	n = 50	Stool	Ampicillin: 78%
Spain (1999)	Garau (15)	Emergency room patients without infection Healthy children, mean age 2.4 years	n = 104 n = 65	Stool Stool	Ciprofloxacin: 24% Ciprofloxacin: 26%
Mexico (1996)	Calva (13)	Healthy children <2 years	n = 20	Stool weekly for 13 weeks	Ampicillin: 100% of children, 90% during the whole period Trimethoprim: 100% of children, 55% during the whole period
The Netherlands (1994)8	London (17)	Healthy volunteers from Weert Healthy volunteers from Roermond	n = 90 (678 strains) n = 92 (670 strains)	Stool weekly for 15 weeks Stool weekly for 15 weeks	Amoxicillin: 8.4% Trimethoprim: 7.5% Amoxicillin: 12.4% Trimethoprim: 9.7%

It shows the resistance pattern that is emerging against E. coli *strains existing in the asymptomatic patients. These trends of resistance against* E. coli *observed by taking a stool sample from patients from different countries. Resistance trends against* E. coli *strains vary spatially.*

Table 2.4 Studies on the prevalence of *E. coli* resistance in symptomatic patients.

Country (year)	First author	Sample size	Study population	Sample material	Prevalence of resistance
North America (2006)	Zhanel (27)	n = 1142	*E. coli* isolates from outpatients from 40 medical centers (30 medical centers throughout the United States, 10 throughout Canada)	Urine	Ampicillin: 37.7% Ciprofloxacin: 5.5% Cotrimoxazole: 21.3% large geographic variability
Spain (2005)	Oteo (29)	n = 7098	*E. coli* isolates from 32 hospital laboratories	Blood	Ampicillin: 59.9% Ciprofloxacin: 19.3%
UK and Ireland (2004)	Reynolds (30)	n = 495	*E. coli* isolates from hospitalized patients with bacteremia	Blood	Amoxicillin: 56.2% Ciprofloxacin: 11.1%
USA (2002)	Karlowsky (24)	n = 286,187	*E. coli* isolates from female outpatients with symptoms of UTI	Urine	Ampicillin:36.4% (1995) 37.0% in 2001; ciprofloxacin: 0.7% (1995) 2.5% in 2001 Cotrimoxazole: 14.8% (1995), 16.1 in 2001
Canada (2000)	Zhanel (26)	n = 1681	*E. coli* isolates from consecutive outpatients with UTI	Urine	Ampicillin: 41.0%; ciprofloxacin: 1.2%; cotrimoxazole: 18.9%
European countries (2000)	Fluit (28)	n = 1918	*E. coli* isolates from hospitalized bacteremia patients	Blood	Ampicillin: 46.7%; ciprofloxacin: 8.1%
England (1999)	Barret (19)	n = 962	isolates from general practice patients with UTI; 65.1% *E. coli*, 23.4% coliform bacteria	Urine	Amoxicillin: 48.3%; ciprofloxacin and norfloxacin: 1.1%; trimethoprim: 24.4%

This shows the trends of resistance against E. coli *strains among symptomatic patients. Here, in this case, blood and urine of concerned patients are used as sample material. Antimicrobial resistance pattern among* E. coli *strains varies globally.*

were taken from patients about whom there is no information regarding their length of stay at the time of specimen collection. There is more variability in the rates of *K. pneumoniae* isolates producing ESBL that were taken from HA infections before rising to 20.9% in 2008. The rate of *K. pneumoniae* isolates positive to produce ESBL taken from CA infections was reported as 5.3% in 2008, which is lower than the rates observed in several years including 2005, 2006, and 2007, and comparable to that observed in 2003 (4.4%). But, usually smaller numbers of *K. pneumoniae* producing ESBL isolates led to correspondingly wider 95% confidence intervals (Table 2.5; Hawser et al., 2010).

K. pneumonia has a high resistance rate to piperacillin-tazobactam (in 2010 it was 37.5% with a statistically nonsignificant decrease to 24.36% in 2015). In the case of ESBL producing isolates, other authors reported higher resistance rates to these antimicrobials (96.9%) (Maina et al., 2013). For each of the analyzed periods, and with the same statistically nonsignificant decreasing pattern, the resistance rate of *K. pneumoniae* to third-generation cephalosporins was very close to the previous one (38.46% in 2010 and 29.72% in 2015). The data reported by our country to European Antimicrobial Resistance Surveillance Network (EARS-Net) is different from the findings in our study: resistance increasing from 44% in 2011 to 73.8% in 2014, so we have the second position after Bulgaria (Holmes et al., 2016). According to our study, which includes 22.22% of the strains that were tested to cephalosporins in 2010, in this study we confirmed the ESBL presence and resistance rate is 29.73%, but the confirmatory tests rate was low. This illustrates an undervaluation about this problem. There are no significant differences between the efficiency of aminopenicillins beta-lactamase inhibitor associations, piperacillin-tazobactam, and third-generation cephalosporins against *K. pneumoniae*, while in the first case, there was a high resistance rate. Any of these antibiotics are an equally reasonable choice for the treatment of *K. pneumoniae*-associated bacteremia according to our findings. In 2010 we found a high resistance rate of *K. pneumoniae* to aminoglycosides (41.17%) but with a significant advancement, although it was statistically nonsignificant in 2015 (21.62%). In 2011, a resistance rate of *K. pneumoniae* to aminoglycosides was somewhat closer to our findings (50%) as reported by

Table 2.5 Percentage of *Klebsiella pneumoniae* isolates from HA and CA IAIs from 2002 to 2008 in Europe.

	Klebsiella HA only			Klebsiella CA only		
Year of isolation	**% ESBL pos**	**95% CI**	**No. of isolates**	**% ESBL pos**	**95% CI**	**No. of isolates**
2002	28.60	15.1–47.2	28	3.70	0–19.8	27
2003	19.00	13.0–27.0	121	4.40	1.0–12.7	68
2004	13.00	8.3–19.8	138	8.40	3.9–16.7	83
2005	14.90	9.8–22.0	134	10.10	5.2–18.3	89
2006	23.80	17.8–30.9	160	11.00	6.4–18.1	118
2007	16.10	10.9–23.0	143	8.90	4.7–15.7	113
2008	20.90	16.1–26.7	225	5.30	1.7–13.2	76

This shows the number of isolates of K. pneumoniae *from 2002 to 2008 taken from hospital-associated (HA) and community-associated (CA) infected individuals. Isolates showing the percentage of extended-spectrum beta-lactamase (ESBL) that are positive with 95% confidence intervals (CIs).*

Romania to EARS-Net, and increased to 67.3% in 2014; both the evolutionary trend and the percentage show a variation from our study (Holmes et al., 2016). One reason for this difference in our study could be the fact that we performed the tests available to check the resistance of antimicrobial agents that were sometimes executed with alternative antibiotics taken from the identical class (for example, gentamicin-amikacin), which could have resulted in changing our data accuracy. In 2010, it was noted that against fluoroquinolones *K. pneumoniae* had a very high resistance rate (47.06%). In 2015 the resistance rate decreased (27.02%). It could be that fluoroquinolones resistance decreased because the pleas of infectious disease specialists for reduction of fluoroquinolones intake were heeded and applied. However, our findings conflict with results taken from EARS-Net, where the increased rate of resistance rate was noted ranging from 30% in 2010 to 66.5% in 2014 (Holmes et al., 2016). In other regions of the globe, high rates of resistance were also found (40.4% Madagascar) (Randrianirina et al., 2010). The concerning growing problem now all over the world is a carbapenem resistance. In 2011, our country reported 10 isolates of *K. pneumoniae* that were resistant to carbapenems to EARS-Net (Albiger et al., 2015) and then in 2014 we reported 81 (31.5%) (Holmes et al., 2016). According to our study, it is observed that against *K. pneumoniae* strains isolated from the blood the carbapenems have excellent activity (no resistance in 2010 and one resistant strain in 2015). However, a major problem for the public health and hospital-acquired infections control could be due to the presence of the carbapenem resistance among these isolates.

2.2.2.2 Global trends

There is a high incidence of multidrug-resistant *K. pneumoniae* in Southern, Eastern, and Central Europe. An increasing number of resistant *K. pneumoniae* strains were reported from Romania among these countries, and these strains were isolated from invasive infections every year. According to the most current report of our country to the surveillance network EARS-Net, we are positioned in first place. Throughout different regions of the world, the rates of antimicrobial resistance are not uniform. Enterobacteriaceae showed an increasing rate of resistance to all antibiotic classes and this has been confirmed through analysis of the temporal antimicrobial trend. Along with *E. Coli*, *K. pneumoniae* is one of the bacteria that displayed this tendency (Rhomberg and Jones, 2009). Percentages of *K. pneumoniae* isolates that were taken from both CA and HA show different trends if observed country-wise (Table 2.6; Hawser et al., 2010).

2.2.3 Streptococcus pneumoniae

A medically significant bacterial pathogen is Streptococcus pneumonia, which causes a number of infections that are easily acquired from the community, including pneumonia, chronic bronchitis, acute bacterial sinusitis, and otitis media. In developing countries, pneumococcal pneumonia is a major cause of morbidity and mortality (Falade and Ayede, 2011; Feldman et al., 2013). Various antibiotic classes have been developing resistance against pneumococci, including penicillins, cephalosporins, and macrolides during the past decades (Appelbaum, 2002).

2.2.3.1 Temporal and global trends

From 2003 to 2005 many invasive pneumococcal isolates reported to ARMed; there were 1298 in total. Out of these total invasive isolates, 25% were reported as penicillin-nonsusceptible *S. pneumoniae* (PNSP), overall either resistant or intermediate to penicillin. PNSP isolates with the lowest proportion

Table 2.6 Percentage of HA and CA Klebsiella isolates in Europe in 2008 by country.

		Klebsiella pneumonia		
Country	Infection source	No. of isolates	% ESBL pos	95% CI
Estonia	HA	0		
	CA	1	0.0	0–83.3
France	HA	13	7.7	0–35.4
	CA	11	0.0	0–30.0
Germany	HA	40	17.5	8.4–32.3
	CA	11	18.2	4.0–48.9
Greece	HA	21	28.6	13.6–50.2
	CA	1	0.0	0–83.3
Italy	HA	25	56.0	37.1–73.4
	CA	2	0.0	0–71.0
Latvia	HA	10	50.0	23.7–76.3
	CA	2	0.0	0–71.0
Lithuania	HA	2	50.0	0–19.2
	CA	4	0.0	0–71.0
Portugal	HA	28	3.6	0–19.2
	CA	2	0.0	0–71.0
Spain	HA	60	13.3	6.7–24.4
	CA	37	5.4	0.6–18.6
Switzerland	HA	4	0.0	0–54.6
	CA	0	–	–
Turkey	HA	15	20.0	6.3–46.0
	CA	0	–	–
United Kingdom	HA	7	14.3	0.5–53.4
	CA	5	0.0	0–48.9

This shows the antimicrobial resistance trends of K. pneumoniae *among inhabitants of Europe in 2008. In this case source of* E. coli *infection can be hospital associated and community associated. This table observed the resistance pattern from both isolates taken from the hospital and community.*

were found in Morocco (17%, n = 42) and Malta (15%, n = 13) according to data reported in 2005. From Algeria (44%, n = 71) and Lebanon (40%, n = 10), frequency of PNSP isolates was found to be higher. A substantial increase and decrease in PNSP isolates were observed in Turkey (ranging from 12.8% to 24.3%; *P* 0.03) and Egypt (ranging from 39% to 17%; *P* 0.002) from 2003 to 2005. During the same time, 1172 pneumococcal isolates (90% of the total) were tested to check its susceptibility to erythromycin. In 2005, the proportion of nonsusceptibility of pneumococcal isolates against erythromycin was found to be lowest both in Morocco (12%, n = 41) and Turkey (10%, n = 98) and the proportions of *S. pneumoniae* isolates that were showing nonsusceptibility against erythromycin were found to be highest in Tunisia (39%, n = 33) and Malta (46%, n = 13). In any one of the countries that were participating in the project from 2003 to 2005, no significant increase or decrease was observed.

In 2005, the proportion of dual nonsusceptibility (defined as penicillin-intermediate or full resistance, together with erythromycin resistance) was found to be highest in Lebanon (20%, n = 10) and Tunisia (24%, n = 33). The proportion of dual nonsusceptibility (Table 2.7) were found to be lowest in Malta (0%, n = 13) and Egypt (3%, n = 121) (Borg et al., 2009).

2.2.4 Mycobacterium tuberculosis

A disease imposing a serious problem associated with public health globally is tuberculosis (TB). This disease inexplicably affects those people who are deficient in resources, particularly those living in

Table 2.7 Number of invasive *S. pneumoniae* isolates and the proportion of penicillin nonsusceptible *S. pneumoniae* (PNSP), erythromycin nonsusceptible *S. pneumoniae* (ENSP), and dual nonsusceptible (DUAL) strains, including 95% CIs as reported by country per year.

Country	Year	No. tested for penicillin	No. tested for erythromycin	PNSP (95% CI) (%)	ENSP (95% CI) (%)	Dual (95% CI) (%)
Algeria	2003	32	27	41 (24–59)	19 (7–39)	11 (3–30)
	2004	113	93	40 (31–49)	22 (14–31)	11 (6–19)
	2005	71	61	44 (32–56)	16 (9–29)	11(5–23)
	Overall	216	181	41 (35–48)	19 (14–26)	11 (7–17)
Cyprus	2003	3	3	0 (0–69)	33 (2–87)	0 (0–69)
	2004	7	7	14 (1–58)	0 (0–44)	0 (0–44)
	2005	16	16	19 (5–46)	13 (2–40)	13 (2–40)
	Overall	26	26	15 (5–36)	12 (3–31)	8 (1–27)
Egypt	2003	49	28	39 (26–54)	32 (17–52)	21 (9–41)
	2004	175	168	27 (21–34)	22 (16–29)	4 (2–9)
	2005	123	121	17 (11–25)	22 (15–31)	3 (1–8)
	Overall	347	317	25 (21–30)	23 (19–28)	5 (3–8)
Lebanon	2003	4	4	50 (9–91)	25 (1–78)	25 (1–78)
	2004	2	2	100 (20–95)	50 (3–97)	50 (3–97)
	2005	10	10	40 (14–73)	20 (4–56)	20 (4–56)
	Overall	16	16	5 (26–74)	25 (8–53)	25 (8–53)
Morocco	2003	38	21	26 (14–43)	5 (0–26)	0 (0–19)
	2004	30	29	3 (0–19)	17 (7–36)	0 (0–15)
	2005	42	41	17 (8–32)	12 (5–27)	5 (1–18)
	Overall	110	91	16 (10–25)	12 (6–21)	2 (0–8)
Turkey	2003	117	105	13 (8–21)	7 (3–14)	3 (1–9)
	2004	149	139	23 (17–31)	9 (5–16)	7 (4–13)
	2005	103	98	24 (17–34)	10 (5–18)	10 (5–18)
	Overall	369	342	20 (16–25)	9 (6–12)	7 (4–10)

This is showing the resistance pattern among the S. pneumoniae *isolates specific to penicillin and erythromycin. It also shows the proportion of both penicillin nonsusceptible* S. pneumoniae *(PNSP) and erythromycin nonsusceptible* S. pneumoniae *(ENSP) strains that included 95% confidence interval (CIs) as reported by country per year.*

Africa and Asia. In developing countries, according to data, it is reported that the ratio of incidence of TB and deaths is more than 90% (Organization, 2006; Organization, 2011). In the world list of 22 high-burden countries related to TB, Ethiopia has a seventh position in the list. In 2011 in the global TB report according to the World Health Organization (WHO), 261/100,000 was reported as the estimated incidence rate, the mortality rate was found to be 35/100,000, and the frequency of all TB forms likely to be reported was 394/100,000 (Organization, 2011).

2.2.4.1 Temporal trends

St. Peter's hospital is TB-specialized hospital that was established in 1961. It is a governmental hospital that is under Federal Democratic Republic of Ethiopia Ministry of Health. It was reported from January 2004 to December 2008, that a total of 376 cases were diagnosed with positive cultures of bacteria that was treated with antituberculosis drugs (at least 1 month, they are cured with first-line drugs that are antituberculosis drugs used previously for their treatment) were recognized. Of the study population, 235 (62.5%) were males and the age range was 9–74 years: mean age, 31.9 and 28 were of median age (9–76). Among these, 87.5% of the study population were in the age range of 15–49. The individuals under study were 64, 74, 69, 83, and 86 across per years, respectively; 102 TB patients (27.1%) showed susceptibility to all the four first drugs of choice used as anti-TB drugs, including streptomycin, ethambutol, isoniazid, and rifampicin. Among 376 cultures positive for *Mycobacterium*, 274 among these (72.9%) showed resistance patterns against at least one drug. Resistance pattern shown by *M. tuberculosis* against any STM (67.3%) was the most commonly found, and this was followed by ETB (43.5%), RIF (46.1%), and INH (56.1%) (Abate et al., 2014). About 71 patients (18.9%) showed resistance against only a single antimicrobial drug, whereas resistance was present by multiple drugs against both polyresistant TB and multidrug resistant TB. So, there is a possibility that it should be 203 in number (203/376 = 54%) that were showing resistance against two or more antimicrobial agents. The highest proportion was found against streptomycin (78.8%), followed by isoniazid among monoresistance. According to our results, there is no single strain that showed resistance against rifampicin. In 29 (7.7%) of the cases, polyresistance was reported. About 15 (53.6%) isoniazid + streptomycin combination was the highest proportion among these. About 174 (46.3%) MDR-TB prevalence (defined as the resistance to at least isoniazid and rifampicin) was found to be high in this study. About 103 (59.2%) and 154 (90.8%) patients included in this study were males and their age range was between 15 and 49 years among the study population. Approximately (80.5%) 140 cases out of MDR-TB cases showed resistance against anti-TB drugs that were used as the first line of treatment. There was no statistically significant association of any drug resistance as well as MDR-TB with any age group. However, any drug resistances were statistically significantly associated with gender. Drug resistance is more commonly seen in male patients ($P < .05$) than in female patients (Abate et al., 2012).

2.2.4.2 Global trends

WHO in different surveys to check drug resistance against microorganisms, in either clinical series or whole-country cohorts, has observed the range of problems related to MDR-TB (Espinal et al., 2001). The number of cases of TB and burden of MDR-TB through various cross-sectional surveys almost certainly are underestimated. The reason behind this is that in high-burden countries they do not consider the actual numerical burden of TB. An exact representation of the burden of MDR-TB that is present worldwide is claimed when the exercise is repeated with a mathematical modeling design by using estimates of drug resistance and number of cases of TB (Table 2.8; Dye et al., 2002). About 105 countries reported cases of XDR tuberculosis in December 2014, of the proportion of patients with multidrug resistant tuberculosis who had also tuberculosis, a representative figure was recorded by 83 countries that was collected from periodic surveys. The average percentage was 9.7% (95% CI, 7.4 to 12.1) when these data were combined. Ten or more cases of XDR tuberculosis for which the data were available in the most recent year was reported by 14 out of total countries. The percentage of patients with multidrug-resistant tuberculosis who had also XDR tuberculosis was found to be maximum in Belarus (29.3% in 2014), Lithuania (24.7% in 2013), Georgia (15.1% in 2014), and Latvia (18.6% in 2014), among these countries (Organization, 2015).

Among first-line drugs used against TB, the highest proportion according to a WHO global surveillance report were found to be monoresistant to streptomycin (14.9%) (Organization, 2010) and the same kind of resistance was reported as well as in other studies that were conducted in the Arsi zone of Ethiopia, Denmark, India, and Turkey (Table 2.8; Gupta et al., 1993; Kart et al., 2002; Senol et al., 2005; Thomsen et al., 2000). Resistance to isoniazid and ethambutol was more common than to other first-time used drugs according to 15-year surveillance conducted in Saudi Arabia (Al-Tawfiq et al., 2005). In different regions, MDR-TB in patients who are retreated varies from 30% to 80% (Pablos-Méndez et al., 1998). And in 2008 in Ethiopia, MDR-TB prevalence was (5.6%—21%) for retreatment cases and (0.9%—2.8%) for new cases (Organization, 2011).

2.2.5 Pseudomonas aeruginosa

Pseudomonas aeruginosa is a universal microorganism that can exist in several diverse environmental sites. Various living sources from which *P. aeruginosa* can be isolated from different sources include

Table 2.8 Estimates of the number of individuals with MDR-TB.

Country	All cases	MDR-TB % (95% CI)	Estimated number of cases	95% CI
England and Wales	6947	(0.5—1.1)	55	29—88
Estonia	935	(10.5—17.6)	131	85—202
Latvia	2783	(7.0—11.0)	250	107—363
Russia	97,223	(4.5—7.6)	5864	3761—9039
United States	15,123	(1.0—1.4)	183	129—275
Peru	54,310	(2.3—3.1)	1666	1068—2570
Mozambique	86,558	(2.4—4.6)	3023	1798—4774
South Africa	215,943	(0.6—2.4)	3267	1098—5809
China (DOTS)	650,502	(2.0—3.7)	18,520	11,305—28,936
China (non- DOTS)	650,502	(6.3—9.0)	49,844	34,515—75,216
Pakistan	273,099	(0—21.6)	26,201	0—62,249
Bangladesh	308,271	(0—3.3)	4351	0—11,217
India	1,864,390	(1.6—5.2)	63,136	25,885—108,340

This shows the distribution of MDR-TB among different countries. Country-wise data of resistance against TB shows the prevalence of TB that includes 95% confidence interval (CIs).

living and nonliving things like animals, humans, and plants. This organism can persist on negligible dietary requirements and can tolerate changes in physical conditions like temperature, pH, etc. So, these capabilities have permitted *P. aeruginosa* to survive in both hospital and community settings. The isolation of this organism can be done from various sources, including antiseptics, equipment used in respiratory therapy, physiotherapy and hydrotherapy pools, sinks, soaps, medicines, and mops in the hospitals (199). Whirlpools, swimming pools, home humidifiers, hot bathtubs, contact lens solutions, vegetables, rhizospheres and soils—all of these are included in the community reservoirs of this organism (Al-Humam et al., 2015; Harris et al., 1984; Lutz and Lee, 2011).

2.2.5.1 Temporal and global trends

Several antipseudomonal drugs showing various rates of resistance against *P. aeruginosa* are presented in Table 2.9 (Flamm et al., 2004; Jones et al., 2004; Karlowsky et al., 2003; Karlowsky et al., 2002a; Obritsch et al., 2004; Rhomberg et al., 2007; Rhomberg and Jones, 2007). Various data reported from several surveillance studies done by the United States since January 2000 only highlights data reported for isolates. Against fluoroquinolones, the highest rates of resistance are reported against *P. aeruginosa*, following resistance to ciprofloxacin and levofloxacin that ranges from 20% to 35%.

2.2.6 Staphylococcus aureus

An important cause of nosocomial infection is *Staphylococcus aureus. S. aureus* is considered the principal source of infections related to surgical sites and respiratory tract, especially the lower portion (Richards et al., 1999a; Richards et al., 1999b). This organism is also responsible and considered the second most important cause of bacteremia that is mainly hospital associated (Wisplinghoff et al., 2004), and cardiovascular infections and pneumonia (Richards et al., 1999a; Richards et al., 1999b). In many United States health care settings, communities and long-term care facilities (Crum et al., 2006), methicillin-resistant *Staphylococcus aureus* (MRSA) is now endemic and even epidemic (Strausbaugh et al., 1996). The proportion of *S. aureus* isolates taken from ICUs that are showing resistance against methicillin has risen from 59.5% to 64.4%, as reported in data collected from the National Nosocomial Infections Surveillance system (Klevens et al., 2007; Klevens et al., 2006). Recent reports also advocate that the main cause of *S. aureus* skin and soft tissue infections is due to community-associated MRSA infections (Carleton et al., 2004; Klein et al., 2007).

2.2.6.1 Temporal trends

Overall rates of MRSA have steadily increased in the United States since 1998 with respect to resistance trends, and the rates of resistance seemed to be still increasing as of March 2005 (53.3%). Among the individual group of the patients strains from intensive care unit patients, nonintensive care unit inpatients, and outpatients, an increase was noted with current MRSA rates of 55%, 59.2%, and 47.9%, respectively. The low rate of increased MRSA was reported among the specimens of ICU patients (Styers et al., 2006). This study, from 1999 to 2005, focused particularly on the overall level of infection of *S. aureus* and the trends of that infection that are typically associated with the community. The study estimated the magnitude of the effect and trends in the incidence and associated mortality rates of infections related to S. aureus and MRSA over a 7-year period (Styers et al., 2006).

Table 2.9 Rates of antimicrobial resistance among *Pseudomonas aeruginosa* isolates from hospitals and ICUs.

Antibiotic	% of strain exhibiting resistance				
	Hospital study 2006 n = (606) (211)	Hospital study 2005 n = (589) (212)	Hospital study 2006 n = (9896) (54)	ICU study 2002 n = (951) (178)	ICU 2000–2002 n = (7500) (95)
Beta-lactams Cefepime	6	5	9	25	12
Ceftazidime	13	10	13	19	17
Piperacillin-tazobactam	11	9	11	10	14
Aztreonam		12	–	32	–
Imipenem	11	7	16	23	22
Meropenem	6	7	–	–	18
Fluoroquinolones Ciprofloxacin	21	22	35	32	33
Levofloxacin	22	22	–	34	32
Aminoglycosides Amikacin	–	–	5	10	–
Tobramycin	8	10	12	16	
Gentamycin	12	12	16		22

This shows the antimicrobial resistance trends among P. aeruginosa *isolates. Isolates were taken from both hospitals and ICUs. Antibiotics showing resistance against* P. aeruginosa *that were particularly observed are Beta-lactams, fluoroquinolones, and aminoglycosides.*

2.2.6.2 Global trends

It is verified that trends for MRSA have followed in each of the nine areas of the US Census Bureau according to geographic distribution analysis of outpatient and inpatient rates of MRSA. Inpatient rates of MRSA were above 50% in all regions, except New England. The Mid-Atlantic (36.3%) and New England (37.6%) areas had the lowest rates of MRSA in outpatients, while the East South Central region had the highest rate (63%), and rates of MRSA in inpatients and outpatients were the same. Regarding the sources of the specimens, rates of MRSA were highest (55.9%) among strains taken from specimens taken from the lower respiratory tract of inpatients. Rates of MRSA were lowest (37.6%) among strains taken from skin and soft tissue specimens of outpatients. The rates of MRSA regarding the sources of specimens were comparatively narrow for both inpatients and outpatients, 48.6%–55.9% and 37.6%–42.8%, respectively (Styers et al., 2006).

2.3 Conclusion

Antimicrobial resistance has become a worldwide challenge. The mechanisms responsible for resistance have created massive financial and clinical burdens on systems related to health care worldwide and resistance mechanisms are considered pandemic. With the passage of time, antimicrobial resistance has become a major problem and to our bad luck is that there are no simple solutions to the problem. Even if lives can be saved, the significant commitment and enforcement that are essential requirements of decisive actions do not remain common in all ages. To our good luck, not all pathogenic strains of bacteria are showing resistance against antibiotics all the time. With empirical treatment with antimicrobial agents administered in the community then many pathogenic strains show a positive response to this treatment. Rather than due to good judgment, sometimes success may be due to luck. Instead of giving the various uncertainties, the best single approach that can be expected to help is that all health care centers and physicians should provide their patients with environments that are totally resistance-free and by adopting strict measures to control infections and to avoid inappropriate use of antibiotics. Various efforts should be done to prevent dumping of antibiotics into the environment through sewer systems, which can lead toward the reversion of antimicrobial resistance and antibiotics again become susceptible to resistant pathogenic strains. Before disposal, complete destruction of antibiotics should be done. In the search for new antimicrobial agents, the microbial bioactive compounds are nowhere near being exhausted despite the negative attitude of big pharma. Similarly, in bacterial pathogens, there is the existence of many drug targets that remain uninvestigated. Many effective new compounds cannot be designed or screened with confidence due to unavailability or lack of current knowledge about the target of inhibitor and inhibitor-resistance interactions. It is acknowledged that at the structural level, more studies conducted about these processes will surely offer new leads in this respect.

References

Aarestrup, F.M., Wegener, H.C., Collignon, P., 2008. Resistance in bacteria of the food chain: epidemiology and control strategies. Expert Review of Anti-infective Therapy 6, 733–750.

Abate, D., Taye, B., Abseno, M., Biadgilign, S., 2012. Epidemiology of anti-tuberculosis drug resistance patterns and trends in tuberculosis referral hospital in Addis Ababa, Ethiopia. BMC Research Notes 5, 462.

Abate, D., Tedla, Y., Meressa, D., Ameni, G., 2014. Isoniazid and rifampicin resistance mutations and their effect on second-line anti-tuberculosis treatment. International Journal of Tuberculosis & Lung Disease 18, 946–951.

Akova, M., 2016. Epidemiology of antimicrobial resistance in bloodstream infections. Virulence 7, 252–266.

Al-Humam, N., Ramadan, R., Al-Hizab, F., Barakat, S., Fadlelmula, A., 2015. Complication of gynecomastia by infection with a novel resistant *Pseudomonas aeruginosa* strain in male goat. Annual Research & Review in Biology 8, 1.

Al-Tawfiq, J.A., Al-Muraikhy, A.A., Abed, M.S., 2005. Susceptibility pattern and epidemiology of *Mycobacterium tuberculosis* in a Saudi Arabian hospital: a 15-year study from 1989 to 2003. Chest 128, 3229–3232.

Albiger, B., Glasner, C., Struelens, M.J., Grundmann, H., Monnet, D.L., Eckmanns, T., 2015. Carbapenemase-producing Enterobacteriaceae in Europe: assessment by national experts from 38 countries, May 2015. Euro Surveillance 20, 30062.

Appelbaum, P.C., 2002. Resistance among *Streptococcus pneumoniae*: implications for drug selection. Clinical Infectious Diseases 34, 1613–1620.

Atkinson, B.A., Lorian, V., 1984. Antimicrobial agent susceptibility patterns of bacteria in hospitals from 1971 to 1982. Journal of Clinical Microbiology 20, 791–796.

Barrett, S., Savage, M., Rebec, M., Guyot, A., Andrews, N., Shrimpton, S., 1999. Antibiotic sensitivity of bacteria associated with community-acquired urinary tract infection in Britain. Journal of Antimicrobial Chemotherapy 44, 359–365.

Bartoloni, A., Bartalesi, F., Mantella, A., Dell'Amico, E., Roselli, M., Strohmeyer, M., Barahona, H.G., Barrón, V.P., Paradisi, F., Rossolini, G.M., 2004. High prevalence of acquired antimicrobial resistance unrelated to heavy antimicrobial consumption. Journal of Infectious Diseases 189, 1291–1294.

Blaettler, L., Mertz, D., Frei, R., Elzi, L., Widmer, A., Battegay, M., Flückiger, U., 2009. Secular trend and risk factors for antimicrobial resistance in *Escherichia coli* isolates in Switzerland 1997–2007. Infection 37, 534.

Bonnet, R., 2004. Growing group of extended-spectrum β-lactamases: the CTX-M enzymes. Antimicrobial Agents and Chemotherapy 48, 1–14.

Bonten, M., Stobberingh, E., Philips, J., Houben, A., 1992. Antibiotic resistance of *Escherichia coli* in fecal samples of healthy people in two different areas in an industrialized country. Infection 20, 258–262.

Borg, M., Tiemersma, E., Scicluna, E., Van De Sande-Bruinsma, N., De Kraker, M., Monen, J., Grundmann, H., members, A.P., collaborators, 2009. Prevalence of penicillin and erythromycin resistance among invasive *Streptococcus pneumoniae* isolates reported by laboratories in the southern and eastern Mediterranean region. Clinical Microbiology and Infection 15, 232–237.

Briñas, L., Zarazaga, M., Sáenz, Y., Ruiz-Larrea, F., Torres, C., 2002. β-Lactamases in ampicillin-resistant *Escherichia coli* isolates from foods, humans, and healthy animals. Antimicrobial Agents and Chemotherapy 46, 3156–3163.

Brumfitt, W., Reeves, D., Faiers, M., Datta, N., 1971. Antibiotic-resistant *Escherichia coli* causing urinary-tract infection in general practice: relation to faecal flora. The Lancet 297, 315–317.

Calva, J.J., Sifuentes-Osornio, J., Céron, C., 1996. Antimicrobial resistance in fecal flora: longitudinal community-based surveillance of children from urban Mexico. Antimicrobial Agents and Chemotherapy 40, 1699–1702.

Cantón, R., González, J., Galán, J., 2012. CTX-M enzymes: origin and diffusion. Frontiers in Microbiology 3, 1–19.

Carleton, H.A., Diep, B.A., Charlebois, E.D., Sensabaugh, G.F., Perdreau-Remington, F., 2004. Community-adapted methicillin-resistant *Staphylococcus aureus* (MRSA): population dynamics of an expanding community reservoir of MRSA. The Journal of Infectious Diseases 190, 1730–1738.

Colodner, R., Rock, W., Chazan, B., Keller, N., Guy, N., Sakran, W., Raz, R., 2004. Risk factors for the development of extended-spectrum beta-lactamase-producing bacteria in nonhospitalized patients. European Journal of Clinical Microbiology and Infectious Diseases 23, 163–167.

Control CfD, Prevention, 2002. Multistate outbreak of *Escherichia coli* O157: H7 infections associated with eating ground beef—United States, June-July 2002. MMWR. Morbidity and mortality weekly report 51, 637.

Crum, N.F., Lee, R.U., Thornton, S.A., Stine, O.C., Wallace, M.R., Barrozo, C., Keefer-Norris, A., Judd, S., Russell, K.L., 2006. Fifteen-year study of the changing epidemiology of methicillin-resistant *Staphylococcus aureus*. The American Journal of Medicine 119, 943–951.

D'costa, V.M., McGrann, K.M., Hughes, D.W., Wright, G.D., 2006. Sampling the antibiotic resistome. Science 311, 374–377.

De Man, P., Verhoeven, B., Verbrugh, H., Vos, M., Van den Anker, J., 2000. An antibiotic policy to prevent emergence of resistant bacilli. The Lancet 355, 973–978.

Domínguez, E., Zarazaga, M., Sáenz, Y., Briñas, L., Torres, C., 2002. Mechanisms of antibiotic resistance in *Escherichia coli* isolates obtained from healthy children in Spain. Microbial Drug Resistance 8, 321–327.

Dunne, E.F., Fey, P.D., Kludt, P., Reporter, R., Mostashari, F., Shillam, P., Wicklund, J., Miller, C., Holland, B., Stamey, K., 2000. Emergence of domestically acquired ceftriaxone-resistant Salmonella infections associated with AmpC β-lactamase. Journal of the American Medical Association 284, 3151–3156.

Dye, C., Espinal, M.A., Watt, C.J., Mbiaga, C., Williams, B.G., 2002. Worldwide incidence of multidrug-resistant tuberculosis. The Journal of Infectious Diseases 185, 1197–1202.

Eckert, C., Gautier, V., Arlet, G., 2005. DNA sequence analysis of the genetic environment of various bla CTX-M genes. Journal of Antimicrobial Chemotherapy 57, 14–23.

Edelstein, M., Pimkin, M., Palagin, I., Edelstein, I., Stratchounski, L., 2003. Prevalence and molecular epidemiology of CTX-M extended-spectrum β-lactamase-producing *Escherichia coli* and *Klebsiella pneumoniae* in Russian hospitals. Antimicrobial Agents and Chemotherapy 47, 3724–3732.

Erb, A., Stürmer, T., Marre, R., Brenner, H., 2007. Prevalence of antibiotic resistance in *Escherichia coli*: overview of geographical, temporal, and methodological variations. European Journal of Clinical Microbiology & Infectious Diseases 26, 83–90.

Espinal, M.A., Laszlo, A., Simonsen, L., Boulahbal, F., Kim, S.J., Reniero, A., Hoffner, S., Rieder, H.L., Binkin, N., Dye, C., Williams, R., Raviglione, M.C., 2001. Global trends in resistance to antituberculosis drugs. World health organization-international union against tuberculosis and lung disease working group on anti-tuberculosis drug resistance surveillance. New England Journal of Medicine 344, 1294–1303.

Falade, A., Ayede, A., 2011. Epidemiology, aetiology and management of childhood acute community-acquired pneumonia in developing countries–a review. African Journal of Medicine and Medical Sciences 40, 293–308.

Feizabadi, M.M., Delfani, S., Raji, N., Majnooni, A., Aligholi, M., Shahcheraghi, F., Parvin, M., Yadegarinia, D., 2010. Distribution of bla TEM, bla SHV, bla CTX-M genes among clinical isolates of *Klebsiella pneumoniae* at Labbafinejad Hospital, Tehran, Iran. Microbial Drug Resistance 16, 49–53.

Feldman, C., Abdulkarim, E., Alattar, F., Al Lawati, F., Al Khatib, H., Al Maslamani, M., Al Obaidani, I., Al Salah, M., Farghaly, M., Husain, E.H., 2013. Pneumococcal disease in the Arabian Gulf: recognizing the challenge and moving toward a solution. Journal of Infection and Public Health 6, 401–409.

Flamm, R.K., Weaver, M.K., Thornsberry, C., Jones, M.E., Karlowsky, J.A., Sahm, D.F., 2004. Factors associated with relative rates of antibiotic resistance in *Pseudomonas aeruginosa* isolates tested in clinical laboratories in the United States from 1999 to 2002. Antimicrobial Agents and Chemotherapy 48, 2431–2436.

Garau, J., Xercavins, M., Rodríguez-Carballeira, M., Gómez-Vera, J.R., Coll, I., Vidal, D., Llovet, T., Ruíz-Bremón, A., 1999. Emergence and dissemination of quinolone-resistant *Escherichia coli* in the community. Antimicrobial Agents and Chemotherapy 43, 2736–2741.

Giske, C.G., Sundsfjord, A.S., Kahlmeter, G., Woodford, N., Nordmann, P., Paterson, D.L., Canton, R., Walsh, T.R., 2008. Redefining extended-spectrum β-lactamases: balancing science and clinical need. Journal of Antimicrobial Chemotherapy 63, 1–4.

Gupta, P., Singhal, B., Sharma, T., Gupta, R., 1993. Prevalence of initial drug resistance in tuberculosis patients attending a chest hospital. The Indian Journal of Medical Research 97, 102–103.

Harris, A.A., Goodman, L., Levin, S., 1984. Community-acquired *Pseudomonas aeruginosa* pneumonia associated with the use of a home humidifier. Western Journal of Medicine 141, 521.

Hawser, S.P., Bouchillon, S.K., Hoban, D.J., Badal, R.E., Cantón, R., Baquero, F., 2010. Incidence and antimicrobial susceptibility of *Escherichia coli* and *Klebsiella pneumoniae* with extended-spectrum β-lactamases in community-and hospital-associated intra-abdominal infections in Europe: results of the 2008 Study for Monitoring Antimicrobial Resistance Trends (SMART). Antimicrobial Agents and Chemotherapy 54, 3043–3046.

Holmes, A.H., Moore, L.S., Sundsfjord, A., Steinbakk, M., Regmi, S., Karkey, A., Guerin, P.J., Piddock, L.J., 2016. Understanding the mechanisms and drivers of antimicrobial resistance. The Lancet 387, 176–187.

Huovinen, P., Toivanen, P., 1980. Trimethoprim resistance in Finland after five years' use of plain trimethoprim. British Medical Journal 280, 72–74.

Ishii, Y., Alba, J., Kimura, S., Yamaguchi, K., 2006. Evaluation of antimicrobial activity of β-lactam antibiotics by Etest against clinical isolates from 100 medical centers in Japan (2004). Diagnostic Microbiology and Infectious Disease 55, 143–148.

Jones, M.E., Draghi, D.C., Thornsberry, C., Karlowsky, J.A., Sahm, D.F., Wenzel, R.P., 2004. Emerging resistance among bacterial pathogens in the intensive care unit–a European and North American Surveillance study (2000–2002). Annals of Clinical Microbiology and Antimicrobials 3, 14.

Kang, C.-I., Kim, S.-H., Kim, D.M., Park, W.B., Lee, K.-D., Kim, H.-B., Oh, M-d, Kim, E.-C., Choe, K.-W., 2004. Risk factors for and clinical outcomes of bloodstream infections caused by extended-spectrum beta-lactamase-producing *Klebsiella pneumoniae*. Infection Control & Hospital Epidemiology 25, 860–867.

Karlowsky, J.A., Draghi, D.C., Jones, M.E., Thornsberry, C., Friedland, I.R., Sahm, D.F., 2003. Surveillance for antimicrobial susceptibility among clinical isolates of *Pseudomonas aeruginosa* and Acinetobacter baumannii from hospitalized patients in the United States, 1998 to 2001. Antimicrobial Agents and Chemotherapy 47, 1681–1688.

Karlowsky, J.A., Kelly, L.J., Thornsberry, C., Jones, M.E., Evangelista, A.T., Critchley, I.A., Sahm, D.F., 2002a. Susceptibility to fluoroquinolones among commonly isolated Gram-negative bacilli in 2000: TRUST and TSN data for the United States. International Journal of Antimicrobial Agents 19, 21–31.

Karlowsky, J.A., Kelly, L.J., Thornsberry, C., Jones, M.E., Sahm, D.F., 2002b. Trends in antimicrobial resistance among urinary tract infection isolates of *Escherichia coli* from female outpatients in the United States. Antimicrobial Agents and Chemotherapy 46, 2540–2545.

Kart, L., Altın, R., Tor, M., Gulmez, I., Oymak, S.F., Atmaca, H.M., Erdem, F., 2002. Antituberculosis drug resistance patterns in two regions of Turkey: a retrospective analysis. Annals of Clinical Microbiology and Antimicrobials 1, 6.

Klein, E., Smith, D.L., Laxminarayan, R., 2007. Hospitalizations and deaths caused by methicillin-resistant *Staphylococcus aureus*, United States, 1999–2005. Emerging Infectious Diseases 13, 1840.

Klevens, R.M., Edwards, J.R., Richards Jr., C.L., Horan, T.C., Gaynes, R.P., Pollock, D.A., Cardo, D.M., 2007. Estimating health care-associated infections and deaths in US hospitals, 2002. Public Health Reports 122, 160–166.

Klevens, R.M., Edwards, J.R., Tenover, F.C., McDonald, L.C., Horan, T., Gaynes, R., System, N.N.I.S., 2006. Changes in the epidemiology of methicillin-resistant *Staphylococcus aureus* in intensive care units in US hospitals, 1992–2003. Clinical Infectious Diseases 42, 389–391.

Kollef, M.H., Fraser, V.J., 2001. Antibiotic resistance in the intensive care unit. Annals of Internal Medicine 134, 298–314.

Kronvall, G., 2010. Antimicrobial resistance 1979–2009 at Karolinska hospital, Sweden: normalized resistance interpretation during a 30-year follow-up on *Staphylococcus aureus* and *Escherichia coli* resistance development. Apmis 118, 621–639.

Lester, S.C., Pla, M.P., Wang, F., Schael, I.P., Jiang, H., O'brien, T.F., 1990. The carriage of *Escherichia coli* resistant to antimicrobial agents by healthy children in Boston, in Caracas, Venezuela, and in Qin Pu, China. New England Journal of Medicine 323, 285–289.

Levy, S.B., Marshall, B., 2004. Antibacterial resistance worldwide: causes, challenges and responses. Nature Medicine 10, S122.

London, N., Nijsten, R., vd Bogaard, A., Stobberingh, E., 1994. Carriage of antibiotic-resistant *Escherichia coli* by healthy volunteers during a 15-week period. Infection 22, 187–192.

Lutz, J., Lee, J., 2011. Prevalence and antimicrobial-resistance of *Pseudomonas aeruginosa* in swimming pools and hot tubs. International Journal of Environmental Research and Public Health 8, 554–564.

Maina, D., Makau, P., Nyerere, A., Revathi, G., 2013. Antimicrobial resistance patterns in extended-spectrum β-lactamase producing *Escherichia coli* and *Klebsiella pneumoniae* isolates in a private tertiary hospital, Kenya. Microbiology Discovery 1, 5.

Minond, D., Saldanha, S., Spicer, T., Qin, L., Mercer, B., Roush, W., Hodder, P., 2011. HTS Assay for Discovery of Novel Metallo-Beta-Lactamase (MBL) Inhibitors.

Naseer, U., Sundsfjord, A., 2011. The CTX-M conundrum: dissemination of plasmids and *Escherichia coli* clones. Microbial Drug Resistance 17, 83–97.

Obritsch, M.D., Fish, D.N., MacLaren, R., Jung, R., 2004. National surveillance of antimicrobial resistance in *Pseudomonas aeruginosa* isolates obtained from intensive care unit patients from 1993 to 2002. Antimicrobial Agents and Chemotherapy 48, 4606–4610.

Organization, W.H., 2006. Global Tuberculosis Control-Surveillance, Planning, Financing. WHO. Report. http://www.who.int/tb/publications/global_report/2006/pdf/full_report_correctedversion.pdf.

Organization, W.H., 2010. Multidrug and Extensively Drug-Resistant TB (M/XDR-TB): 2010 Global Report on Surveillance and Response. Multidrug and extensively drug-resistant TB (M/XDR-TB): 2010 global report on surveillance and response.

Organization, W.H., 2011. Fluorescent Light-Emitting Diode (LED) Microscopy for Diagnosis of Tuberculosis: Policy Statement. World Health Organization, Geneva.

Organization, W.H., 2015. Global Tuberculosis Report 2015. World Health Organization.

Pablos-Méndez, A., Raviglione, M.C., Laszlo, A., Binkin, N., Rieder, H.L., Bustreo, F., Cohn, D.L., Lambregts-van Weezenbeek, C.S., Kim, S.J., Chaulet, P., 1998. Global surveillance for antituberculosis-drug resistance, 1994–1997. New England Journal of Medicine 338, 1641–1649.

Paterson, D.L., Bonomo, R.A., 2005. Extended-spectrum β-lactamases: a clinical update. Clinical Microbiology Reviews 18, 657–686.

Pitout, J.D., Hanson, N.D., Church, D.L., Laupland, K.B., 2004. Population-Based laboratory surveillance for *Escherichia coli*–producing extended-spectrum β-lactamases: importance of community isolates with bla CTX-M genes. Clinical Infectious Diseases 38, 1736–1741.

Randrianirina, F., Vaillant, L., Ramarokoto, C.E., Rakotoarijaona, A., Andriamanarivo, M.L., Razafimahandry, H.C., Randrianomenjanahary, J., Raveloson, J.R., Hariniana, E.R., Carod, J.-F., 2010. Antimicrobial resistance in pathogens causing nosocomial infections in surgery and intensive care units of two hospitals in Antananarivo, Madagascar. The Journal of Infection in Developing Countries 4, 074–082.

Rao, S.P., Rama, P.S., Gurushanthappa, V., Manipura, R., Srinivasan, K., 2014. Extended-spectrum beta-lactamases producing *Escherichia coli* and *Klebsiella pneumoniae*: a multi-centric study across Karnataka. Journal of Laboratory Physicians 6, 7.

Reacher, M.H., Shah, A., Livermore, D.M., Wale, M.C., Graham, C., Johnson, A.P., Heine, H., Monnickendam, M.A., Barker, K.F., James, D., 2000. Bacteraemia and antibiotic resistance of its pathogens reported in England and Wales between 1990 and 1998: trend analysis. British Medical Journal 320, 213–216.

Rhomberg, P.R., Deshpande, L.M., Kirby, J.T., Jones, R.N., 2007. Activity of meropenem as serine carbapenemases evolve in US medical centers: monitoring report from the MYSTIC program (2006). Diagnostic Microbiology and Infectious Disease 59, 425–432.

Rhomberg, P.R., Jones, R.N., 2007. Contemporary activity of meropenem and comparator broad-spectrum agents: MYSTIC program report from the United States component (2005). Diagnostic Microbiology and Infectious Disease 57, 207–215.

Rhomberg, P.R., Jones, R.N., 2009. Summary trends for the meropenem yearly susceptibility test information collection program: a 10-year experience in the United States (1999-2008). Diagnostic Microbiology and Infectious Disease 65, 414–426.

Richards, M.J., Edwards, J.R., Culver, D.H., Gaynes, R.P., 1999a. Nosocomial infections in medical intensive care units in the United States. Critical Care Medicine 27, 887–892.

Richards, M.J., Edwards, J.R., Culver, D.H., Gaynes, R.P., 1999b. Nosocomial infections in pediatric intensive care units in the United States. Pediatrics 103, e39.

Rodríguez-Bano, J., Navarro, M.D., Romero, L., Martínez-Martínez, L., Muniain, M.A., Perea, E.J., Pérez-Cano, R., Pascual, A., 2004. Epidemiology and clinical features of infections caused by extended-spectrum beta-lactamase-producing *Escherichia coli* in nonhospitalized patients. Journal of Clinical Microbiology 42, 1089–1094.

Ruppe, E., Hem, S., Lath, S., Gautier, V., Ariey, F., Sarthou, J.L., Monchy, D., Arlet, G., 2009. CTX-M beta-lactamases in *Escherichia coli* from community-acquired urinary tract infections, Cambodia. Emerging Infectious Diseases 15, 741–748.

Senol, G., Komurcuoglu, B., Komurcuoglu, A., 2005. Drug resistance of *Mycobacterium tuberculosis* in Western Turkey: a retrospective study from 1100-bed teaching hospital. Journal of Infection 50, 306–311.

Singer, R.S., Finch, R., Wegener, H.C., Bywater, R., Walters, J., Lipsitch, M., 2003. Antibiotic resistance—the interplay between antibiotic use in animals and human beings. The Lancet Infectious Diseases 3, 47–51.

Strausbaugh, L.J., Crossley, K.B., Nurse, B.A., Thrupp, L.D., Committee, S.L.-T.C., 1996. Antimicrobial resistance in long-term–care facilities. Infection Control & Hospital Epidemiology 17, 129–140.

Styers, D., Sheehan, D.J., Hogan, P., Sahm, D.F., 2006. Laboratory-based surveillance of current antimicrobial resistance patterns and trends among *Staphylococcus aureus*: 2005 status in the United States. Annals of Clinical Microbiology and Antimicrobials 5, 2.

Tadesse, D.A., Zhao, S., Tong, E., Ayers, S., Singh, A., Bartholomew, M.J., McDermott, P.F., 2012. Antimicrobial drug resistance in *Escherichia coli* from humans and food animals, United States, 1950–2002. Emerging Infectious Diseases 18, 741.

Tenover, F.C., Hughes, J.M., 1996. The challenges of emerging infectious diseases: development and spread of multiply-resistant bacterial pathogens. Journal of the American Medical Association 275, 300–304.

Thomsen, V., Bauer, J., Lillebaek, T., Glismann, S., 2000. Results from 8 yrs of susceptibility testing of clinical *Mycobacterium tuberculosis* isolates in Denmark. European Respiratory Journal 16, 203–208.

Tsay, R.-W., Siu, L., Fung, C.-P., Chang, F.-Y., 2002. Characteristics of bacteremia between community-acquired and nosocomial *Klebsiella pneumoniae* infection: risk factor for mortality and the impact of capsular serotypes as a herald for community-acquired infection. Archives of Internal Medicine 162, 1021–1027.

Tzouvelekis, L., Tzelepi, E., Tassios, P., Legakis, N., 2000. CTX-M-type β-lactamases: an emerging group of extended-spectrum enzymes. International Journal of Antimicrobial Agents 14, 137–142.

Von Baum, H., Marre, R., 2005. Antimicrobial resistance of *Escherichia coli* and therapeutic implications. International Journal of Medical Microbiology 295, 503–511.

Vorland, L., Carlson, K., Aalen, O., 1985. Antibiotic resistance and small R plasmids among *Escherichia coli* isolates from outpatient urinary tract infections in northern Norway. Antimicrobial Agents and Chemotherapy 27, 107–113.

Walusansa, A., 2017. College of Health Sciences School of Biomedical Sciences Department of Medical Microbiology. Makerere University.

Wisplinghoff, H., Bischoff, T., Tallent, S.M., Seifert, H., Wenzel, R.P., Edmond, M.B., 2004. Nosocomial bloodstream infections in US hospitals: analysis of 24,179 cases from a prospective nationwide surveillance study. Clinical Infectious Diseases 39, 309–317.

Zhanel, G.G., Hisanaga, T.L., Laing, N.M., DeCorby, M.R., Nichol, K.A., Weshnoweski, B., Johnson, J., Noreddin, A., Low, D.E., Karlowsky, J.A., 2006. Antibiotic resistance in *Escherichia coli* outpatient urinary isolates: final results from the North American Urinary Tract Infection Collaborative Alliance (NAUTICA). International Journal of Antimicrobial Agents 27, 468–475.

Zhanel, G.G., Karlowsky, J.A., Harding, G.K., Carrie, A., Mazzulli, T., Low, D.E., Hoban, D.J., 2000. A Canadian national surveillance study of urinary tract isolates from outpatients: comparison of the activities of trimethoprim-sulfamethoxazole, ampicillin, mecillinam, nitrofurantoin, and ciprofloxacin. Antimicrobial Agents and Chemotherapy 44, 1089–1092.

CHAPTER

Antibiotics' presence in hospitals and associated wastes

3

Kanwal Rehman[1], Waqar Pervaiz[1], Francis Victor[1], Basit Ateeq[1], Muhammad Sajid Hamid Akash[2]
[1]*Department of Pharmacy, University of Agriculture, Faisalabad, Pakistan;* [2]*Department of Pharmaceutical Chemistry, Government College University Faisalabad, Faisalabad, Pakistan*

3.1 Introduction

The development of modern antibiotics plays a significant role in increasing expected lifespan and in improving human quality of life by virtue of their selective and lethal action against pathogenic microorganisms (Nicolaou and Rigol, 2018) In developing countries where the conditions of sanitation are poor, antibiotics decrease the rate of morbidity and mortality by changing the outcome of bacterial infections (Rossolini et al., 2014). Antibiotics not only save patients' lives but they have also played a vital role in achieving major advances in the field of surgery and medicine. They have successfully treated and prevented food-borne and other poverty-related bacterial infections in patients who are receiving chemotherapeutic treatments (Ventola, 2015). Despite antibiotics being a powerful weapon against pathogenic microorganisms, the war against such pathogens has not been won yet and it may never be. The rapid emergence of antimicrobial drug resistance is occurring worldwide and is endangering the efficacy of antibiotics (Sengupta et al., 2013). Development of antimicrobial drug resistance is becoming an increasing threat to global health (Khalid et al., 2016). Unregulated distribution, widespread use of broad-spectrum antibiotics, and lack of interest by pharmaceutical industries in new drug development due to economic and regulatory barriers are among the major reasons behind the global antibiotic resistance crisis (Sturm et al., 1997). Appropriate and effective use of antibiotics in hospitals is an essential aspect of preventing the prevalence of antimicrobial drug resistance and this is directly in the hands of health care professionals working in hospital settings (Organization, 2014). Antibiotics are extensively used in hospitals for prophylaxis and treatment of various bacterial infections and diseases (Devarajan et al., 2016). Antimicrobial-drug resistance in hospitals is driven a by number of factors like poor hospital hygiene, selective pressures created by inappropriate use of broad-spectrum antibiotics for longer periods, and mobile genetic elements that encode bacterial resistance by genetic mutations and acquisition of resistant genes followed by the selection of resistant variants (Weinstein, 2001). Antibiotics that are administered to humans are partially metabolized in the digestive tract and are then discharged in the hospital and communal effluents, which ultimately contaminates environmental water bodies (Kummerer, 2004). These effluents contain a significant amount of human and environmental bacterial flora, antibiotic-resistant bacteria, and antibiotic-resistant genes. All these microbial contaminants make up an environment that harbors antibiotic resistant genes, which pose a global

Antibiotics and Antimicrobial Resistance Genes in the Environment. https://doi.org/10.1016/B978-0-12-818882-8.00003-6

public threat with serious health and economic consequences (Control and Prevention, 2013). Such types of environments act as resistance hot spots that support the proliferation of antibiotic-resistant genes and emergence of new resistant strains possibly by the mechanisms involving horizontal gene transfer, gene mutation, and recombination (Berendonk et al., 2015).

3.2 History of antibiotics

There is well-documented evidence that the ancient civilizations used some sort of antimicrobial agents for the management of microbial infections. However, the golden era of antibiotics began with the discovery of penicillin by Alexander Fleming in 1928 (Sengupta et al., 2013). At that time, penicillin was in limited use; it was first prescribed in the 1940s for the treatment of serious infections and widely used by the public starting in 1942. This discovery not only revolutionized modern medicine but has also saved the lives of millions of people (Control and Prevention, 2013). Most of the antibiotics in this golden era were discovered by using Selman Waksman's strategy, which involves whole-cell screening of natural product extracts to find natural sources from which new antibiotics were discovered (Lewis, 2012). By the mid-1960s, the discovery of new sources of antibiotics was becoming difficult by using the simple and effective strategies adopted by Selman Waksman. Moreover, the development of resistance to these early antibiotics as a result of horizontal gene transfer and chromosomal mutation also pose problems in the development of new antibiotics (Brown et al., 2006). All these problems led to the emergence of a new era in the field of antibiotics, called the medicinal chemistry era. In this era new antibiotics were discovered by creating synthetic analogues of natural sources of antibiotics that were used in the golden era. These cycles of innovations in the medicinal chemistry era led to the development of new and improved antibiotics with respect to their pharmacology, antimicrobial spectrum, and their ability to overcome resistance (Cho et al., 2014). From the late 1960s to the early 1980s, many new antibiotics were discovered and introduced to solve the problem of antibiotic resistance, but after that the pool of antibiotic reservoir began to dry up and very few antibiotics were discovered (Spellberg and Gilbert, 2014). The resistance era started in the 1980s, in which the emphasis was mainly on the target-based discovery of new antibacterial agents, as shown in Fig. 3.1. However, this model of drug discovery failed to provide a sufficient number of new antibiotics to cope with the threat of ever-increasing resistance (Davies and Davies, 2010). Discovery of each class of new antibiotic was followed by an emerging wave of resistance. A timeline of the introduction of new antibiotics in their respective era, along with the resistance developed, is illustrated in Fig. 3.1 (Wijaya, 2018).

3.3 Emerging trends of antibiotics in hospitals

With this rapid decrease in the number of newly discovered antibiotics, the medical community is trying its best to maintain the existing collection of antibiotics because emerging antimicrobial resistance is the mainspring toward the great threat to this postantibiotic era (Allerberger et al., 2009). Antibiotics are often prescribed in an irrational manner, which is the leading cause of developing resistance in microorganisms. In order to prevent this emerging resistance, the following main trends regarding the use of antibiotics in hospitals have been observed over the last few years. Firstly, how antibiotics are prescribed within the hospitals; secondly, how antibiotics are used as a quality metrics;

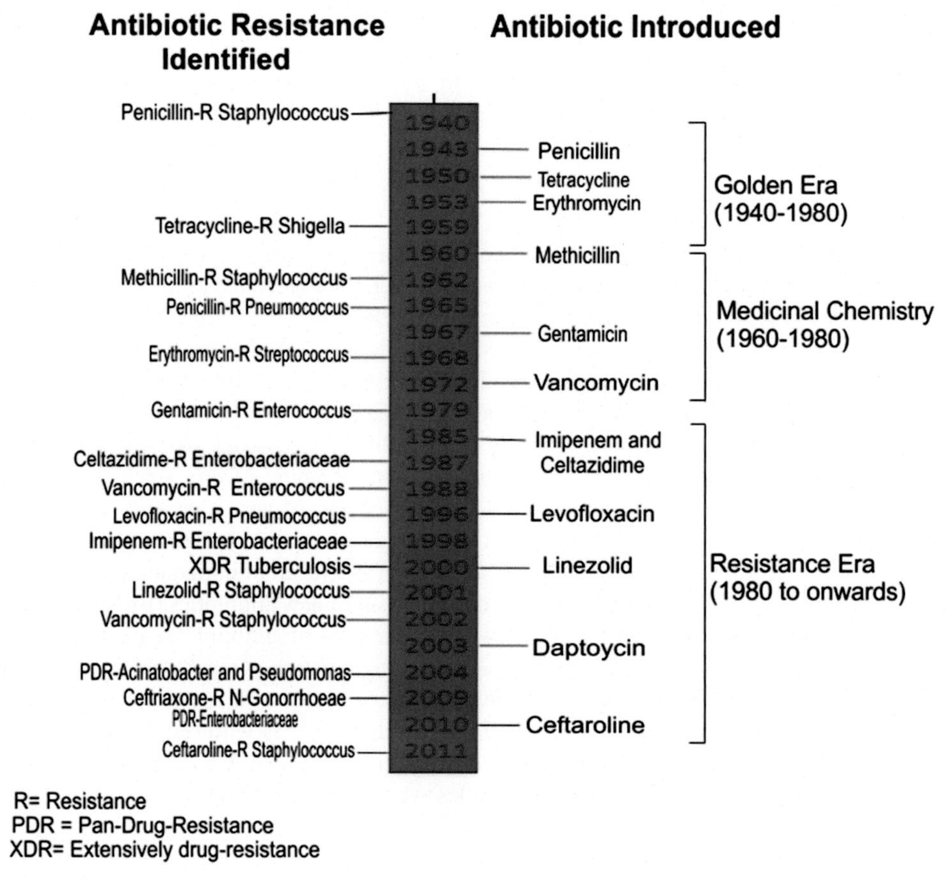

FIGURE 3.1

Timeline of development of antibiotic and their emerging resistance.

thirdly, using various methods for estimating overall antibiotic consumption; and, finally, the use of antibiotic stewardship programs (Jacob and Gaynes, 2010). These trends are used in various hospitals to overcome the harmful consequences related to the use and misuse of antibiotics (Morris, 2014).

3.4 Prescribing pattern of antibiotics

Inappropriate pattern of antibiotic prescribing by physicians is considered as one of the main factors responsible for escalating the rates of antimicrobial resistance (Krockow et al., 2018). The emerging antimicrobial resistance is a worldwide problem that is estimated to cost more than 10 million lives by 2050 (Davey et al., 2013). Penicillins (penicillin, co-amoxiclav, and amoxicillin), cephalosporins (cefixime, cephalexin, and ceftriaxone), and macrolides (particularly azithromycin) are among the most commonly prescribed antibiotics in hospital settings (Hashemi et al., 2013). Use of all antibiotics contributes one way or another in the development of antimicrobial resistance, and despite all the efforts being made, the use of antibiotics in hospitals is increasing day by day. For example, in the United Kingdom the overall use of antibiotics in hospitals increased by 5.7% from 2013 to 2014. Globally, half of the antibiotics that were prescribed to hospital inpatients are considered to be inappropriate (Nash et al., 2002).

Inappropriate prescribing involves various factors, including, but not limited to, prescribing antibiotics when they are not medically required (e.g., prescribing antibiotics for treating viral illnesses), prescribing antibiotics for inappropriate period of time, prescribing antibiotics in incorrect dosage due to lack of information, and/or prescribing antibiotics in conditions that may rectify without drug treatment (e.g., some bacterial infections of throat). Moreover, excessive use of broad-spectrum antibiotics as compared to narrow-spectrum antibiotics, which are effective against a wide variety of different bacterial species, is another major determinant of inappropriate prescribing of antibiotics. All these patterns of antibiotic prescribing are responsible for spreading and emerging antibiotic drug resistance. This problem cannot be prevented entirely but its rapid prevalence can be decreased. For this purpose, various strategies have been adapted to overcome the problems associated with inappropriate use of antibiotics in hospitals. Use of various antibiotic monitoring systems and hospital formularies, which monitor the overall use of antibiotics in hospital, are often seen to reduce antibiotic prescription rates (Okeke et al., 1999). Adoption of a National Essential Drug List also proved helpful as it limits the antibiotics available to prescribers. Using advance and effective antibiotic treatment protocols and educating health care professionals regarding appropriate use of antibiotics is also very important because most of them are unaware of deleterious effects associated with inappropriate use of antibiotic (Leekha et al., 2011). However, implementation of the above-mentioned protocols and strategies does not guarantee proper and rational use of antibiotics in hospitals because of a number of other factors like irregular drug supply, unqualified drug sellers, financial constraints, and availability of drugs from unofficial sources (Mercurio, 2009).

3.5 Antibiotics as quality metrics

The use of antibiotics in hospitals is presumed to be an essential feature when determining the quality of care given to the patients in health care settings and to evaluate various trends adopted by the hospitals regarding antibiotic use and its relationship with the emerging resistance in microorganisms toward such antibiotics. For this purpose, the Infectious Diseases Society of America (IDSA) established various guidelines after 9 years of data collection (2000–09) on therapeutic use of antibiotics and updated them in 2010 (America, 2011). Such guidelines provide recommendations that serve as evidence to quantify the extent of care provided to the patient in health care settings and are based on four important components such as the selection of antibiotic, onset time of antibiotic, risk factors for resistant microbes to such antibiotic, and the severity of illness against which such antibiotics are used (Mandell et al., 2007). These guidelines are implemented to achieve positive patient outcomes by minimizing the use of antibiotics in hospitals and to reduce the chances of multidrug resistance in microorganisms (Arnold et al., 2006; Chastre et al., 2003). The attributes such as availability, validity, sensitivity, and stability are used to quantify the use of antibiotics, but these attributes vary from country to country and over time. Various methods have been developed to calculate the use of antibiotics, of which the method of defined daily dose (DDD) is widely used. In this method, antibiotic use is calculated from purchase data, which is easily available in hospitals. But now other methods such as dispensing data of pharmacy, billing data of patients, and bar-coded administration data are also used for this purpose. However, the results obtained from all these methods do not give the same measurement of antibiotic use and show marked differences for a network of hospitals. Despite the fact that the method of DDD was solely developed to calculate the use of antibiotics, it shows some limitations as well. Firstly, the estimate of antibiotic use obtained from this approach is merely of apparent use of antibiotics in hospitals and not the actual use. Secondly, this method is not appropriate when antibiotic use is measured in the pediatric population (Maier et al., 2006; Ansari et al., 2009; Muller et al., 2006; Polk et al., 2007; Bestehorn et al., 2009) Due to these limitations, new approaches such as recommended daily dose and prescribed daily dose, were also developed to calculate the use of antibiotics (Bestehorn et al., 2009; Maier et al., 2006; Polk et al., 2007). However, these approaches show significance only for individual hospital settings and are not used for comparison of antibiotic use between different hospitals and countries because they show the same limitations that are observed in previous methods.

3.6 Measurement of antibiotic consumption

Antibiotic consumption within a hospital usually refers to the mean quantities of antibiotics being consumed at the patient level, hospital setting, or within an entire institution or organization. Bed utilization of antibiotics, their mean daily doses, and overall duration of antimicrobial therapy are a few parameters on the basis of which antibiotic consumption is measured (Morris, 2014). Approaches like grams of antimicrobial therapy, cost of therapy, DDDs, and days of therapy (DOT) are a few parameters that are used to measure antibiotic consumption.

3.7 Grams of antimicrobial therapy

Investigators from all over the world have shown a great interest in the overall quantity of antibiotics consumed by the patients due to their impact on patients' health (Schirmer et al., 2012). But this approach has some limitations as grams of antibiotics used cannot give any significant information regarding drug potency. Moreover, it is not a well-established fact that there is any significant relationship between grams of antibiotics used and development of antimicrobial resistance. It is also unclear that this approach has any clear-cut benefit over other approaches used (Polk et al., 2007; Rosenthal et al., 2008).

3.8 Cost of antimicrobial therapy

Cost of therapy (COT) is one of the easiest approaches utilized by hospitals to measure the consumption of antibiotics (Goff et al., 2012). Many providers are hesitant to report the cost of therapy because antibiotics show time-to-time fluctuations in their procurement cost. Moreover, procurement cost also varies from hospital to hospital. For example, procurement cost is usually less in government hospitals as compared to private hospitals. In addition to these, the general prescribing trend of antibiotics has shifted toward more expensive and broad-spectrum antibiotics, which has also affected the overall cost of antimicrobial therapy (Morris, 2014). Despite being a simple and easy approach, COT can't be used effectively for measuring the consumption of antibiotics because of a great degree of variability in costs of antibiotics from country to country and complexity in purchasing agreements between different health institutions.

3.9 Antimicrobial defined daily dose

Defined daily dose is simply defined as specific quantity of a drug received by the patient on any day during the course of its therapy. DDD is the most acceptable approach used worldwide for measuring the consumption of antibiotics. This approach was developed in the 1970s and is further modified by the World Health Organization Collaborating Center for Drug Statistics and Methodology (Organization, 2006). The main advantage of this approach over others is that it is relatively easy to adapt by the hospitals for reporting the overall consumption of antibiotics within a hospital setting. Most of the pharmacy departments established within a hospital have mechanisms to calculate with relative ease the overall quantity of antibiotics prescribed, dispensed, or consumed, thus giving the complete picture of antibiotic consumption (Dellit et al., 2007). However, in the case of renal and pediatric patients, DDD shows some limitations while standardizing the dose (Rosenthal et al., 2008).

3.10 Antimicrobial days of therapy

Antimicrobial DOT is another approach used by hospitals for measuring antibiotic consumption. This approach is relatively easy to understand and provides more clinical significance to health care professionals while calculating antibiotic consumption, especially when antibiotics are prescribed or dispensed for a specified period of time. Further advances in this approach led to the development of a new concept called "antimicrobial exposure days" (Kubin et al., 2012; Polk et al., 2007). As compared to other approaches used, DOT is more pertinent to pediatric populations. However, it is difficult to analyze DOT data accurately in the absence of an electronic pharmacy database. Therefore, pharmacy departments of hospitals should possess an efficient electronic database system for accurately calculating antibiotic consumption within a hospital using DOT approach. One disadvantage of this approach is that it overestimates the use of drugs when they are given in multiple doses per day (Polk et al., 2007)

3.11 Antibiotic stewardship program

Antibiotic stewardship program (ASP) is a new approach offering the opportunity for health care professionals to improve overall health care (McGowan, 2012). As microorganisms are developing resistance against various antibiotics, the medical community has introduced this program for combating emerging antimicrobial resistance and for controlling the use of antibiotics in different health care settings (Akpan et al., 2016). The term "antibiotic stewardship" was first used in 1996 when Gerding and McGowan said that the appropriate use of antibiotics can be used as an effective tool for preventing and overcoming the trends observed in emerging antimicrobial resistance (McGowan and Gerding, 1996). ASP optimizes the antimicrobial therapy to maximize the patient outcomes by reducing adverse effects. The main focus of ASP is to minimize overall use of antibiotics, reduce antimicrobial drug resistance and cost of therapy (Fraser et al., 1997; Solomon et al., 2001). Data regarding total number of antibiotics purchased, dispensed, and administered in a particular health care setting is analyzed under ASP for estimating the overall use of antibiotics. However, the parameter that provides the most accurate and relevant information regarding the use of antibiotics is the number of antibiotics administered to the patients (Schirmer et al., 2012). The key strategies used under ASP to access the use of antibiotics include prospective audit with feedback and all possible interventions, by applying formulary restrictions and preauthorization. In addition to these educational programs and guidelines, clinical pathways, cycling of antibiotics, and IV to oral conversion are also a part of this program. ASP is applicable worldwide with slight differences in strategies and policies adapted by different health care settings (Bruce et al., 2009; McGowan and Gerding, 1996). However, most of the hospitals don't have basic infrastructure and facilities to properly implement ASP, and the main problem associated with ASP is that the antibiotic selection becomes homogenized (Jacob and Gaynes, 2010).

3.12 Antibiotics and hospital-associated wastes

Antibiotics, being an important group of pharmaceuticals, are among the most commonly prescribed medications in hospitals for the treatment and prevention of various bacterial ailments. Much of the antibiotics administered to hospital inpatients are partially metabolized in the body while the rest are added to hospital effluents via excretion. Similarly, unused antibiotics are also dumped into hospital effluents. All these ultimately contribute to the residues of antibiotics in hospital-associated wastes (Kummerer, 2004; Rooklidge, 2004). Moreover, hospital wastewaters are a complex mixture of biological and chemical substances, including human and environmental bacterial flora, antimicrobial-resistant bacteria and antimicrobial resistant genes, diagnostic laboratory and research activity wastes, toxic metals, medicines and their metabolites, chemicals, disinfecting agents, detergents, radioactive markers, iodinated contrast media, and nutrients (Verlicchi et al., 2010). Rapid industrial and economic development as well as urban growth have resulted in an increase per capita waste production. This ultimately leads to an increased amount of hospital-associated wastes, particularly residues of antibiotics, antimicrobial-resistant bacteria, antimicrobial-resistant genes, and toxic metals in hospital effluents. These hospital-associated wastes are potentially very dangerous because they make up an environment that serves as hot spots for developing antimicrobial resistance and spreading resistance genes (Michael et al., 2013; Schwartz et al., 2003). Absence of definite policies, rules, and regulations regarding the production and disposal of hospital wastes and improper training of health care personnel associated with the management of hospital wastes contribute to the increasing hazards and increased production of hospital-associated wastes (Askarian et al., 2004). Moreover, pouring of chemical substances into hospital sewerage systems, mixing of hospital wastes with domestic wastes, inadequate treatment of hospital wastes, and lack of incinerators at hospital levels are responsible for emerging antimicrobial resistance in bacteria, thus making bacterial ailments more resilient and more difficult to treat (Levy and Marshall, 2004).

Recent studies proved that antibiotic concentrations found in hospital effluents correlate with the total volume of antibiotics prescribed in hospital settings. There is a positive correlation between concentration of antibiotics in hospital wastewaters and volume of antibiotics prescribed (Watkinson et al., 2009). For example, ciprofloxacin, which was among the most commonly prescribed antibiotics in hospitals, was found in higher concentration in hospital wastewaters as compared to other antibiotics. High ciprofloxacin residual levels in hospital wastewaters can cause genotoxicity in bacterial strains (Diwan et al., 2010). However, in addition to prescription volume, there are other factors also involved that play a significant role in the detection of residual amount of antibiotics found in hospital wastewaters. These factors include time of collection of samples, antibiotics metabolism, temperature sensitivity, environmental stability, rate of flowing water, etc. (Kemper, 2008). For example, ceftriaxone and other β-lactam antibiotics were not detected in higher concentration in hospital wastewaters regardless of been major antibiotics prescribed in the hospitals. Reasons behind their nondetection are easy degradation, their high metabolic rates, and the process of decarboxylation (Kummerer, 2004). Moreover, it was reported that variations in antibiotics residue levels in hospital wastewaters were detected when samples were collected at different times of the day at the same place. These variations in antibiotic concentration levels may occur due to variations in antibiotic administration during the course of the day, wastewater flow rates, and pharmacokinetic parameters of the drug like metabolic half-life, drug metabolism, and drug excretion. In addition to the above-mentioned factors, other processes that influence the antibiotic concentration detected in hospital wastewaters include hydrolysis, photolysis, and drug sorption (Brown et al., 2006). However, it is not a well-established fact that only the presence of antibiotics in hospital effluents is responsible for development of resistance in bacterial strains. Therefore, while discussing this topic, factors like bacterial load in hospital effluents, extent of their exposure to antibiotics, and availability of favorable environment are also important and need to be considered as well. It is widely acknowledged that the antibiotics are often diluted in the recipient waters and the exposure of bacteria strains to subtherapeutic levels of antibiotics over longer duration provides an ideal condition for the transfer of antimicrobial-resistant genes located on mobile genetic elements like plasmids, transposons, and integrons (Kummerer, 2004). The development of antimicrobial resistance in bacterial strains is explained either by acquired resistance or chromosomal resistance. However, acquired resistance developed by acquisition of genetic material is responsible for development of resistance against many antibiotics and antibiotic classes. This type of resistance can easily be transferred from one bacterial strain to another possibly by the mechanisms of conjugation, transformation, or transduction (Aminov, 2011).

3.13 Conclusion

There are substantial quantities of antibiotics existing in hospital wastewater. Noteworthy associations have been found among the measures of utilization of antibiotics and amounts or levels of antibiotics found in wastewater of hospitals for various drugs. Presence of antibiotics in the wastes of hospitals are the major sources of exposure of antibiotic resistance that need to be addressed on priority basis.

References

Akpan, M.R., Ahmad, R., Shebl, N.A., Ashiru-Oredope, D., 2016. A review of quality measures for assessing the impact of antimicrobial stewardship programs in hospitals. Antibiotics (Basel) 5.

Allerberger, F., Gareis, R., Jindrák, V., Struelens, M.J., 2009. Antibiotic stewardship implementation in the EU: the way forward. Expert Review of Anti-infective Therapy 7, 1175–1183.

America IDSo, 2011. Standards, Practice Guidelines, and Statements Developed and/or Endorsed by IDSA.

Aminov, R.I., 2011. Horizontal gene exchange in environmental microbiota. Frontiers in Microbiology 2, 158.

Ansari, F., Erntell, M., Goossens, H., Davey, P., Group, E.I.H.C.S., 2009. The European surveillance of antimicrobial consumption (ESAC) point-prevalence survey of antibacterial use in 20 European hospitals in 2006. Clinical Infectious Diseases 49, 1496–1504.

Arnold, F.W., McDonald, L.C., Smith, R.S., Newman, D., Ramirez, J.A., 2006. Improving antimicrobial use in the hospital setting by providing usage feedback to prescribing physicians. Infection Control & Hospital Epidemiology 27, 378–382.

Askarian, M., Vakili, M., Kabir, G., 2004. Results of a hospital waste survey in private hospitals in Fars province, Iran. Waste Management 24, 347–352.

Berendonk, T.U., Manaia, C.M., Merlin, C., Fatta-Kassinos, D., Cytryn, E., Walsh, F., Bürgmann, H., Sørum, H., Norström, M., Pons, M.-N., 2015. Tackling antibiotic resistance: the environmental framework. Nature Reviews Microbiology 13, 310.

Bestehorn, H., Steib-Bauert, M., Kern, W., 2009. Comparison of defined versus recommended versus prescribed daily doses for measuring hospital antibiotic consumption. Infection 37, 349–352.

Brown, K.D., Kulis, J., Thomson, B., Chapman, T.H., Mawhinney, D.B., 2006. Occurrence of antibiotics in hospital, residential, and dairy effluent, municipal wastewater, and the Rio Grande in New Mexico. The Science of the Total Environment 366, 772–783.

Bruce, J., MacKenzie, F.M., Cookson, B., Mollison, J., Van der Meer, J.W., Krcmery, V., Gould, I.M., 2009. Antibiotic stewardship and consumption: findings from a pan-European hospital study. Journal of Antimicrobial Chemotherapy 64, 853–860.

Chastre, J., Wolff, M., Fagon, J.-Y., Chevret, S., Thomas, F., Wermert, D., Clementi, E., Gonzalez, J., Jusserand, D., Asfar, P., 2003. Comparison of 8 vs 15 days of antibiotic therapy for ventilator-associated pneumonia in adults: a randomized trial. JAMA 290, 2588–2598.

Cho, H., Uehara, T., Bernhardt, T.G., 2014. Beta-lactam antibiotics induce a lethal malfunctioning of the bacterial cell wall synthesis machinery. Cell 159, 1300–1311.

Control CfD, Prevention, 2013. Office of Infectious Disease Antibiotic Resistance Threats in the United States, 2013. Centers for Disease Control and Prevention, Atlanta, GA.

Davey, P., Brown, E., Charani, E., Fenelon, L., Gould, I.M., Holmes, A., Ramsay, C.R., Wiffen, P.J., Wilcox, M., 2013. Interventions to improve antibiotic prescribing practices for hospital inpatients. Cochrane Database of Systematic Reviews.

Davies, J., Davies, D., 2010. Origins and evolution of antibiotic resistance. Microbiology and Molecular Biology Reviews 74, 417–433.

Dellit, T.H., Owens, R.C., McGowan, J.E., Gerding, D.N., Weinstein, R.A., Burke, J.P., Huskins, W.C., Paterson, D.L., Fishman, N.O., Carpenter, C.F., 2007. Infectious Diseases Society of America and the Society for Healthcare Epidemiology of America guidelines for developing an institutional program to enhance antimicrobial stewardship. Clinical Infectious Diseases 44, 159–177.

Devarajan, N., Laffite, A., Mulaji, C.K., Otamonga, J.-P., Mpiana, P.T., Mubedi, J.I., Prabakar, K., Ibelings, B.W., Poté, J., 2016. Occurrence of antibiotic resistance genes and bacterial markers in a tropical river receiving hospital and urban wastewaters. PLoS One 11, e0149211.

Diwan, V., Tamhankar, A.J., Khandal, R.K., Sen, S., Aggarwal, M., Marothi, Y., Iyer, R.V., Sundblad-Tonderski, K., Stålsby-Lundborg, C., 2010. Antibiotics and antibiotic-resistant bacteria in waters associated with a hospital in Ujjain, India. BMC Public Health 10, 414.

Fraser, G.L., Stogsdill, P., Dickens, J.D., Wennberg, D.E., Smith, R.P., Prato, B.S., 1997. Antibiotic optimization: an evaluation of patient safety and economic outcomes. Archives of Internal Medicine 157, 1689–1694.

Goff, D.A., Bauer, K.A., Reed, E.E., Stevenson, K.B., Taylor, J.J., West, J.E., 2012. Is the "low-hanging fruit" worth picking for antimicrobial stewardship programs? Clinical Infectious Diseases 55, 587–592.

Hashemi, S., Nasrollah, A., Rajabi, M., 2013. Irrational antibiotic prescribing: a local issue or global concern? EXCLI journal 12, 384.

Jacob, J.T., Gaynes, R.P., 2010. Emerging trends in antibiotic use in US hospitals: quality, quantification and stewardship. Expert Review of Anti-infective Therapy 8, 893–902.

Kemper, N., 2008. Veterinary antibiotics in the aquatic and terrestrial environment. Ecological Indicators 8, 1–13.

Khalid, L., Mahsood, N., Ali, I., 2016. The public health problem of OTC antibiotics in developing nations. Research in Social and Administrative Pharmacy 12, 801–802.

Krockow, E.M., Colman, A.M., Chattoe-Brown, E., Jenkins, D.R., Perera, N., Mehtar, S., Tarrant, C., 2018. Balancing the risks to individual and society: a systematic review and synthesis of qualitative research on antibiotic prescribing behaviour in hospitals. Journal of Hospital Infection.

Kubin, C.J., Jia, H., Alba, L.R., Furuya, E.Y., 2012. Lack of significant variability among different methods for calculating antimicrobial days of therapy. Infection Control & Hospital Epidemiology 33, 421–423.

Kummerer, K., 2004. Resistance in the environment. Journal of Antimicrobial Chemotherapy 54 (2), 311–320 (Find this article online).

Leekha, S., Terrell, C.L., Edson, R.S., 2011. General principles of antimicrobial therapy. Mayo Clinic Proceedings 156–167. Elsevier.

Levy, S.B., Marshall, B., 2004. Antibacterial resistance worldwide: causes, challenges and responses. Nature Medicine 10, S122.

Lewis, K., 2012. Antibiotics: recover the lost art of drug discovery. Nature 485, 439.

Maier, L., Steib-Bauert, M., Kern, P., Kern, W., 2006. Trends in antibiotic use at a university hospital: defined or prescribed daily doses? Patient days or admissions as denominator? Infection 34, 91–94.

Mandell, L.A., Wunderink, R.G., Anzueto, A., Bartlett, J.G., Campbell, G.D., Dean, N.C., Dowell, S.F., File Jr., T.M., Musher, D.M., Niederman, M.S., 2007. Infectious Diseases Society of America/American Thoracic Society consensus guidelines on the management of community-acquired pneumonia in adults. Clinical Infectious Diseases 44, S27–S72.

McGowan, J.E., 2012. Antimicrobial stewardship—the state of the art in 2011 focus on outcome and methods. Infection Control & Hospital Epidemiology 33, 331–337.

McGowan, J.J., Gerding, D., 1996. Does antibiotic restriction prevent resistance? New Horizons (Baltimore, Md.) 4, 370–376.

Mercurio, B.C., 2009. Health in the developing world: the case for a new funding and support agency. Asian Journal of WTO & International Health Law and Policy 4, 27–64.

Michael, I., Rizzo, L., McArdell, C., Manaia, C., Merlin, C., Schwartz, T., Dagot, C., Fatta-Kassinos, D., 2013. Urban wastewater treatment plants as hotspots for the release of antibiotics in the environment: a review. Water Research 47, 957–995.

Morris, A.M., 2014. Antimicrobial stewardship programs: appropriate measures and metrics to study their impact. Current Treatment Options in Infectious Diseases 6, 101–112.

Muller, A., Monnet, D.L., Talon, D., Hénon, T., Bertrand, X., 2006. Discrepancies between prescribed daily doses and WHO defined daily doses of antibacterials at a university hospital. British Journal of Clinical Pharmacology 61, 585–591.

Nash, D.R., Harman, J., Wald, E.R., Kelleher, K.J., 2002. Antibiotic prescribing by primary care physicians for children with upper respiratory tract infections. Archives of Pediatrics & Adolescent Medicine 156, 1114–1119.

Nicolaou, K.C., Rigol, S., 2018. A brief history of antibiotics and select advances in their synthesis. Journal of Antibiotics 71, 153.

Okeke, I.N., Lamikanra, A., Edelman, R., 1999. Socioeconomic and behavioral factors leading to acquired bacterial resistance to antibiotics in developing countries. Emerging Infectious Diseases 5, 18.

Organization, W.H., 2006. WHO Collaborating Centre for Drug Statistics Methodology: ATC Classification Index with DDDs and Guidelines for ATC Classification and DDD Assignment. Norwegian Institute of Public Health, Oslo, Norway.

Organization, W.H., 2014. Antimicrobial Resistance: Global Report on Surveillance. World Health Organization.

Polk, R.E., Fox, C., Mahoney, A., Letcavage, J., MacDougall, C., 2007. Measurement of adult antibacterial drug use in 130 US hospitals: comparison of defined daily dose and days of therapy. Clinical Infectious Diseases 44, 664–670.

Rooklidge, S.J., 2004. Environmental antimicrobial contamination from terraccumulation and diffuse pollution pathways. The Science of the Total Environment 325, 1–13.

Rosenthal, V.D., Maki, D.G., Mehta, A., Álvarez-Moreno, C., Leblebicioglu, H., Higuera, F., Cuellar, L.E., Madani, N., Mitrev, Z., Dueñas, L., 2008. International nosocomial infection control consortium report, data summary for 2002–2007, issued January 2008. American Journal of Infection Control 36, 627–637.

Rossolini, G.M., Arena, F., Pecile, P., Pollini, S., 2014. Update on the antibiotic resistance crisis. Current Opinion in Pharmacology 18, 56–60.

Schirmer, P.L., Mercier, R.C., Ryono, R.A., Nguyen, N., Lucero, C.A., Oda, G., Holodniy, M., 2012. Comparative assessment of antimicrobial usage measures in the Department of Veterans Affairs. Infection Control & Hospital Epidemiology 33, 409–411.

Schwartz, T., Kohnen, W., Jansen, B., Obst, U., 2003. Detection of antibiotic-resistant bacteria and their resistance genes in wastewater, surface water, and drinking water biofilms. FEMS Microbiology Ecology 43, 325–335.

Sengupta, S., Chattopadhyay, M.K., Grossart, H.-P., 2013. The multifaceted roles of antibiotics and antibiotic resistance in nature. Frontiers in Microbiology 4, 47.

Solomon, D.H., Van Houten, L., Glynn, R.J., Baden, L., Curtis, K., Schrager, H., Avorn, J., 2001. Academic detailing to improve use of broad-spectrum antibiotics at an academic medical center. Archives of Internal Medicine 161, 1897–1902.

Spellberg, B., Gilbert, D.N., 2014. The future of antibiotics and resistance: a tribute to a career of leadership by John Bartlett. Clinical Infectious Diseases 59, S71–S75.

Sturm, A.W., Van der Pol, R., Smits, A., Van Hellemondt, F., Mouton, S., Jamil, B., Minai, A., Sampers, G., 1997. Over-the-counter availability of antimicrobial agents, self-medication and patterns of resistance in Karachi, Pakistan. Journal of Antimicrobial Chemotherapy 39, 543–547.

Ventola, C.L., 2015. The antibiotic resistance crisis: part 1: causes and threats. Pharmacy and Therapeutics 40, 277.

Verlicchi, P., Galletti, A., Petrovic, M., Barceló, D., 2010. Hospital effluents as a source of emerging pollutants: an overview of micropollutants and sustainable treatment options. Journal of Hydrology 389, 416–428.

Watkinson, A., Murby, E., Kolpin, D.W., Costanzo, S., 2009. The occurrence of antibiotics in an urban watershed: from wastewater to drinking water. The Science of the Total Environment 407, 2711–2723.

Weinstein, R.A., 2001. Controlling antimicrobial resistance in hospitals: infection control and use of antibiotics. Emerging Infectious Diseases 7, 188.

Wijaya, L., 2018. How to manage and control healthcare associated infections. In: IOP Conference Series: Earth and Environmental Science. IOP Publishing, p. 012105.

CHAPTER

Current trends of antimicrobials used in food animals and aquaculture

4

Muhammad Ibrahim[1], Fiaz Ahmad[2], Bushra Yaqub[1], Ayesha Ramzan[1], Ayesha Imran[1], Muhammad Afzaal[3], Safdar Ali Mirza[4], Iqra Mazhar[3], Muhammad Younus[5], Qaisar Akram[5], Muhammad Sulman Ali Taseer[5], Ali Ahmad[5], Sarfraz Ahmed[6]

[1]*Department of Biochemistry, Bahauddin Zakariya University, Multan, Pakistan;* [2]*Central Cotton Research Institute, Multan, Pakistan;* [3]*Sustainable development study center GC University Lahore, Punjab, Pakistan;* [4]*Department of Botany, GC University, Lahore, Punjab, Pakistan;* [5]*Department of Basic Sciences, University of Veterinary and Animal Sciences, Narowal, Punjab, Pakistan;* [6]*Department of Basic Siences, University of Veterinary and Animal Sciences Lahore, Narowal, Punjab, Pakistan*

4.1 Introduction

Antimicrobials are used globally both for humans and animals to obviate and treat contagious diseases (O'neill, 2014). Furthermore, in some countries, antimicrobials are used in animal breeding as growth promoters (Flórez et al., 2017). Antimicrobial agents are one of the medicinal innovations of humanity that allows us to cure both human and veterinary infections of microbes. Since the 1940s, several antimicrobials have contributed significantly for prevention, restriction, and cure of contagious diseases in animals. Low- and subtherapeutic antimicrobial dosage plays a very important role in improving feeding proficiency, stimulating animal growth, disease avoidance, and control (Magouras et al., 2017).

There are four ways in which substances expressing antimicrobial activity are used in animals. Therapeutic usage of antimicrobials is considered to prevent existing microbial diseases, usually used for individual animal cure. It involves testing of each infected animal, which involves laboratory examination, determining the microbes and antimicrobial sensitivity testing. Antimicrobials are administered either orally or via inoculation only to animals showing symptoms of that particular illness. The dosage that is injected is related to both the type of animal and the severity of illness. Metaphylaxis includes prior medication to the whole animal group that might lessen the numbers of sick or deceased animals. It might also reduce the antimicrobial dosage required for the treatment of huge numbers of the symptomatically sick populace, therefore treatment expenses are also reduced. Antimicrobial prophylactic application exists for individual and animal groups. It is generally used for operative prophylaxis in animals. In cattle, the prophylactic intramammary injection of antimicrobial agents at the end of the suckling phase prevents mastitis. In swine and cattle husbandry, antimicrobial prophylactic usage occurs at significant time periods like weaning. Antimicrobial prophylaxis usage is critical in numerous pigs and cattle herds. In its absence, continual breathing and enteral illnesses in

Antibiotics and Antimicrobial Resistance Genes in the Environment. https://doi.org/10.1016/B978-0-12-818882-8.00004-8

the byres and piggeries cannot be effectively controlled. Growth promotion also involves antimicrobial usage in food animals. Antimicrobial growth promoters were first endorsed in the mid-1950s.It was revealed that small and subtherapeutic dosage of antimicrobials like penicillin, procaine, and tetracycline (1/10 to 1/100 the quantity of curative dosage), given to animals in food, could increase the food/mass ratio for chickens, pigs, and cows. All substances used to stimulate growth are certified on the base of European Union (EU)—wide rules (guideline 70-524-EWG). These regulations narrate the usage of the particular substances in various animals in accordance to animal's age, maximal and minimal antimicrobial consumption in mg/kg food. Formerly, just four substances were permitted in the EU, having certified growth promotions with antimicrobial functions. These were flavophospholipol, monensin—Na, salinomycin—Na, and avilamycin. In 1996, the glycopeptide-avoparcin usage as a growth promoter was prohibited. Cross-resistance to glycopeptides (vancomycin; teicoplanin), macrolides (erythromycin; clarithromycin) and streptogramins (dalfo/quinupristin) was the major reason for banning them (Ungemach, 1999; Schwarz et al., 2001).

The widespread and inappropriate utilization of antimicrobials in food animals are contributing factors for the emergence and spread of antimicrobial resistance (AMR). Diseases have become untreatable due to the resistance against therapeutic agents. This also poses a risk to public health through potential transfer of resistance genes to human pathogens. Both pathogenic and commensal microbes are exposed to antimicrobials and in response AMR develops. It has been detected that microbes develop resistance by any of the four mechanisms: through drug inactivation or its modification, alteration in the drug target site, modification in the metabolic pathways to overcome drug effects, and by minimizing entry and promoting active efflux of the drugs (Sharma et al., 2018). Microbes can develop antimicrobial resistance by mutating existing genes (vertical gene transfer) or by obtaining new genes from the environment, other spp., or strains (horizontal gene transfer) (Jeters et al., 2009). Resistance between bacterial spp. has been seen through antibiotic-resistant genes and includes among the primary genes leading to AMR: blaTEM genes for the antibiotics (penicillin, amoxicillin, ampicillin) (Bailey et al., 2011); van for glycopeptides (avoparcin, vancomycin) (Leavis et al., 2003); erm gene cluster for macrolides (erythromycin, tylosin, tilmicosin, kitasamycin, oleandomycin) (Ramos et al., 2012); vatD, vatE, erm gene cluster, satA for streptogramins (virginiamycin, quinupristin-dalfopristin) (Ramos et al., 2012); sul genes for sulfonamides (sulfisoxazole, sulfadimethoxine, sulfamethazine) (Cain and Hall, 2012); tet genes for tetracyclines (chlortetracycline, oxytetracycline, doxycycline) (Ramos et al., 2012); rgpA—F, mbrA—D genes for polypeptides (bacitracin) (Cain and Hall, 2012); and cmaA, floR, fexA, fexB, cfr, cat gene for amphenicols (chloramphenicol (Cain and Hall, 2012).

4.2 Global consumption of antimicrobial trends in food animals

Global utilization of antimicrobials in the production of food animals has been estimated at 63,151 (±1560) tons in 2010 and it is estimated to increase by 67% to 105,596 (±3605) tons by 2030.Two-thirds (66%) of global antimicrobial consumption growth (67%) is due to the increasing number of animals raised for food production. The remaining third (34%) is due to a shift in farming practices. It is expected that the larger section of animals to be raised by 2030 will be via intensive farming. Roughly 46% of antimicrobial consumption growth by 2030 in Asia is likely due to shifts in production systems. By 2030, antimicrobial consumption in Asia is predicted to be 51,851 tons. It represents 82% of the current global consumption of antimicrobials in food animals in 2010.

In 2010, China (23%), the United States (13%), Brazil (9%), Germany (3%), and India (3%) were the five countries having substantial shares of global antimicrobial consumption in food animal production. By 2030, it is expected that this ranking will be China (30%), the United States (10%), Brazil (8%), India (4%), and Mexico (2%). Five countries with the greatest projected percentage increases in antimicrobial consumption by 2030 are expected to be Myanmar (205%), Indonesia (202%), Nigeria (163%), Peru (160%), and Vietnam (157%). At the present time, China and Brazil are among the large-scale consumers of antimicrobials. But these are not the countries with the rapid projected increases in antimicrobial consumption. This shows that these two countries have already begun moving toward more escalated livestock production systems using antimicrobials to sustain animal health and increase productivity. Antimicrobial consumption for animals in the BRICS (Brazil, Russia, India, China and South Africa) countries is supposed to grow by 99% by 2030.

4.3 Frequent trends of use of antimicrobials in the treatment of infectious and contagious diseases in food animals

4.3.1 Use of antimicrobials in pigs

Pig weaning is a slow progression that starts at almost 3 months of age and shows the transfer of piglet dependence from lactate to other foodstuffs. But in a majority of developed nations pig weaning is a rapid progression taking place earlier in life, at the age of 19–25 days. It is frequently related to increased risk of stomach dysfunction and dysentery. Pigs eating feed containing antimicrobials showed no symptoms of gastrointestinal impairment at any point, while naturally lactate controls developed dysentery (Li et al., 2013). With antimicrobial usage, there was high day-by-day mass growth in contrast with lactate controls. The streptomycin repeatedly used for pig weaning from delivery to 28 days of age resulted in 8% mass increase compared to natural pigs of 56 days weaning time. Similarly, Aureomycin, Terramycin (oxytetracycline), and penicillin enhanced the swine growth by 10% (Li, 2017). Antimicrobials used are listed in Table 4.1. Specifically, in chicken and pig farming, antimicrobial usage has become a basic part of animal feeds. In the United States, more than a thousand experiments were performed from 1950 to 85. From the results, it was observed that antimicrobials were the efficient agents for feed and growth improvement in adult pigs, as well as the whole growth-finished phase, and reduced death and disease conditions mostly in juvenile pigs. The death rate can be doubly high in the farm environment compared to research locations where the amenities are usually clean, the disease load is less, and the atmosphere is less traumatic (Luecke et al., 1951). The advantageous effects of antimicrobials on body mass growth were accompanied by improved feed consumption effectiveness, enhanced desire for food, and extra common exterior of the fur hair and hide. Thus, antimicrobials have been used as a feed preservative globally for many years. The extent of the improvement in growth rate is based on the type of antimicrobial, nourishment stage, farm atmosphere, and swine conditions (Cromwell, 2002).

4.3.2 Use of antimicrobials in goats and sheep

Research has provided information on antimicrobial use in livestock spp., including cattle and swine, but there is little information on drug use practices for sheep and goats in different countries (Acar et al., 2000; Menzies, 2000). In these countries they are considered as minor spp. They are food-producing animals that do not have a large economic footprint. They are not often targeted for drug

Table 4.1 Antimicrobial use for treatment of diseases in pigs.

Therapeutic areas	Diseases	Causative microbes	Antimicrobial use
Gastrointestinal tract	Edema disease	*Escherichia coli*	Trimethoprim, sulfonamides, aminopenicillins
	Pneumonia (bronchopneumonia) (Porcine enzootic - pneumonia)	*Streptococci* *Pasteurella multocida* *Mycoplasma hyopneumoniae*	Benzylpenicillin, trimethoprim, sulfonamides Tiamulin, lincomycin, tetracyclines
Reproductive tract	Balanoposthitis	*Actinobaculum suis*	Benzylpenicillin, aminopenicillins
	Metritis (endometritis)	Coliforms, Gram-positive bacteria	Trimethoprim, sulfonamides, benzylpenicillin
Kidneys and urinary tract	Urinary tract infection (Cystopyelonephritis)	*Actinobaculum suis*	Benzylpenicillin, aminopenicillins
	Inflammation of the bladder (Cystitis)	*E. coli*	Trimethoprim, sulfonamides, aminopenicillins
Respiratory tract	Progressive atrophic rhinitis (Soontornvipart, Kohout et al.)	Toxigenic *P. multocida*	Tetracyclines, trimethoprim
	Pleuropneumonia	*Actinobacillus pleuropneumoniae*	Benzylpenicillin, tiamulin, tetracyclines
	Dysentery	*Brachyspira hyodysenteriae*	Tiamulin (The Finnish Food Safety Authority, 2018)
Central nervous system	Meningitis	*Streptococcus suis*	Benzylpenicillin, aminopenicillins
Musculoskeletal system	Lameness	*S. suis*	Sulfonamides, cephalosporins, aminoglycosides
	Arthritis	*S. suis*, *Mycoplasma hyosynoviae*	Penicillins, lincosamide, tylosin
Mammary gland	Postpartum dysgalactia syndrome (PPDS)	Gram-negative bacteria (mostly *E. coli*)	Trimethoprims, sulfonamides, aminopenicillins
	Chronic mastitis	Gram-positive bacteria	Benzylpenicillin, aminopenicillins (De Briyne et al., 2014a)
	Acute mastitis	Gram-negative bacteria	Trimethoprim, sulfonamides, aminopenicillins (The Finnish Food Safety Authority, 2018)

development and approval. Additionally, the market for drug use is small, resulting in limited financial commitments from pharmaceutical companies. With sparse clinical data generated in North America supporting drug use, it is difficult to license drugs for use in these species. Veterinarians and sheep producers therefore have a limited selection of licensed drugs (Navarre and Marley, 2006). It is thought that much of the drugs used is extra-label drug use (ELDU), which means usage is not in accordance with information mentioned on its label, package insert, or product monograph. Penicillin, tetracycline, oxytetracycline, and florfenicol are extra-label drugs generally used (Fajt, 2001).Antimicrobials used in goats and sheep are listed in Table 4.2.

4.3.3 Use of antimicrobials in cattle and cows

Cattle production is the third largest animal farming in the world (approximately 65 million globally), after swine and poultry (Food et al., 2014). China (6.7 million), Brazil (9.6 million), the United States (US) (11.4 million), the 28 member countries of the European Union (7.5 million), and India (4.5 million) are the fundamental cattle-producing countries in the world, resulting in an excess of one billion cattle population in 2015 (Ali et al., 2016). Cattle raising at massive levels normally involves moving animals from cow-calf systems (a permanent herd used to produce young beef), to back grounding (postweaning intermediate feeding, normally forage-based diets) and feedlot (a building where livestock are fattened for market, usually with high-energy grain-based diets). For the treatment and prevention of diseases in cattle and cows, antimicrobials can be administered in live cattle at any developmental stage (Podberscek, 2009). In the husbandry environment, cattle can be more prone to endemic pathogens. These pathogens are normally neglected, causing severe damage to animal health, and affecting herd growth and farm productivity. The chances of transmission of diseases cause significant economic pressure for antimicrobial use against bovine infectious diseases (Radostits et al., 2006; Van Epps and Blaney, 2016). A number of the antimicrobials being used for the treatment of infections in cattle and cows are mentioned in Table 4.3. Commonly used antimicrobials for treatment of diseases in cattle and cows are trimethoprim/sulfonamides, oxytetracycline, benzylpenicillin, and polymyxin B (The Finnish Food Safety Authority, 2018).

4.3.4 Use of antimicrobials in horse

Meat is one of the major sources of nutrients in human food, for its contribution of high-biological-value protein. Recently, there has been an interest in meat from alternative sources, other than bovine, swine, and poultry. The main producers of horsemeat are China, Kazakhstan, Mexico, Russia, and Argentina, while Mongolia, Switzerland, Italy, Kazakhstan, and Russia are the largest consumers (Vanegas Azuero and Gutiérrez, 2016). It is very challenging to estimate the antimicrobial use in horses. Deciding the volume of antimicrobials to be used in horses in most countries is difficult, if not impossible. As a result, antimicrobials administration strategy and estimation of antimicrobials usage is difficult. There are insufficient answers to the questions of "how much" and "how are" these antimicrobials are administered in horses (Weese, 2015). The major spp. of bacteria detected at the onset of instillation were *Staphylococcus aureus*, *Streptococcus equi* subsp. *Zooepidemicus*, *Acinetobacter lwoffii*, *Staphylococcus xylosus*, *Staphylococcus vitulinus*, *Enterobacter agglomerans*, *Flavimonas oryzihabitans* and *Staphylococcus sciuri* (HIDAKA et al., 2015). Aminoglycosides (e.g., gentamicin, or amikacin) are concentration-dependent bactericidal drugs, therefore the higher the drug

Table 4.2 Antimicrobials use for treatment of diseases in goats and sheep.

Therapeutic areas	Diseases	Causative microbes	Antimicrobial use
Gastrointestinal tract	Colibacillosis	Enterotoxigenic *Escherichia coli*	Broad-spectrum antimicrobials
	Salmonella dysentery	*Salmonella typhimurium*	Broad-spectrum antimicrobials
	Abomasitis	*Clostridium* spp.	Oral penicillins
	Coccidiosis	*Eimeria* spp.	Salinomycin, decoquinate, amprolium, and sulfonamides
Reproductive tract	Enzootic abortion of ewes	*Chlamydophila abortus*	Tetracycline, oxytetracycline, tylosin
	Campylobacter abortion	*Campylobacter jejuni*, *Campylobacter fetus* spp.	Oxytetracycline, sulfamethazine penicillin G, streptomycin, tetracycline and tylosin
	Listeria abortion	*Listeria monocytogenes*	Oxytetracycline
	Toxoplasma abortion	*Toxoplasma gondii*	Decoquinate
	Salmonella abortion	*S. typhimurium*, *salmonella abortusovis*, *salmonella dublin*	Broad-spectrum antimicrobial
	Leptospira abortion	*Leptospira hardjo*, *leptospira Pomona*	Penicillin G, streptomycin, tetracyclines
	Brucella ovis ram epididymitis	*Brucella ovis*	Dihydrostreptomycin with oxytetracycline
Respiratory tract	Pneumonic pasteurellosis	*Mannheimia haemolytica*, *Pasteurella multocida*	Tilmicosin, oxytetracycline, ceftiofur, florfenicol
	Pasteurella septicemia (Sheep)	*Bibersteinia trehalosi*,	Tilmicosin
	Necrotic laryngitis	*Fusobacterium necrophorum*	Penicillin G, oxytetracycline
	Mycoplasma pneumonia	*Mycoplasma arginine*, *Mycoplasma ovipneumoniae*	Tylosin, oxytetracycline
Kidneys	Enterotoxaemia	*Clostridium perfringens* type C and D	Oral virginiamycin, Penicillin G
Urinary tract	Leptospirosis	*Leptospira interrogans*	Dihydrostreptomycin, oxytetracycline
	Cystitis	*Corynebacterium renale*, other spp.	Broad-spectrum antimicrobials
Musculoskeletal system and foot	Contagious foot rot	*Dichelobacter Nodosus*, *F. necrophorum*	Oxytetracycline

Table 4.2 Antimicrobials use for treatment of diseases in goats and sheep.—cont'd

Therapeutic areas	Diseases	Causative microbes	Antimicrobial use
	Foot scald	*F. necrophorum*	Zinc sulfate foot bath
	Polyarthritis	*Chlamydophila pecorum*	Oxytetracycline
	Polyarthritis (goats)	*Mycoplasma mycoides* (other *Mycoplasma* spp.)	Oxytetracycline, tylosin
Mammary gland	Gangrenous mastitis	*Staphylococcus aureus, M. haemolytica*	Tilmicosin
	Contagious agalactia	*Mycoplasma agalactiae, Mycoplasma mycoides*	Tetracyclines, tylosin
	Subclinical and clinical mastitis	*S. aureus, M. haemolytica*	Tilmicosin, cloxacillin, cephapirin,
Central nervous system	Bacterial meningitis	Many spp.	Broad-spectrum antimicrobials
	Listeriosis	*L. monocytogenes*	Oxytetracycline, penicillin G
Oral cavity	Tooth root abscess	Many spp.	Oxytetracycline, florfenicol
	Actinobacillosis	*Actinobacillus lignieresii*	Sodium iodide
	Actinomycosis	*Actinomyces bovis*	Sodium iodide, sulfadimethoxine, isoniazid
Eyes	Pinkeye (infectious keratoconjunctivitis)	*Chlamydia psittaci, Mycoplasma conjunctivae, Neisseria*	Spiramycin, oxytetracycline, tiamulin
Skin	Secondary infection of contagious ecthyma (Benkendorff)	*S. aureus*	Tilmicosin, oxytetracycline, ampicillin
	Dermatomycosis (lumpy wool in sheep	*Dermatophilus congolensis*	Oxytetracycline

concentration, the greater the bactericidal effect use against skin, subcutaneous tissue, eye, and urinary tract infections in horses (Papich, 2001; Williams and Pinard, 2013; Carapetis et al., 2017). Beta-lactam antibiotics such as penicillins, potentiated aminopenicillins, and cephalosporins are slowly bactericidal and used for the treatment of urinary tract, skin, subcutaneous tissue, and respiratory tract infections (Papich, 2001; Gordon and Radtke, 2017b; Wilson, 2001). A summary of the major antimicrobials used in horses is provided in Table 4.4.

4.3.5 Use of antimicrobials in poultry

Poultry is one of the world's most popular food industries. Poultry refers to the breeders and production animals of broilers, chickens, and turkeys. Chicken is the most frequently farmed spp., producing more than 90 billion tons of chicken meat per year (Agyare et al., 2018). Most countries use a wide variety of

Table 4.3 Antimicrobials used for treatment of diseases in cattle and cows.

Therapeutic areas	Diseases	Causative microbes	Antimicrobial use
Gastrointestinal tract	Coccidiosis	*Eimeria*	Trimethoprim, sulfonamides
	Diarrhea (neonatal diarrhea) Diarrhea (in preweaning calves)	*Escherichia coli* Several viruses, bacteria	Trimethoprim, sulfonamides
Respiratory tract	Bovine respiratory disease (BRD)	*Mannheimia haemolytica*, *Pasteurella multocida*, *Histophilus somni*, *Ureaplasma* sp. *Mycoplasma* sp.	Oxytetracycline, Benzylpenicillin, macrolides
	Ovine respiratory disease (pneumonia)	*M. haemolytica*, *Pasteurella multocida* mycoplasma	Benzylpenicillin, oxytetracycline
Reproductive tract	Acute metritis	*Trueperella pyogenes*, *E. coli*, *streptococci*, *staphylococci*	Benzylpenicillin, oxytetracycline
Urinary tract	Cystitis	*Corynebacterium renale*, *E. coli*	Benzylpenicillin
Musculoskeletal system	Arthritis	*T. pyogenes*, *E. coli*, *Mycoplasma bovis* other bacteria	Benzylpenicillin, oxytetracycline
	Cellulitis, Bursitis	*T. pyogenes*, *Escherichia coli*, *Streptococci*, *Staphylococci*	Benzylpenicillin
Skin	Interdigital phlegmon	*Fusobacterium necrophorum*, *Dichelobacter* (former *Bacteroides nodosus*)	Benzylpenicillin, Oxytetracycline
	Digital dermatitis	*Treponema* spp.	Oxytetracycline
Eye	Infectious keratoconjunctivitis	*Listeria monocytogenes* *Moraxella* spp mycoplasma	Benzylpenicillin, polymyxin B + oxytetracycline (applied locally)
	Uveitis	*L. monocytogenes*	Polymyxin B + oxytetracycline, benzylpenicillin
Others infections	Systemic infections of newborn ruminants (omphalitis, polyarthritis, meningitis, sepsis)	Several bacterial spp. (*E. coli*, *T. pyogenes*, *Streptococci*, *Staphylococci*)	Trimethoprim, sulfonamides, oxytetracycline, Benzylpenicillin + enrofloxacin
	Umbilical infections	*T. pyogenes*, *streptococci*, *staphylococci*	Benzylpenicillin

Table 4.3 Antimicrobials used for treatment of diseases in cattle and cows.—cont'd

Therapeutic areas	Diseases	Causative microbes	Antimicrobial use
	Listeriosis	*L. monocytogenes*	Benzylpenicillin, oxytetracycline
	Tick-borne fever	*Anaplasma phagocytophilum*	Oxytetracycline
	Necrobacillosis	*Fusobacterium necrophorum*	Benzylpenicillin, oxytetracycline (The Finnish Food Safety Authority, 2018; De Briyne et al., 2014a)

antimicrobials to grow poultry (Sahoo et al., 2010; Landers et al., 2012; Boamah et al., 2016). In order to meet the demand, initially scientists began to look for ways to produce more meat at a relatively cheaper level, resulting in the use of antimicrobial agents (Dibner and Richards, 2005). Poultry diseases always involve an entire flock falling ill, which prompts a decision on administration of medicine that must be taken. Several factors affect the decision and the most important of them is the cause behind the disease. Before initiating the treatment, dead or euthanized broilers, chickens, and turkeys, samples of their organs or blood or bacterial samples must be sent for testing to obtain diagnosis. Conducting a field diagnosis is difficult, and antimicrobials are all too often prescribed for precautionary reasons. Phenoxymethyl-penicillin, amoxicillin, and trimethoprim/sulfonamides are mostly used for treating gastrointestinal tract infections, arthritis, and other systemic infections (The Finnish Food Safety Authority, 2018). A summary of the major antimicrobials used in poultry is listed in Table 4.5.

4.3.6 Use of antimicrobials in cats and dogs

The issue of eating dogs and cats is highly emotive, especially in countries like the United Kingdom and United States. In these countries, the idea of consuming a cat or a dog is considered as heinous and amoral, as in United Kingdom and United States cats and dogs are mainly kept as pet animals. Regions where there are records of dog eating include Southeast Asia and Indochina, North and Central America, parts of Africa, and the islands of the Pacific. During the Stone Age and Bronze Age, dog eating was apparently also common in Europe. Still less has been written or discovered about the eating of domestic cats. It has a briefer history than dog eating and the level of consumption of cat meat is also comparatively low. Nowadays, the consumption of dogs and cats still occurs in a number of countries, including China, Thailand, Cambodia, and Vietnam. In 1996, it was proclaimed that dog meat was still being eaten in parts of Eastern Switzerland. It has been estimated that in Asia, around 13–16 million dogs and 4 million cats are eaten each year (Podberscek, 2009). A number of the antimicrobials being used for the treatment of infections in cats and dogs are mentioned in Table 4.6. Commonly used antimicrobials in dogs and cats are beta lactams (particularly cephalexin and amoxicillin-clavulanate in dogs). In dogs, trimethoprim-sulfonamides are the second-most used antimicrobials after beta lactams. In cats, macrolide-lincosamides, azithromycin, and erythromycin are the second-most common class of antimicrobials after beta lactam use for treating skin, ear, eyes, and oral cavity and gastrointestinal tract infections (Wael and Husein, 2011; Nuttall, 2016; Winer et al.,

Table 4.4 Antimicrobials used for treatment of diseases in horse.

Therapeutic areas	Diseases	Causative microbes	Antimicrobial use
Skin, subcutaneous tissue	Superficial pyoderma	*Staphylococcus* spp	Aminoglycosides, Enrofioxacin.
	Wounds and abscesses	*Streptococcus* (Vanegas Azuero and Gutiérrez, 2016) *Staphylococcus aureus*	Sulfamethoxazole-trimethoprim, cephalexin, clindamycin, erythromycin (Bowen et al., 2017)
	Cellulitis	*Staphylococcus* and *Streptococcus*	Penicillin-aminoglycoside, trimethoprim-sulfadiazine
	Lymphangitis	*Corynebacterium*, *Pseudo tuberculosis*, *Histoplasma*, *Farciminosum*	Analgesic drugs (morphine) (Fjordbakk et al., 2008), ceftiofur, cefazolin, rifampin, penicillin G, ampicillin, amikacin, enrofloxacin (Wilson, 2001)
Eyes	Corneal ulcers	*Streptococcus equi*, *Zooepidemicus*, *Pseudomonas*, *Aeruginosa*, *Staphylococcus* spp.	Aminoglycosides (solution of gentamycin or tobramycin), Fluoroquinolones (Williams and Pinard, 2013)
Respiratory tract	Sinusitis	*S. equi*, *Streptococcus*, *Zooepidemicus*	Penicillin, TMS, and/or metronidazole (Gordon and Radtke, 2017a)
	Pneumonia	*S. zooepidemicus*, *Rhodococcus equi*, *Klebsiella*, pneumonia.	Erythromycin or azithromycin penicillin G, ceftiofur, TMS, ampicillin-gentamicin
Urinary tract	Cystitis	*E. coli*, *Streptococcus* sp and *Staphylococcus* sp.	Aminoglycosides, gentamicin-penicillin G, or ampicillin (Wilson, 2001)
	Acute Pleuropneumonia	Gram-positive aerobes: (*S. Zooepidemicus*) Gram-negative aerobes	Penicillin G or ampicillin-gentamicin, metronidazole
Mammary gland	Mastitis	*S. zooepidemicus*, *E.coli*, *Klebsiella*, Pneumoniae	Penicillin, gentamicin, amikacin, cephalothin
Urinary tract	Cystitis	*E. coli*, *Streptococcus* sp, and *Staphylococcus* sp.	Aminoglycosides, gentamicin-penicillin G or ampicillin

Table 4.4 Antimicrobials used for treatment of diseases in horse.—cont'd

Therapeutic areas	Diseases	Causative microbes	Antimicrobial use
Musculoskeletal system	Septic arthritis	*R. equi*, *Streptococcus*, *Zooepidemicus*	Cefazolin or cephalothin amikacin, gentamicin, oxacillin
	Osteomyelitis	Enterobacteriaceae, *Streptococcus* sp, *Staphylococcus* sp.	Cefazolin or cephalothin amikacin, gentamicin, oxacillin
Oral cavity and gastrointestinal tract	Inflammatory bowel disease (IBD)	Etiology unclear	Dexamethasone, omeprazole (Boshuizen et al., 2018)
	Acute colitis	*Clostridium difficile*, *C. perfringens* (lesser extent)	Acute colitis
	Diarrhea	*Salmonella* sp.	Gentamicin, metronidazole (Wilson, 2001)

2016; Gómez-Poveda and Moreno, 2018). Fluoroquinolones are used commonly in both cats and dogs for treating skin, reproductive tract, and ear infections (Hölsö et al., 2005; Wael and Husein, 2011; Nuttall, 2016; Adel and Khadidja, 2017).

4.3.7 Use of antimicrobials in rabbits

Rabbits are small mammals used as food animals. Colibacillosis is a widespread ailment in rabbit's reproduction. It has become one of the chief contagious illnesses that cause danger in the rabbit farming industry (Milon et al., 1999). In commercial farming, the European rabbit (*Oryctolagus cuniculus*), a lactate female can produce a total quantity of milk corresponding to her body mass via four to five pairs of mammae in 35 days (Rosell and de la Fuente, 2018). This attempt predisposes to diseases that affect the mammae throughout lactation. It includes mostly mammary gland microbial diseases like mastitis. In rabbits, diseases by *Pasteurella multocida* are also widespread. Pasteurellosis in them present rhinitis with blood-stained nasal discharge, pneumonia, serous otitis media, pyometra, orchitis, pustule, and septic infection (Langan et al., 2000). Respiratory anthrax is another lethal illness in rabbits in the absence of earlier treatment with antimicrobials. Rabbits are extremely vulnerable to diseases caused by *Bacillus anthracis* spores via intranasal instillation; they succumb within 2–4 days post illness. For the prevention and cure of diseases, different antimicrobials including penicillin, ampicillin, linezolid, chloramphenicol, rifampin, vancomycin, ciprofloxacin, levofloxacin, moxifloxacin, doxycycline, amoxicillin, clindamycin, and meropenem are used (Li et al., 2013). Antimicrobials used for rabbits are listed in Table 4.7.

4.4 Aquaculture

Aquaculture includes all forms of culturing aquatic animals and plants in marine, fresh, and brackish environments (Pillay and Kutty, 2005). Aquaculture is one of the most hopeful alternatives for

Table 4.5 Antimicrobials used for treatment of diseases in poultry.

Therapeutic area	Disease	Causative microbe	Antimicrobial use
Gastrointestinal tract	Necrotic enteritis	*Clostridium perfringens*	Phenoxymethyl-penicillin* amoxicillin*, tylosin, trimethoprim-sulfonamides, tetracycline
Musculoskeletal system	Tenosynovitis in broiler breeders	*Staphylococcus aureus*	Phenoxymethyl-penicillin* amoxicillin*, trimethoprim
	Arthritis (in turkeys)	*S. aureus*	Phenoxymethyl-penicillin*, amoxicillin*, trimethoprim-sulfonamides, tetracycline
Other infections	Erysipelas	*Erysipelothrix rhusiopathiae*	Phenoxymethyl-penicillin*, amoxicillin*, trimethoprim-sulfonamides, tetracycline
	Pasteurella infection (in adult chickens and turkeys)	*Pasteurella multocida*	Phenoxymethyl-penicillin*, amoxicillin*, trimethoprim-sulfonamides, tetracycline
	Colibacillosis (a systemic infection)	*Escherichia coli*	Amoxicillin*, tetracycline (De Briyne et al., 2014a; The Finnish Food Safety Authority, 2018)

proficiently and sustainably increasing the production of animal proteins (Liao and Chao, 2009). To meet the protein demands of growing global populace, aquaculture is considered to be the fastest protein food production sector and accounts for 50% of overall food supply (Okocha et al., 2018). Aquaculture include fishes, catfish , Atlantic salmon, rainbow trout, tilapia, Pacific oyster, Eastern oyster, Pacific white shrimp, yellow perch, and bluegill sunfish, mollusks, etc. Nowadays, freshwater pool aquaculture farming major product is finfish. For freshwater farming, diverse sources of water supply like containers, pools, streams, and channels are used. Aquaculture farming generally consists of cage culture for sea finfish and freshwater pools or brackish water for crustaceans (Hall, 2011). Whilst many cages and pools benefit from natural water exchange used for provision of oxygen and waste disposal, simultaneously the fish and crustaceans are exposed to illness-causing microbes present in the water. To improve aquaculture farming, new techniques have been developed including closed recirculating aquaculture systems that decrease the risk of disease vectors and wastes (Martins et al., 2010).

Table 4.6 Antimicrobials used for treatment of diseases in cats and dogs.

Therapeutic areas	Diseases	Causative microbes	Antimicrobial use
Skin	Lesion and superficial skin inflammation	*Staphylococcus pseudintermedius*	Fluoroquinolones
	Superficial skin infection (hair follicle infection, impetigo)	*Staphylococcus intermedius*	Amoxicillin, clavulinic acid, cefotraxione, ciprofloxacin (Wael and Husein, 2011)
	Deep skin infection (canine pyoderma)	*Staphylococcus intermedius*	Clavulanate, amoxicillin, lincomycin, clindamycin, cefovecin, cefpodoxime, marbofloxacin, difloxacin, orbifloxacin, enrofloxacin, azithromycin, tobramycin, pradofloxacin, rifampin, amikacin, netilmicin, chloramphenicol, gentamicin (Ferran et al., 2016)
	Wounds and abscess	*Pasteurella multocida*, *Staphylococcus* spp., obligate anaerobes	Amoxicillin, cephalosporins, Fluoroquinolones, metronidazole, clavulunate (Roy et al., 2007; Little and Kennedy, 2010)
Ear	Otitis externa	*Staphylococcus*, *Malassezia*, *Pseudomonas*	Polymixin B, fusidic acid, florfenicol, gentamicin, enrofloxacin, neomycin marbofloxacin, cefadroxil, amoxicillin, miconazole, clindamycin, lincomycin, fluoroquinolones (Nuttall, 2016)
Respiratory tract and thoracic cavity	Pneumonia	*Streptococcus canis*, *Mycoplasma* spp., Chlamydia felis, *Bronchiseptica*, *Zooepidemicus*, *Bordetella*	Doxycycline, fluoroquinolone' penicillin, clindamycin ampicillin-sulbactam
	Bronchitis		

Continued

Table 4.6 Antimicrobials used for treatment of diseases in cats and dogs.—cont'd

Therapeutic areas	Diseases	Causative microbes	Antimicrobial use
		Chlamydophila spp., *Escherichia coli*, *Mycoplasma* spp., *streptococcus* spp, *Bronchiseptica*, *Bordetella*	Doxycycline, amoxicillin, clavulanate, amikacin
	Pyothorax	*Fusobacterium*, *Prevotella*, *Clostridium*, *Bacteroides*, *Peptostreptococcus*, *Streptococcus* spp., *Mycoplasma* spp. *Porphyromonas*, *Pasteurella* spp, *actinomyces*	Enrofloxacin, marbofloxacin, penicillin clindamycin(Lappin et al., 2017)
Oral cavity and gastrointestinal tract	Gingivitis, periodontitis	*Streptococcus*, *staphylococcus*, *Enterococcus*, *actinomyces* sp., *lactobacillus* sp. (mostly in dogs) (Pieri et al., 2012)	Tetracyclines, clindamycin amoxicillin, clavulanate metronidazole (Hale and FAVD)
	Root abscess	*Pasteurella*, anaerobes	Chloramphenicol, azithromycin, metronidazole, amoxicillin, (Winer et al., 2016)
	Inflammatory bowel disease	*Campylobacter jejuni*, *Clostridium difficile*, *Clostridium perfringens*, *Salmonella* (Honneffer et al., 2014)	Tylosin, oxytetracycline, metronidazole (Simpson and Jergens, 2011)
	Anal sac inflammation	Gram-positive cocci, gram-negative cocci, gram-positive rods, gram-negative rods (Frankel et al., 2008)	Cefovecin, enrofloxacin, orbifloxacin
	Diarrhea	*Enterobacteriaceae*, *streptococcus* Gamma-, beta-Proteobacteria, *Bacilli*, *Collinsella*, *Clostridium* (Suchodolski et al., 2015)	Metronidazole (De Briyne et al., 2014a)

Table 4.6 Antimicrobials used for treatment of diseases in cats and dogs.—cont'd

Therapeutic areas	Diseases	Causative microbes	Antimicrobial use
Reproductive tract	Prostatitis	*Staphylococcus* spp., *Klebsiella* spp., *E. coli*, *Pseudomonas* spp., *Pasteurella* spp., *Mycoplasma* spp., *Ureaplasma* spp. (Nižański et al., 2014)	Tetracycline fluoroquinolone, sulfamethoxazole, chloramphenicol, trimethoprim (Sykes, 2013; Adel and Khadidja, 2017)
	Pyometra	*E. coli*	Sulfadoxine, amoxicillin, trimethoprim, clavulanic acid (Fieni et al., 2014)
Urinary tract	Urinary tract infections (UTIs)	*E. coli* (52.5%), *Staphylococcus* spp., *Enterococcus* spp.	Amoxicillin, cephalexin, sulfamethoxazole, clavulanic acid, enrofloxacin (Wong et al., 2015)
Musculoskeletal system	Arthritis	*Staphylococcus aureus*, *staphylococcus* spp. *Pseudomonas aeruginosa* (Marchandeau et al., 2014)	Amoxicillin-clavulanic acid, cephalosporin, clindamycin, enrofloxacin amikacin, azathioprim (Soontornvipart et al., 2003)
Eyes	Conjunctivitis	*Enterococcus* spp., *Micrococcus* spp., *Pseudomonas* spp. *Pasteurella* spp., *staphylococci*, *Mycoplasma*, *Bacillus* spp., (Płoneczka-Janeczko et al., 2017)	Penicillins, fusidic acid, cephalosporins, aminoglycosides, oxytetracycline, polymyxin, erythromycin (Gómez-Poveda and Moreno, 2018)
	Melting keratitis	*Pseudomonas aeruginosa*, Staphylococcus and Streptococcus spp.	Atropine, cyprofloxacincollyre, hyaluronic acid, NAC 10% (Ion et al., 2015)
Other	Leptospirosis	*Leptospira* (Miotto et al., 2018)	Penicillins, doxycycline
	Lyme borreliosis	*Borrelia* spp.	minocycline, amoxicillin clarithromycin, ceftriaxone, erythromycin, cefotaxime, doxycycline (Littman et al., 2018)

Table 4.7 Antimicrobials used for treatment of diseases in rabbits.

Therapeutic area	Disease	Causative microbe	Antimicrobial use
Respiratory tract	Pasteurellosis	*Pasteurella multocida* (*coccobacillus*)	Enrofloxacin, (1−3), (1−6) β-glucans (Palócz et al., 2014)
Gastrointestinal tract	Epizootic Rabbit Enteropathy	*Escherichia coli*, *Haemophilus paracuniculus* , *Proteus mirabilis*, *Citrobacter* spp. and *Klebsiella* spp.	Lincomycin, spectinomycin and neomycin, tylosin, apramycin, bacitracin, tiamulin (Puón-Peláez et al., 2018)
	Diarrhea	*E. coli*, *Eimeria* spp., adenovirus, coronavirus, *salmonella* spp., *Yersinia* spp.	Sulfonamides, tetracyclines, and neomycin (Banerjee et al., 1987)
	Intestinal Coccidiosis	*Eimeria* spp.	Prophylactic (bifuran, sulfa drugs, amprosol) (Bhat et al., 2010)
Eyes	Myxomatosis	*Myxoma* virus	Vaccination (Marchandeau et al., 2014)
	Phacoclastic uveitis	*Encephalitozoon cuniculi*	Surgery (phacoemulsification) (enucleation) (Sandmeyer et al., 2011)
Skin and Musculoskeletal system	Pododermatitis	*Pasteurella* sp. or *Staphylococcus aureus.*	Antiseptic products: salicylic acid mupirocin, neomycin
	Abscesses	*Pasteurella* sp. or *S. aureus*	Cephalosporin or azithromycin (Esther van Praag)
Ear	Otitis media/interna	*Pasteurella multocida*, *Streptococcus* spp., *E. coli*, *Enterococcus* spp., *Pseudomonas* spp.	Chloramphenicol and penicillin, ciprofloxacin, enrofloxacin-marbofloxacin-penicillin chloramphenicol

4.5 Global aquaculture trends

Worldwide aquaculture has developed significantly over the past 50 years to around 52.5 million tons (68.3 million, counting sea-going plants), in 2008 worth US$98.5 billion (US$106 billion, counting oceanic plants) and contributing to around 50% of the world's aquatic food supply. Asia dominates this production, accounting for 89% by capacity and 79% by cost; among Asian countries, China is the leading producer (32.7 million tons in 2008). The speedy development in this region is due to different factors that include preexisting aquaculture practices, populace and financial development, relaxed regulatory framework, and expanding export opportunities. Countries contributing more in aquaculture than wild-caught fish are China, India, Vietnam, Bangladesh, and Egypt. The top 15 aquatic culture–producing countries in 2010 by percentage of total worldwide production appear in Fig. 4. 2. It has been reported that China is contributing more than 60% of the worldwide aquaculture production and also using bulk antimicrobials to guarantee sufficient production and disease management. During the 1980s–1990s, development in aquaculture was rapid in Europe and North America, but since this time it has stagnated, probably owing to administrative restrictions on sites and other competitive factors, though they have continued the growth of markets for fish and seafood (Bostock et al., 2010).

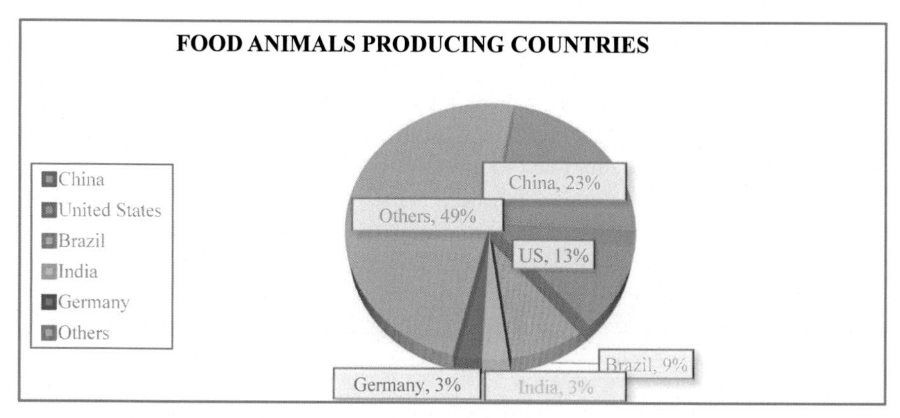

FIGURE 4.1

Top five countries and percentages of their shares in global antimicrobial consumption in food animals in 2010 (Van Boeckel, Brower et al., 2015).

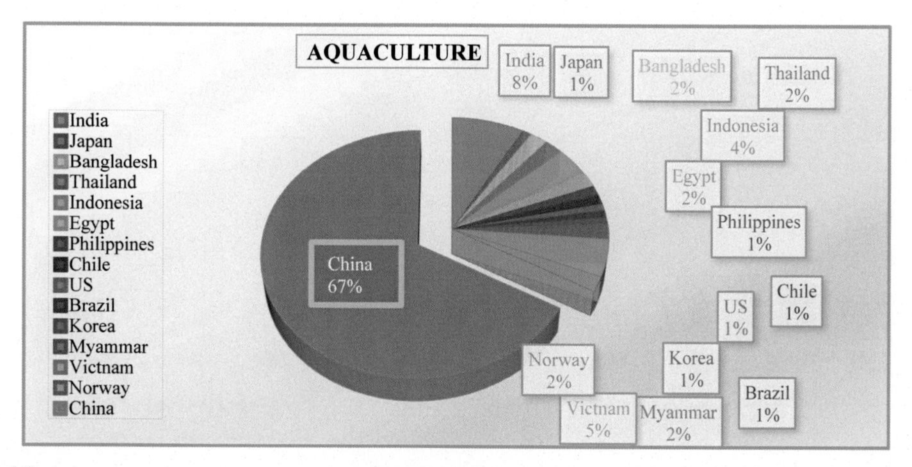

FIGURE 4.2

Top 15 aquatic culture–producing countries in 2010 by percentage of total worldwide production (FAOSTAT., Accessed Nov 24, 2014. http://www.faostat.fao.org.)

By 2015 we outreached to a point where the marine food consumed worldwide was ~160 million metric tons, Mmt and was grown in farms rather than taken from natural sources. This 80 Mmt of farmed marine food consisted of fish, shellfish, shrimps, and seaweed, with approximately 90% farmed in Asia. By 2050, it is expected that worldwide aquaculture production will double, with well-managed fisheries predicted to demise over this time period. Undoubtedly, aquaculture will be a major contributor to the protein supply for the future overall diet (Stentiford, February 2, 2017). Fig. 4.3 indicates aquatic culture expanding to meet world fish demand.

4.6 Need for aquaculture

In the last 60 years, swine, poultry, and cattle production has increased globally while the poultry production surpassed the others. Around 1985, aquaculture was the only animal producing industry globally. Prior to this era, aquaculture was considered as a noncommercial matter, a traditional way of life and a source of nourishment for its producers. Increased demand for a healthy choice of protein, enriched seafood feed production, reduction of numbers of wild fish, and advanced farming techniques have led to high-density fish production in recent years (Cole et al., 2009). Up to 80% reduction in global finfish and shellfish stocks (collectively as "fish") occur in order to provide food to increasing

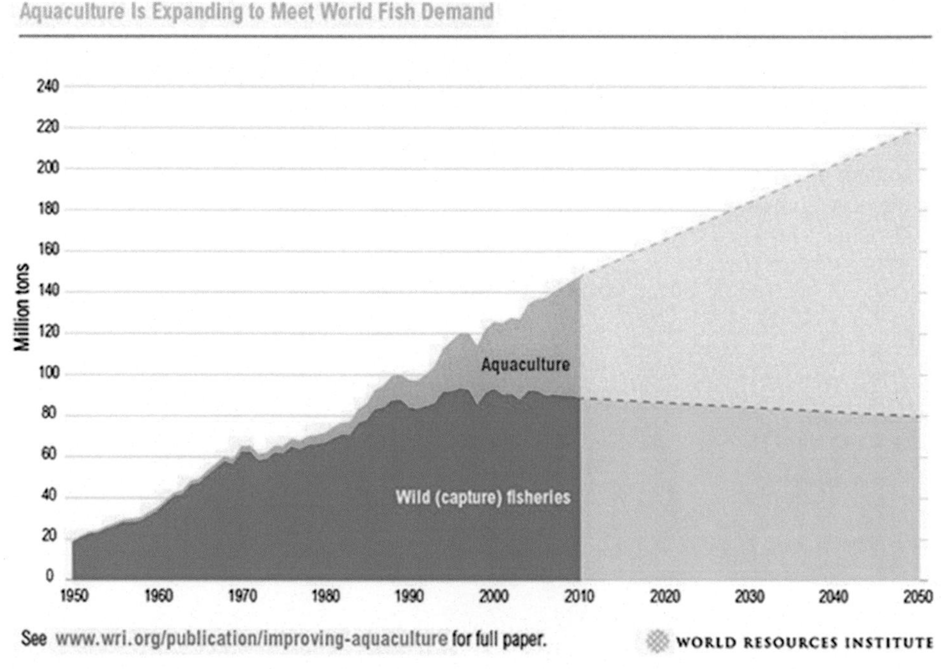

FIGURE 4.3

Aquatic culture expanding to meet world fish demand.

Adapted from (Waite, R. June 2014. Aquaculture Is Expanding to Meet World Fish Demand. *World Resources Institute: https://www.wri.org/resources/charts-graphs/aquaculture-expanding-meet-world-fish-demand).*

human population day by day (Pauly and Zeller, 2016, 2017). Due to its capability to provide sensible, secure, reliable, and alternative food producing systems, our dependency on aquaculture has increased to reduce this extinction. Aquaculture systems produced 70.5 million tons of food fish and 26.1 million tons of aquatic algae in 2014. According to aquaculture production figures, a significant contribution of aquaculture resulted in increasing the total fish consumption from 1962 to 2002 from 5% to 49%, respectively. It is predicted that the production of aquaculture in Europe will reach 4 million tons by 2030 (Pauly and Zeller, 2017). This globally increased production of aquaculture has resulted in new and improved farmed spp., more than 580 species in total (consisting of 362 finfish and 62 crustaceans), with a wide range of growth and maximum production conditions (Pauly and Zeller, 2017; Naylor et al., 2000).

4.7 Legislation concerning antimicrobial use in aquaculture

Antimicrobial usage in aquaculture farming is governed by a range of factors that include laws and policies by the particular management union, the unique microbes that exist and their antimicrobial sensitivities, the medication period, the illness condition of the host, and the system framework, like saltiness, temperature, light phase, etc. Statistics related to the antimicrobial dosages used in aquaculture are limited, as only some countries scrutinize the amount of antimicrobials used (Sapkota et al., 2008). Specifically in Europe, North America, and Japan, policies regarding antimicrobial consumption are stringent and fewer antimicrobials are approved for use in aquaculture farming. In 2001, the European Veterinary Medicinal Products Directive, as amended and codified in Directive (2001)/82/EC, excluded the prophylactic usage of antimicrobials for aquaculture (Committee, 2004; Watts et al., 2017). In spite of productivity rate >20-fold, Norway instituted strict regulations for antimicrobial usage, with 99% decrease between 1987 and 2013 in combination with improved vaccinations; their outstanding stewardship has been certified (Watts et al., 2017).

4.8 Antimicrobial agents used in aquaculture

Aquaculture leads to the endorsement of conditions that facilitate spreading the number of illnesses and harms. A broad variety of compounds are used in aquaculture farming that include antimicrobials, pesticides, hormones, anesthetics, a variety of pigments, mineral deposits and vitamins, though not all of them are antimicrobials. As with livestock animal production, antimicrobials are also used in aquaculture in attempting to cure ailments (Burka et al., 1997). Antimicrobial treatment patterns also differ among countries and among individual aquaculture farming within the same country. The most important reason to follow antimicrobial usage is to control contagious illnesses in breeding areas, to avert losses in farming; to limit the introduction of microbes to new facilities when larvae, fry, or brood stock are moved; to limit the disease spreading to natural fish via the reproduction sewage or when cultured fish are stocked out; and to avoid the strengthening of microbes previously widespread in a watershed (Phillips et al., 2004). There are inadequate statistics concerning antimicrobial usage in global aquaculture. For many of the farmed genuses, we are deficient in sufficient information regarding pharmacokinetics (Sapkota et al.) and pharmacodynamics of drug administration. The drugs available for the cure of widespread transmittable illnesses are becoming increasingly inadequate and costly and, in several situations, not available due to the emergence of drug resistance that is shocking and erases the previous 60 years' medicinal developments (Serrano, 2005). In aquaculture farming antimicrobials have been used predominantly for curative purposes and as prophylactic agents (Serrano, 2005; Shao, 2001). Drugs are rarely used as growth promoters in aquaculture farming. Prophylactic treatments when used are typically confined for breeding, the immature or larval stages of aquatic animal farming. Prophylactic usage is more typically found in lower-level farming units that can't afford or get access to the recommendation of veterinary professionals.

4.9 Route of antimicrobial usage in aquaculture

In the marine environment antimicrobials are typically used during the fatten phase. Antimicrobials in aquaculture farming are delivered via feed medication, bioencapsulation, immersion baths, dip, flush, or in exceptional cases, intramuscularly or intraperitoneally (Smith et al., 2008). Feed medication is not completely digested by fish, and most of the time is inadequately digested and metabolized, with the result being that there is a continuous release of antimicrobials into the surroundings. Medicated feed also primarily affects the gut flora of fish (Navarrete et al., 2008). Consistent presentation to antimicrobials leads to the variety of resistant microbes and an increase in antimicrobial-resistant genes (ARGs) transfer. These circumstances eventually imply that the feces of feed-medicated fish are affluent in ARGs (Martinez and Baquero, 2000).

4.10 WHO list of antimicrobials used in aquaculture

The World Health Organization (WHO) catalog is a classification of 260 antimicrobials. The catalog was proposed as a reference for the people and veterinary health establishment to prioritize threat measurement with respect to increase in antimicrobial resistance. Two criteria are measured for incorporation in the catalog: first, for the treatment of genuine human illnesses the antimicrobial should be the only way or one of the few accessible therapies and, second, it ought to be used to treat illnesses caused either by microbes that might be transferred to human via nonhuman origin (Magouras et al., 2017) or human illnesses caused by microbes that could gain resistant genes from nonhuman origin. "Crucially significant" antimicrobials met both criteria. "Most significant" antimicrobials met any one of the criteria, and "significant" antimicrobials are those that do not meet any criterion but still are considered important antimicrobials. The WHO catalog included six widespread classes of antibiotics like aminoglycosides, macrolides, penicillins, quinolones, sulfonamides, and tetracyclines and are frequently used in aquaculture farming and cultivation.

4.11 Unregulated use of antimicrobials in aquaculture

Unregulated use of antimicrobials in aquaculture industry might cause individual well-being and foodstuff security concern. A result of the utilization of the antimicrobials in food animals is the occurrence of medicine residues, even in a smaller amount, in the eatable tissues of the treated creature. Antimicrobials used in accordance to brand instructions must not result in residues at butchery. The explanation for residues occurrence in eatable tissues of animals suggests multiple reasons; noncompliance of prescribed extraction period; dispersion of excess quantity of drug at a particular inoculation site; utilization of antimicrobial-polluted tools, or failure to appropriate hygienic apparatus used for mixing or managing medicines; assimilation error; inadvertent feed with chemical spill or feed medication; animal characteristics like age, pregnancy, inborn ailment, and hypersensitivities; chemical reactions among medicines; changes in heat for water spp.; ecological pollution; and inappropriate drugs usage (Okocha et al., 2018).

Drug residues in aquaculture foodstuffs can result in resistant bacterial growth and be poisonous to customers, which can lead to morbid conditions or death. For example, chloramphenicol residues increase the possibility of cancer and in lower concentration may produce aplastic anemia; other lethal effects include hypersensitivity by penicillin, mutagenicity and nephropathy by gentamicin, and immunopathology and carcinogenic effects by sulfamethazine, oxytetracycline, and furazolidone (Beyene, 2016).

4.12 Use of antimicrobials in fish

Fish as foodstuff contribute about 17% of overall animal proteins; out of this, half originate from aquaculture farming (Troell et al., 2014). The aquaculture farmed fish spp. included salmon, turbot, marine bass, marine bream, trout, tuna, sole, halibut, cod, and European eel. In Ireland, Norway, and Scotland, marine enclosed aquaculture farming is mainly restricted to salmon, and marine bass and marine bream in Italy, Greece, and Spain. Recently, a rise in the tilapia and mullet farming have been observed in Egypt. The salmon production in Chile is continually affected by microbes such as bacteria, parasites, fungi, and viruses responsible for progression of illnesses, numerous of which resulted in the loss of millions of finfish and thus major farming loss (Asche et al., 2009). Over the last 30 years, the facultative intracellular bacterium *Piscirickettsia salmonis*, causative agent of Salmonid Rickettsial Syndrome (SRS) consistently overwhelmed the salmon farming (Rozas and Enriquez, 2014). The bacterium is responsible for >80% of finfish deaths that happened due to contagious illnesses in the three chief fish group cultured in Chilean industry, i.e., *Salmo salar* (Atlantic salmon), *Oncorhynchus kisutch* (coho salmon), and *O. mykiss* (rainbow trout) (Makrinos and Bowden, 2017). Between 2007 and 17, salmon farming industry utilized >5500 tons of antimicrobials with each ton of salmon production receiving a standard of 500 g antimicrobials in accordance to a statement by the national fisheries service (Sernapesca). The two broad-spectrum antibiotics frequently administered in salmon farming are florfenicol and oxytetracycline. In 2017, 393.9 tons of antibiotics were used, among them 92.2% of florfenicol, 6.7% of oxytetracycline, and the remaining 1% correspond to erythromycin and amoxicillin (Lozano et al., 2018). The antimicrobials used for fishes appear in Table 4.8.

4.13 Use of antimicrobials in crustaceans

Crustaceans have been an essential diet source for people for many years. Almost all crustacean diseases have viral etiology except two or three. For the cure of such diseases, there are no well-known antimicrobials. The alternatives available include devastation or separation of tainted stocks and amendment of on-farm husbandry measures. Due to the unavailability of antiviral drugs for shrimp diseases, a few defensive procedures are brood stock screening programs and decontamination of farmed tools or pool services. The few antimicrobials and preventive measures adopted for crustaceans are presented in Table 4.9. Hence, in few circumstances excess of chloride and lime might be the single solution for infected stock devastation and pool disinfection purposes. For the earlier identification of

Table 4.8 Antimicrobials use for treatment of diseases in fishes.

Affected spp.	Diseases	Causative microbes	Antimicrobial use
Fin fish	Tenacibaculosis	*Tenacibaculum maritimum*	Oral broad-spectrum antibacterials
Fin fish	Vibriosis	*Vibrio anguillarum*, *Vibrio ordalii*	Caprylic acid with broad-spectrum antibacterials
Fin fish	Epitheliocystis	*Chlamydia* spp.	Broad-spectrum antibacterials
Fin fish	Botulism	*Clostridium botulinum*, *Clostridium argentinense*, *Clostridium butyricum*	Guanidine hydrochloride (spickler, July 2007: http://www.cfsph.iastate.edu/DiseaseInfo/factsheets.php #46)
Salmonid	Bacterial gill disease (BGD)	*Flavobacterium branchiophila*	Florfenicol, oxytetracycline, chloramine T
Salmonid	Piscirickettsiosis	*Piscirickettsia salmonis*	Broad-spectrum antibacterials
Salmonid	Furunculosis	*Aeromonas salmonicida*	Sulfadiazine, trimethoprim, old quinolones (1st and 2nd generations) oxolinic acid, flumequine (Sekkin and Kum, 2011)
Salmonid	Infectious salmon anemia (ISA)	Isa virus	Virus inactivated by disinfectants (chloramine-T, iodophors, Virkon S) (spickler, March 2010 : http://www.cfsph.iastate.edu/DiseaseInfo/factsheets.php #36)
Salmonid	Infectious hematopoietic necrosis(IHN)	Rhabdoviridae	Virus inactivated by disinfectants (Spickler, July 2007 : http://www.cfsph.iastate.edu/DiseaseInfo/factsheets.php #45)
Rainbow trout	Rainbow trout-Gastro Enteritis (RTGE)	*Candidatus arthromitis*	Broad-spectrum antibacterials
Rainbow trout, salmonids, Catfish	Enteric red mouth disease (ERM)	*Yersinia ruckeri*	Sulfadiazine, trimethoprim, old quinolones
Rainbow trout, Redfin perch	Epizootic hematopoietic necrosis (EHN)	*Iridoviridae ranavirus*	Virus inactivated by disinfectants (Spickler,

Table 4.8 Antimicrobials use for treatment of diseases in fishes.—cont'd

Affected spp.	Diseases	Causative microbes	Antimicrobial use
			July 2007 : http://www.cfsph.iastate.edu/DiseaseInfo/factsheets.php #46)
Trout	Red mark syndrome	*Flavobacterium psychrophilum*	Broad-spectrum antibacterials (Sekkin and Kum, 2011)
Trout	Viral hemorrhagic septicemia	*Novi rhabdovirus*	Antivirals (spickler, March 2010: http://www.cfsph.iastate.edu/DiseaseInfo/factsheets.php #36)
Turbot	Furunculosis	*Aeromonas salmonicida*	Sulfamerazine (Snieszko, 1954)
Catfish	Hemorrhagic septicemia	*Aeromonas veronii*	Amoxicillin, gentamicin, ofloxacin (Deemagarn and Tohmee, 2014)

Table 4.9 Antimicrobials used for treatment of diseases in crustaceans.

Affected spp.	Disease	Causative microbe	Antimicrobial use
Crustacean	Vibriosis	*Luminous Vibrio* spp. (*V. harveyi*)	Oxytetracycline, quinolones
Crustacean	Necrotizing hepatopancreatitis (NHP)	Intracellular proteobacteria	Oxytetracycline
Crustacean	Gaffkaemia	*Aerococcus viridans*	Oxytetracycline
Prawns	Vibrio infections	*Vibrio Parahaemolyticus*	Oxytetracycline
Prawns	Bacterial shell disease	*Vibrio anguillarium*, *Pseudomonas* spp., *aeromonas* spp., *Vibrio* spp.,	Oxytetracycline
Shrimp	Vibrio infections	*V. Parahaemolyticus*	Oxytetracycline, furazolidine, prefuran
Shrimp	Bacterial shell disease	*V. anguillarium*, *aeromonas* spp.,	Prefuran, oxytetracycline
Shrimp	Protozoan infections	*Zoothammium* spp.	Prefuran
Shrimp	Larval mycosis	*Lagenidium* spp.	Treflan (Alderman et al., 1998)

necrotizing hepatopancreatitis caused by *Vibrio* spp., shrimps are treated with oxytetracycline via medication feed. *Vibrio* spp. show resistance to antibiotics like chloramphenicol, furazolidone, oxytetracycline, and streptomycin (Rodgers and Furones, 2009).

4.14 Use of antimicrobials in mollusks

Mollusks like snails, oysters, and bivalves are also a food source for humans. Mollusks as a foodstuff may also contribute to the prevention of diseases by providing key nutrients, immunostimulatory compounds, and other secondary metabolites (Benkendorff, 2010). Mostly mollusks are affected by viral or parasitic diseases and it is not feasible to exploit any antimicrobial in open seawater culturing. The mollusk parasitic illnesses are often intracellular (e.g., *Bonamia ostreae*) and no antimicrobials are available for cure. It might be probable to monitor behavioral changes in several stocks, especially brood stock and larvae in hatcheries, and collection of complete evidence (e.g., aquaculture features like hotness and salinity etc.) may also be useful for manipulation of restricted surrounding conditions. This is in fact possible in a controlled hatchery situation, where infections can break out very rapidly in vulnerable stock. Larvae stage feed behaviors may also present an earlier sign of health troubles. Indication of weakness includes gaping shells in immature or adult stages, which can also be used to forecast possible harms, as diminished growth in motile genuses, e.g., scallops, clams. However, particular substitutes are needed for mollusks. These usually include reduction in stock mass, changes in saltiness, and lower water temperature, as well as avoidance of the transfer of shellfish from known enzootic areas (Rodgers and Furones, 2009). Preventive measures adopted for mollusks are presented in Table 4.10.

4.15 Future perspectives

Demand for animal protein by humans is rising globally at an uncontrollable rate, which leads to wide usage of antimicrobials at greater extent for disease prevention and growth promotion in food animals. Patterns of antimicrobial consumption in middle-income and high-income countries differ in many respects. Mapping the antimicrobial consumption in livestock provides a baseline estimate of its global importance. Low-income countries lack in knowledge based on consumption of antimicrobials results to intense AMR in food animals. Globally, intensive livestock farming has increased food production at a low cost per unit produced, but at an unrecognized price paid in increased antimicrobial resistance (Van Boeckel et al., 2015).

The unique advantages of antimicrobials use in food animals are definite targeting of pathogens, well-known mechanisms of activity, and preferable stability for administration, for the prevention and treatment of bacterial and parasitic diseases, for the improvement of animal food production, and protection of the environment and public health. Absence of antimicrobials use in food-producing animals may cause deleterious effects on production of food derived from animals and, thus, on public health. Contrary to that, it is also important to administer antimicrobials to animals in ways that avoid the negative impacts. In the near future, it is expected that no new class of antimicrobials is going to be administered in food animals. Keeping in view the development of antimicrobial resistance in food animals, our aim is to act against AMR by taking preventive measures such as vaccination, advanced farm management, implementation of improved farming systems, upgraded techniques used for better hygiene on farms, and cautious and sagacious use of antimicrobial agents.

Table 4.10 Preventive measures adopted for mollusks.

Affected spp.	Diseases	Causative microbes	Treatment/Control
Oyster	Iridovirosis (oyster velar virus disease)	Iridoviridae	No treatment, brood stock screening, stock destruction and pond disinfection
Oyster	Herpesvirosis (oyster Herpes-like virus disease)	Herpesviridae	No treatment or control
Oyster	Bonamiosis	*Bonamia ostreae*, *Bonamia* sp.	No treatment, reduced stocking densities, lower water temperature
Oyster	Perkinsosis	*Perkinsus marinus*, *Perkinsus olseni*, *Perkinsus* spp,	No treatment, development of resistant oyster stocks (Rodgers and Furones, 2009)
Oyster	Roseovarius oyster disease	*Roseovarius crassostreae*	No treatment, selective breeding of oysters
Oyster	Haplosporidiosis	*Haplosporidium costale*, *Haplosporidium nelsoni*	No treatment, sterilization and filtration of inflow water (Rodgers and Furones, 2009)
Mollusks	QX disease (Marteiliosis)v	*Marteiliare fringens*, *Marteilia maurini*, *Marteilia sydneyi*,	No treatment. high salinity (Maloy et al., 2007; Zannella et al., 2017)
Mollusks	Brown ring disease *Tapes philippinarum*	*Vibrio tapetis*	Nitrofurans (Rodgers and Furones, 2009)
Mollusks	Pacific oyster nocardiosis	*Nocardiacrass ostreae*	No treatment, sterilization and filtration of inflow water (Zannella et al., 2017)

Therefore, more research is required to understand how antimicrobials are being used and how to cope with their inappropriate uses in nature (Hao et al., 2014). Moreover, research must also be done to identify the resistance-causing genes in microbes and food animals. Identification of mobile genetic elements and modes of spreading of these elements would needs also to be investigated and researched. Raising livestock without using antimicrobial agents is impossible, therefore, qualitative and quantitative analysis of antimicrobials must be done. More advanced dosage schemes can help brighten the future of antimicrobial trends in food animals (Hao et al., 2014).

4.16 Conclusion

With extensive animal production, microbial and pathogenic diseases became more chronic, caused by *Actinobacillus pleuropneumoniae*, *E. coli*, *Clostridium welchii*, *S. aureus*, *S. pneumonia*, *Salmonella*, and others. More than a hundred antimicrobials, including β-lactams, aminoglycosides, tetracyclines, amphenicols, macrolides, sulfonamides, fluoroquinolones, lincosamides, polypeptides, and polyene, have been used for the production of food for animals and aquaculture throughout the world. These antimicrobials have played a crucial role in prevention, treatment, and control of animal diseases caused by microorganisms. Due to certain advantages, such as exact targeting of pathogens, well-known mechanisms of activity and desired stability, antimicrobials have justified their usage in food animals and aquaculture, thus playing a main part in the prevention and treatment of bacterial and parasitic diseases. The improper use of antimicrobials is the main cause of development of antimicrobial resistance. As a result of AMR in food animals and aquaculture, costs/charges to treat antimicrobial-resistant infections in humans have also increased. It is well acknowledged that the problems relating to antimicrobial use in animal food and aquaculture are of global concern. But still, pharmacological research on food animals and aquaculture drugs has helped to lessen the possibility of noxious resistance and sporadic public health and environmental concern. As well, advanced and more productive medicines are required for future successful animal production.

References

Acar, J., Casewell, M., Freeman, J., Friis, C., Goossens, H., 2000. Avoparcin and virginiamycin as animal growth promoters: a plea for science in decision-making. Clinical Microbiology and Infection 6, 477–482.

Adel, A., Khadidja, M., 2017. Canine prostatic disorders. Veterinary Medicine-Open Journal 2, 83–90.

Agyare, C., Boamah, V.E., Zumbi, C.N., Osei, F.B., 2018. Antibiotic Use in Animal Production and its Effects on Bacterial Resistance.

Alderman, D., Hastings, T., Baticados, M., Paclibare, J., Bell, T., Lightner, D., Dierberg, F., Kiattisimkul, W., Graslund, S., Bengtsson, B., 1998. Experimental study on prevention and treatment of disease of shrimp post larvae (Pl2-Pl15) in nursery ponds. Journal of Fisheries and Aquatic Science 1, 139–155.

Ali, H., Rico, A., MURSHED-E-Jahan, K., Belton, B., 2016. An Assessment of Chemical and Biological Product Use in Aquaculture in Bangladesh.

Asche, F., Hansen, H., Tveteras, R., Tveterås, S., 2009. The salmon disease crisis in Chile. Marine Resource Economics 24, 405–411.

Bailey, J.K., Pinyon, J.L., Anantham, S., Hall, R.M., 2011. Distribution of the bla TEM gene and bla TEM-containing transposons in commensal Escherichia coli. Journal of Antimicrobial Chemotherapy 66, 745–751.

Banerjee, A.K., Angulo, A.F., Dhasmana, K.M., Kong, A.S.J., 1987. Acute diarrhoeal disease in rabbit: bacteriological diagnosis and efficacy of oral rehydration in combination with loperamide hydrochloride. Laboratory Animals 21, 314–317.

Benkendorff, K., 2010. Molluscan biological and chemical diversity: secondary metabolites and medicinal resources produced by marine molluscs. Biological Reviews 85, 757–775.

Beyene, T., 2016. Veterinary drug residues in food-animal products: its risk factors and potential effects on public health. Journal of Veterinary Science & Technology 7, 1–7.

Bhat, T., Jlthendran, K.P., Kurade, N., 2010. Rabbit Coccidiosis and its Control: A Review.

Boamah, V.E., Agyare, C., Odoi, H., Dalsgaard, A., 2016. Practices and Factors Influencing the Use of Antibiotics in Selected Poultry Farms in Ghana.

Boshuizen, B., Ploeg, M., Dewulf, J., Klooster, S., DE Bruijn, M., Picavet, M.-T., Palmers, K., Plancke, L., DE Cock, H., Theelen, M., 2018. Inflammatory bowel disease (IBD) in horses: a retrospective study exploring the value of different diagnostic approaches. BMC Veterinary Research 14, 21.

Bostock, J., Mcandrew, B., Richards, R., Jauncey, K., Telfer, T., Lorenzen, K., Little, D., Ross, L., Handisyde, N., Gatward, I., 2010. Aquaculture: global status and trends. Philosophical Transactions of the Royal Society B: Biological Sciences 365, 2897–2912.

Bowen, A.C., Carapetis, J.R., Currie, B.J., Fowler Jr., V., Chambers, H.F., Tong, S.Y., 2017, November. Sulfamethoxazole-trimethoprim (cotrimoxazole) for skin and soft tissue infections including impetigo, cellulitis, and abscess. In: Open forum infectious diseases, Vol. 4. Oxford University Press, US, p. ofx232. No. 4.

Burka, J.F., Hammell, K.L., Horsberg, T., Johnson, G.R., Rainnie, D., Speare, D.J., 1997. Drugs in salmonid aquaculture—a review. Journal of Veterinary Pharmacology and Therapeutics 20, 333—349.

Cain, A.K., Hall, R.M., 2012. Evolution of a multiple antibiotic resistance region in IncHI1 plasmids: reshaping resistance regions in situ. Journal of antimicrobial chemotherapy 67, 2848—2853.

Carapetis, J.R., Bowen, A.C., Currie, B.J., Tong, S.Y.C., Fowler JR., V.,, Chambers, H.F., 2017. Sulfamethoxazole-trimethoprim (cotrimoxazole) for skin and soft tissue infections including impetigo, cellulitis, and abscess. Open Forum Infectious Diseases 4.

Cole, D.W., Cole, R., Gaydos, S.J., Gray, J., Hyland, G., Jacques, M.L., POWELL-Dunford, N., Sawhney, C., Au, W.W., 2009. Aquaculture: environmental, toxicological, and health issues. International Journal of Hygiene and Environmental Health 212, 369—377.

Committee, S., 2004. Directive 2001/82/EC of the European Parliament and of the Council of 6 November 2001 on the Community code relating to veterinary medicinal products. Official Journal L 311, 1—66.

Cromwell, G.L., 2002. Why and how antibiotics are used in swine production. Animal Biotechnology 13, 7—27.

De Briyne, N., Atkinson, J., Pokludová, L., Borriello, S., 2014a. Antibiotics used most commonly to treat animals in Europe. The Veterinary Record 175, 325.

Deemagarn, T., Tohmee, N., 2014. Study on Hemorrhagic Septicemia in Freshwater Fish.

Dibner, J., Richards, J., 2005. Antibiotic growth promoters in agriculture: history and mode of action. Poultry Science 84, 634—643.

Fajt, V.R., 2001. Label and extralabel drug use in small ruminants. Veterinary Clinics of North America: Food Animal Practice 17, 403—420.

FAOSTAT., F. Accessed 24 Nov 2014. . http://www.faostat.fao.org. Food and Agriculture Organization of the United Nations. 2014.

Ferran, A.A., Liu, J., Toutain, P.-L., BOUSQUET-Mélou, A., 2016. Comparison of the in vitro activity of five antimicrobial drugs against Staphylococcus pseudintermedius and *Staphylococcus aureus* biofilms. Frontiers in Microbiology 7, 1187.

Fieni, F., Topie, E., Gogny, A., 2014. Medical treatment for pyometra in dogs. Reproduction in Domestic Animals 49, 28—32.

Fjordbakk, C., Arroyo, L., Hewson, J., 2008. Retrospective study of the clinical features of limb cellulitis in 63 horses. The Veterinary Record 162, 233—236.

Flórez, A.B., Vázquez, L., Mayo, B., 2017. A functional metagenomic analysis of tetracycline resistance in cheese bacteria. Frontiers in Microbiology 8, 907.

Food, Trade, A. O. O. T. U. N. & Division, M., 2014. Food Outlook: Biannual Report on Global Food Markets. October 2014. Food and Agriculture Organization of the United Nations.

Frankel, J.L., Scott, D.W., Erb, H.N., 2008. Gross and cytological characteristics of normal feline anal-sac secretions. Journal of Feline Medicine & Surgery 10, 319—323.

Gómez-Poveda, B., Moreno, M.A., 2018. Antimicrobial prescriptions for dogs in the Capital of Spain. Frontiers in veterinary science 5, 309.

Gordon, D.L., Radtke, C.L., 2017a. Treatment of chronic sinusitis in a horse with systemic and intra-sinus antimicrobials. Canadian Veterinary Journal 58, 289.

Gordon, D.L., Radtke, C.L., 2017b. Treatment of chronic sinusitis in a horse with systemic and intra-sinus antimicrobials. The Canadian Veterinary Journal 58, 289—292.

Hall, S.J., 2011. Blue Frontiers: Managing the Environmental Costs of Aquaculture. WorldFish.

Hao, H., Cheng, G., Iqbal, Z., Ai, X., Hussain, H.I., Huang, L., Dai, M., Wang, Y., Liu, Z., Yuan, Z., 2014. Benefits and risks of antimicrobial use in food-producing animals. Frontiers in microbiology 5, 288.

Hidaka, S., Kobayashi, M., Ando, K., Fujii, Y., 2015. Efficacy and safety of lomefloxacin on bacterial extraocular disease in the horse. Journal of Veterinary Medical Science 14—0507.

Hölsö, K., Rantala, M., Lillas, A., Eerikäinen, S., Huovinen, P., Kaartinen, L., 2005. Prescribing antimicrobial agents for dogs and cats via university pharmacies in Finland—patterns and quality of information. Acta Veterinaria Scandinavica 46, 87.

Honneffer, J.B., Minamoto, Y., Suchodolski, J.S., 2014. Microbiota alterations in acute and chronic gastrointestinal inflammation of cats and dogs. World Journal of Gastroenterology 20, 16489—16497.

Ion, L., Ionascu, I., Birtoiu, A., 2015. Melting Keratitis in dogs and cats. Agriculture and Agricultural Science Procedia 6, 342—349.

Jeters, R.T., Wang, G.-R., Moon, K., Shoemaker, N.B., Salyers, A.A., 2009. Tetracycline-associated transcriptional regulation of transfer genes of the Bacteroides conjugative transposon CTnDOT. Journal of Bacteriology 191, 6374—6382.

Landers, T.F., Cohen, B., Wittum, T.E., Larson, E.L., 2012. A review of antibiotic use in food animals: perspective, policy, and potential. Public Health Reports 127, 4—22.

Langan, G.P., Lohmiller, J.J., Swing, S.P., Wardrip, C.L., 2000. Respiratory diseases of rodents and rabbits. Veterinary Clinics: Small Animal Practice 30, 1309—1335.

Lappin, M.R., Blondeau, J., Boothe, D., Breitschwerdt, E.B., Guardabassi, L., Lloyd, D.H., Papich, M.G., Rankin, S.C., Sykes, J.E., Turnidge, J., Weese, J.S., 2017. Antimicrobial use guidelines for treatment of respiratory tract disease in dogs and cats: antimicrobial guidelines working group of the International Society for Companion animal infectious diseases. Journal of Veterinary Internal Medicine 31, 279—294.

Leavis, H.L., Willems, R.J., Top, J., Spalburg, E., Mascini, E.M., Fluit, A.C., Hoepelman, A., DE Neeling, A.J., Bonten, M.J., 2003. Epidemic and nonepidemic multidrug-resistant *Enterococcus faecium*. Emerging Infectious Diseases 9, 1108.

Li, J., 2017. Current status and prospects for in-feed antibiotics in the different stages of pork production—a review. Asian-Australasian Journal of Animal Sciences 30, 1667.

Li, Y., Cui, X., Solomon, S.B., Remy, K., Fitz, Y., Eichacker, P.Q., 2013. B. anthracis edema toxin increases cAMP levels and inhibits phenylephrine-stimulated contraction in a rat aortic ring model. American Journal of Physiology - Heart and Circulatory Physiology 305, H238—H250.

Liao, I.-C., Chao, N.-H., 2009. Aquaculture and food crisis: opportunities and constraints. Asia Pacific Journal of Clinical Nutrition 18, 564—569.

Little, S., Kennedy, M., 2010. Feline Infectious Peritonitis. WINN Feline Foundation.

Littman, M.P., Gerber, B., Goldstein, R.E., Labato, M.A., Lappin, M.R., Moore, G.E., 2018. ACVIM consensus update on Lyme borreliosis in dogs and cats. Journal of Veterinary Internal Medicine 32, 887—903.

Lozano, I., DÍAZ Pérez, N., MUÑOZ Mimiza, S., Riquelme, C., 2018. Antibiotics in Chilean Aquaculture: A Review.

Luecke, R., THORP Jr., F., Newland, H., Mcmillen, W., 1951. The growth promoting effects of various antibiotics on pigs. Journal of Animal Science 10, 538—542.

Magouras, I., Carmo, L.P., Stärk, K.D., SCHÜPBACH-Regula, G., 2017. Antimicrobial usage and-resistance in livestock: where should we focus? Frontiers in Veterinary Science 4, 148.

Makrinos, D.L., Bowden, T.J., 2017. Growth characteristics of the intracellular pathogen, Piscirickettsia salmonis, in tissue culture and cell-free media. Journal of Fish Diseases 40, 1115—1127.

Maloy, A.P., Ford, S.E., Karney, R.C., Boettcher, K.J., 2007. Roseovarius crassostreae, the etiological agent of Juvenile Oyster Disease (now to be known as Roseovarius Oyster Disease) in *Crassostrea virginica*. Aquaculture 269, 71—83.

Marchandeau, S., Pontier, D., Guitton, J.-S., Letty, J., Fouchet, D., Aubineau, J., Berger, F., Léonard, Y., Roobrouck, A., Gelfi, J., Peralta, B., Bertagnoli, S., 2014. Early infections by myxoma virus of young rabbits (*Oryctolagus cuniculus*) protected by maternal antibodies activate their immune system and enhance herd immunity in wild populations. Veterinary Research 45, 26-26.

Martinez, J., Baquero, F., 2000. Mutation frequencies and antibiotic resistance. Antimicrobial Agents and Chemotherapy 44, 1771—1777.

Martins, C., Eding, E.H., Verdegem, M.C., Heinsbroek, L.T., Schneider, O., Blancheton, J.-P., D'orbcastel, E.R., Verreth, J., 2010. New developments in recirculating aquaculture systems in Europe: a perspective on environmental sustainability. Aquacultural Engineering 43, 83–93.

Menzies, P., 2000. Antimicrobial drug use in sheep and goats. Antimicrobial Therapy in Veterinary Medicine 3, 591–601.

Milon, A., Oswald, E., DE Rycke, J., 1999. Rabbit EPEC: a model for the study of enteropathogenic Escherichia coli. Veterinary Research 30, 203–219.

Miotto, B.A., Guilloux, A.G.A., Tozzi, B.F., Moreno, L.Z., DA Hora, A.S., Dias, R.A., Heinemann, M.B., Moreno, A.M., Filho, A.F.D.S., Lilenbaum, W., Hagiwara, M.K., 2018. Prospective study of canine leptospirosis in shelter and stray dog populations: identification of chronic carriers and different Leptospira species infecting dogs. PLoS One 13 e0200384-e0200384.

Navarre, C., Marley, S., 2006. Antimicrobial drug use in sheep and goats. In: Antimicrobial Therapy in Veterinary Medicine, fourth ed. Blackwell Publ, Ames, Iowa, pp. 519–528.

Navarrete, P., Mardones, P., Opazo, R., Espejo, R., Romero, J., 2008. Oxytetracycline treatment reduces bacterial diversity of intestinal microbiota of Atlantic salmon. Journal of Aquatic Animal Health 20, 177–183.

Naylor, R.L., Goldburg, R.J., Primavera, J.H., Kautsky, N., Beveridge, M.C., Clay, J., Folke, C., Lubchenco, J., Mooney, H., Troell, M., 2000. Effect of aquaculture on world fish supplies. Nature 405, 1017.

Niżański, W., Levy, X., Ochota, M., Pasikowska, J., 2014. Pharmacological treatment for common prostatic conditions in dogs – Benign prostatic Hyperplasia and Prostatitis: an update. Reproduction in Domestic Animals 49, 8–15.

Nuttall, T., 2016. Successful management of otitis externa. Practice 38, 17–21.

O'neill, J., 2014. Antimicrobial resistance: tackling a crisis for the health and wealth of nations. Review on Antimicrobial Resistance 1, 1–16.

Okocha, R.C., Olatoye, I.O., Adedeji, O.B., 2018. Food safety impacts of antimicrobial use and their residues in aquaculture. Public Health Reviews 39, 21.

Palócz, O., Gál, J., Clayton, P., Dinya, Z., Somogyi, Z., Juhász, C., Csikó, G., 2014. Alternative treatment of serious and mild Pasteurella multocida infection in New Zealand White rabbits. BMC Veterinary Research 10, 276-276.

Papich, M.G., 2001. Current concepts in antimicrobial therapy for horses. Proceedings of the Annual Convention of the AAEP 47, 94–102.

Pauly, D., Zeller, D., 2016. Catch reconstructions reveal that global marine fisheries catches are higher than reported and declining. Nature Communications 7, 10244.

Pauly, D., Zeller, D., 2017. Comments on FAOs state of world fisheries and aquaculture (SOFIA 2016). Marine Policy 77, 176–181.

Phillips, I., Casewell, M., Cox, T., DE Groot, B., Friis, C., Jones, R., Nightingale, C., Preston, R., Waddell, J., 2004. Does the use of antibiotics in food animals pose a risk to human health? A critical review of published data. Journal of Antimicrobial Chemotherapy 53, 28–52.

Pieri, F., Daibert, A.P., Bourguignon, E., Moreira, M., 2012. Periodontal Disease in Dogs.

Pillay, T.V.R., Kutty, M.N., 2005. Aquaculture: Principles and Practices. Blackwell publishing.

Płoneczka-Janeczko, K., Bania, J., Bierowiec, K., Kiełbowicz, M., Kiełbowicz, Z., 2017. Bacterial diversity in feline Conjunctiva based on 16S rRNA gene Sequence analysis: a Pilot study. BioMed Research International.

Podberscek, A.L., 2009. Good to pet and eat: the keeping and consuming of dogs and cats in South Korea. Journal of Social Issues 65, 615–632.

Puón-Peláez, X.-H., Mcewan, N.R., OLVERA-Ramirez, A.M., 2018. Epizootic Rabbit Enteropathy (ERE): A Review of Current Knowledge.

Radostits, O.M., Gay, C.C., Hinchcliff, K.W., Constable, P.D., 2006. Veterinary Medicine E-Book: A Textbook of the Diseases of Cattle, Horses, Sheep, Pigs and Goats. Elsevier Health Sciences.

Ramos, S., Igrejas, G., Rodrigues, J., CAPELO-Martinez, J.-L., Poeta, P., 2012. Genetic characterisation of antibiotic resistance and virulence factors in vanA-containing enterococci from cattle, sheep and pigs subsequent to the discontinuation of the use of avoparcin. The Veterinary Journal 193, 301–303.

Rodgers, C., Furones, M., 2009. Antimicrobial agents in aquaculture: practice, needs and issues. Options Méditerranéennes 86, 41–59.

Rosell, J., de la Fuente, L., 2018. Mastitis on rabbit farms: Prevalence and risk factors. Animals 8, 98.

Roy, J., Messier, S., Labrecque, O., Cox, W.R., 2007. Clinical and in vitro efficacy of amoxicillin against bacteria associated with feline skin wounds and abscesses. Canadian Veterinary Journal 48, 607.

Rozas, M., Enriquez, R., 2014. Piscirickettsiosis and Piscirickettsia salmonis in fish: a review. Journal of Fish Diseases 37, 163–188.

Sahoo, K.C., Tamhankar, A.J., Johansson, E., Lundborg, C.S., 2010. Antibiotic use, resistance development and environmental factors: a qualitative study among healthcare professionals in Orissa, India. BMC Public Health 10, 629.

Sandmeyer, L.S., Bauer, B.S., Grahn, B.H., 2011. Diagnostic ophthalmology. The Canadian Veterinary Journal 52, 327–328.

Sapkota, A., Sapkota, A.R., Kucharski, M., Burke, J., Mckenzie, S., Walker, P., Lawrence, R., 2008. Aquaculture practices and potential human health risks: current knowledge and future priorities. Environment International 34, 1215–1226.

Schwarz, S., Kehrenberg, C., Walsh, T., 2001. Use of antimicrobial agents in veterinary medicine and food animal production. International Journal of Antimicrobial Agents 17, 431–437.

Sekkin, S., Kum, C., 2011. Antibacterial drugs in fish farms: application and its effects. In: Recent Advances in Fish Farms. IntechOpen.

Serrano, P.H., 2005. Responsible Use of Antibiotics in Aquaculture. Food & Agriculture Org.

Shao, Z.J., 2001. Aquaculture pharmaceuticals and biologicals: current perspectives and future possibilities. Advanced Drug Delivery Reviews 50, 229–243.

Sharma, C., Rokana, N., Chandra, M., Singh, B.P., Gulhane, R.D., Gill, J.P.S., Ray, P., Puniya, A.K., Panwar, H., 2018. Antimicrobial resistance: its surveillance, impact, and alternative management strategies in dairy animals. Frontiers in veterinary science 4, 237.

Simpson, K.W., Jergens, A.E., 2011. Pitfalls and progress in the diagnosis and management of canine Inflammatory bowel disease. Veterinary Clinics: Small Animal Practice 41, 381–398.

Smith, P.R., LE Breton, A., Horsberg, T.E., Corsin, F., 2008. Guidelines for Antimicrobial Use in Aquaculture. Guide to Antimicrobial Use in Animals, pp. 207–218.

Snieszko, S.F., 1954. Fish furunculosis. The Progressive Fish-Culturist 16, 143-143.

Soontornvipart, K., Kohout, P., Proks, P., 2003. Septic Arthritis in Dogs: A Retrospective Study of 20 Cases, 2000-2002.

Stentiford, G., February 2, 2017. Solving the $6 billion per year global aquaculture disease problem. Marine Science.

Suchodolski, J.S., Foster, M.L., Sohail, M.U., Leutenegger, C., Queen, E.V., Steiner, J.M., Marks, S.L., 2015. The fecal microbiome in cats with diarrhea. PLoS One 10 e0127378-e0127378.

Sykes, J.E., 2013. Canine and Feline Infectious Diseases. Elsevier Health Sciences.

The Finnish Food Safety Authority, E., The Faculty of Veterinary Medicine at the University of Helsinki, 2018. Recommendations for the Use of Antimicrobials in the Treatment of the Most Significant Infectious and Contagious Diseases in Animals, pp. 1–56. https://www.ruokavirasto.fi/globalassets/viljelijat/elaintenpito/elainten-laakitseminen/hallittu_laakekekaytto/mikrobilaakekaytonperiaatteet/mikrobilaakkeiden_kayttosuositukset_en.pdf.

Troell, M., Naylor, R.L., Metian, M., Beveridge, M., Tyedmers, P.H., Folke, C., Arrow, K.J., Barrett, S., Crépin, A.-S., Ehrlich, P.R., 2014. Does aquaculture add resilience to the global food system? Proceedings of the National Academy of Sciences 111, 13257–13263.

Ungemach, F., 1999. Use of antibiotics in veterinary medicine-consequences and prudent use. Tierarztliche Praxis Ausgabe Grobtiere Nutztiere 27, 335–340.

Van Boeckel, T.P., Brower, C., Gilbert, M., Grenfell, B.T., Levin, S.A., Robinson, T.P., Teillant, A., Laxminarayan, R., 2015. Global trends in antimicrobial use in food animals. Proceedings of the National Academy of Sciences 112, 5649–5654.

Van Epps, A., Blaney, L., 2016. Antibiotic residues in animal waste: occurrence and degradation in conventional agricultural waste management practices. Current Pollution Reports 2, 135–155.

Vanegas Azuero, A.M., Gutiérrez, L.F., 2016. Horse meat: production, consumption and nutritional value. CES Medicina Veterinaria y Zootecnia 11, 86–103.

Wael, M., Husein, M., 2011. Diagnosis of recurrent pyoderma in dogs by traditional and molecular based diagnostic assays and its therapeutic approach. The Journal of American Science 7, 120–134.

Waite, R., June 2014. Aquaculture Is Expanding to Meet World Fish Demand. World Resources Institute. https://www.wri.org/resources/charts-graphs/aquaculture-expanding-meet-world-fish-demand.

Watts, J., Schreier, H., Lanska, L., Hale, M., 2017. The rising tide of antimicrobial resistance in aquaculture: sources, sinks and solutions. Marine Drugs 15, 158.

Weese, J.S., 2015. Antimicrobial use and antimicrobial resistance in horses. Equine Veterinary Journal 47, 747–749.

Williams, L.B., Pinard, C.L., 2013. Corneal ulcers in horses. Compendium: Continuing Education for Veterinarians 35, E4.

Wilson, W.D., 2001. Rational selection of antimicrobials for use in horses. Proceedings of the Annual Convention of the AAEP 75–93.

Winer, J.N., Arzi, B., Verstraete, F.J.M., 2016. Therapeutic management of feline chronic Gingivostomatitis: a Systematic review of the literature. Frontiers in Veterinary Science 3, 54-54.

Wong, C., Epstein, S.E., Westropp, J.L., 2015. Antimicrobial Susceptibility patterns in urinary tract infections in dogs (2010-2013). Journal of Veterinary Internal Medicine 29, 1045–1052.

Zannella, C., Mosca, F., Mariani, F., Franci, G., Folliero, V., Galdiero, M., Tiscar, P.G., Galdiero, M., 2017. Microbial diseases of bivalve mollusks: infections, immunology and antimicrobial defense. Marine Drugs 15, 182.

CHAPTER

5

Major natural sinks for harboring microorganisms with altered antibiotic resistance versus major human contributing sources of antibiotic resistance: a detailed insight

Arshia Amin[1], Tehseen Zahra[1], Hafsa Raja[1], Mehroze Amin[2], Erum Dilshad[1], Muhammad Naveed[3], Iftikhar Ahmed[4]

[1]*Department of Bioinformatics and Biosciences, Capital University of Science and Technology, Islamabad, Pakistan;* [2]*Institute of Biochemistry and Biotechnology, University of Central Punjab, Lahore, Punjab, Pakistan;* [3]*Department of Biotechnology, University of Central Punjab, Lahore, Punjab, Pakistan;* [4]*National Culture Collection of Pakistan (NCCP), Bio-resources Conservation Institute (BCI), National Agriculture Research Center (NARC), Islamabad, Pakistan*

5.1 Introduction

One of the major ecological disorders threatening indigenous biodiversity is biological invasions. An anticipated increased rate of environmental degradation will have significant impacts on the brain structure of ecosystems worldwide. Invasions of biology have numerous overarching ecosystem impacts, possibly disturbing interactions with species and worldwide environmental procedures. We are very dependent on our ability to effectively forecast and handle the impact of biologic invasions on the ecosystems of complicated biology and its spatiotemporal dynamics. Biological invasions have many overarching effects on the ecosystems, potentially disturbing interactions with species and global environmental practices. The effects of biological invasions in ecosystems of complex biology and their spatiotemporal dynamics are very dependent on our capacity to efficiently predict and manage. We give insight into, and practical application of, traditional (stable isotopes, genetic populations) and emerging (met barcoding, citizen science) methods and techniques for biological invasions. We also present several models and machine-learning methods presently available that can be used to predict new or documented interactions, thus making it possible to predict network and ecosystem stability more robustly and efficiently. Finally, we address the significance of methodological developments in the creation of science and societal problems in the research of local and international species history. One of the most critical ecological disturbances threatening indigenous biodiversity is biological invasion. An anticipated rise in species extinction will have a significant impact on the

Antibiotics and Antimicrobial Resistance Genes in the Environment. https://doi.org/10.1016/B978-0-12-818882-8.00005-X

world's ecosystems' composition and operation. The objective of our research is to determine, using a meta-analysis, which environmental characteristics influence the biodiversity loss of invasive species globally. We examined property role, such as invaders' trophic and geographic situation, invaded taxonomic groups, the sort of habitats invaded, and whether invasive species are classified as one of the most hazardous invasive species for biodiversity loss. (Fig. 5.1)

5.1.1 Biological invasion is affected by climate change

The main drivers of environmental change for conservation, agriculture, and human health are the biological invasions. Ecological changes such as hydrology, fire regimes, webs of food stuffs, the use of soil nutrients, and nutrient cycling include intrusive alien species effects (Prins and Gordon, 2014) (Ziska et al., 2011) which facilitate future invasions and related changes in the impact of trajectory development, including through hybrids (Vlachopoulos, O'Rourke and Nichols, 2011) and changes in the development of species response (Richardson, 2011) (Allsopp et al., 2014) (Sahni and Elkayam, 2012). The invasion method has therefore received much attention. Biological disasters, which distinguish invasion stages and procedures involved in individual research, are now commonly recognized.

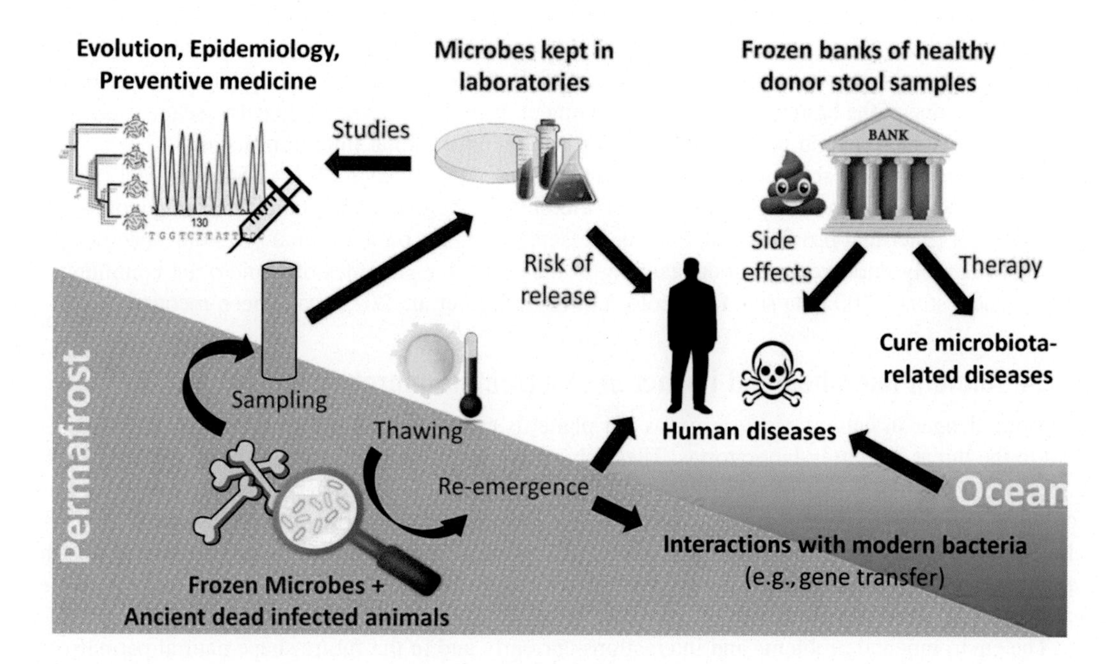

Figure 5.1 Environmental origin of antibiotic resistance genes.

5.1.2 Importance of permafrost microbials among other microbes

Bacteria permafrost. In the permafrost of sustainable ancient microorganisms, important figures are known. There's been an error. Therefore, the permafrost represents a stable and unique physico-chemical complex that keeps an unparalleled record of lifetime of a familiar habitat.

A series of permafrost stressors must be addressed by microbial communities. Temperatures below zero stabilize the secondary nucleic acid structure and lower membrane fluidity and protein structural flexibility. Communities are subjected to ~2 mGy soil-mineral radiation per year (Margesin et al., 2008). Reactive oxygen, which could damage DNA, RNA, proteins, and lipids at low temperatures, may increase (Castro-Sowinski, 2019). The high solvent concentrations of liquid water are present in permafrost to approximately 10°C (Margesin et al., 2008). Then live cells are likely to be spread patchily and be contained within a thin brine or pocket of the soil matrix in order to overcome the temperature effects (Margesin et al., 2008; Chénard and Lauro, 2017). Permafrost from late Pleistocene in the Arctic and subarctic persists. Stresses in the society are likely to build up over time, requiring counteractive adaptation in severe environmental circumstances for a long-term survival. However, we do not understand very much about the ecological methods used by microbial societies to address the problems posed by permafrost expenditure over thousands of years.

5.1.3 Microorganisms in cold environment

The world's largest portion of cold ecosystem is exposed during the year to het below 5°C. Cold ecosystems range from high mountains to deep sea, polar zones and mountains. There are highly diverse cold ecosystems. This encompasses aquatic and aquatic ecology (90% below 5°C), sea ice, lakes, snow and natural habitat, cold pools, permafrost, poor soils, even wind and rain. Current cold house surveys are focused on the earth's cryosphere permafrost (Dick, 2018; Kallmeyer, 2017) and the appreciations of the cryosphere as one of the biomes in the cold environment, including ice sheets, glaciers, sea ice, lake ice, and dry soil. Studies have in recent years concentrated on cold habitat microorganisms.

The integration of fresh evolving techniques (Aliyu et al., 2017; A.J., 2017; Margesin, 2017) have lately achieved greater knowledge of the role of microbial living in cool habitats. The metagenomics of an ecological tester that provides data on gene presence and thus on functional potential, specifically, have considerably improved our understanding of the collective genomes of a microbial community. For instance, some 1400 data sets from cold settings (Aliyu et al., 2017) have been recognized.

5.1.4 Ancient microbes and impact of resurrected microbes

A second danger in natural ecosystems, as our planet is currently undergoing climate change, exists next to the threat current in laboratories. These changes can be seen in the atmosphere, in the oceans, and in the cryosphere (loss of weight of ice, covering snow). These changes take place in the environment. Potential risks for people, society, our economies, in specific ecosystems are posed by climate change. The largest increase in temperature has been observed in areas with medium and high latitudes, where Arctic temperatures are rising rapidly, two times faster than in other places around the earth.

The environmental problems and interactions currently and in the future shape natural permafrost microbial communities. There has been increasing evidence in recent years that permafrost represents a gigantic reservoir that contains ancient microbes or bacteria that can return to life when environmental conditions change and release them again. Up to 108 cells/g dry land (Fuller et al., 2004) can be used in permafrost that is feasible. The permafrost thawing enhances the resurrection of different dormant vectors (Miller and Whyte, 2011) like fungi, which are sustainable for thousands of years. For example, a viable sample of a giant virus called *Pithovirus sibericum* was found in an ice core harvested from Siberian permafrost in 2014, which is 30,000 years old and was restored in the laboratory. Of interest, its natural ameboid host has been found to remain infected (Borcard et al., 2018).

The reappearance of ancient microbes and viruses is a source of new relationships between current natural groups. These risks come from the reappearance of pathogens in laboratories and/or the retention of permafrost pathogens that are made available after the thawing of climate change. Microbes in deep rock (Spijkerboer, 2013) in ice sheets' permafrost have been discovered in Omni (Bell, 2012) on this earth. They are found in profound rocks or in strata of sediment (Kallmeyer, Pockalny, Adhikari, Smith, and D'Hondt, 2012). This assessment concentrated on the capacity to recover long-sleeping microbial pathogens. In specific, we examined the danger of reemergence of ancient microbes in study equipment and ice sheets and permafrost, the temporal adaptation and the associated virulence of diseases and hosts, and then the feasible and potential pathogens for the resurrection of further studies into the area of the microbial/pathogenic environment.

5.1.5 Mechanisms of antibiotic resistance

Antimicrobial agents are substances that can kill or inhibit microorganism growth. The word *antimicrobial* refers to antibiotics, other chemicals and compounds for use in killing microorganisms. The word *"antibiotic"* relates to substances that are manufactured naturally by certain fungi and bacteria and are synthetically created. Bacteriostatic or bactericidal medicines may also be categorized. By avoiding reproduction of microorganisms, bacteriostatic substances, such as tetracyclines and sulfonamides, operate.

There are various mechanisms that inhibit or kill microorganisms through antibiotics. Since these processes are disrupted or rendered ineffective, it is essential to know the various ways in which antibiotics work. Furthermore, while a good antibiotic should destroy the microorganism it targets, it would not ideally harm human cells. It should also be taken into account. The cell wall, which is present in the bacteria (or prokaryotic cells), but absent in people (or eukaryotic cells), is one of the most common mechanisms of action. The cell wall is a layer of peptidoglycan and inner cell membrane in bacteria categorized as gram positive. Gram-negative bacteria also contain an inner membrane and a peptidoglycan cell wall, but also have an external membrane. This external cell membrane can sometimes assist in preventing the adverse antibiotic bacteria that disturb the cell wall composing peptidoglycan (Ehret et al., 2011). The environmental origin of antibiotic resistance genes is shown is Fig. 5.2.

5.1.6 Drug resistance

Drug resistance is the decrease in the effective treatment of a disease or condition by a medication like an antimicrobial or an antineoplastic.

In particular, drugs that target only specific bacterial molecules (mostly proteins) develop antibiotic resistance. Since the medicines are so particular, any changes in these molecules can interfere with or

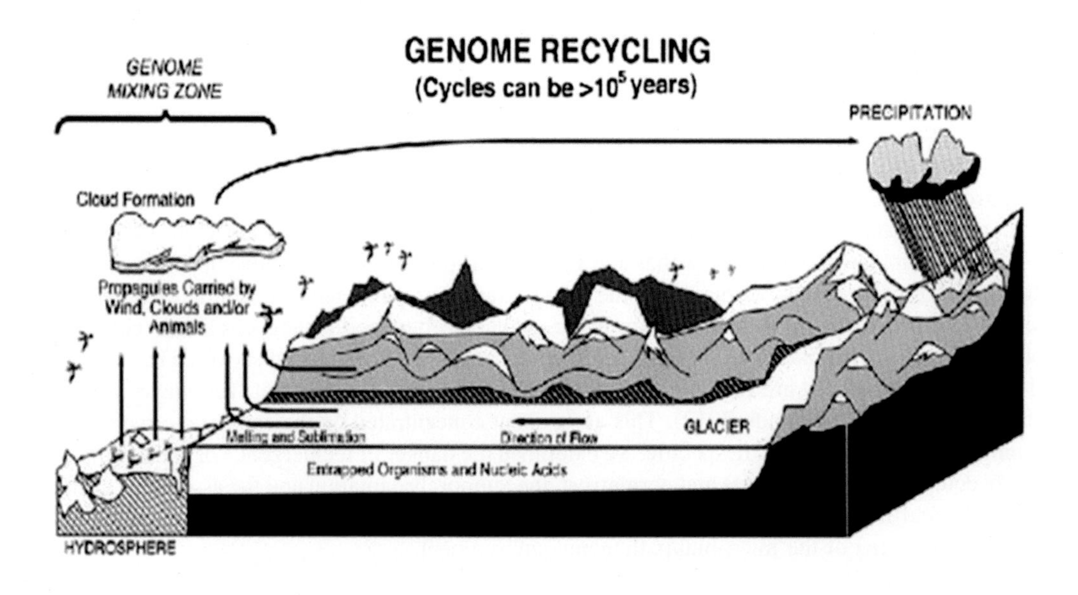

Figure 5.2 Reuse of the genome by glacial ice.

Source Rogers, S. O., Theraisnathan, V., Ma, L. J., Zhao, Y., Zhang, G., Shin, S.-G., Starmer, W. T., 2004. Comparisons of protocols for decontamination of environmental ice samples for biological and molecular examinations. Applied and Environmental Microbiology, 70 (4), 2540–2544.

negate their damaging impact, resulting in resistance to antibiotics. In addition, concern is increasing regarding the abuse of antibiotics in animal production, which alone is three times the amount dispensed to individuals in the European Union, leading to the growth of super-resistant bacteria.

Bacteria are not only able to alter the antibiotic enzyme but can also alter and thus neutralize the antibiotic through the application of enzymes. For example, *Staphylococcus aureus* and *Enterococci* resistant to vancomycin, *Streptococcus* resistant to macrolides, and *Acinetobacter baumannii* and *Pseudomonas aeruginosa* resistant to aminoglycosides are instances of antibiotic microbes.

This problem is only one component of chemical resistance; other issues are the strength of physical factors such as temperature, stress, noise, radiation, and magnetism that are physical variables that affect microbial life.

Antibiotic resistance is when one or more antibiotics enable bacteria to survive and to develop. This causes infection through the resistant bacteria. A particular form of antimicrobial drug resistance is bacterial antibiotic resistance. Although the resistance to antimicrobial drugs used to treat infections with these microbes can also be achieved in other microbes such as viruses and fungi, this chapter focuses upon antibiotic-resistant bacteria.

Resistance in nature is generally created. However, bacterial exposure to antibiotics is more common, and because of the routine use of antibiotics, resistance is growing more rapidly. Common diseases, like bacterial pneumonia, without effective antibiotics, would once again be life-threatening. Complex procedures would be much difficult and infection deaths would be much greater, like open heart activities.

The world's greatest risk to public health is bacterial antibiotic resistance. In the United States, two million individuals each year develop antibiotic-resistant diseases, with at least 23,000 deaths from those diseases, according to the U.S. Centers for Disease Control and Prevention. The medical system and people also suffer from resistant infections. The research has shown that long-term clinical visits, longer recovery times, and increased medical costs can lead to resistant diseases. Resistance to antibiotics has shown that the health care system has an annual burden of $20 billion. Alternative medicines can be less efficient, cause more side effects (be more poisonous) and be more costly if used to treat infections that are resistant. Medical treatments like organ transplants, chemotherapy, and significant operations become more risky without efficient methods of treating infections.

Existing natural communities are faced with new relationships owing to the reemergence of ancestral microbes. These dangers stem from the reemergence of pathogens held in laboratories or pathogens retained in permafrost that are becoming accessible at thawing owing to climate change.

We now see evidence of the recurrence in natural environments of some types of bacteria that did not have any natural hosts for centuries.

Pathogens appear to be adapted locally and temporarily to their cohosts, but when pathogens from a different setting or time enter the host community, the level to which a new host interferes with the pathogen depends on specific genotypical associations, the time lag between both the host and the pathogen, and the relationships with native or current host and pathogen species.

5.1.7 Glacial ice formation plus significance

Microorganisms can be found surviving even in some of the most extreme and harshest environments our earth has ever witnessed. Having been isolated from the highest temperatures or the coldest environments that entailed sea ice, subglacial lakes, permafrost, and glacial ice, microbes survived and continued to be active in extreme weather conditions. This is a significantly interesting phenomenon as these are comparable conditions with extraterrestrial environmental conditions. Study of life in the coldest conditions of glaciers in the Arctic and Antarctic environments is particularly significant in light of the history of ice on Earth. A frosty spread has been found on Europa (a moon of Jupiter) and ongoing investigations point toward hints of ice on Mars (Chyba, 2000). Ice in the polar areas on earth has been amassing for a few million years. The transition from snow to ice is a perplexing phenomenon constrained by various factors and takes quite a while. Snowfall is the initial phase in this procedure, followed by the transformation of snow to firn (wet snow). Frosty ice is shaped when the interconnecting air entries between the grains (precious stone or a gathering of gems) in firn are fixed, leaving gas in air pockets (WSB, 1994).Therefore, glacial ice holds an important record of the past atmosphere alongside a ceaseless record of microbial life. Amid glacial ice, study related to microbes helps understanding the mechanism of survival about these organisms in extreme weather conditions or environment and to know in what way these microbes are compared to contemporary microorganisms with today's environments.

5.1.8 Glacial ice and the study of climate

Atmospheric elements are trapped by ice sheets and ice caps of the earth in chronological order with different times. Study dealing with chemical and physical composition of the resulting data and such ice cores may help understanding of the climatic environment of past times and global climate changes. Connection of the adjustment in atmosphere (e.g., increments in temperature) with the barometrical arrangement (e.g., CO_2 levels) amid those occasions may assist us with understanding the present and future atmosphere changes, particularly with the current worldwide environmental changes that are occurring. Different parts of the ice centers help us to comprehend distinctive parts of the past atmosphere. The O18/O16 proportion is subject to the temperature, environmental chemistry is reflected by anion-to-cation concentrations, sulfates in the ice give a record of the volcanic ejections, dust concentrations demonstrate the turbidity of the atmosphere, following component concentrations can be utilized in the measurement of environmental emissions, and contractions and expansions of localized vegetation can be determined with the measurement of nitrate concentration levels (WSB, 1994).

There exist some additional records in the environment that provide helpful information regarding climate changes in the past, e.g., in tree rings, coral reefs, and sea sediments. Yet, most of these methodologies do have repercussions. Proxy records, as compared to other records, from ice cores are meticulous and continuous records of different paleoclimate factors inclusive of accurate and precise records of primitive atmosphere's gas concentrations (WSB, 1994). Utilizing the ice centers recovered from various places of the world, researchers have been able to recreate past atmosphere records up to a few hundred thousand years. Ice centers from Greenland, Antarctica, and Tibet have assembled the atmospheric records revisiting the last frosty interglacial cycle (Dansgaard et al., 1993; Grootes et al., 1993; Lorius et al., 1985; Thompson et al., 1997). Studies at the Vostok ice center have recreated the atmospheric record for as long as 420,000 years, which covers four icy cycles (Petit et al., 1999). The study of the ice from polar districts gives an increasingly complete image of the past atmosphere record contrasted with the study of ice from other geographic areas. The volume and size of polar glaciers do have a key role to keep the global climate patterns regulated while subtropical and tropical glaciers don't (Thompson et al., 2000).

Studies dealing with climate conditions in the past help us to understand the evolution of life on earth. Our planet has witnessed climatic fluctuations with notable changes in varying temperature, ending up with ice eras over long spans of time. The cycles called glacial and interglacial cycles have been a typical event from the beginning of time, however, three such occasions (2.7–2.8 billion years prior, 2.3–2.5 billion years back, and 600–800 million years prior) have been sufficiently large to have assumed a critical role in the advancement of life. The two times of worldwide glaciation that happened 600–800 million years and 2.3–2.5 billion years prior brought about a condition called "snowball earth" when the earth was totally shrouded in ice (Hoffman et al., 1998; Kirschvink et al., 2000). These times of incredibly low temperatures and broad ice spread were trailed by times of quick warming that were caused by the arrival of enormous quantities of CO_2 into the atmosphere from volcanic sources, and the ending of a portion of these cold periods concurred with vast shooting star impacts. Nonetheless, the state of snowball earth may have taken four to 30 million years to turn around and such delayed times of solidifying conditions and changed geochemistry may have had genuine ramifications for the different organic biological systems of the time. The time of warming after the second snowball earth may have been in charge of the Cambrian blast, a period when all the creature phyla quickly advanced (Hoffman et al., 1998). Studies dealing with chemical and physical properties relating to ice cores have helped in understanding the past climates and the ways they have impacted the evolution on the earth.

5.1.9 Preservation of life in ice

Being a protective matrix of environmental microorganisms, ice has provided a notable record of microbial evolution and of primitive biodiversity (Ma et al., 2000). Living beings caught in ice may stay reasonable in a lethargic state, or are dynamic and ready to complete a low dimension of digestion. Thymidine and leucine in investigations of *E.coli* at 150 and 700°C demonstrated metabolic movement in cells at 150°C yet not at 700°C (Carpenter et al., 2000). Fluid water is available in ice at low temperatures, and is fundamental for the metabolic action of the cells. Water is available at the interface of ice precious stones and furthermore as a film on air bubbles, and has appeared to exist in ice up to temperatures of 600°C (Ostroumov and Siegert, 1996; Price, 2000). The nonappearance of fluid water (water action just about zero) might be in charge of the nonattendance of metabolic movement in *E.coli* at 700°C. Glacial ice has steady temperatures, yet higher than 600°C, making it perfect for long haul protection of microorganisms (Willerslev et al., 2004). The low temperatures in these conditions will stop cell division yet hold low dimensions of metabolic movement. Nonstop catabolism in cells with diminished anabolism rates leads to the creation of free radicals in charge of harming DNA and proteins (Aldsworth et al., 1999). Low metabolic action in these conditions keeps the arrival of free radicals, along these lines securing the phenotypes. The absence of cell division in these conditions prevents hereditary mutations brought about by dynamic replication while the low metabolic movement takes into account DNA repair of any unconstrained mutations. Some corrosive dissolvable proteins (SASP) have been seen in lethargic cells like bacterial endospores and help lessen the DNA harm by binding to the DNA (Willerslev et al., 2004). Microorganisms stuck amid environmental ice that struggled through freeze-thaw cycles are likely to continue in a viable or dormant state over a long span of time (Rogers et al., 2004) (Figure 5.2).

Genome reuse through glacial ice. Microorganisms transported into the atmosphere by different methods like breezes and volcanism are stored onto the glaciers by precipitation. These living beings get entangled in the icy mass with the arrangement of ice. These living beings are caught in glacial ice for thousands to a huge number of years. The softening of the icy mass discharges the captured microorganisms into the hydrosphere (lakes, streams, waterways, and seas). This helps the microorganisms discharged from glacial ice to collaborate with the surviving populaces of similar species in the genome blending zone and gives a chance to create a fleeting quality stream (Rogers et al., 2004). The existence of what we call viable microbes within ice thousands of years before and danger of release of old pathogens strains is an intriguing reason for monitoring and study of such ice cores.

Studies dealing with organisms that are extricated from environmental ice may possibly provide insight toward the process of evolution, while in the case of pathogens, studies might be significant in the wake of precautions and health safety. Environmental ice carries a huge and diverse mix of microbes, and with the increased rate of glacial melting, ancient microbes have already been released in huge numbers. The release of ancient microbes may lead to these microbes entering the contemporary population from which they have been separated for hundreds of thousands of years. The temporal gene flow that may occur because of this mixing of the ancient and modern genotypes has been termed as "genome recycling" (Fig. 5.1). The rate of glacial retreat has increased in the last 1000 years and recent increases in temperatures because of global climate change have accelerated this process (Rogers et al., 2004). The melting of environmental ice releases an estimated 1017 to 1021 viable microbes every year (Smith et al., 2004). The release of pathogenic microbes is of even greater concern as they may have an increased virulence because of immunity deficiency or defense to these strains in the host. The appearance of a few influenza types and calicivirus decades apart may be related to the melting of environmental ice.

5.1.10 Microorganisms in cold environments

Microorganisms are isolated and identified from glacier samples of ice across various regions. In glacial ice and snow, the microorganism global concentration is likely a consequence of atmospheric rotation and wind currents. Different kinds of solidified circumstances, such as permafrost, ocean ice, glacial ice, and gradual addition of ice from subglacial lakes have been differentiated and separated between bacteria and fungi. Cryopegs are unfrozen layers of soil, covered by a constantly solidified soil. The particles broken up in the water bring down the freezing point, preventing the water from freezing and creating pockets with the temperature range from 9 to 11°C (Ozerskaya et al., 2004). Ozerskaya et al. (2004) revealed the disconnection of 40 parasitic strains from two cryopegs in Kolyma swamp district of Siberia. The fungi had a place within the genera *Alternaria, Aureobasidium, Cladosporium, Geomyces, Penicillium, Ulocladium, Valsa*, and *Verticillium*. More genera were recovered from the MEA medium and at 25°C instead of 40°C (David Gilichinsky et al., 2005). Cryogenic methods isolated bacteria, fungi, and yeast near the East Siberian Sea. The most abundant bacteria were the psychrobacteria the genera *Psychrobacterium, Arthrobacterium, Frigoribacterium, Subtercola, Microbacterium, Rhodococcus, Erwinia, Paenibacillus*, and *Bacillus*, from Siberian cryopegs (Bakermans et al., 2003). Fungi were restricted to the soils in the Arctic where water due to low temperatures is not available organically.

The vast majority of the dirt bacteria and fungi were oligotrophs that can survive in the low supplement conditions (Bergero et al., 1999). Fifty four contagious strains were retrieved from Arctic soils by Bergero et al. (1999) and were placed within 21 genera. The most commonly known genera were *Acremonium, Cladosporium, Geomyces, Mortierella, Phoma*, and *Thelebolus*. The psychrooligotrophic fungi had been from nine species with genera *Geomyces, Phoma, Mortierella*, and *Thelebolus*. Most of the fungi had sterile mycelia dull or hyaline that showed fungal survival processes. Permafrost is another cold condition that is very concentrated with microorganisms. Microorganisms in these situations cease to exist below zero temperatures and foundation radiation (Hindmarsh, 1992). It has been shown that a flimsy film of water encompasses the dirt and ice particles in permafrost even at below zero temperatures. The fluid water may encourage the exchange of supplements and poisonous synthetic substances, and may be fundamental for the survival of organisms in these conditions. Research on permafrost lands in Antarctic and Arctic areas has shown variations in their chemical and physical properties. Antarctic permafrost soils are lower in organic matter, temperature, higher pH, and not oxidized environment.

5.1.10.1 Impact and origin of revived microbes in natural populations

In earth, microbes are pervasive (Ghai et al., 2011) including their traces in sediment strata or deep rock (Gadd, 2010); they have also been found impounded in ice glaciers and permafrost (Bidle et al., 2007). Recent ecological studies pertaining to resurrection (Orsini et al., 2013) in their special condition has hinted that microbes that are long dormant are possible to be revived even today, after millions of years (Barras, 2017).

5.1.10.2 Microbes revived in the laboratory or conserved in research facilities

Hypothetically transferable today, ancient microorganisms or viruses have remained largely under study among diagnostic and research facilities across the globe. Although the development of pathogens in space is most likely a lot higher risk than development of pathogens after some time, there are imperative cases to consider: one example of such a case is in 2003, a SARS virus was removed from a three-dimensional laboratory in Singapore; in 2007, a break in biosecurity caused diseases in the feet and mouths of chickens in the United Kingdom (Von Bubnoff, 2005; Hamilton et al., 2015). The risks connected with the research of antiquated pathogens arise not only from solidified pathogenic species held in the laboratory but also from the allocation and prospective recreation of the genome sequence of such pathogens. For example, the genome sequence of the Spanish flu virus of 1918, a wild virus that killed 50 million people around the world, was reproduced from viral RNA retrieved from unfortunate tissues in 1918. An advancement of studies was then aimed at studying in vitro and creature models of the destructiveness of the virus, using viral builds that contain the qualities of 1918. As a distinct difference to contemporary human flu H1N1 vitrics, the 1918 pandemic virus was observed to be incredibly harmful, causing 100% demise in mice and showing a high growth phenotype in human bronchial epithelial cells (Tumpey et al., 2005). This sort of study, breathing life back into old lethal viruses, is still profoundly debatable. While a few analysts suggest that such examinations give key novel bits of knowledge into virus science and pathogenesis, critical data about how to anticipate and control future pandemics is as yet missing (Drancourt and Raoult, 2005) (Taubenberger et al., 2012). The processes and related dangers of revived pathogens from the past can be contrasted and the danger of rising and reemerging zoonotic infections worldwide because of spatial invasion of nonnative species setting up in another range and spreading into new populaces (Blackburn et al., 2011; Dunn and Hatcher, 2015). Restored pathogens from an earlier time and rising pathogens because of spatial attack are both a wellspring of new host pathogen cooperation. In the two cases, the result of these new communications is unusual. Spatial or temporal invasion by external pathogens may face dangerous consequences for animals, humans, or plants, e.g., crayfish plague, Zika, tick-borne diseases, and amphibian decline etc. yet such pathogens are possible to be nonviral to native habitants, hence causing no such real threat that matters.

5.1.10.3 Old microbes from in the permafrost

Another next risk of the ecosystem in our planet earth is occurring already as long as our planet is subjected to the mechanisms for climate change. These modifications are not a secret in our atmosphere today. After all, there are potential hazards to people and society, economics, and biological communities for environmental changes. In mid- to high-scale areas, ice temperatures have been seen to have increased greatly, doubling the global rate (Schuur and Abbott, 2011; Slenning, 2010). Ice temperatures are increasing rapidly. These lines are formed by current and future problems for the environment and by common microbial communities in the permafrost. In particular, a recorded factor can be used to rekindle torpid propagule and sensible vectors that may lead to future problems (Fig. 5.2). The permafrost fills up as a distinctive bank and includes a wide array and other sensible vectors (seeds, eggs, bulbs or spores from crops and spineless beings). The risk of diseases may be increased (Revich and Podolnaya, 2011; Revich, Tokarevich, and Parkinson, 2012). There has been growing evidence over the last few years that permafrost is a huge store of ancient microorganisms that are able to come back into life if the circumstances alter and free them again. The permafrost-caught measure of microorganisms can reach 108 cells/g of dry soil (Legendre et al., 2014), which is still sensible. The defrosting of the permafrost increases the regeneration of diverse torpid vectors, including bacteria that may remain sensible for a few million years (Vishnivetskaya et al., 2000). For example, in 2014, in a 30,000 year-old ice center collected from Siberian permafrost and revived in the research center, a practical example of a monster virus, called the *Pithovirus Sibericum*, was found. Oddly enough, this virus has been found. One of those viruses is a distant relative of Gemini viruses, a known plant virus, although the second virus is recognized as a collection of RNA viruses that are pathogenic to beneficial arthropods such as humpback bees as well as restore and bugs. Surprisingly, the viruses were unbroken and remained irresistible 700 years after ice. In this respect, there is a risk, especially as the Russian Arctic is affected by environmental changes in particular. The developed plant *Nicotiana Benthamiana*, a relative of tobacco located in North Australia and unmistakably not the main host of that virus, has been reported to be infective (Holmes, 2014) by the plant virus in particular (Ng et al., 2014). These results show that potentially irresistible pathogens can be released from the layers of ancient permafrost submitted to freeze with potential ramifications for humans, animals, and plants. Similarly, for their contemporary abuse (i.e., mining and boring (Revich et al., 2012; Vorobyova et al., 1997; Vishnivetskaya et al., 2000; Legendre et al., 2014), the weight of the wealthy mineral resources of the cold areas is increasing. Such occasions will lead to flooding and soil disruption, which can release onto the surface soil and vegetation bacterial spores or viruses devoured by brushing-living things, as well as increase the risk of contamination in individuals coming in touch with contaminated animal products (half-cooked meat, bone (Revich et al., 2012). Although the risk of potential ice-captured pathogens is small, and the hotter temperatures associated with ice dissolution could somewhat undermine the spread of viruses between modern populations (Holmes, 2014), the prospective pathogen to reduce viral nucleic acids (Legendre et al., 2014) is not excluded from posing future hazards on human well-being. In Russia, an epizootic cycle in *Bacillus anthracis* caused the loss somewhere between 1897 and 1925 of 1.5 million deer led to the creation of more than 13,000 graves with the lives of man-made animals. These bodies can return with ice defrost, which can cause *B. anthracis* to continue among steers and reindeers and also infect individuals who process contaminated animal items, or ingest tainted meat incorrectly cooked (Revich et al., 2012). Many people were forced to migrate, while others were isolated. This disease is due to the freezing of reindeer carcasses that have never been living since an epidemic 75 years ago Fig. 5.3.

Halted in permafrost, community DNA represent an important reservoir of genes that may be obtained from lateral gene transfer by the existing microbes in the liquefaction or thaw (Legendre et al., 2014). The community DNA as discovered in 100 ka to 8 MA ice samples has been metagenomically studied and diverse orthologic diversities have been searched for metabolism genes. Where some microbes have been removed from the same sample, they can be cultivated in the laboratory. Their inquiry proposes a channel to a wide-scale horizontal exchange of quality, which could possibly spill out microbial phylogeny and speed up the rhythm of microbial growth in previous soil liquefying (Legendre et al., 2014). In bacterial strains exuded by permafrost from Siberian and Antarctic areas from 5000 to 30,000 years ago, the quality of encoding obstruction against prevalent or existing semi-synthetic antitoxins, as well as mobile parts that participated in their flat exchanges, were surprisingly detected (Perron et al., 2015; Petrova et al., 2009; Petrova et al., 2014). Taking these results together, the hypothesis is strengthened that there existed a repository of oppositional characteristics in a range of bacterial kinds of animals that prior to the anthropogenic use of antimicrobial substances and that evidence has evolved to close antitoxin formation in indigenous habitats (Perron et al., 2015; Petrova et al., 2009; Petrova et al., 2014). These results also confirm the evolving group of

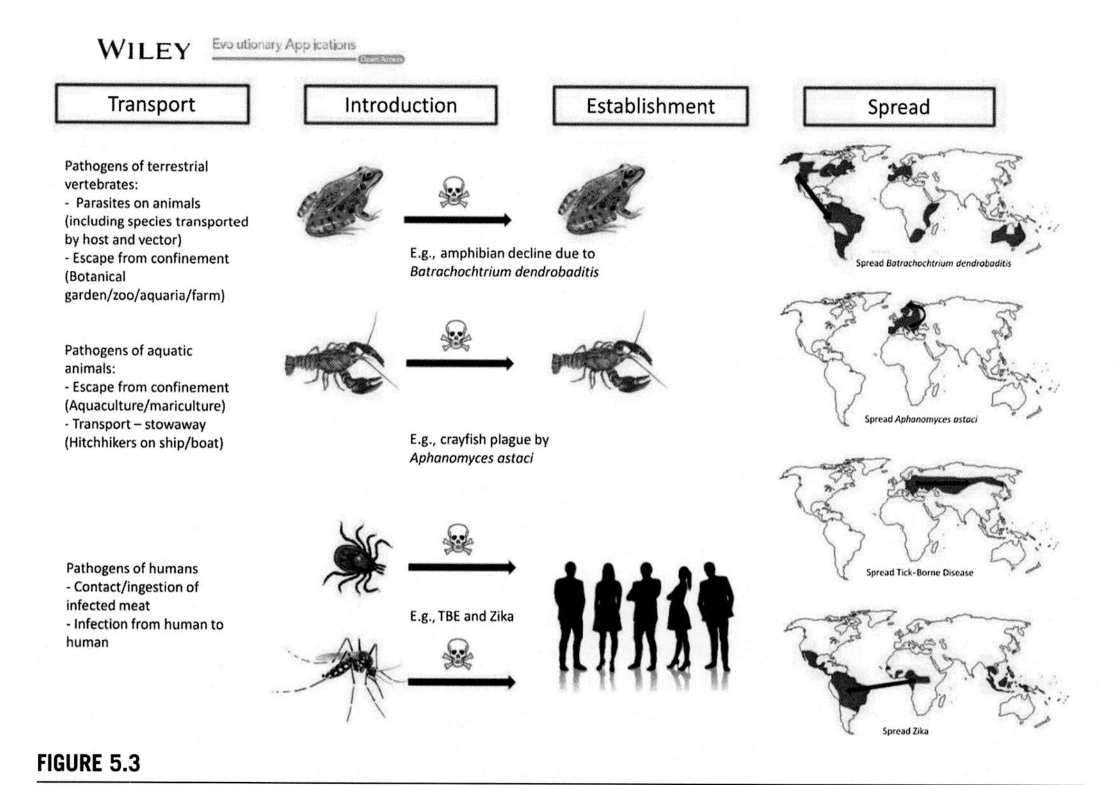

FIGURE 5.3

Overview of the different trajectories of how frozen pathogens or microbes can be introduced into current populations.

evidence that pathogens are the obstruction characteristics that can be translated into pathogens and thus enormously impact the development of multidrug-safe bacteria in clinical environments (Bhullar et al., 2012). These results also strengthen the nonpathogens.

In addition, organisms protect their hosts from pathogenic contaminations to make them increasingly unpredictable. In the host populations there is a collaboration among these "guarded" organisms and pathogenic agents. Ford, Williams, Paterson, and King (Ford et al., 2017) experimentally codeveloped a host protective microorganism (*Enterococcus faecalis*) and a nonevolved (*Caenorhabditis elegans*) pathogen (*Staphylococcus aureus*). They discovered that all the pathogen populations recreated in 10 submissions have been adapted locally to protect the microorganism, and the cautious organism has been disadvantaged locally. Paralleling hostpathogenic coevolution, this instance shows that pathogenic neighborhood adaptation is more frequently acknowledged than neighboring adjustments as pathogens react more quickly to reciprocal adjustment than hosts (due to shorter ages, larger populations, and greater movement rates) (Ford et al., 2017). (Ford et al., 2017; Kawecki and Ebert, 2004) proved by conducting time change experiments that both the protective and pathogen organisms had been late encountered in transformative history in well-being against enemy populations. Guarded microorganisms from what has come were substantially higher in pathogens smothering than previous protection organisms, while in future pathogenic ones, cautious microorganisms were considerably higher than previous pathogens to smother. The well-being of a key animal variety should be better than its current enemy against its current enemy since the current opponent has begun to respond to the versatile modifications of the main species. This shows that there are links between nonnative pathogens (elevation from an alternative region or an alternating moment) and local hosts. Protective microorganisms, besides pathogens, may also join the network and modify the pathogen and host associations. The virulence of pathogens can be reduced in beneficial ways, but it could still be reduced in adverse manner, depending on the time span between the host and the pathogens (Ford et al., 2017; Kawecki and Ebert, 2004; Dybdahl and Lively, 1998).

Considerations that determine the danger of ancestral pathogens developing infectious illnesses.

Increases and communications of infectious diseases generally follow routes involving a resource (i.e., environment, vector, or optional host) for local host interaction (e.g., pathogen-associated health-related hazards arising from the combination of environmental and socioeconomically variables that affect pathogenic microbial potentials and virulence). There is a strong dependence on the level of contact as well as the overflow between stores and the local host (Lambin et al., 2010) on the danger of increased disease. Contact between the non-tamed animals and humans or individuals for the provision of life for transmission of pathogens is crucial for zoonosis representing most emerging infectious disorders. Numerous cynanthropic species (i.e., species, animals or plants living in close proximity to people, such as rodents, fluttering creatures, bats, and some other species of mammals, have been reported to transmit zoonotic diseases and are now and again used as repository hosts for such pathogens. Urbanization-related anthropogenic weights commonly enhance their respect for animals and individuals and thus promote disease development (Hassell et al., 2017). Domesticated animal cultivation is also a key factor in the interplay of human animal life. The casual domesticated animals are common in urban African regions and are often characterized as being in close proximity to humans through low biosecurity and mixed species. Evidence of late zoonotic growth in Asia (for instance, avian influenza diseases; World Health Organization Global Influenza Program Surveillance Network, 2005), and the pursuit of usually stable zoonosis (for instance, cow-like tuberculosis; Gortazar (Gortazar et al., 2011) involves domesticated animals that are used as tin shelters, linking individuals with life and epidemiological conditions. The chain of disease is further supposed to be weak. The pathogenic components that affect host susceptibility to disease area and sexual orientation, and other immunocompetence and immunological history (Hassell et al., 2017) are not all specific (Bento et al., 2017), and are "dead end" in hosts can assume to be controlling contamination. Ancient pathogens, such as the 1918 flu virus, are examples of previous pandemics would not be indistinguishable. Certainly, for example, due to the 1918 flu outbreak, the vast majority of these diseases are now somewhat invulnerable to the infection of 1918 because the consequence is, to some extent, human influenza diseases. In the case of certain diseases with reemerged viruses that contain some genes present in the 1918 influenza virus, continuous flu drugs and vaccines have been used among mice (Von Bubnoff, 2005).

5.1.10.4 Permafrost evolutions

The permafrost triggers 20% of land (Ershov, 1998) while much of biosphere remains below 5°C constantly (Ashcroft, 2002). In the glacier ground layers, permanent and perennial lake water microbes have lately been discovered around ice sediment regions (Psenner and Sattler, 1998).The disengagement of microorganisms, together with permafrost, are an immobile contending territory and are concerned with whether microbes are as per time as shops. It is fascinating that many representatives of ancient microscopic bodies are nowadays subatomic microbes. Now and again, geo-intelligent information demonstrates the seniority of these life forms, while molecular information presents proof for their innovation (Greenblatt et al., 2004). Living (or if nothing else feasible) microbes evidently happen somewhere down in strong solidified ground (permafrost) wide open to the harsher elements areas (Gilichinsky and Wagener, 1995).

5.1.10.5 DNA evolution

Bacteria are the most primitive organisms to have emerged billion of years ago on this planet, as a part of their survival they've sharpened their skills by making their genome flexible which enables them to be protected from the negative impacts of toxic chemicals. Bacteria are identified as potent originators for the dissemination of antimicrobial-resistant genetic apparatus (Woodford et al., 2011). They are competent enough for the transmission and upholding of genes responsible for antimicrobial resistance as part of mobile genetic elements (MGE; plasmids, transposons and integrons). The transposons and integrons, because of their genomic plasticity have played a major role in the fitness quotient and robustness of bacteria enabling them to survive in various environments. Integrons, that are transported by plasmids or enclosed in transposons, perform the task of resistance gene dissemination plays an important role in the revealing of Super Bugs (Xu et al., 2011). Since its earliest assessment in 1989 (Stokes and Hall, 1989) molecular mechanisms involved in the mobility of integrons, their excision and integration for gene cassettes, is currently being scrutinized (Hall et al., 1999; Mazel, 2006). Establishing role of MGEs in genomic evolution justifies the predictions of Barbara McClintok that transposons play a major role in the genomic diversity and evolution. Owing to their capacity to relocate between host genomes, MGEs play a vital function of acting as vehicles for resistance gene acquisition and their successive propagation.

5.2 Resurrection of ecological research avenues

5.2.1 Temporal adaptation of pathogen infectivity and host susceptibility

Just before fully grasping the results of association of novel swarm pathogens it is vital to get insights of pathogen-host mechanism of interaction on many temporal and spatial adaptation scales (Penczykowski et al., 2016). A study of examples of cross-sectional host-pathogen adjustment over time was usually seen as an increasingly fast strategy to induce coevolutionary dynamics (Penczykowski et al., 2016; Burdon and Thrall, 2014; Gandon et al., 2008). This is not always evident, however, because time has brief, slack contrasts covering these members (Gandon et al., 2008). This is not always evident.

Contrasts in host weakness and pathogens' infectivity trigger strong pressures and often result after some time in the increased arms competition (Decaestecker et al., 2013; Decaestecker et al., 2007; Futuyma and Agrawal, 2009; Schmid Hempel, 2011). The comparative speed at which they coevolve is one of the most essential components of adaptation in two opponent species. Because pathogens are usually smaller than hosts, have a much shorter age and larger numbers and greater motion rates, their development potential is commonly higher (Greischar and Koskella, 2007; Kawecki and Ebert, 2004). This has led to thinking that pathogens should be adjusted more often than their hosts. Pathogens often, however, are agamic while their hosts expressly imitate them, which can speed up the host's response to pathogens. The host's inheritable variety is the fuel for host obstruction and pathogenic infectiveness proportional coevolutionary components (Decaestecker et al., 2013; Decaestecker et al., 2007; Futuyma and Agrawal, 2009; Schmid Hempel, 2011; Greischar and Koskella, 2007; Kawecki and Ebert, 2004). It is crucial to notice that there are two types of host-pathological coevolutionary components: (1) weapon competition elements, which are a growth of certain breadths of fresh transformations and thus a mild operation, and (2) adverse recurrence-dependent codevelopmental elements. The coevolutionary Red Queen Dynamics (RQDs) are connected with circumstances of infectiveness that are already in place after a period of time and are suggested for work in all biology, such as developing and maintaining sexual proliferation, and shaping the structure of ordinary population and network (Decaestecker et al., 2007; Futuyma and Agrawal, 2009; Schmid Hempel, 2011; Greischar and Koskella, 2007; Kawecki and Ebert, 2004; Dybdahl and Lively, 1995; George Priya Doss et al., 2008; Wood et al., 2007). Because of the timeframe needed to observe long-term genotype recurrence and adaptation changes the attempts to consider long-term effects on transformative and biological exemplars of natural modifications. One strategy to avoiding this problem is by applying a "resurrection biology" strategy (for example, the water insect Daphnia and its normal asides), while other stages are sediment by distinct layers and represent distinct moments around some pool (Decaestecker et al., 2013). Such revitalizing population may be used for time-shift studies when, at some stage, future or modern times, pathogens are confronted to the host's past. The investigator can easily observe long-term aspects of host-pathogenic coevolution through investigation of these special associations. Basically, if the well-being of a contemporary population (for example, the host) is greater than the well-being of the past population, but less than that of things that come into existence in a fixed rival population (for example, the pathogen), the host populace reacts over a certain time period to determining the pressures of the pathogens (Decaestecker et al., 2007; Futuyma and Agrawal, 2009; Schmid Hempel, 2011; Greischar and Koskella, 2007; Kawecki and Ebert, 2004; Dybdahl and Lively, 1995; George Priya Doss et al., 2008; Wood et al., 2007; Decaestecker et al., 2013; Buckling and Rainey, 2002; Koskella, 2013). The time allocation for testing coevolution, especially in the context of rapid developmental changes (Schmid Hempel, 2011) is a fundamental consideration in these studies (Decaestecker et al., 2007). Thus, it's best to conduct time change tests across times that go through the anticipated window over which coevolution is expected to happen (Penczykowski et al., 2016). If we anticipate that host and pathogen alleles are being cycled as Dybdahl and Lively (1995) guessed, we will be able to imagine the infectivity of blends including previous pathogens and hosts. Since the pathogen development has become late in the host, mixes of near future pathogens with previous hosts have been believed to be the most infectious (Gandon et al., 2008). In this line, the best adapted strains to later (future) hosts are revived pathogens from the past, as they will need some chance in order to adapt to the usually fresh host. Nevertheless, this can sometimes lead to a greater ineffectiveness of pathogens in the factorless pathogen-host couples. However, it's a little less predictable.

Recently a strong method has been designed to investigate bacterial genome evolutions at molecular level, synthetically and with ancient motifs, and to restore an ancient quality. This paleo-experimental strategy involves expelling the cutting-edge type of quality from a surviving being (for example, *E. coli*), a surviving being. Further study installations are used to monitor the versatility of the hybrid genome. The family variation of tufB, for instance, a fundamental quality encoding factor, Tu, is derived from a 700 million-year-old and incorporated in a sophisticated *E. coli* genome rather than local tufB quality. Interestingly, exploratory progress has allowed bacteria to recover wellness by aggregating the tufB advertiser transformations, leading to an extended articulation of the ancient EF-Tu protein. These changes can provide the simplest route for hereditary compensatory changes, in particular in high-conservation proteins. These investigations into paleo-experimental evolvement allow for the reenactment and restoration of the history of biomolecules of ancient times by labs, and offer a detailed understanding of the primary roles of determinism and contingency (Kacar et al., 2017).

5.2.2 Virulence in revitalized host-pathogen interactions

Because of anthropogenic and other climate changes, primitive age pathogens have possibility to reemerge. That may bring temporal or spatial mismatches among pathogens and hosts (antagonists), scaled up outbursts in diseases toward hosts and consequently host shifts, faced with certain pathogens (Penczykowski et al., 2016). Old pathogens may reemerge due to climate change and other kinds of anthropogenic changes. This can contribute to spatial or temporal discrepancies between antagonists (host and pathogens), leading to host changes and the fast spread of disease among sensitive plant communities not historically subjected to specific pathogens (Penczykowski et al., 2016). Also, it can prompt adjusted harmfulness impacts. To appraise the impact of developed pathogens (from various areas or distinctive time focuses) in regular populaces, it is consequently critical to gauge their harmfulness upon presentation. There are three stages in the adaptation of a pathogen to another host: (1) coincidental contaminations, (2) advancement of harmfulness not long after effective attack, and (3) the development of ideal destructiveness in a cotransformative connection between the host and the pathogen. These three stages show how the destructiveness of the pathogen developing. In the main stage (coincidental contaminations), pathogens taint and cause malady in a host that isn't a piece of the typical transmission course, however, is an impasse for the pathogen. These unplanned diseases could happen when pathogens are restored from the permafrost or when pathogens extend their contamination zone (e.g., when pathogens are presented in another territory by means of extraordinary species, or by means of air travel or transporting). Pathogens will in general be locally and transiently adjusted to their cohappening hosts, yet when pathogens from an alternate situation or diverse time enter the host network, the dimension of destructiveness in these new host pathogen's cooperation is flighty (Levin and Eden, 1990) (Ebert and Bull, 2008). Much of the time, infectivity and destructiveness are, by and large, higher in built-up relationship than in novel ones (Ebert, 1994; Lively, 1989).There can, be that as it may, be impressive variation around this normal (Ebert, 1998; Kaltz et al., 1999) A mix might be harmful by and large, yet may infrequently be exceedingly destructive (Ebert, 1994). To enter the second stage, an incidental contamination of one individual spreads to another host, at that point another, framing a constant transmission chain that effectively attacks the new host populace. At first, this spread will establish a plague, in which the quantity of tainted hosts increases. Stage 2 likewise applies to the root of novel variations of a pathogen in the old host. These are extraordinary freaks of existing pathogens that can begin pestilences (Ebert and Bull, 2008).The attacking pathogen will by and large not be all around adjusted to the new host, and thus, there will be quick development of the pathogen and most likely the host. Destructiveness might be a long way from ideal in this underlying foundation in the new host, yet their favorable position outweighs other imperfect characteristics, with the goal that they can spread in any case. Subsequent to persevering for quite a while in another host populace, the pathogen should approach balance destructiveness. This is the third stage in the adaptation of a pathogen to a host, the development of ideal harmfulness. The pathogen may then achieve a determination limit, in which exchange offs among the pathogen's different wellness parts oblige further advancement of destructiveness and transmission (Anderson & May 1982; Ewald, 1980).

Anderson and May (1982) discovered proof for an exchange off between pathogen-induced host mortality (i.e., destructiveness) and host-induced pathogen mortality (i.e., host recuperation). The exchange off display is adaptable and makes it conceivable to foresee changes in ideal destructiveness for various conditions. The exchange off destructiveness predicts that transmission stage generation and host abuse are adjusted, with the end goal that the pathogen's lifetime transmission achievement is augmented. After an average time stint the host is dead in highest lifeline transmission attainment, from various environment pathogens (Jensen et al., 2006). Amid the coevolutionary weapons contest, hosts advance to lessen the harm that pathogens cause to their wellness (Ebert and Bull, 2008). A few segments may affect the two accomplices, while others may not. Consequently, it is critical to determine the bases of destructiveness while talking about host-pathogen coevolution (Ebert and Bull, 2008; Ebert, 1994; Lively, 1989; Ebert, 1998; Kaltz et al., 1999; Anderson & May 1982; Ewald, 1980; Jensen et al., 2006; Masri et al., 2015). The test setup comprised five special developmental medicines: (i) the host control, when the host is adapted without regard to the general research center; (ii) the host one-sided adaptation, when the host is adapted to the tribal pathogen. (iii) pathogen adapting to the ancestral host populace and control of pathogen, where the pathogen adapted to general laboratory conditions without the need for a host; and (iv) host-pathogen coevolution where both the host and the pathogen coadapted to their ceaseless coevolutionary enemy. One of the finest and most advanced uses of destructive development was innovative—weakening live infections (Ebert and Bull, 2008). The normal method for constructing a live anticorps was to modify a damaging pathogen to grow in culture before the appearance of a hereditary structure. As the virus evolved to better grow in society, it also frequently evolved ineffectively in the normal host and subsequently decreased its harmfulness (Ebert, 1998). The outcome is a change between the capacity to grow in one lot and that of another (neighborhood adaptation). This technique did not generally succeed, yet it was sufficiently vigorous to prevail as a rule. Development can reverse the low destructiveness of decreased immunizations. If the anticorps strain is allowed to spread between hosts once more, the development of variants that grow better and recover elevated virulence can be further determined on a regular level (Ebert and Bull, 2008).

Studies about interactions of pathogen-host temporal and spatial measures are becoming important as increasing evidence shows such scales varying because of elements entailing fragmentation of habitat, dispersal, and climate changes (Guppy and Withers, 1999).

As surmised, enriched nutrients in environment and increasing temperatures are said to be the reasons behind size of population, increasing pathogen-host interactions with regard to pathogen transmission, and intra-host competitiveness can make the coevolutionary race or virulence results among pathogens and host intensified (Aalto et al., 2015; Reyserhove et al., 2017).

5.3 Major man-made sources of antibiotics resistance

5.3.1 Pharmaceuticals

A huge number of active compounds are presently used in bulk quantities in pharmaceuticals to provide cures or prevent human and animal illnesses; though, they have also been known over the past 20 years as microcontaminants in soil and water ecologies (Boxall, 2004; Carvalho and Santos, 2016). Amongst the numerous pharmaceutical goods, the occurrence of antibiotics in soil and those in water ecosystems produce certain concerns, as their cumulative use and resulting progress of multidrug-resistant bacteria present severe health risks both for humans and animals (Thiele-Bruhn, 2003; Kümmerer, 2009). Some antibiotics are in use for many therapeutic actions, for example, cancer drugs (e.g., actinomycin D, anthracyclines, bleomycin, anthracenones, and epothilones) or also some pesticides (such as oxytetracycline and also streptomycin). Antibiotics at present in use are regular, synthetically synthesized and also semi-synthetic molecules. Those that are natural are formed by bacteria and fungi (e.g., benzylpenicillin and gentamicin) are with bacteriostatic or bactericide effects. Semisynthetic composites are chemically changed natural antibiotics by introducing a preservative into the drug preparation, which are extra stable and less decomposable. Much research has focused on the expansion of advanced technologies for the removal of antibiotics in environment, e.g., supplementary oxidation and coagulation, etc. These supplementary treatments obviously increase management costs and, for this purpose, are presently underutilized. Wastewater from clinics and the pharmaceutical productions, and prohibited drug disposal and aquaculture (Justino et al., 2016) can also be an important source of aquatic contamination (Kümmerer, 2009). The presentation of dung and sludge in the soil as enrichers, along with irrigation with improved water, can enhance chances of the blowout of antibiotics and antibiotic-resistance genes in the dirt (Kumar et al., 2005; Tasho and Cho, 2016). It has also been emphasized that antibiotics may be recycled by plants or other soil organisms impregnated with animal feed or irrigation wastewater. Though the opposing impacts of antibiotic consumption existing in plants are not well recognized and studied, they may cause allergic or poisonous reactions and/or may induce antibiotic resistance in humans (Kumar et al., 2005). Lastly, from the exterior soil, their precise intrinsic features, antibiotics can be leaked to shallower layers and groundwater (Díaz-Cruz et al., 2003).

5.3.2 Persistence of antibiotics in natural environment

Amount of antibiotics in natural surroundings such as soil or water varies (nanograms to kgs/g of soil). Increased amounts are commonly found in areas with robust anthropogenic pressures such as hospital wastes, rivers, and wastewater effluents and soils preserved with dung or soils used for cattle. Environmental residual absorptions of antibiotics occur not only due to their incessant release into the location, but also to their extraordinary intrinsic tenacity. In detail, some antibiotics, such as penicillins, degrade easily, though others, such as fluoroquinolones (e.g., ciprofloxacin), macrolides (e.g., tylosin) and tetracyclines, are significantly more tenacious, which results in their lengthier stay in the atmosphere, dispersing further and collecting in advanced concentrations (Hamscher et al., 2002). Due to their nonstop and continuous introduction into the atmosphere, sea or soil organisms are recurrently open to these chemicals (González-Pleiter et al., 2013). In addition, as they are active even in very small concentrations, they can have a poisonous effect. The instantaneous occurrence of numerous antibiotics with new pharmaceuticals and/or other xenobiotics may result in a synergistic result, a phenomenon well identified in pharmacology (González-Pleiter et al., 2013). Small concentrations can lead to horizontal transmission of resistance genes in *E. coli*, as established for the broad-spectrum antibiotic tetracycline usually used to treat both humans and animals. This absorption is in the range usually found in soils. As a consequence, thoughtful concern has been raised about the likely role of subtherapeutic tetracycline meditations in endorsing antibiotic resistance.

Ecological danger assessments of antibiotics, through the request of a formula of toxic action approach, should create better use of ecotoxicological variables directing microorganisms, particularly bacteria. In fact, a thoughtful amount of antibiotic complexes and bacteria in the environment is vital for a satisfactory danger assessment of these molecules. For competence reasons, risk management has also been practiced for antibiotics, though it has not yet been completely recognized whether these examinations are useable for antibiotics. Antibiotics and their handling products display an extensive range of physicochemical and biological things reliant on the abiotic goods of the environment (i.e., neutral, cationic, anionic, or zwiteric compounds), with positive and negative signs. Similarly absorption, antibiotic activity, toxicity, and light reactivity can also vary with different pH. Though many medications observed in the environment were accepted earlier than when these guidelines were created, consequently there is a rising interest in making available ecological risk data with respect to products previously found on the market (Grenni, 2018).

5.3.3 Environmental side effects

Antibiotics are openly intended to have an influence on indigenous microorganisms and bacterial reactions to them depend on the concentration (Bernier and Surette, 2013). At extraordinary concentrations, antibiotics yield antimicrobial actions in vulnerable cells and can give bactericidal and bacteriostatic results, even though lethal concentrations infrequently occur external to therapeutic uses. Bacteriostatic agents constrain the development of bacterial cells but do not execute them, while bactericidal agents destroy bacteria. So, these categories are not unconditional, as the bushing effect of each drug varies from the test scheme and the species tested (French, 2006). The harmful effect of antibiotics on natural microbial populations could be the vanishing or inhibition of few microbial groups involved in significant ecosystem roles due to bactericidal and bacteriostatic effects (direct effect). Nevertheless, antibiotics can act as a discriminating force in some microbial populations, which can change resistance, producing genetic and phenotypic inconsistency and manipulating numerous physiological actions (Ding and He, 2010).

5.3.4 Direct environmental side effects

Most biogeochemical cycles are facilitated completely by microorganisms. Antibiotics can act as an ecological element in the environment and are sometimes responsible for vanishing or inhibition of some bacterial groups (Allen et al., 2010). The effects can even be found in nontarget organisms with significant ecological functions (Woegerbauer et al., 2015). Many studies have shown that antibiotics can influence the growth and enzymatic activities of indigenous bacterial communities and, eventually, result in disturbance of ecological functions such as biomass manufacturing and nutrient treatment, leading to the cost of nonsteady functions (Ng et al., 2014; Holmes, 2014; Legendre et al., 2014). Antibiotics, including broad-spectrum antibiotics, have a discriminatory effect on numerous microbial groups that range from a group can be as large as fungi or bacteria or a narrower one such as a single species. As a result, the discriminatory antibiotic effect changes the relative species richness. These properties depend on the type of microbial community (Sørensen, 2009), soil or water composition (i.e., soil texture, adsorption capacity, pH, water content, temperature (Ashbolt et al., 2013), and antibiotic concentrations. Transformation and discharge of antibiotics result in fluctuations in microbial community structure after the accumulation of antibiotics in soil and aquatic environments (Aminov and Mackie, 2007; Hou et al., 2014).

For example: Sulfonamides have been set up to tempt an alteration in microbial diversity by decreasing not only microbial biomass but also the association between bacteria and fungi (Underwood et al., 2011).

As regards the nitrogen cycle, nitrification and denitrification are known to be achieved by various prokaryotes and nitrification, in particular ammonium and archaea oxidizing bacteria (AOB and AOA) (Roose-Amsaleg and Laverman, 2016).

Environmentally important concentrations of fluoroquinolones and sulfonamides have been shown to partly constrain denitrification and have been shown to remove from the ground pig manure containing antibiotic tylosin, altering the behavior of nitrogen facilitated by these microbial communities (Roose-Amsaleg and Laverman, 2016).

5.3.5 Indirect effects

Indirect effects involve deviations in bacterial ecology, growth of resistance, and pharmaceutical biodegradation. Low concentrations (i.e., nanograms per liter or kg soil) of antibiotics may have long-term indirect effects on species microbials or consortia that are not directly bothered by their occurrence (e.g., through population dynamics). "Low concentrations" means nonlethal and subinhibitory, which are beneath the so-called minimum inhibitory concentration, the lowest concentration of medications that, under recognized in vitro conditions, constrains the visible development of a bacterium population target.

These concentrations can act in three different ways:

- Select resistance (by enriching preexisting resistant bacteria and by selecting de novo resistance) (Ashbolt et al., 2013);
- Making genetic and phenotypic inconsistency (accrue the rate of adaptive evolution, including the progress of resistance);
- As signaling molecules (inducing many physiological activities, including virulence, biofilm creation, and gene expression) (Müller, 2002).

It has been well known since the early ages of antibiotic usage that bacterial resistance has been designated at low concentrations of antibiotics (French, 2006). Bacterial resistance is a natural niche (homeostatic response) of bacteria against products that search to prevent their growth. It refers to the skill of a microorganism to live and increase, regardless of the occurrence of a biocidal molecule as an antibiotic (Martinez, 2009). Many elements that help to resist higher concentrations of antibiotics have also been revealed to have other practical roles, e.g., cell homeostasis, signal trafficking of molecules and metabolic enzymes. At nonlethal concentrations, bacteria may use antibiotics as extracellular chemicals to activate different cellular responses and can be well-thought-out friendly signals that coordinate and control the working of the community microbial (Abrudan et al., 2015). In ordinary environments, microbes are classically found in polymicrobial communities, involving and exchanging a variety of helpful compounds that serve as cell signals (Pal et al., 2015).

Nonlethal points of antibiotics can change the expression of genes involved in a diversity of bacterial purposes such as metabolism, guideline, virulence, DNA repair, and stress response (Świeciło, 2013). In addition, heavy metal pollution can also coselect resistance to bacterial antibiotics in the atmosphere and antibiotic resistance genes and antibiotic resistance (MRA) genes are often create together in the similar genetic element signals (Pal et al., 2015). In any case, the coselection possibility of biocides and metals is specific to certain antibiotics; for example, resistance genes to quaternary ammonium compounds (QAC) and class 1 integrons (resistance genes for almost all antibiotic families, that include beta-lactams, aminoglycosides, chloramphenicol, phosphomycin, macrolides, rifampicin, and quinolones) are more joint in bacteria exposed to detergents and biocides. Plasmids provide limited chances for biocides and metals to encourage horizontal transfer of antibiotic resistance from end-to-end coselection, while there is plenty of potential for secondary selection through biocide/acid resistance (BMRG) genes (Cornforth and Foster, 2015; Friman et al., 2015). Antibiotics are intended to be obstinate to biodegradation and many of them (e.g., quinolones, diaminopyrimidine) are said to have a great persistence in the soil (DT50 > 100 d) Likewise, ciprofloxacin and oxolinic acid are measured as justly persistent in water (DT50 >90 d) (Amorim et al., 2014). The key process of degradation of an antibiotic depends on its chemical structure. For example, some antibiotics, including fluoroquinolones, are photosensitive molecules, so photodegradation has been described as their foremost course of transformation. In other cases, some bacterial strains or populations proficient in degrading some antibiotics have been acknowledged, as in the case, more or less, of quinolones and sulfonamides (Adamek et al., 2016).

5.3.6 Events to reduce the discharge of antibiotic and antibiotic-resistance genes

To diminish antibiotic resistance, strengthened by the usage of antibiotics in veterinary medicine, since 2006 the European Union has barred the use of antibiotics as growth promoters since low-dose administration (minimum inhibitory concentration) to moderate the metabolism of the commensal like bacterial flora can endorse the extent of this occurrence (D'Costa et al., 2011). Unluckily, in other countries (e.g., United States, Canada, and Asia), they are still extensively used as growth promoters. Many countries have also limited the use of antibiotics in aquaculture, especially antibiotics used in the cure of human infections (Chattopadhyay, 2014). Though national programs to control antimicrobial resistance, and rationalize the use of antibiotics in case of humans, are decreasing the amount of antibiotics cast-off in human therapy, their whole elimination is not possible. Consequently, the quantity of antibiotics released into the environment from human and veterinary medication will probably remain at extraordinary levels in the future. The efficiency of reducing the use of antibiotics in lessening the number of resistance genes is contentious. Some authors have shown that the decrease of antibiotics in the environment can also decrease the amount of resistance genes and their transfer to humans, but others show that, although resistance is declining, their diminution is slow and strong populations persist (Cabello et al., 2016). In addition, the fact that some resistance genes in human pathogens are created in environments that are not considered by previous antibiotic contamination (Pallecchi, 2008) proposes that antibiotic-resistance genes may continue even in the deficiency of selective antibiotic pressure. Measureable and qualitative data on the profusion of resistance genes in different environments (soil, water) and an improved understanding of the interactions between antibiotics and ecological bacteria are essential for risk calculations (Grenni, 2018) (Sørensen, 2009).

5.4 Conclusion

It has been noted that the wider problem of antibiotic resistance and antimicrobial resistance has many parallels to that of climate change, as they are both, arguably, an interdisciplinary, complex, global "tragedy of the commons." Many hundreds of research years have gone into producing our current understanding of the mechanisms of antimicrobial-resistance selection and transmission. However, despite this deep knowledge base, many rather fundamental questions about the spread of antimicrobial and antibiotic resistance in environment are unanswered. A recent report that conducted a systematic observational analysis of antibacterial resistance research funding showed that only 3% of research projects on antibiotic resistance proposed to tackle issues that relate to the environment.

The combined removal of pollutants that are potential selective agents, disinfection, and deactivation of the genetic material may be a useful strategy to reduce the pollution of environments with resistance factors.

The discharge of antibiotics and resistance genes into natural ecosystems is a new development in evolutionary terms. There is specific apprehension regarding their influence on nontarget bacteria and their linked ecological purposes. These contaminants can primarily or secondarily affect microbial communities. To lessen the impact of resistance there must be a separation between human-related and environmental bacteria. Interaction among both can disturb the structure and purpose of ecological microbial populations. Resistance genes are considered as non "degradable contaminants" and this even goes for automatic replication. The primary effects of antibiotics have been established on natural microbial communities but the indirect effects of the antibiotics in ecosystems are still mostly unknown. In addition to regulating the use of antibiotics, studies are desired to recover their degradation in natural environments. Biodegradation (and mineralization) of antibiotics has been professed, which requires microbial version and selection procedures that happen during comparatively long periods of contact in an unspoiled environment.

References

A.J., B, 2017. Microbiomes of the Built Environment: A Research Agenda for Indoor Microbiology, Human Health, and Buildings. National Academies Press.

Aalto, S.L., Decaestecker, E., Pulkkinen, K., 2015. A three-way perspective of stoichiometric changes on host–parasite interactions. Trends in Parasitology 31 (7), 333–340.

Abrudan, M.I., Smakman, F., Grimbergen, A.J., Westhoff, S., Miller, E.L., Van Wezel, G.P., Rozen, D.E., 2015. Socially mediated induction and suppression of antibiosis during bacterial coexistence. Proceedings of the National Academy of Sciences 112 (35), 11054–11059.

Achtman, M., Zurth, K., Morelli, G., Torrea, G., Guiyoule, A., Carniel, E., 1999. *Yersinia pestis*, the cause of plague, is a recently emerged clone of Yersinia pseudotuberculosis. Proceedings of the National Academy of Sciences 96 (24), 14043–14048.

Adamek, E., Baran, W., Sobczak, A., 2016. Assessment of the biodegradability of selected sulfa drugs in two polluted rivers in Poland: effects of seasonal variations, accidental contamination, turbidity and salinity. Journal of Hazardous Materials 313, 147–158.

Aldsworth, T.G., Sharman, R.L., Dodd, C.E.R., 1999. Bacterial suicide through stress. Cellular and Molecular Life Sciences CMLS 56 (5–6), 378–383.

Aliyu, G., Ezati, N., Iwakun, M., Peters, S., Abimiku, A., 2017. Diagnostic system strengthening for drug resistant tuberculosis in Nigeria: impact and challenges. African Journal of Laboratory Medicine 6 (2), 1–6.

Allen, H.K., Donato, J., Wang, H.H., Cloud-Hansen, K.A., Davies, J., Handelsman, J., 2010. Call of the wild: antibiotic resistance genes in natural environments. Nature Reviews Microbiology 8 (4), 251.

Allsopp, N., Colville, J.F., Verboom, G.A., 2014. Fynbos: Ecology, Evolution, and Conservation of a Megadiverse Region. Oxford University Press, USA.

Aminov, R.I., Mackie, R.I., 2007. Evolution and ecology of antibiotic resistance genes. FEMS Microbiology Letters 271 (2), 147–161.
Amorim, C.L., Moreira, I.S., Maia, A.S., Tiritan, M.E., Castro, P.M.L., 2014. Biodegradation of ofloxacin, norfloxacin, and ciprofloxacin as single and mixed substrates by Labrys portucalensis F11. Applied Microbiology and Biotechnology 98 (7), 3181–3190.
Anderson, R.M., May, R.M., 1982. Coevolution of hosts and parasites. Parasitology 85 (2), 411–426.
Ashbolt, N.J., Amézquita, A., Backhaus, T., Borriello, P., Brandt, K.K., Collignon, P., others, 2013. Human health risk assessment (HHRA) for environmental development and transfer of antibiotic resistance. Environmental Health Perspectives 121 (9), 993–1001.
Ashcroft, F., 2002. Life at the Extremes: The Science of Survival. Univ of California Press.
Bakermans, C., Tsapin, A.I., Souza-Egipsy, V., Gilichinsky, D.A., Nealson, K.H., 2003. Reproduction and metabolism at- 10 C of bacteria isolated from Siberian permafrost. Environmental Microbiology 5 (4), 321–326.
Barras, C., 2017. Wakey, wakey. New Scientist 234 (3126), 34–37.
Bell, E., 2012. Life at Extremes: Environments, Organisms, and Strategies for Survival, vol. 1. Cabi.
Bento, G., Routtu, J., Fields, P.D., Bourgeois, Y., Du Pasquier, L., Ebert, D., 2017. The genetic basis of resistance and matching-allele interactions of a host-parasite system: the Daphnia magna-Pasteuria ramosa model. PLoS Genetics 13 (2), e1006596.
Bergero, R., Girlanda, M., Varese, G.C., Intili, D., Luppi, A.M., 1999. Psychrooligotrophic fungi from arctic soils of franz joseph land. Polar Biology 21 (6), 361–368.
Bernier, S.P., Surette, M.G., 2013. Concentration-dependent activity of antibiotics in natural environments. Frontiers in Microbiology 4, 20.
Bhullar, K., Waglechner, N., Pawlowski, A., Koteva, K., Banks, E.D., Johnston, M.D., Wright, G.D., 2012. Antibiotic resistance is prevalent in an isolated cave microbiome. PLoS One 7 (4), e34953.
Bidle, K.D., Lee, S., Marchant, D.R., Falkowski, P.G., 2007. Fossil genes and microbes in the oldest ice on Earth. Proceedings of the National Academy of Sciences 104 (33), 13455–13460.
Blackburn, T.M., Pyšek, P., Bacher, S., Carlton, J.T., Duncan, R.P., Jarošik, V., Richardson, D.M., 2011. A proposed unified framework for biological invasions. Trends in Ecology & Evolution 26 (7), 333–339.
Borcard, D., Gillet, F., Legendre, P., 2018. Numerical Ecology with R.
Boxall, A.B.A., 2004. The environmental side effects of medication. EMBO Reports 5 (12), 1110–1116.
Buckling, A., Rainey, P.B., 2002. Antagonistic coevolution between a bacterium and a bacteriophage. Proceedings of the Royal Society of London – Series B: Biological Sciences 269 (1494), 931–936.
Burdon, J.J., Thrall, P.H., 2014. What have we learned from studies of wild plant-pathogen associations?—the dynamic interplay of time, space and life-history. European Journal of Plant Pathology 138 (3), 417–429.
Cabello, F.C., Godfrey, H.P., Buschmann, A.H., Dölz, H.J., 2016. Aquaculture as yet another environmental gateway to the development and globalisation of antimicrobial resistance. The Lancet Infectious Diseases 16 (7), e127–e133.
Carpenter, E.J., Lin, S., Capone, D.G., 2000. Bacterial activity in South Pole snow. Applied and Environmental Microbiology 66 (10), 4514–4517.
Carvalho, I.T., Santos, L., 2016. Antibiotics in the aquatic environments: a review of the European scenario. Environment International 94, 736–757.
Castro-Sowinski, 2019. The ecological role of micro-organisms in the antarctic environment. Microbiology.
Chattopadhyay, M.K., 2014. Use of antibiotics as feed additives: a burning question. Frontiers in Microbiology 5, 334.
Chénard, C., Lauro, F.M., 2017. Microbial Ecology of Extreme Environments. In Microbial Ecology of Extreme Environments. https://doi.org/10.1007/978-3-319-51686-8.
Chyba, C., 2000. Energy for microbial life on Europa. Nature 406 (6794), 368.

Cornforth, D.M., Foster, K.R., 2015. Antibiotics and the art of bacterial war. Proceedings of the National Academy of Sciences 112 (35), 10827–10828.

D'Costa, V.M., King, C.E., Kalan, L., Morar, M., Sung, W.W.L., Schwarz, C., others, 2011. Antibiotic resistance is ancient. Nature 477 (7365), 457.

Diaz-Cruz, M.S., de Alda, M.J.L., Barcelo, D., 2003. Environmental behavior and analysis of veterinary and human drugs in soils, sediments and sludge. TRAC Trends in Analytical Chemistry 22 (6), 340–351.

Dansgaard, W., Johnsen, S.J., Clausen, H.B., Dahl-Jensen, D., Gundestrup, N.S., Hammer, C.U., others, 1993. Evidence for general instability of past climate from a 250-kyr ice-core record. Nature 364 (6434), 218.

Decaestecker, E., De Gersem, H., Michalakis, Y., Raeymaekers, J.A.M., 2013. Damped long-term host–parasite Red Queen coevolutionary dynamics: a reflection of dilution effects? Ecology Letters 16 (12), 1455–1462.

Decaestecker, E., Gaba, S., Raeymaekers, J.A.M., Stoks, R., Van Kerckhoven, L., Ebert, D., De Meester, L., 2007. Host–parasite 'Red Queen'dynamics archived in pond sediment. Nature 450 (7171), 870.

Dick, G., 2018. Genomic Approaches in Earth and Environmental Sciences.

Ding, C., He, J., 2010. Effect of antibiotics in the environment on microbial populations. Applied Microbiology and Biotechnology 87 (3), 925–941.

Drancourt, M., Raoult, D., 2005. Palaeomicrobiology: current issues and perspectives. Nature Reviews Microbiology 3 (1), 23.

Dunn, A.M., Hatcher, M.J., 2015. Parasites and biological invasions: parallels, interactions, and control. Trends in Parasitology 31 (5), 189–199.

Dybdahl, M.F., Lively, C.M., 1995. Host–parasite interactions: infection of common clones in natural populations of a freshwater snail (*Potamopyrgus antipodarum*). Proceedings of the Royal Society of London – Series B: Biological Sciences 260 (1357), 99–103.

Dybdahl, M.F., Lively, C.M., 1998. Host-parasite coevolution: evidence for rare advantage and time-lagged selection in a natural population. Evolution 52 (4), 1057–1066.

Ebert, D., 1994. Virulence and local adaptation of a horizontally transmitted parasite. Science 265 (5175), 1084–1086.

Ebert, D., 1998. Experimental evolution of parasites. Science 282 (5393), 1432–1436.

Ebert, D., Bull, J.J., 2008. The evolution and expression of virulence. Evolution in Health and Disease 2, 153–167.

Ehret, G.B., Munroe, P.B., Rice, K.M., Bochud, M., Johnson, A.D., Chasman, D.I., others, 2011. Genetic variants in novel pathways influence blood pressure and cardiovascular disease risk. Nature 478 (7367), 103.

Ershov, E.D., 1998. Foundations of Geocryology. Moscow State University, Moscow, Russia, pp. 1–575.

Ewald, P.W., 1980. Evolutionary biology and the treatment of signs and symptoms of infectious disease. Journal of Theoretical Biology 86 (1), 169–176.

Ford, S.A., Williams, D., Paterson, S., King, K.C., 2017. Co-evolutionary dynamics between a defensive microbe and a pathogen driven by fluctuating selection. Molecular Ecology 26 (7), 1778–1789.

French, G.L., 2006. Bactericidal agents in the treatment of MRSA infections—the potential role of daptomycin. Journal of Antimicrobial Chemotherapy 58 (6), 1107–1117.

Friman, V.-P., Guzman, L.M., Reuman, D.C., Bell, T., 2015. Bacterial adaptation to sub lethal antibiotic gradients can change the ecological properties of multitrophic microbial communities. Proceedings of the Royal Society B: Biological Sciences 282 (1806), 20142920.

Fuller, B.J., Lane, N., Benson, E.E., 2004. Life in the Frozen State. CRC press.

Futuyma, D.J., Agrawal, A.A., 2009. Macroevolution and the biological diversity of plants and herbivores. Proceedings of the National Academy of Sciences 106 (43), 18054–18061.

Gadd, G.M., 2010. Metals, minerals and microbes: geomicrobiology and bioremediation. Microbiology 156 (3), 609–643.

Gandon, S., Buckling, A., Decaestecker, E., Day, T., 2008. Host−parasite coevolution and patterns of adaptation across time and space. Journal of Evolutionary Biology 21 (6), 1861−1866.
George Priya Doss, C., Rajasekaran, R., Sudandiradoss, C., Ramanathan, K., Purohit, R., Sethumadhavan, R., 2008. A novel computational and structural analysis of nsSNPs in CFTR gene. Genomic Medicine 2 (1−2), 23−32.
Ghai, R., Pašić, L., Fernández, A.B., Martin-Cuadrado, A.-B., Mizuno, C.M., McMahon, K.D., et al., 2011. New abundant microbial groups in aquatic hypersaline environments. Scientific Reports 1, 135.
Gilichinsky, D., Wagener, S., 1995. Microbial life in permafrost: a historical review. Permafrost and Periglacial Processes 6 (3), 243−250.
Gilichinsky, D., Rivkina, E., Bakermans, C., Shcherbakova, V., Petrovskaya, L., Ozerskaya, S., et al., 2005. Biodiversity of cryopegs in permafrost. FEMS Microbiology Ecology 53 (1), 117−128.
González-Pleiter, M., Gonzalo, S., Rodea-Palomares, I., Leganés, F., Rosal, R., Boltes, K., Fernández-Piñas, F., 2013. Toxicity of five antibiotics and their mixtures towards photosynthetic aquatic organisms: implications for environmental risk assessment. Water Research 47 (6), 2050−2064.
Gortazar, C., Vicente, J., Boadella, M., Ballesteros, C., Galindo, R.C., Garrido, J., de La Fuente, J., 2011. Progress in the control of bovine tuberculosis in Spanish wildlife. Veterinary Microbiology 151 (1−2), 170−178.
Greenblatt, C.L., Baum, J., Klein, B.Y., Nachshon, S., Koltunov, V., Cano, R.J., 2004. Micrococcus luteus-survival in amber. Microbial Ecology 48 (1), 120−127.
Greischar, M.A., Koskella, B., 2007. A synthesis of experimental work on parasite local adaptation. Ecology Letters 10 (5), 418−434.
Grenni, P., Ancona, V., Caracciolo, A.B., 2018. Ecological effects of antibiotics on natural ecosystems: A review. Microchemical Journal 136, 25−39.
Grootes, P.M., Stuiver, M., White, J.W.C., Johnsen, S., Jouzel, J., 1993. Comparison of oxygen isotope records from the GISP2 and GRIP Greenland ice cores. Nature 366 (6455), 552.
Guppy, M., Withers, P., 1999. Metabolic depression in animals: physiological perspectives and biochemical generalizations. Biological Reviews 74 (1), 1−40.
Hall, R.M., Collis, C.M., Kim, M.J., Partridge, S.R., Recchia, G.D., Stokes, H.W., 1999. Mobile gene cassettes and integrons in evolution. Annals of the New York Academy of Sciences 870, 68−80. https://doi.org/10.1111/j.1749-6632.1999.tb08866.x.
Hamilton, K., Visser, D., Evans, B., Vallat, B., 2015. Identifying and reducing remaining stocks of rinderpest virus. Emerging Infectious Diseases 21 (12), 2117.
Hamscher, G., Sczesny, S., Höper, H., Nau, H., 2002. Determination of persistent tetracycline residues in soil fertilized with liquid manure by high-performance liquid chromatography with electrospray ionization tandem mass spectrometry. Analytical Chemistry 74 (7), 1509−1518.
Hassell, J.M., Begon, M., Ward, M.J., Fèvre, E.M., 2017. Urbanization and disease emergence: dynamics at the wildlife livestock human interface. Trends in Ecology & Evolution 32 (1), 55−67.
Hindmarsh, R.C.A., 1992. The growth and decay of ice. Studies in polar research series. Geological Magazine 129 (3), 376−377.
Ho, S.Y.W., Phillips, M.J., Cooper, A., Drummond, A.J., 2005. Time dependency of molecular rate estimates and systematic overestimation of recent divergence times. Molecular Biology and Evolution 22 (7), 1561−1568.
Hoffman, P.F., Kaufman, A.J., Halverson, G.P., Schrag, D.P., 1998. A Neoproterozoic snowball earth. Science 281 (5381), 1342−1346.
Holmes, E.C., 2014. Freezing viruses in time. Proceedings of the National Academy of Sciences of the United States of America 111 (47), 16643−16644.
Hou, L., Yin, G., Liu, M., Zhou, J., Zheng, Y., Gao, J., Tong, C., 2014. Effects of sulfamethazine on denitrification and the associated N_2O release in estuarine and coastal sediments. Environmental Science & Technology 49 (1), 326−333.

Jaenicke, R., 1996. Stability and folding of ultrastable proteins: eye lens crystallins and enzymes from thermophiles. The FASEB Journal 10 (1), 84–92.
Jensen, K.H., Little, T., Skorping, A., Ebert, D., 2006. Empirical support for optimal virulence in a castrating parasite. PLoS Biology 4 (7), e197.
Johnson, L.R., Mangel, M., 2006. Life histories and the evolution of aging in bacteria and other single-celled organisms. Mechanism of Ageing and Development 127 (10), 786–793.
Justino, C.I.L., Duarte, K.R., Freitas, A.C., Panteleitchouk, T.S.L., Duarte, A.C., Rocha-Santos, T.A.P., 2016. Contaminants in aquaculture: Overview of analytical techniques for their determination. TRAC Trends in Analytical Chemistry 80, 293–310.
Kacar, B., Ge, X., Sanyal, S., Gaucher, E.A., 2017. Experimental evolution of *Escherichia coli* harboring an ancient translation protein. Journal of Molecular Evolution 84 (2–3), 69–84.
Kallmeyer, J., 2017. Life at Vents and Seeps, Vol. 5. Walter de Gruyter GmbH & Co KG.
Kallmeyer, J., Pockalny, R., Adhikari, R.R., Smith, D.C., D'Hondt, S., 2012. Global distribution of microbial abundance and biomass in subseafloor sediment. Proceedings of the National Academy of Sciences 109 (40), 16213–16216.
Kaltz, O., Gandon, S., Michalakis, Y., Shykoff, J.A., 1999. Local maladaptation in the anther-smut fungus Microbotryum violaceum to its host plant Silene latifolia: evidence from a cross-inoculation experiment. Evolution 53 (2), 395–407.
Kawecki, T.J., Ebert, D., 2004. Conceptual issues in local adaptation. Ecology Letters 7 (12), 1225–1241.
Kirschvink, J.L., Gaidos, E.J., Bertani, L.E., Beukes, N.J., Gutzmer, J., Maepa, L.N., Steinberger, R.E., 2000. Paleoproterozoic snowball Earth: extreme climatic and geochemical global change and its biological consequences. Proceedings of the National Academy of Sciences 97 (4), 1400–1405.
Koskella, B., 2013. Phage-mediated selection on microbiota of a long-lived host. Current Biology 23 (13), 1256–1260.
Kumar, K., Gupta, S.C., Baidoo, S.K., Chander, Y., Rosen, C.J., 2005. Antibiotic uptake by plants from soil fertilized with animal manure. Journal of Environmental Quality 34 (6), 2082–2085.
Kümmerer, K., 2009. Antibiotics in the aquatic environment—a review—part I. Chemosphere 75 (4), 417–434.
Lambin, E.F., Tran, A., Vanwambeke, S.O., Linard, C., Soti, V., 2010. Pathogenic landscapes: interactions between land, people, disease vectors, and their animal hosts. International Journal of Health Geographics 9 (1), 54.
Legendre, M., Bartoli, J., Shmakova, L., Jeudy, S., Labadie, K., Adrait, A., et al., 2014. Thirty-thousand-year-old distant relative of giant icosahedral DNA viruses with a pandoravirus morphology. Proceedings of the National Academy of Sciences 111 (11), 4274–4279.
Levin, B.R., Eden, C.S., 1990. Selection and evolution of virulence in bacteria: an ecumenical excursion and modest suggestion. Parasitology 100 (S1), S103–S115.
Levy, M., Miller, S.L., 1998. The stability of the RNA bases: implications for the origin of life. Proceedings of the National Academy of Sciences 95 (14), 7933–7938.
Lively, C.M., 1989. Adaptation by a parasitic trematode to local populations of its snail host. Evolution 43 (8), 1663–1671.
Lorius, C., Jouzel, J., Ritz, C., Merlivat, L., Barkov, N.I., Korotkevich, Y.S., Kotlyakov, V.M., 1985. A 150,000-year climatic record from Antarctic ice. Nature 316 (6029), 591.
Ma, L.-J., Rogers, S.O., Catranis, C.M., Starmer, W.T., 2000. Detection and characterization of ancient fungi entrapped in glacial ice. Mycologia 92 (2), 286–295.
Margesin, R., 2017. Psychrophiles: from biodiversity to biotechnology. Springer.
Margesin, R., Schinner, F., Marx, J.-C., Gerday, C., 2008. Psychrophiles: From Biodiversity to Biotechnology. Springer.

Martinez, J.L., 2009. Environmental pollution by antibiotics and by antibiotic resistance determinants. Environmental pollution 157 (11), 2893–2902.

Masri, L., Branca, A., Sheppard, A.E., Papkou, A., Laehnemann, D., Guenther, P.S., et al., 2015. Host–pathogen coevolution: the selective advantage of Bacillus thuringiensis virulence and its cry toxin genes. PLoS Biology 13 (6), e1002169.

Mazel, D., 2006. Integrons: agents of bacterial evolution. Nature Reviews Microbiology 4, 608–620. https://doi.org/10.1038/nrmicro1462.

Miller, R.V., Whyte, L., 2011. Polar Microbiology: Life in a Deep Freeze. American Society for Microbiology Press.

Ng, T.F.F., Chen, L.-F., Zhou, Y., Shapiro, B., Stiller, M., Heintzman, P.D., et al., 2014. Preservation of viral genomes in 700-y-old caribou feces from a subarctic ice patch. Proceedings of the National Academy of Sciences 111 (47), 16842–16847.

Nickle, D.C., Learn, G.H., Rain, M.W., Mullins, J.I., Mittler, J.E., 2002. Curiously modern DNA for a"250 million-year-Old"Bacterium. Journal of Molecular Evolution 54 (1), 134–137.

Orsini, L., Schwenk, K., De Meester, L., Colbourne, J.K., Pfrender, M.E., Weider, L.J., 2013. The evolutionary time machine: using dormant propagules to forecast how populations can adapt to changing environments. Trends in Ecology & Evolution 28 (5), 274–282.

Ostroumov, V.E., Siegert, C., 1996. Exobiological aspects of mass transfer in microzones of permafrost deposits. Advances in Space Research 18 (12), 79–86.

Ozerskaya, S.M., Ivanushkina, N.E., Kochkina, G.A., Fattakhova, R.N., Gilichinsky, D.A., 2004. Mycelial fungi in cryopegs. International Journal of Astrobiology 3 (4), 327–331.

Pal, C., Bengtsson-Palme, J., Kristiansson, E., Larsson, D.G.J., 2015. Co-occurrence of resistance genes to antibiotics, biocides and metals reveals novel insights into their co-selection potential. BMC Genomics 16 (1), 964.

Pallecchi, L., Bartoloni, B.A., Paradisi, F., Rossolini, G.M., 2008. Antibiotic resistance in the absence of antimicrobial use: mechanisms and implications. Expert review of anti-infective therapy 6 (5), 725–732.

Parkes, R.J., Cragg, B.A., Wellsbury, P., 2000. Recent studies on bacterial populations and processes in subseafloor sediments: a review. Hydrogeology Journal 8 (1), 11–28.

Penczykowski, R.M., Laine, A.-L., Koskella, B., 2016. Understanding the ecology and evolution of host–parasite interactions across scales. Evolutionary Applications 9 (1), 37–52.

Perron, G.G., Whyte, L., Turnbaugh, P.J., Goordial, J., Hanage, W.P., Dantas, G., Desai, M.M., 2015. Functional characterization of bacteria isolated from ancient arctic soil exposes diverse resistance mechanisms to modern antibiotics. PLoS One 10 (3), e0069533.

Petit, J.-R., Jouzel, J., Raynaud, D., Barkov, N.I., Barnola, J.-M., Basile, I., et al., 1999. Climate and atmospheric history of the past 420,000 years from the Vostok ice core, Antarctica. Nature 399 (6735), 429.

Petrova, M., Gorlenko, Z., Mindlin, S., 2009. Molecular structure and translocation of a multiple antibiotic resistance region of a Psychrobacter psychrophilus permafrost strain. FEMS Microbiology Letters 296 (2), 190–197.

Petrova, M., Kurakov, A., Shcherbatova, N., Mindlin, S., 2014. Genetic structure and biological properties of the first ancient multiresistance plasmid pKLH80 isolated from a permafrost bacterium. Microbiology 160 (10), 2253–2263.

Price, P.B., 2000. A habitat for psychrophiles in deep Antarctic ice. Proceedings of the National Academy of Sciences 97 (3), 1247–1251.

Prins, H.H.T., Gordon, I.J., 2014. Invasion Biology and Ecological Theory: Insights from a Continent in Transformation. Cambridge University Press.

Psenner, R., Sattler, B., 1998. Life at the freezing point. Science 280 (5372), 2073–2074.

Revich, B.A., Podolnaya, M.A., 2011. Thawing of permafrost may disturb historic cattle burial grounds in East Siberia. Global Health Action 4 (1), 8482.

Revich, B., Tokarevich, N., Parkinson, A.J., 2012. Climate change and zoonotic infections in the Russian Arctic. International Journal of Circumpolar Health 71 (1), 18792.

Reyserhove, L., Samaey, G., Muylaert, K., Coppé, V., Van Colen, W., Decaestecker, E., 2017. A historical perspective of nutrient change impact on an infectious disease in Daphnia. Ecology 98 (11), 2784–2798.

Richardson, D.M., 2011. Fifty Years of Invasion Ecology: The Legacy of Charles Elton. John Wiley & Sons.

Rogers, S.O., Theraisnathan, V., Ma, L.J., Zhao, Y., Zhang, G., Shin, S.-G., Starmer, W.T., 2004. Comparisons of protocols for decontamination of environmental ice samples for biological and molecular examinations. Applied and Environmental Microbiology 70 (4), 2540–2544.

Roose-Amsaleg, C., Laverman, A.M., 2016. Do antibiotics have environmental side-effects? Impact of synthetic antibiotics on biogeochemical processes. Environmental Science and Pollution Research 23 (5), 4000–4012.

Sahni, G., Elkayam, U., 2012. Cardiovascular disease in pregnancy. In: An Issue of Cardiology Clinics-E-Book, vol. 30. Elsevier Health Sciences.

Schmid Hempel, P., 2011. Evolutionary Parasitologythe Integrated Study of Infections, Immunology, Ecology, and Genetics.

Schuur, E.A.G., Abbott, B., 2011. Climate change: high risk of permafrost thaw. Nature 480 (7375), 32.

Slenning, B.D., 2010. Global climate change and implications for disease emergence. Veterinary Pathology 47 (1), 28–33.

Smith, A.W., Skilling, D.E., Castello, J.D., Rogers, S.O., 2004. Ice as a reservoir for pathogenic human viruses: specifically, caliciviruses, influenza viruses, and enteroviruses. Medical Hypotheses 63 (4), 560–566.

Sørensen, I.K., 2009. Consolidated Presentation of the Joint Scientific Opinion of the GMO and BIOHAZ Panels on the "Use of Antibiotic Resistance Genes as Marker Genes in Genetically Modified Plants" and the Scientific Opinion of the GMO Panel on "Consequences of the Opinion on the Use of Antibiotic Resistance Genes as Marker Genes in Genetically Modified Plants on Previous EFSA Assessments of Individual GM Plants". EFSA-Q-2009–20000589 and EFSA-Q-2009-00593.

Spijkerboer, T., 2013. Fleeing Homophobia: Sexual Orientation, Gender Identity and Asylum. Routledge.

Stokes, H.W., Hall, R.M., 1989. A novel family of potentially mobile DNA elements encoding site-specific gene-integration functions: integrons. Molecular Microbiology 3, 1669–1683. https://doi.org/10.1111/j.1365-2958.1989.tb00153.x.

Tasho, R.P., Cho, J.Y., 2016. Veterinary antibiotics in animal waste, its distribution in soil and uptake by plants: a review. The Science of the Total Environment 563, 366–376.

Taubenberger, J.K., Baltimore, D., Doherty, P.C., Markel, H., Morens, D.M., Webster, R.G., Wilson, I.A., 2012. Reconstruction of the 1918 influenza virus: unexpected rewards from the past. mBio 3 (5), e00201–e00212.

Thiele-Bruhn, S., 2003. Pharmaceutical antibiotic compounds in soils—a review. Journal of Plant Nutrition and Soil Science 166 (2), 145–167.

Thompson, L.G., Yao, T., Davis, M.E., Henderson, K.A., Mosley-Thompson, E., Lin, P.-N., Bolzan, J.F., 1997. Tropical climate instability: the last glacial cycle from a Qinghai-Tibetan ice core. Science 276 (5320), 1821–1825.

Thompson, L.G., Yao, T., Mosley-Thompson, E., Davis, M.E., Henderson, K.A., Lin, P.-N., 2000. A high-resolution millennial record of the South Asian monsoon from Himalayan ice cores. Science 289 (5486), 1916–1919.

Tumpey, T.M., Basler, C.F., Aguilar, P.V., Zeng, H., 2005. Characterization of the reconstructed 1918 Spanish influenza pandemic virus. Science 310, 77–80.

Underwood, J.C., Harvey, R.W., Metge, D.W., Repert, D.A., Baumgartner, L.K., Smith, R.L., Barber, L.B., 2011. Effects of the antimicrobial sulfamethoxazole on groundwater bacterial enrichment. Environmental Science & Technology 45 (7), 3096–3101.

Veiga-Crespo, P., Poza, M., Prieto-Alcedo, M., Villa, T.G., 2004. Ancient genes of *Saccharomyces cerevisiae*. Microbiology 150 (7), 2221–2227.

Vishnivetskaya, T., Kathariou, S., McGrath, J., Gilichinsky, D., Tiedje, J.M., 2000. Low-temperature recovery strategies for the isolation of bacteria from ancient permafrost sediments. Extremophiles 4 (3), 165–173.

Vlachopoulos, C., O'Rourke, M., Nichols, W.W., 2011. McDonald's Blood Flow in Arteries: Theoretical, Experimental and Clinical Principles. CRC press.

Von Bubnoff, A., 2005. The 1918 flu virus is resurrected. Nature 437 (7060), 794–795.

Vorobyova, E., Soina, V., Gorlenko, M., 1997. The deep cold biosphere: facts and hypothesis. FEMS Microbiology Reviews 15 (3), 4.

Vreeland, R.H., Rosenzweig, W.D., Powers, D.W., 2000. Isolation of a 250 million-year-old halotolerant bacterium from a primary salt crystal. Nature 407 (6806), 897.

Willerslev, E., Hansen, A.J., Poinar, H.N., 2004. Isolation of nucleic acids and cultures from fossil ice and permafrost. Trends in Ecology & Evolution 19 (3), 141–147.

Woegerbauer, M., Zeinzinger, J., Gottsberger, R.A., Pascher, K., Hufnagl, P., Indra, A., et al., 2015. Antibiotic resistance marker genes as environmental pollutants in GMO-pristine agricultural soils in Austria. Environmental Pollution 206, 342–351.

World Health Organization Global Influenza Program Surveillance Network, 2005. Evolution of H5N1 avian influenza viruses in Asia. Emerging infectious diseases 11 (10), 1515.

Wood, C.L., Byers, J.E., Cottingham, K.L., Altman, I., Donahue, M.J., Blakeslee, A.M.H., 2007. Parasites alter community structure. Proceedings of the National Academy of Sciences 104 (22), 9335–9339.

Woodford, N., Turton, J.F., Livermore, D.M., 2011. Multi resistant Gram negative bacteria: the role of high-risk clones in the dissemination of antibiotic resistance. FEMS Microbiology Reviews 35, 736–755. https://doi.org/10.1111/j.1574-6976.2011.00268.x.

WSB, P., 1994. Transformation of snow to ice. In: The Physics of Glaciers, third ed., pp. 8–26.

Xu, H., Broersma, K., Miao, V., Davies, J., 2011. Class 1 and class 2 integrons in multidrug-resistant gram-negative bacteria isolated from the Salmon River, British Columbia. Canadian Journal of Microbiology 57, 460–470. https://doi.org/10.1139/w11-029.

Ziska, L.H., Blumenthal, D.M., Runion, G.B., Hunt, E.R., Diaz-Soltero, H., 2011. Invasive species and climate change: an agronomic perspective. Climatic Change 105 (1–2), 13–42.

CHAPTER

Dissemination of antibiotic resistance in the environment

6

Saima Saima[1], Marium Fiaz[1], Rabeea Zafar[1,2], Iftikhar Ahmed[3], Muhammad Arshad[1]

[1]*Institute of Environmental Sciences and Engineering (IESE), School of Civil and Environmental Engineering (SCEE), National University of Sciences and Technology (NUST), Islamabad, Pakistan;* [2]*Department of Environmental Design, Health & Nutritional Sciences, Faculty of Sciences, Allama Iqbal Open University, Islamabad, Pakistan;* [3]*National Culture Collection of Pakistan (NCCP), Bio-resource Conservation Institute (BCI), National Agricultural Research Centre (NARC), Islamabad, Pakistan*

6.1 Background

Antibiotic resistance is a well-established and most urgent threat to global health now. It is the ability of a bacteria to fight against the effects of the antibiotics. Among the three major threats to human health in the 21st century, antibiotic resistance is considered one of them, and a global effort is required to contain it (WHO, 2014). Antibiotics were first discovered in 1928 as products of fungi and bacteria. Later, antibiotic-resistant bacteria started appearing but the problem was dismissed as of little importance; later in the 1950s multidrug-resistant bacteria (bacteria able to resist more than two classes of antibiotics) started creating problems (Berglund, 2015). As a result, diseases that were once considered completely curable can now make a comeback because of antibiotic resistance.

Soon after, ARGs were discovered and it was thought that they emerged to protect target bacteria from antibiotics. But the recent evidence suggests that they emerged due to the sub-inhibitory concentration of antibiotics in our environment. These sub-inhibitory concentrations are 200 times below the minimal inhibitory concentration (MIC) and acceptable by environmental quality standards but they can still cause selection of antibiotic resistance (induce SOS response) and may even cause development of multidrug resistant opportunistic pathogens. Development of resistance against antibiotics has been strongly linked to antibiotic overuse (Berglund, 2015). In a study by Klein et al. (2018) an increase of 65% in overall consumption of antibiotics over the 15 years from 2000 to 15 was reported, which has many contributing factors, one of them being increase in bacterial infections, especially in rapidly developing rural areas in developing countries. Large amount of the antibiotics released (by humans and animals) remain in unaltered form and persist, and wastewater treatment plants remove little of it because they are not designed for that purpose.

As has already been established, major factors in the rise and spread of antibiotic resistance are overconsumption of antibiotics, which comes in the form of noncompliance of patients to full

Antibiotics and Antimicrobial Resistance Genes in the Environment. https://doi.org/10.1016/B978-0-12-818882-8.00006-1

prescription of antibiotics, prescription of drugs by physicians without first establishing the infection to be bacterial, and use as growth promoters in agriculture (Berglund, 2015). The problem is even more aggravated in developing countries where self-medication is common. Large-scale prophylactic use of antibiotics in fishery and farming industries also adds to the problem (Phillips et al., 2004). Antibiotics exert pressure for selection of antibiotic resistance on the microflora. Animal husbandry, aquaculture facilities, pharmaceutical manufacturing effluents, and municipal wastewater systems are considered the hotspots for emergence of antibiotic resistance—not just the medical settings. These hotspots are characterized by very high loads of bacterial contaminants and antibiotics that give way for emergence of antibiotic-resistant bacteria (ARBs) and antibiotic-resistant genes (ARGs) that are later released in the environment.

Continuous exposure of anthropogenically generated ARGs to environmental bacteria causes development and proliferation of resistance in environmental bacteria. Alternatively, environmental bacteria are considered to have a gene pool that has not yet been explored fully. Many of the genes they harbor may have the ability to be used as ARGs if they find their way into a pathogenic bacteria. Sánchez-Osuna et al. (2019) suggested that novel resistance mechanisms against newer antibiotics (even synthetic) may already be present in the microbial pangenome, which spread in a very brief time period upon clinical introduction of the drugs.

The increased dissemination of antibiotic resistance is considered to be because of three major mechanisms: genetic mutation and recombination, horizontal gene transfer and spread of ARBs owing to the selective pressure of antibiotics, and micropollutants like biocides and heavy metals. The concentration of chemical contaminants decreases due to degradation, dilution, or sorption, but ARGs are able to persist in environment and multiply inside a host, which earns them the easy-to-get-hard-to-lose nature.

When it comes to ARG transfer from environment to a pathogenic bacteria, horizontal gene transfer (HGT) is of high importance. It may occur via conjugation, transduction, OR transformation. The phenomenon of conjugation occurs more often and facilitates HGT with transfer of plasmids and integrative conjugative elements. It has been observed in soil, marine sediments, wastewater, activated sludge, and seawater (Davison, 1999).

During genetic transfer the element of importance is integron, it allows catching and expression of gene cassettes, which have antibiotic resistant genes embedded in them. Integrons can carry genes like intI1 that are responsible for encoding enzymes such as integrases. The gene cassettes are excised and integrated into the integron with integrases and may also change the order of the genes inside the cassette, which affects their expression (Cambray et al., 2010).

6.2 Discovery and development of antibiotics

Not so long ago in human history infections caused by bacteria could be cured using herbal therapies and folk medicines only. These were often insufficient and led to serious illness and high mortality rates. Interest in antimicrobial therapy was incited by the revelations of microbial existence by Louis Pasteur and Robert Koch in the late nineteenth century and proof that they were responsible for most of the serious diseases, including anthrax and cholera. Thus there was a revolution in the treatment of serious bacterial infections as microbiologists were able to identify cause of the infection and were provided a target for therapy. The revolution in the industry was with the interest to manufacture "magic bullets"—chemicals that can selectively kill only infectious microbes and not harm the human body—which led to the discoveries of the first synthetic antimicrobial "sulfonamides," introduced in the 1930s. A second revolution in medicine came with Alexander Fleming's discovery of penicillin, an antimicrobial substance produced by microbes themselves. The enormous use of penicillin during World War II sparked interest in producing more natural antibiotics. Due to the screening methodology developed by Waksman, scientists at the time (the 1950s) were able to discover more than half of the antibiotics known to this day, referred to as the golden age of antibiotics.

Much of the progression late in the drug formulation field included making drugs with better pharmacodynamics and pharmacokinetic properties with large range of activity of the newer derivatives of natural antibiotics (Fig. 6.1).

6.3 Classification of antibacterial drugs

Antibiotics are generally classified into two categories. The first one includes synthetic drugs like quinolones and sulfonamides, and the second category includes those derived from microorganisms such as penicillins. In recent years the discovery of semi-synthetic drugs has blurred the difference between natural and synthetic drugs. Antimicrobials, either natural or synthetic, work by targeting and disrupting the vital processes in the bacterial cells that are not present or different in mammalian cells.

Antibiotics can be categorized depending upon their different properties, such as inhibitory effect, range of activity, and bacterial target. There are bactericidal and bacteriostatic compounds; the former kill the infection causing bacteria, whereas the latter stop the growth or division of the bacterial cell. Some drugs have a broad spectrum of activity, and thus are used against a wide range of gram-positive and gram-negative bacteria, whereas others have a relatively small range of clinical activity. Mechanisms of action and targets vary for antibacterial drugs (Fig. 6.2). The antibacterial drugs mostly target the bacterial cell wall synthesis and membrane integrity, synthesis of protein, folic acid metabolism, and replication and transcription DNA.

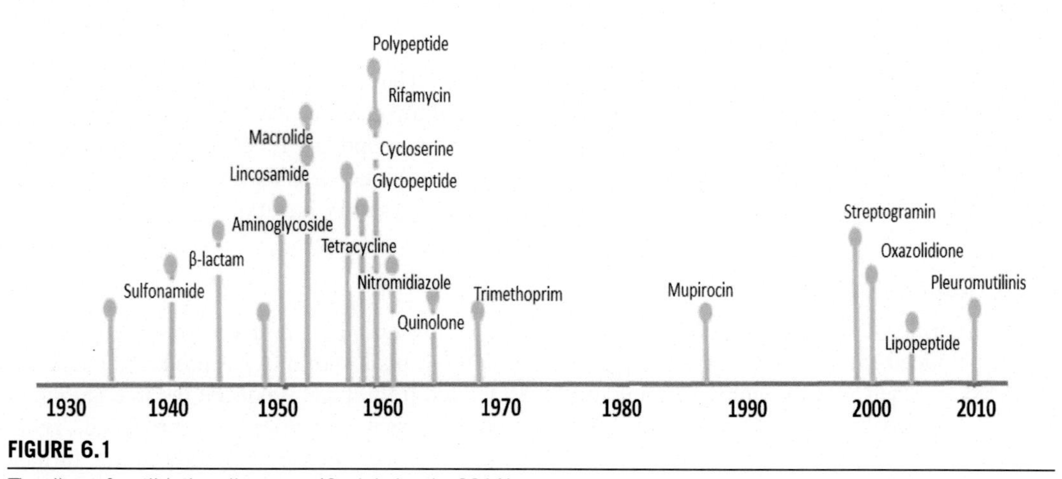

FIGURE 6.1

Timeline of antibiotics discovery (Cruickshank, 2011).

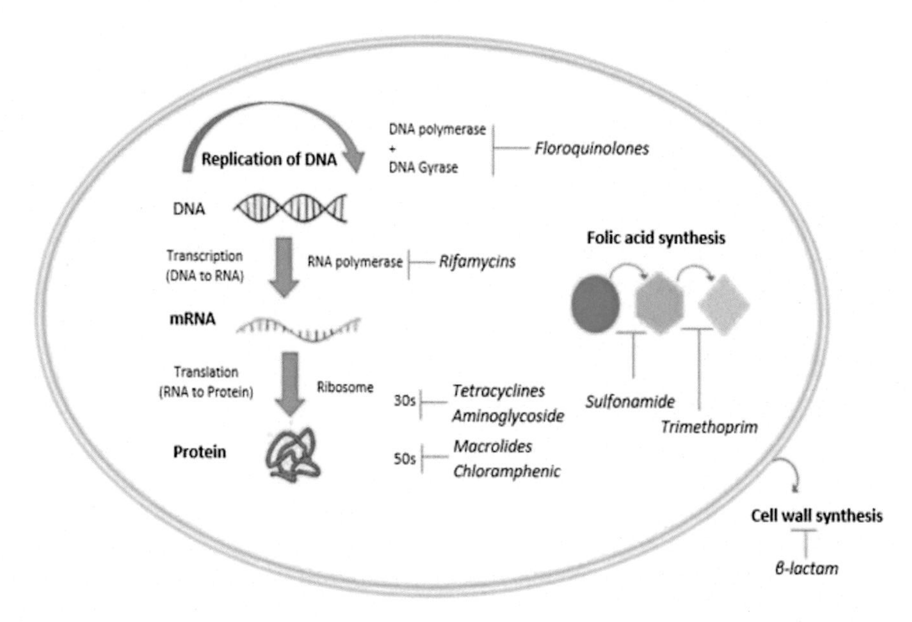

FIGURE 6.2

Bacterial targets of various antibiotics.

(Adapted from Chanda S, Rakholiya K (2011) Combination therapy: synergism between natural plant extracts and antibiotics against infectious diseases. Microbiol Book Series *520-529).*

6.3.1 Beta-lactams and fluoroquinolones

Beta-lactam antibiotics were used in World War II and proved to be very useful and are one of the most extensively used antibiotics. All beta-lactam antibiotics have a beta-lactam ring structure. Naturally, bacteria have a peptidoglycan layer (cell wall) around them, which gives them protection and structure. The cell wall is made with the help of an enzyme called transpeptidase, which catalyzes the cycle of cross-linking of amino acids for making the protein layer.

These antibiotics work by inactivating this enzyme. The transpeptidase forms a covalent penicilloyl-enzyme complex because of the stereo-chemical alikeness of the beta-lactam ring to the amino acid it was supposed to bind to. This is the reason the transpeptidase is called penicillin-binding protein (PBP). The enzyme-complex of transpeptidase with antibiotic stops the enzymatic activity and results in a weak cross-linked wall that ultimately causes cell lysis (Wilke et al., 2005).

6.3.2 Cephalosporins

Cephalosporins were produced from a fungi called *Cephalosporium* when it was observed that the fungus produced antibiotics that resembled penicillin. Later, cephalosporin C was produced but never marketed, but then 7-aminocephalosporanic acid was derived from it and produced in large amounts. From 7-aminocephalosporanic acid cephalosporins with different properties were produced. Their mode of action is like that of the penicillins. But the extra atom in the ring allowed for further semi-synthetic modifications (Greenwood, 2000; Walsh, 2003).

6.3.3 Third-generation cephalosporins, cephamycins and carbapenems

The third-generation cephalosporins are broad-spectrum antibiotics (Walsh, 2003). These compounds persist against many beta-lactamases and have improved activity against many gram-negative bacteria. They are unlike earlier generations because they have a capability to reach the central nervous system. The examples of these antibiotics are, cefpodoxime, ceftazidime, cefotaxime etc. With each generation of cephalosporins their range of activity increased; this is attributed to better permeation of the drugs to the bacterial cell, increased affinity toward PBP, and decreased affinity toward hydrolysis by beta-lactamases.

Cephamycins were first derived from *Streptomyces* spp. in 1972 and were found to be similar to cephalosporins as they shared the same cephem nucleus. Some examples of these include cefoxitin, letamoxef, and flomoxef (Greenwood, 2000).

Carbapenems are also beta-lactam ring—containing antibiotics and are very effective against beta-lactamases. Due to a broader range of activity they are very active against gram-negative and -positive bacteria. Examples of them include meropenem, imipenem, ertepam, and others (Greenwood, 2000; Walsh, 2003).

6.3.4 Fluoroquinolones

Fluoroquinolones are also broad-spectrum antibiotics, which were developed by modifying the first-generation quinolone called nalidixic acid, a narrow-spectrum quinolone. Many infections caused in both humans and animals by gram-positive or -negative bacteria are treated with fluoroquinolones. Ciprofloxacin is one of the most used fluoroquinolones and became available in the market in 1987 (Fàbrega et al., 2009). DNA replication in a bacterial cell occurs with the help of the enzymes topoisomerase and DNA gyrase. Negative supercoils are added in the bacterial DNA because of gyrase and this also helps maintain the coiling density of the DNA during replication. It also relieves the torsional stress of the coiled DNA and catalyzes the cleavage reaction by ATP binding and hydrolyses. Removal of the knots and tangles in DNA and decatenation of the daughter DNA after replication is one of the primary functions of topoisomerase IV (Levine et al., 1998).

Quinolones act by increasing DNA-cleavage complexes by increased production of topoisomerases. The cleavage complexes of DNA are formed when double-stranded breaks are introduced in the DNA strands that are four base pairs apart, and so to maintain genomic integrity, topoisomerases bind by the tyrosine residues to the 5′ end termini of the cleaved DNA. The integrity of the DNA is compromised as permanent breaks are introduced and relegation is prevented, the cell then is unable to survive and dies. The antibiotic also acts by decreasing the amount of DNA-cleavage complexes as it attaches to the DNA and enzyme complex and prevents the enzyme from untangling and decatenating the replicating DNA, hence causing mitotic failure and ultimately cell death (Aldred et al., 2013; Hooper, 1999).

6.4 Antibiotic resistance development

Many antibiotics are produced by bacteria in environment; resistance to them had emerged years before humans started using them. The antibiotic-producing bacteria and those susceptible to them coexisted in their own natural habitats without assisting in the selection of dangerous resistant pathogenic bacteria so that those that still responded to most antibiotics were found in nature. Once humans started using high antibiotic concentrations in clinical settings more than 70 years ago, this led to development of highly resistant bacterial strains that were also human pathogens. Because of the selective pressure exerted by the antibiotics in environment, the bacteria in the natural habitat either attained resistance against them by HGT, mostly through conjugation or through de novo mutations, which further caused selection of resistance in pathogenic strains when they came in contact with each other in various environmental compartments, specifically aquatic, such as wastewater treatment plants (Fig. 6.3).

In the beginning, all major human pathogenic bacteria responded to antibiotics; the antibiotics produced in the 1930s were very successful and caused a remarkable decrease in human morbidity and death rate. Although antibiotic resistance was documented soon after the discovery of penicillin, the production of new drug derivatives of higher potency in the 1940s—1960s did not let resistance to rule out antibiotic therapy as it kept up with emergence of resistance. Therefore the news of bacterial resistance did not cause much distress. Conversely, the initial success of antibiotics caused the misbelief that the problem of bacterial infections has been overcome and would soon be eradicated. This led to the "innovation gap" of 40 years from the mid-1960s to 2000 when no major efforts were made to invent newer classes of antibiotics. Thus the infections caused by multidrug-resistant pathogens became inevitable and the already-present drugs started losing efficacy (Fig. 6.4).

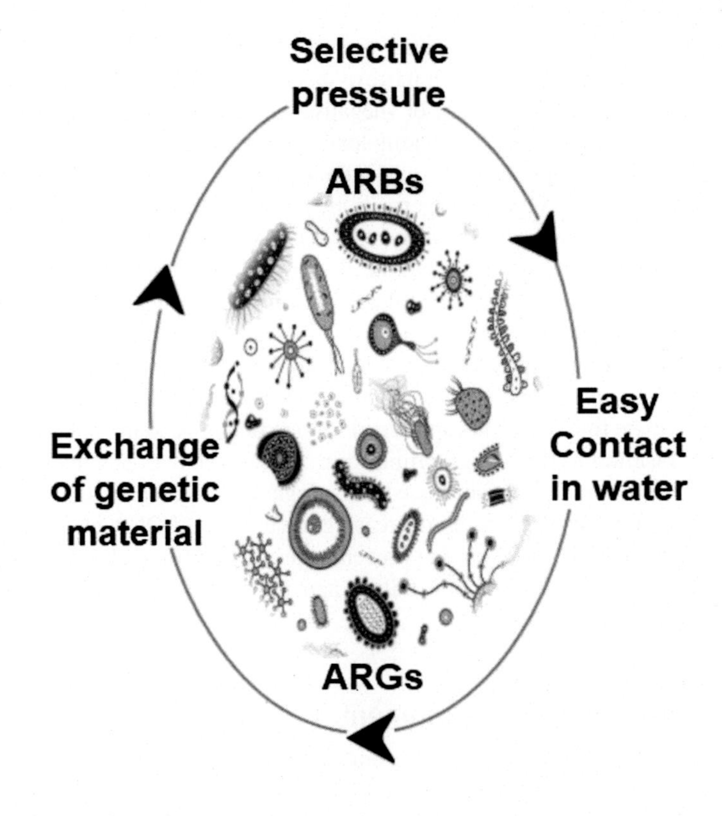

FIGURE 6.3

Antibiotic selective pressure causes development of antibiotic resistance in aquatic environment where because of opportunities for easy contact, exchange and recombination of genetic material occurs.

6.5 Causes of antibiotic resistance

6.5.1 Overuse

Antibiotic use is directly related to development of resistance and spread among pathogenic strains. Antibiotics remove the competing strains that are sensitive to them and leave behind the resistant strains (Berglund, 2015).

6.5.2 Incorrect prescription

Incorrect prescription of antibiotics boosts emergence of resistance in bacteria. Studies have shown that the antibiotic prescribed and or duration of treatment is wrong 30%–50% of the time (CDCP, 2013). Studies also report that 30%–60% drugs prescribed in intensive care units are not necessary (Luyt et al., 2014).

6.5.3 Use in other sectors

The use of antibiotics as prophylactics is very common in livestock, poultry, and fish farming industries for growth promotion. Humans are exposed to these antibiotics when the products from these industries are consumed. It was first observed that resistant bacteria can transfer from farm animals to humans when higher resistant rates were observed in intestinal bacteria of both farm animals and farmers about 35 years ago. Also when the livestock excrete waste, 95% of the antibiotics are released in unchanged form, which goes into the natural environment (Elmund et al., 1971).

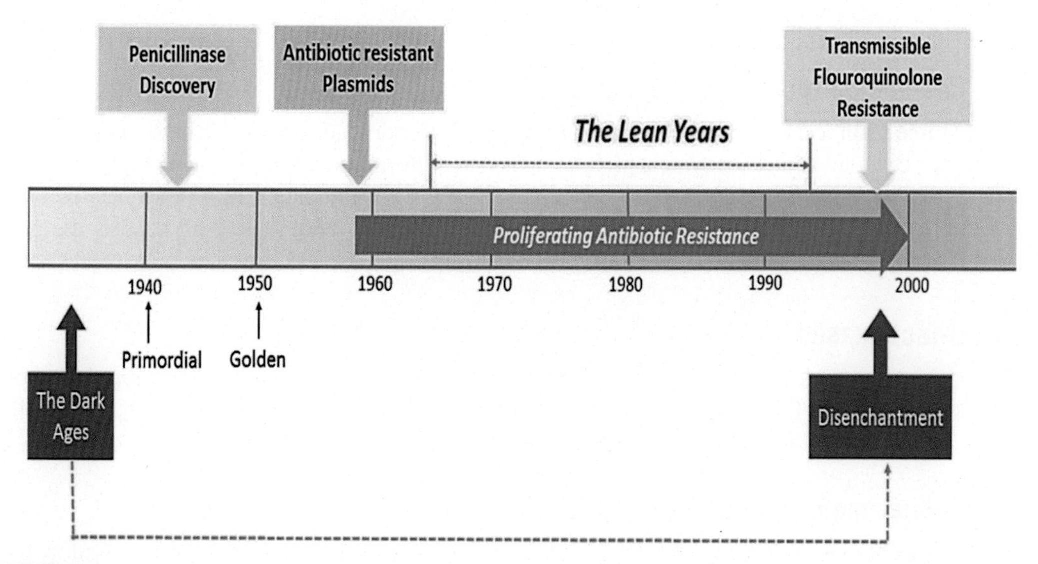

FIGURE 6.4

Major events in the history of antibiotics.

(Adapted from Davies J, Davies D (2010) Origins and evolution of antibiotic resistance. Microbiology and Molecular Biology Reviews *74:417-433 doi:10.1128/MMBR.00016-10).*

6.5.4 Newer antibiotics

Since antibiotics are cheap and are used for short periods of time, it has been deemed unwise to spend too much money on research and labs to develop new ones. Instead, investors have focused more on making drugs for chronic diseases like diabetes, asthma, or psychiatric disorders. According to the Office of Health and Economics in London, a new antibiotic will have a net worth of $50 million whereas a drug that treats a neuromuscular disease will have a net worth up to one billion. There are reports that 15 of the 18 leading pharmaceutical industries have abandoned antibiotic fields (Infectious Disease Society of America, 2010). Also microbiologists advise restraint in antibiotic consumption and to preserve their use for when it's really needed. Because even if new antibiotics are developed, the bacteria will eventually find a way to resist them as the phenomenon exists naturally. Because of lack of newer, effective antibiotics it's getting harder to control the rising multidrug-resistant strains with the older, less effective antibiotics.

6.6 Mechanism of antibiotic resistance

Bacteria processes have exceptional genetic flexibility due to which they have an ability to respond to any threat to their existence in the environment, including antibiotics. Potential pathways for antibiotic resistance are depicted in Fig. 3. Bacteria that share the same ecological niche develop abilities and techniques to continue in the presence of harmful compounds in the environment by adopting mainly two major genetic strategies:

1. Mutations in the genes—due to the mechanism of action of the compound.
2. Horizontal gene transfer—due to acquisition of foreign DNA.

6.6.1 Resistance mechanism against B-lactams

There are three main resistance mechanisms that bacteria have adopted to resist the detrimental effects of b-lactams. Beta-lactamase (enzymes) production is one of the main resistance mechanisms found in resistant gram-negative bacteria, beta-lactamases hydrolyze the beta-lactam ring in all beta-lactam-containing antibiotics. Second is the modification of target site of antibiotic. And third is the prevention of drugs to enter the cell by various mechanisms.

6.6.2 B-lactamases

The two major schemes of classification of beta-lactamases are: (1) Ambler classification, based on molecular structure of the enzyme; and (2) Bush-Jacoby classification, based on function of the enzyme. The one most commonly used is the Ambler classification.

6.6.3 Ambler molecular classification

In this scheme of classification protein homology, in particular amino acid similarity, is used as the basis for dividing the groups of enzymes. There are four classes in this scheme of classification, categorized as A, C, D, and B. The classes C, A, and D are serine group enzymes as they use serine for hydrolysis of the target antibiotic. Class B beta-lactamases are also called metallo beta-lactamases as they use zinc ions for hydrolyzing beta-lactams.

The largest group of enzymes among these four is class A. The enzymes in this group have highly similar sequences, which are considered to have derived from a single ancestral gene. Despite having high similarities they have different enzymatic properties and substrate profiles. Class A enzymes prefer penicillin for their substrate. These enzymes can be periplasmic or cell bound and plasmid or chromosomal gene mediated. They can be produced by both gram-positive and gram-negative bacteria (Ambler, 1980).

The class B of metallo beta-lactamases is a smaller class of enzymes and needs zinc as a metal cofactor. They are inhibited by metal chelators only and not by beta-lactamase inhibitors like clavulanic acid (Bush, 1988).

Class C beta-lactamases were first discovered in 1981 in *Escherichia coli* K-12 strain. These enzymes are like class-A enzymes except that their binding site is more open and can accept bulkier molecules of penicillins, oxyaminocephalosporins, cephamycins, and monobactams. Clavulanic acid doesn't aid in their inhibition, unlike in the case of class-A enzymes. Carbapenem resistance has been related with AmpC production coupled with loss of porin. They are both chromosome and plasmid mediated (Philippon et al., 2002).

Class D enzymes were discovered in 1981; they don't share much homology with class A and C enzymes. Their subfamily is named using OXA nomenclature because it was observed that they have a strong hydrolytic activity against semi-synthetic penicillins such as oxacillin. It was later discovered that they can hydrolyze cephalosporins, beta-lactam/beta-lactamase inhibitor combinations, and carbapenems as well (Leonard et al., 2013).

6.6.4 Extended-spectrum beta-lactamases

Extended-spectrum beta-lactamases (ESBLs) are active against penicillins, first, second, and third-generation cephalosporins and aztreonam. They cannot hydrolyze cephamycins (cefoxitin) and carbapenems (imipenem). They are inhibited by clavulanic acid.

6.6.5 ESBL diversity

There are three main types of ESBLs, namely TEM, SHV, and CTX-M. There are others as well, but they are relatively rare.

6.6.6 TEM

TEM-1 was first detected in a clinical isolate of *E. coli*. Point mutations in TEM enzymes are clustered in five points in the enzymes and are adjacent to seven evolutionary conserved elements. These conserved elements are located near the active site and increase the size of enzymes to accommodate the oxyamino components of cephalosporins to allow for broad-spectrum resistance (Stürenburg and Mack, 2003).

6.6.7 SHV

In 1972, a sulfydryl variable beta-lactamase was identified and then SHV was derived from it as it is the description of the biochemical property of the enzyme. In the beginning it was chromosomally mediated and had narrow range activity. But later due to mutations its activity range increased. Currently, there are 40 SHV enzymes, each varying in number and amino acid mutation. These enzymes are now mobilized as they are mediated by plasmid (Stürenburg and Mack, 2003).

6.6.8 CTX-M

The word CTX-M is derived from the name of the cephalosporin, cefotaxime, indicating its hydrolyzing ability. They were detected on the chromosome of environmental bacteria, *Klyvera,* and are not related to TEM or SHV (Peirano and Pitout, 2010). These enzymes are mediated on plasmid as well and can be found in association with SHV, TEM, and OXA beta-lactamases. Presently, there are 120 CTX-M enzymes, which are divided into five phylogenetic groups. They are less effective at hydrolysis of penicillins but show better activity against cephalosporins (Bonnet, 2004).

6.6.9 Target site alteration

Penicillin binding proteins, also called transpeptidases, are enzymes that catalyze the making of peptidoglycan layer around bacterial cells; they have been reported to have undergone mutations so that they are less likely to bind to the beta-lactam antibiotic. This happens by mutations in the genes responsible for encoding PBPs. Mutations in PBPs have been reported in pathogenic strains such as *Neisseria meningitides* (Ropp et al., 2002).

6.6.10 Reduced permeability

Hydrophilic compounds or antibiotics are allowed inside the bacterial cell by proteins called porins. They make pathways in the outer membrane of the cell wall and make way for substances to enter. Mutations in the genes encoding porins can cause their underexpression or loss, which helps bacteria gain resistance (Zgurskaya and Nikaido, 2000).

6.6.11 Efflux pump

Efflux pumps are proteins of transport, responsible for expulsion of potentially harmful substances including antibiotics. Most bacteria with overly expressed efflux pumps are multidrug resistant. Mutations in the genes encoding efflux pumps cause their overexpression. There are five systems of efflux pumps among gram-negative bacteria. Efflux pump—encoding genes can be found on plasmids and chromosomes. It is reported that resistant as well as susceptible bacteria carry genes for different efflux pumps and that they are overly expressed once the bacteria is exposed to more than one class of antibiotics as well as other toxic substances because of the extended substrate range of the efflux pump system. Overexpression of efflux pump has been reported in *Pseudomonas aeruginosa* and many other bacteria over the years (Lomovskaya et al., 2001).

6.6.12 Resistance mechanism against fluoroquinolones

The bacterial resistance to fluoroquinolones is either target-mediated, plasmid-mediated, or chromosome-mediated mutations.

6.6.13 Mutations in target

Resistance to fluoroquinolones (FQs) is often target mediated. To reduce the affinity of antibiotic for the target, mutations occur in gyrase and topoisomerase IV–encoding genes. About ≤10-fold FQs resistance occurs due to one mutation in the type II enzyme–encoding genes and for higher level resistance (10–100 fold) mutations in both type of enzymes occur (Hooper, 1999). About 90% of the mutations occurring are in the amino acid serine. The rest of the mutations occur in the acidic residues that hold the metal-water bridge for antibiotic. Mutations in serine are more common because with the latter the catalytic activity of the cell is reduced about 5–10 times (Aldred et al., 2013). Mutations have been reported in gyrA genes of environmental bacterial isolates from WWTPs (Conte et al., 2017).

6.6.14 Resistance mediated by plasmid

Plasmid-mediated quinolone resistance confers low level of resistance but is an emerging problem because of ease of transfer. Unlike target-mediated quinolone resistance it can be transmitted to others by horizontal gene transfer. Plasmid-mediated resistance to quinolones is by three gene families.

The first are Qnr genes, which encode penta-peptide repeat proteins that decrease the binding of topoisomerases to DNA and can also attach to them by themselves so that there are lesser enzyme complexes for the antibiotic to attack. There are about 100 variants of Qnr divided into at least five families. QnrA and QnrD genes were reported in environmental bacteria isolated from hospital wastewater (Wang et al., 2018).

Another gene, aac(6′)-Ib-cr, encodes for a protein that acetylates the piperazinyl amine of the antibiotic and decreases its activity. The third protein is efflux pump; so far only three have been identified that confer resistance to fluoroquinolones. QepA, an efflux pump that was plasmid mediated, was reported in a clinical *E.coli* isolate by Yamane et al. (2007).

6.6.15 Chromosome-mediated resistance

Chromosome-mediated quinolone resistance is achieved by downregulation of porin expression or by upregulation of efflux pump, which are chromosome encoded. By these mechanisms the cellular concentration of antibiotic is decreased. Jaffe et al. (1982) reported that downregulation of porin (OmpC and OmpF) by mutation altered the permeability of cefoxitin in *E. coli* and decreased its effectiveness.

6.6.16 Antibiotic resistance consequences

How antibiotic resistance affects us depends up on the level of resistance, type and place of infection, and availability of effective treatment alternatives. As resistance develops in the infectious strains there are risks of increase in severity of diseases, length of disease time period, mortality rate, and health care costs.

Most of the currently available antibiotics have lost their effectiveness, with only 15 out of 44 showing some activity and only five of them pass the phase 3 against some infection-causing gram-negative bacteria, and the absence of newer antibiotics doesn't help the situation much (PEW charitable trust, 2017). Antimicrobial resistance is not only a problem in developing countries but it is spread worldwide. Resistance has been reported to have increased in strains responsible for causing community- and hospital-associated infections, which is a matter of concern because community-based infections could be transferred around even by normal contact. According to Golkar et al. (2014) more people in the United States have been victims of methicillin-resistant *Staphylococcus aureus* over a year than Parkinson's disease, homicide, emphysema, and HIV/AIDS altogether. One bacteria may have different resistance mechanisms. Furthermore, a study claims that in the European Union, the United States, and Thailand infections caused by resistant bacteria claim up to 25,000, 23,000, and 38,000 lives each year, respectively (WHO, 2014).

6.7 Antibiotic-resistant genes

Resistant genes in bacteria encode resistance mechanisms against drugs that are detrimental to their survival. But that is not considered their initial function; ARGs used to have regulatory functions, but with time, due to external pressure, they adopted their roles of providing protection to their host. Environmental bacteria may carry a pool of genes that may have the ability to be used as resistance genes if ensnared by pathogenic bacteria. These genes may encode a novel resistance mechanism to antibiotics that are our last resort. In the beginning it was believed that ARGs evolved due to contaminated environment to shield bacteria from harsh conditions, but then it was revealed that they existed even prior to the use of antibiotics and that we only found out about them recently (D'Costa et al., 2011). It will be right to call ARGs the "emerging environmental pollutants" because they are ubiquitous in various environmental compartments (Table 6.1).

6.8 Dissemination of antibiotic-resistant genes in environment

For bacteria to ensnare ARGs, the mechanism most important is HGT. It may occur through one of the following three processes:

1. Conjugation
2. Transduction
3. Transformation

6.8.1 Conjugation

Conjugation is the exchange of genetic material by cell-to-cell connection via pilus formation. It was first discovered in the 1950s and has been proven to have occurred in water, soil, seawater sediments, activated sludge, and sewage wastewater. There have been reports of DNA transfer among a vast range of hosts, even to eukaryotes (Davison, 1999). The important genetic material commonly getting transferred includes integrative conjugative elements and plasmids.

6.8.2 Integrons

Integrons are mobile genetic elements that provide a platform for capture and expression of gene cassettes that have ARGs. Their basic feature is a gene, intI, which encodes for an enzyme called tyrosine recombinase, which excises and integrates genes into the cassette. It can also reshuffle the

Table 6.1 ARGs and their ARBs with their resistance profiles reported from various water samples in different countries.

ARBs resistance profile	ARGs reported	ARBs reported	Sample	Country	References
Ampicillin, cefotaxime, cefoxitin, cefpodoxime, ceftazidime, ceftriaxone, cefuroxime, chloramphenicol, ciprofloxacin, streptomycin, co-trimoxazole, gentamycin, nalidixic acid, trimethoprim, streptomycin, tetracycline	*blaTEM, blaCTX-M, sul1, sul2, tetA, tetB, strA, aphA2, cat1, dhfr1, dhfr7*	*Escherichia coli*	Estuary water samples	India	Divya and Hatha (2019)
Streptomycin, chloramphenicol, kanamycin, ampicillin, ampicillin/sulbactam, neomycin	*Sul1, tetA*	*E. coli, Klebsiella pneumonia, Enterobacter cloacae, Pseudomonas aeruginosa*	Raw water, treated water before distribution, water after distribution	United States	Bergeron et al. (2017)
Aminoglycosides, beta-lactams, tetracyclines, vancomycin, sulfonamides, fluoroquinolones, chloramphenicol, macrolide-mincosamide-streptogramin	*tetO, tetX, tetM, sul1, sul2, sul3, ermB, ermAereA, oqxB, qnrA, qnrD*	*Arcobacter* spp., *Acinetobacter* spp., *Dechloromonas* spp., *Escherichia* spp., *Aeromonas* spp.	Hospital wastewater samples	China	Wang et al. (2018)
Cefotaxime, ceftazidime, carbapenems, tigecycline, ciprofloxacin	*blaCTX-M, blaSHV, blaGES,aac(6′)-Ib-cr, oqx*AB, *qnr*S	*Klebsiella oxytoca, K. pneumonia, E. coli*	Hospital effluent, river water, sanitary effluent (industrial +domestic wastewater), wastewater treatment plant	Brazil	Conte et al. (2017)

Continued

Table 6.1 ARGs and their ARBs with their resistance profiles reported from various water samples in different countries.—cont'd

ARBs resistance profile	ARGs reported	ARBs reported	Sample	Country	References
Amikacin, amoxicillin clavulanate/ sulbactam, ampicillin, aztreonam, cefazolin, cefpodoxime, cefotaxime, ceftazidime, cefoxitin, levofloxacin, cefepime, cefuroxime, chloramphenicol, colistin, ciprofloxacin, aztreonam, norfloxacin, moxifloxacin	*blaCTX-M, blaSHV, blaOXA, blaTEM*	*E. coli, Escherichia vulneris, Serratia odorifera, Enterobacter cloacae, Enterobacter aerogens, Citrobacter braakii, Citrobacter freundii, Citrobacter koseri, Citrobacter farmeri, K. pneumoniae, K. oxytoca*	Wastewater treatment plants	Spain	Ojer-Usoz et al. (2014)

order of genes, influencing the expression of gene cassette. The commonly found integrons are class 1 integrons in most clinical strains. Expression of intI gene is induced by SOS response. SOS response is triggered by b-lactam, fluoroquinolone, and trimethoprim antibiotics. So this means that a bacteria carrying an integron in a population if exposed to these antibiotics will have high chances of expression of relevant ARGs. Integrons could be mobile or chromosomal. Mobile integrons can be transferred to other bacteria via plasmids (Cambray et al., 2010). Studies have reported presence of integrons in various aquatic environments, showing their ubiquity in the natural environment.

6.8.3 Transduction

Transduction is the transfer of genetic material to bacteria through bacteriophages. It is easier for phage particles to survive environmental degradation than naked DNA, and also due to their small size they are well suited for DNA dissemination. This can commonly occur in marine environment. According to various studies, bacteriophages in freshwater reservoirs, plants, natural lakes, and marine environment carried plasmids, chromosome markers, and also transferred virulence genes.

6.8.4 Transformation

Transformation occurs when bacteria takes up genetic material from the environment released by another bacteria after its decease. Successful transformation can occur if degradation and dilution of

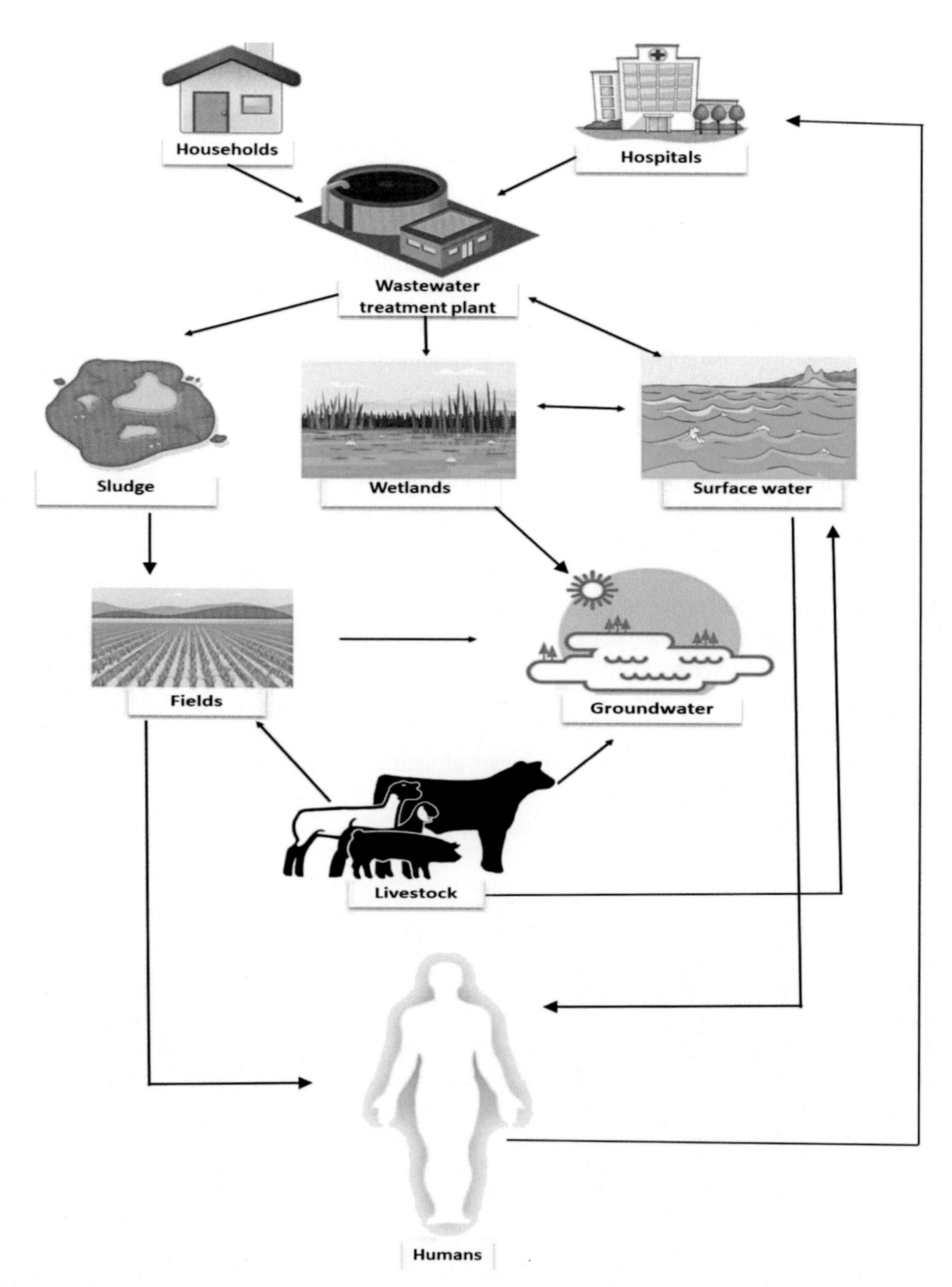

FIGURE 6.5

Antibiotics and their metabolites are released from various sources and ultimately reach wastewater treatment plants (WWTPs), where multiple processes occur and ARGs and ARBs are developed. The sludge and water from these treatment plants, when reused as fertilizers and for irrigation purposes, cause contamination of crops, soil, food animals, and groundwater. Humans become the ultimate receptors of ARGs by various direct and indirect exposure routes.

bacterial DNA is prevented by its adhesion to particles of sediment or soil. It can also occur in biofilms where the DNA of lysed bacteria is captured by other nearby bacterial cells. Studies have proven that natural transformation occurs in seawater, river bodies, groundwater, and soil (Davison, 1999). Transformation has been considered a reason for spread of penicillin-resistant genes.

6.8.5 Pathways for dissemination of ARGs in environment

Bacterial strains resistant to antibiotics have been reported from almost every environment on earth. The summary of dissemination and exposure is presented in Fig. 5. We know that dissemination of ARGs can occur and HGT via conjugation is one of the main facilitators for development of antibacterial resistance in nonpathogenic bacteria present in humans, animals, and various environmental matrices. The pathogenic bacteria discharged into the environment, e.g., via sewage or agricultural runoff, ultimately become resistant by acquiring resistant genes from the nonpathogenic bacteria present in the environment, which act as a source of resistance proliferation. Moreover, high concentrations of antibiotics in the environment promote natural selection and could cause development of ARGs. This will also assist in establishing the environment, specifically aquatic environment, as a pool of antibiotic-resistant genes and their further propagation to pathogenic bacteria through direct and indirect pathways. Drug formulation facilities are major point sources for discharge of antibiotics with quantities much higher than any other source. Antibiotics can enter our environment through various pathways, and it's established that because of the selective pressure put on the microflora wherever antibiotics go, resistant bacteria and ARGs follow. So there are points where ARGs, resistant bacteria, and environmental microflora can mix. These points, called hotspots, are where ARGs increase in number and resistant strains develop and further evolve. There are various ways in which humans are exposed to these ARGs, so they can become a part of their microbiome.

References

Aldred, K.J., McPherson, S.A., Turnbough Jr., C.L., Kerns, R.J., Osheroff, N., 2013. Topoisomerase IV-quinolone interactions are mediated through a water-metal ion bridge: mechanistic basis of quinolone resistance. Nucleic Acids Research 41, 4628–4639.

Ambler, R.P., 1980. The structure of β-lactamases. Philosophical Transactions of the Royal Society of London B Biological Sciences 289, 321–331.

Bergeron, S., Raj, B., Nathaniel, R., Corbin, A., LaFleur, G., 2017. Presence of antibiotic resistance genes in raw source water of a drinking water treatment plant in a rural community of USA. International Biodeterioration & Biodegradation 124, 3–9.

Berglund, B., 2015. Environmental dissemination of antibiotic resistance genes and correlation to anthropogenic contamination with antibiotics. Infection Ecology & Epidemiology 5, 28564.

Bonnet, R., 2004. Growing group of extended-spectrum β-lactamases: the CTX-M enzymes. Antimicrobial Agents and Chemotherapy 48, 1–14.

Bush, K., 1988. Beta-lactamase inhibitors from laboratory to clinic. Clinical Microbiology Reviews 1, 109–123.

Cambray, G., Guerout, A.-M., Mazel, D., 2010. Integrons. Annual Review of Genetics 44, 141–166.

CDCP, 2013. Office of Infectious Disease Antibiotic Resistance Threats in the United States. Atlanta, GA. Retrieved from. https://www.cdc.gov/drugresistance/pdf/ar-threats-2013-508.pdf.

Chanda, S., Rakholiya, K., 2011. Combination therapy: synergism between natural plant extracts and antibiotics against infectious diseases. Microbiol Book Series 520–529.

Conte, D., Palmeiro, J.K., Nogueira, K.S., Rosa De Lima, T.M., Cardoso, M.A., Pontarolo, R., Pontes, F.L.D., Dalla-Costa, L.M., 2017. Characterization of CTX-M enzymes, quinolone resistance determinants, and antimicrobial residues from hospital sewage, wastewater treatment plant, and river water. Ecotoxicology and Environmental Safety 136, 62–69.

Cruickshank, M., 2011. Antimicrobial Stewardship in Australian Hospitals. Australian Commission on Safety & Quality in Health Care, Sydney.

Davies, J., Davies, D., 2010. Origins and evolution of antibiotic resistance. Microbiology and Molecular Biology Reviews 74, 417–433. https://doi.org/10.1128/MMBR.00016-10.

Davison, J., 1999. Genetic exchange between bacteria in the environment. Plasmid 42, 73–91.

D'Costa, V.M., King, C.E., Kalan, L., Morar, M., Sung, W.W., Schwarz, C., Froese, D., Zazula, G., Calmels, F., Debruyne, R., Golding, G.B., Poinar, H.N., Wright, G.D., 2011. Antibiotic resistance is ancient. Nature 477, 457–461.

Divya, S.P., Hatha, A.A.M., 2019. Screening of tropical estuarine water in south-west coast of India reveals emergence of ARGs-harboring hypervirulent *Escherichia coli* of global significance. International Journal of Hygiene and Environmental Health 222, 235–248 (d).

Elmund, G.K., Morrison, S., Grant, D., Nevins, M., 1971. Role of excreted chlortetracycline in modifying the decomposition process in feedlot waste. Bulletin of Environmental Contamination and Toxicology 6, 129–132.

Fàbrega, A., Madurga, S., Giralt, E., Vila, J., 2009. Mechanism of action of and resistance to quinolones. Microbial Biotechnology 1, 40–61.

Golkar, Z., Bagasra, O., Pace, D.G., 2014. Bacteriophage therapy: a potential solution for the antibiotic resistance crisis. The Journal of Infection in Developing Countries 2, 129–136.

Greenwood, D., 2000. Antimicrobial Chemotherapy. Oxford University Press, Oxford; New York.

Hooper, D.C., 1999. Mode of action of fluoroquinolones. Drugs 2, 6–10.

Infectious Diseases Society of America, 2010. The 10×20 Initiative: pursuing a global commitment to develop 10 new antibacterial drugs by 2020. Clinical Infectious Diseases 50, 1081.

Jaffe, A., Chabbert, Y.A., Semonin, O., 1982. Role of porin proteins OmpF and OmpC in the permeation of beta-lactams. Antimicrobial Agents Chemother 22, 942–948. https://doi.org/10.1128/aac.22.6.942.

Klein, E.Y., et al., 2018. Global increase and geographic convergence in antibiotic consumption between 2000 and 2015. Proceedings of the National Academy of Sciences 115, E3463–E3470.

Leonard, D.A., Bonomo, R.A., Powers, R.A., 2013. Class D β-lactamases: a reappraisal after five decades. Accounts of Chemical Research 46, 2407–2415.

Levine, C., Hiasa, H., Marians, K.J., 1998. DNA gyrase and topoisomerase IV: biochemical activities, physiological roles during chromosome replication, and drug sensitivities. Biochimica et Biophysica Acta (BBA) - Gene Structure and Expression 1400, 29–43.

Lomovskaya, O., Warren, M.S., Lee, A., et al., 2001. Identification and characterization of inhibitors of multidrug resistance efflux pumps in *Pseudomonas aeruginosa*: novel agents for combination therapy. Antimicrobial Agents Chemother 45, 105–116.

Luyt, C.-E., Bréchot, N., Trouillet, J.-L., Chastre, J., 2014. Antibiotic stewardship in the intensive care unit. Critical Care 18, 480.

Ojer-Usoz, E., González, D., García-Jalón, I., Vitas, A.I., 2014. High dissemination of extended-spectrum β-lactamase-producing *Enterobacteriaceae* in effluents from wastewater treatment plants. Water Research 56, 37–47.

Peirano, G., Pitout, J.D., 2010. Molecular epidemiology of *Escherichia coli* producing CTX-M β-lactamases: the worldwide emergence of clone ST131 O25: H4. International Journal of Antimicrobial Agents 35, 316–321.

PEW Charitable Trusts, 2017. Antibiotics Currently in Global Clinical Development. Retrieved from. https://www.pewtrusts.org/-/media/assets/2019/03/antibiotics-currently-in-global-clinical-development.pdf?la=en&hash=078238EF15FACD9753ED2C4EBAB58F16B664B59E.

Philippon, A., Arlet, G., Jacoby, G.A., 2002. Plasmid-determined AmpC-type β-lactamases. Antimicrobial Agents Chemother 46, 1–11.

Phillips, I., Casewell, M., Cox, T., De Groot, B., Friis, C., Jones, R., Nightingale, C., Preston, R., Waddell, J., 2004. Does the use of antibiotics in food animals pose a risk to human health? A critical review of published data. Journal of Antimicrobial Chemotherapy 53, 28–52.

Ropp, P.A., Hu, M., Olesky, M., Nicholas, R.A., 2002. Mutations in ponA, the gene encoding penicillin-binding protein 1, and a novel locus, penC, are required for high-level chromosomally mediated penicillin resistance in *Neisseria gonorrhoeae*. Antimicrobial Agents Chemother 46, 769–777.

Sánchez-Osuna, M., Cortés, P., Barbé, J., Erill, I., 2019. Origin of the mobile di-hydro-pteroate synthase gene determining sulfonamide resistance in clinical isolates. Frontiers in Microbiology 9, 3332.

Stürenburg, E., Mack, D., 2003. Extended-spectrum β-lactamases: implications for the clinical microbiology laboratory, therapy, and infection control. Journal of Infection 47, 273–295.

Walsh, C., 2003. Antibiotics: Actions, Origins, Resistance. American Society for Microbiology (ASM).

Wang, Q., Wang, P., Yang, Q., 2018. Occurrence and diversity of antibiotic resistance in untreated hospital wastewater. The Science of the Total Environment 621, 990–999.

WHO, 2014. Antimicrobial Resistance: Global Report on Surveillance. World Health Organization. http://apps.who.int/iris/bitstream/10665/112642/1/9789241564748_eng.pdf?ua=1.

Wilke, M.S., Lovering, A.L., Strynadka, N.C., 2005. *β*-Lactam antibiotic resistance: a current structural perspective. Current Opinion in Microbiology 8, 525–533.

Yamane, K., et al., 2007. New plasmid-mediated fluoroquinolone efflux pump, QepA, found in an *Escherichia coli* clinical isolate. Antimicrobial Agents Chemother 51, 3354–3360. https://doi.org/10.1128/aac.00339-07.

Zgurskaya, H.I., Nikaido, H., 2000. Multidrug resistance mechanisms: drug efflux across two membranes. Molecular Microbiology 37, 219–225.

CHAPTER

Long-range transport of antibiotics and AMR/ARGs

7

Safdar Ali Mirza[1], Iram Liaqat[2], Muhammad Umer Farooq Awan[1], Muhammad Afzaal[3]

[1]*Department of Botany, GC University, Lahore, Punjab, Pakistan;* [2]*Department of Zoology, GC University, Lahore, Pakistan;* [3]*Sustainable Development Study Center (Env.Science) GC University, Lahore, Punjab, Pakistan*

7.1 Introduction

Antibiotics are a universally applied to control microbes. Human modulations made microbial behavior complicated. The startling situation of resistance to antibiotics makes for complications in the applied field of clinical microbiology. The foundation of modern clinical microbiology was built just after the discovery of penicillin by Alexander Fleming in 1928. Several gram-negative antibiotic-resistant bacterial species challenge health care. Cellular modifications include antibiotic degradation, metabolic pathways and rapid transport of antibiotic out of the cell. Multidrug resistance is more common in clinically significant pathogens and threatens to delay treatment of diseases.

Bacterial resistance occurs by reduction in drug channels through the outer membrane into cell. The outer membrane of gram-negative bacteria' is a petrifying barrier to polar molecules. The beta-barrel proteins (porins) traverse the bacterial outer membrane and help the uptake of small polar nutrients, which is the main entry pathway for many antibiotics (Nikaido, 2003; Davin-Regli et al., 2008). Numerous clinical investigations have related antibiotic resistance to altered expression of porins or their constrained function due to point mutation(s) (Pages et al., 2008). The World Health Organization (WHO) along with the Alliance for the Prudent Use of Antibiotics (APUA), organized the Global Policy for Containment of Antimicrobial Resistance (Sredkova, 1998).

The application of antibiotics is not restricted to human use but also used in animal husbandry and aquaculture. One of the terrifying issues is the accidental release of antibiotics in environment and their immense threat to human health and to environment. Antibiotics can enter the food chain, even more concerning are Antibiotic Resistance Genes (ARGs) that may relocate among human pathogens and environmental bacteria (Du and Liu, 2012; Bengtsson-Palme and Larsson, 2015; Li et al., 2015; Martinez et al., 2015; Van et al., 2015). The phenomenon of antibiotic resistance has been under consideration already for a few decades (D'Costa et al., 2011; Wright and Poinar, 2012). ARGs have increased in recent years with increased release of antibiotics and other pollutants into the environment (Yin et al., 2013; Bengtsson-Palme et al., 2014; Czekalski et al., 2014; Yang et al., 2017). As such, antibiotics and their individual and combined influence on the environment have become a key subject in the field of environmental sciences (Pruden et al., 2006).

Antibiotics and Antimicrobial Resistance Genes in the Environment. https://doi.org/10.1016/B978-0-12-818882-8.00007-3

Single applied antibiotics and their combined influence on the environment (ARGs, antibiotic resistant bacteria [ARBs], etc.) has become a key subject in environmental sciences. In the aquatic environment, the behavior of rivers and lakes is different due to diverse hydraulic characteristics. Hydraulic characteristics can gradually reduce the concentration of pollutants in river sediments (Pruden et al., 2006; Reuther, 2009). Among aquatic environments, lakes are a significant source of fresh drinking water containing about 90% of fresh water worldwide while rivers have 2% fresh drinking water (McConnel and Abel, 2013). Rivers have drawn much attention for the transportation of antibiotics and ARGs (Reuther, 2009; Storteboom et al., 2010; Chen et al., 2013; Rodriguez-Mozaz et al., 2015). As compared to rivers, lakes have more time to retain contaminants that indicate the slow discharge of pollutants and potentially add ARGs to larger levels than rivers (Reuther, 2009; Lyandres, 2012; Czekalski et al., 2014).

7.2 Historical perspective of antibiotics

The beliefs of ancient people about endemic diseases were often superstitious and related to punishment from God where sacrifices and lustrations to appease the anger of the gods were required for the management and prevention of disease. In the history of medicine the application of antibiotics is one of the flourishing forms for broad-spectrum chemotherapy.

The use of antibiotics is not restricted to modern medical approaches; antibiotics have been used directly and indirectly for thousands of years, as evidenced from traces of well-known antibiotics i.e. tetracycline found in human skeletal bones from ancient Sudanese Nubia back to 350–550 A.D. (Bassett et al., 1980; Nelson et al., 2010). This indicates that ancient people were aware of tetracycline-containing materials in the diet. Further examples of ancient antibiotic contacts are from a histological investigation of femoral mid-shafts of the late Roman period skeletons from the Dakhla Oasis, Egypt. These indicated distinct fluorochrome labeling consistent with the existence of tetracyclines in their food supplements (Cook et al., 1989) traces of application of additional antibiotics in ancient populations are hard to find. Anecdotal evidence may point to these occurrences. Historically, in Jordan people have strong beliefs about red soil that contains antibiotic-like similarities; the latest investigation showed the vast diversity of bacteria producing antibiotics in soil used for the treatment of skin infection (Falkinham et al., 2009).

Traditional Chinese medicine is another example of use of some antimicrobial materials for thousands of years in traditional Chinese medicines; the antimalarial drug *qinghaosu* (artemisinin) from *Artemisia* plant extract is the best example of this. For thousands of years traditional Chinese herbalists have used plants for treating many illnesses (Cui and Su, 2009).

7.3 Invisible organisms causing diseases

Even before microorganisms were observed, various investigators believed in their existence and also their responsibility for diseases. Lucretius (about 98–55 B.C.) the Roman philosopher and the physician Girolamo Fracastoro (1478–1553) a physician of Verona explained that infectious diseases were caused by invisible microorganisms he called them "seminaria" or "seeds". The Ancient Roman scholar Varro in the second century B.C. later studied the principle of infection caused by invisible organisms. A millennium later in the 13th century Roger Bacon suggested that invisible living creatures produced disease. Were Kircher (1659) observed microscopic worms in the blood of plague patients and Von Plenciz (1762) concluded that each disease was caused by invisible microorganisms (Surinder, 2012).

7.4 Phylogenetic analysis of antibiotic resistance genes

Phylogenetic analysis and establishment of a phylogenetic tree for evolutionary history of species is one the modern tools in applied microbiology. The genetic investigation of ARGs could help in discovering phylogenetic relationships. This investigation suggests the lasting presence of genes and existence of these genes resistant to a variety of antibiotics in nature (Aminov and Mackie, 2007; Kobayashi et al., 2007). The structure-based example of phylogenetic analysis of serine and metallo-β-lactamases indicates that both enzymes were produced about two billion years ago with some serine β-lactamases present on plasmids (Hall and Barlow, 2004; Garau et al., 2005). The phylogeny of housekeeping genes and β-lactamase is similar in *Klebsiella oxytoca* indicating that for over 100 million years these genes have been evolving in this host (Fevre et al., 2005). Another phylogenetic study of β-lactamases in the metagenomic clones derived from the 10,000-year-old "cold-seep" sediments clearly indicates the presence of these enzymes in ancient evolution (Song et al., 2005).

7.5 Antibiotic resistance

Previous studies explained the possible resistance of arsenicals applied for the treatment of syphilis without warning of high frequency of arsenic-resistant infections. In the case of penicillin it is hoped that extensive application of penicillin will eventually not result in disease resistance to penicillin (Rollo et al., 1952) but the application is restricted only to spirochete bacterium *T. palladium* (Cha et al., 2004) and no other disease-causing bacteria as well as the enterobacteriaceae, which become resistant not only to original penicillin but also to semi-synthetic penicillins, cephalosporins and carbapenems (Kumarasamy et al., 2010).

Bacterial multidrug resistance is also an emerging problem. ECDC/EMEA joint working group presented data that shows that in European Union countries the death rate has increased due to multidrug-resistant bacterial infections i.e. 25,000 patients die every year, while in the United States 63,000 patients die annually from the infections of hospital-acquired bacteria. Disease cure and management have a great influence on the financial system of any country and it is not only restricted to health issues but also has effects on development. Expected economic expenditures due to infections caused by multidrug-resistant bacteria in the European Union showed losses of 1.5 billion EUR per year (ECDC/EMEA Joint Working Group, 2009; Rustam and Aminov, 2010).

The previous health economic data indicated the estimated cost of only six species of antibiotic-resistant bacteria was about 1.3 billion dollars in 1992 and in 2006 it increased to 1.87 billion dollars; this cost was much higher than treating influenza per year. Due to multidrug resistance it is very difficult to understand the exact mode of action of antibiotic resistance.

Latest works in the field of antimicrobial resistance recommend that not all interactions of bacteria with antibiotics can be elucidated within the frames of the classical bullet-target concept. Present work on novel antibiotic resistance mechanisms apply the "kin selection" concept, since this resistance mechanism operates at the population/system level (Lee et al., 2010).

Antibiotic resistance complications describe briefly the interrelationship between human history and global evolution. It is also necessary to know the wax and wane in the antibiotics era, which is very helpful to understand different aspects of survival of microbes under different conditions. Microorganisms apply the same natural method to defend from immense attack of antibiotics constantly applied by humankind. All these tools make the foundation of genetic engineering and molecular genetics for the benefit of mankind in industrial and clinical applications. These techniques have been essential to reshape recent technology ranging from the production of recombinant proteins to construction of entire metabolic pathways. Microbial evolution is one of the critical aspects of antimicrobial and antibiotic resistance phenomena; during the history of microbial evolution microbial world has adopted diversified metabolic mechanisms in response to selection under different conditions. It is strongly suggested as critical in selection of antibiotics in targeting the pathogens and considering the limitations to avoid faster emergence of novel resistance mechanisms.

7.6 Distribution of antibiotic resistance genes in environment

Water is one of the most consumed liquids all over the world. Water covers about 71% of the earth's surface, while 3% of the earth's water is fresh lakes being a significant resource of fresh waterworldwide. Water has directly and indirectly higher ecological risk for human health and ecosystems in case of contamination. Recently, ARGs and antibiotics are considered as promising pollutants. The distribution and occurrence of ARGs and antibiotics in global freshwater lakes are at a critical level of pollution. It is very important to collect data for the classification and identification of ARGs and antibiotics in fresh water to minimize the health risks of ecosystems. One study reported 57 different antibiotics at least once in the studied lakes (Bengtsson-Palme and Larsson, 2015).

Now a days antibiotics are also being applied extensively in aquaculture and in animal farms. The great risk to human health and ecosystems is accidental release of antibiotics into the environment. These antibiotics become part of food web and ultimately these ARGs can transport human pathogens (Du and Liu, 2012; Bengtsson-Palme and Larsson, 2015; Li et al., 2015; Martinez et al., 2015; Van et al., 2015) resistance to antibiotics are not the latest phenomenon (D'Costa et al., 2011; Wright and Poinar, 2012) but fast and extensive increase of ARGs has occurred in current years with large quantity of release of antibiotics along with other pollutants into the environment (Yin et al., 2013; Czekalski et al., 2014; Bengtsson-Palme and Larsson, 2015; Yang et al., 2017).

Water resources such as lakes and rivers are main sources of ARGs and antibiotics due to inclusion of wastewater effluents from treatment plants (Zhang et al., 2009; Liu et al., 2012; Zhu et al., 2013; Lavilla et al., 2014). Lakes and rivers have different behaviors due to variations in hydraulic characteristics. The quantity of pollutants in river sediments reduces downstream due to hydraulic characteristics (Pruden et al., 2006; Reuther, 2009)

Rivers are nature's gift and perform a major role in the management of freshwater ecosystems. Rivers always get most consideration in aquatic environments for their frequent transportation of ARGs and antibiotics (Pruden et al., 2006; Storteboom et al., 2010; Chen et al., 2013; Rodriguez-Mozaz et al., 2015). As compared to rivers, in lakes the contaminants stay for a long time due to less flow of water and the pollutants slowly discharge around the lakes; this type of transport of ARGs and antibiotics is more critical than in rivers (Reuther, 2009; Lyandres, 2012). Several comparative studies showed that lakes have great potential for the transportation of ARGs and antibiotics than rivers

(Czekalski et al., 2014). Interestingly, if we talk about transportation of antibiotics and ARGs, the rivers contain higher diversity of the microworld, while in lakes there are more chances for origination of new species of microorganisms that might be more dangerous than the preexisting bacteria and viruses. No complete review of antibiotics and ARGs in lakes has yet been conducted (Yuy and Wenjuan, 2018).

7.7 Antimicrobial resistance in environment

One of the greatest intimidations to human health is the increasing antibiotic resistance in bacteria. The WHO has warned pharmacologists and microbiologists to develop strategies to cope with this emerging issue. Several studies have shown that the antibiotic resistance is multifactorial and multi-directional anthropogenic phenomena. The sources and distribution of antibiotic resistance is present in broad range. The application of antibiotics has been increasing day by day in every biotic field; not only is human health directly affected but animals and agriculture are experiencing terrible effects as well. Antimicrobial resistant (AMR) bacteria may be found on soils, on surface waters, and in human and animal waste streams. Strong supervision and comprehensive assessments of sources of contaminants can be helpful in controlling antimicrobial resistance. Lack of awareness and overuse of antimicrobials are two main hurdles in controlling the biocontaminants in the environment. Encouragement and implementation of excellent practices to manage environmental contamination with antimicrobials should be of high priority. The sources and distribution of resistant organisms play important roles in the environment with great influence on human health, food security and animal health and will increase as serious threats in the future. The development of antimicrobial resistance is linked up with number anthropogenic activities such as mismanagement in discharges from pharmaceuticals, agriculture, and human waste contamination management.

Antimicrobials are extensively being used in different fields such as agriculture, apiculture, aquaculture, poultry and livestock. In the field of plant production the application of antimicrobial systems also contaminates soil, which is also a big source of retention of antimicrobial resistance and treatments applied for the control of different diseases. Antimicrobial resistance is a very complicated phenomena; survival of resistant microorganisms depends on various abiotic and biotic factors, including edaphic factors, solar radiation, temperature, rainfall and residue retention capacity. There are several biotic factors, but the most important factors are rate of bacterial proliferation, die off, exchange of resistance genes and dispersion. Thus, resistance developing in the environment may be clinically relevant across all divisions. Nevertheless, there exist many flaws in the studies relating antimicrobial resistance ecology, environmental contamination with antimicrobial residues i.e. the importance of the public health threat posed by AMR bacteria in the environment and the effects of antimicrobial residues on soil ecosystem services, such as biogeochemical cycles, are still unknown.

7.8 Need for antimicrobial environmental protection

Protection of environment from antimicrobial and antibiotic resistance urgentiy needs characterization of residues and AMR bacteria from agronomic sources to mitigate the risks associated with human health. The key responsibilities in research are the determination of the limitations directly and indirectly in influencing public health expenses posed by environmental contamination with ARGs, AMR organisms and antimicrobial residues.

In addition, it is necessary to collect data to execute the research, but unfortunately at national level there is lack of recorded data regarding antimicrobial contamination and AMR bacteria in the environment. The record of microflora is helpful for experiment design and detecting antimicrobial resistance. Programs and tools for scientific measurement of destruction still do not exist.

In the management of antimicrobial resistance systems in the environment, it is necessary to have strong scientific supervision and assessment of antimicrobial resistance and synchronized inspections in the animal, human and food sectors to find pathways for the separation of antimicrobial residues.

7.8.1 Mechanisms of antibiotic resistance

Exact pathways or mechanisms of antibiotic resistance are very difficult to find because of the multifactorial and multidirectional aspects that influence the survival of pathogenic bacteria. Rapidly increasing resistance of bacterial pathogens is one of the major threats for human health. Once it was considered that specific places are the pools of antibiotic resistance, such as the environment of hospitals and some public places (such as public toilets), but rapid increase in the world's population and extensive use of unwanted antibiotics have changed the concept; more recently, the multidrug-resistant organisms are often found in community settings, indicating that the stock of antibiotic-resistant bacteria is now out of the hospital.

Extensive application of antibiotics is the reason for bacterial resistance; even multidrug-resistant organisms have created an alarming situation in clinical microbiology, the response of bacteria toward similar antibiotic attacks for a long time gave opportunity to bacterial adaptation and finding a means of survival; we can call this the pinnacle of evolution and the phenomenon of "survival of the fittest." The vast genetic alterations in the expression of genes occurred in pathogenic bacteria, which are big tool for them to resist antibiotics and transfer of these genes to the residues.

To understand the genetic and biochemical basis of resistance it is necessary to develop strategies to cope with the resistance and novel therapeutic approaches against multidrug-resistant organisms. Antimicrobial resistance is a huge threat and scientists are working to find major mechanisms of antibiotic resistance that they come across in clinical practice. The antimicrobial and antibiotic resistance management system is intricate; the journey starts from collection and assessment of data followed by the discovery and commercialization. The regular system of management of antimicrobial compounds for the treatment of infectious diseases has changed the therapeutic models. Definitely, antibiotics have become the most significant intervention in clinical microbiology necessary for the improvement of complex medical approaches, such as solid organ transplantation, management of patients with cancer, cutting-edge surgical procedures, etc.

Disease management is always affecting economics. Many countries are spending billions of dollars every year in controlling infectious diseases; the United States is spending an estimated 20 billion dollars annually to manage the damage cause by multidrug resistant organisms (DiazGranados et al., 2005; Cosgrove, 2006; Sydnor and Perl, 2011). The data presented by the US Centers for Disease Control and Prevention reports that in the US at least 23,000 people die every year due to antibiotic-resistant organism infections. Recent investigations have presented data showing estimated deaths of about 300 million by 2050 due to antibiotic resistance; the ultimate loss is about $100 trillion (£64 trillion) to the world economy.

Antimicrobial resistance is not new phenomena and this is the consequence of variation and genetic recombination among organisms within the same environment. Secondly against natural antimicrobial molecules bacteria, for their survival, develop some mechanisms to cope with the action of antimicrobials.

References

Aminov, R.I., Mackie, R.I., 2007. Evolution and ecology of antibiotic resistance genes. FEMS Microbiology Letters 271, 147–161.

Bassett, E.J., Keith, M.S., Armelagos, G.J., Martin, D.L., Villanueva, A.R., 1980. Tetracycline-labeled human bone from ancient Sudanese Nubia (A.D. 350). Science 209, 1532–1534.

Bengtsson-Palme, J., Larsson, D.G.J., 2015. Antibiotic resistance genes in the environment: prioritizing risks. Nature Reviews Microbiology 13, 396.

Bengtsson-Palme, J., Boulund, F., Fick, J., Kristiansson, E., Larsson, D.G.J., 2014. Shotgun metagenomics reveals a wide array of antibiotic resistance genes and mobile elements in a polluted lake in India. Frontiers in Microbiology 5, e648.

Cha, J.Y., Ishiwata, A., Mobashery, S., 2004. A novel β-lactamase activity from a enicillin-binding protein of *Treponema pallidum* and why syphilis is still treatable with penicillin. Journal of Biological Chemistry 279, 14917–14921.

Chen, B.W., Liang, X.M., Huang, X.P., Zhang, T., Li, X.D., 2013. Differentiating anthropogenic impacts on ARGs in the Pearl River Estuary by using suitable gene indicators. Water Research 47, 2811–2820.

Cook, M., Molto, E., Anderson, C., 1989. Fluorochrome labelling in Roman period skeletons from Dakhleh Oasis, Egypt. American Journal of Physical Anthropology 80, 137–143.

Cosgrove, S.E., 2006. The relationship between antimicrobial resistance and patient outcomes: mortality, length of hospital stay, and health care costs. Clinical Infectious Diseases 42 (Suppl. 2), S82–S89 [PubMed: 16355321].

Cui, L., Su, X.Z., 2009. Discovery, mechanisms of action and combination therapy of artemisinin. Expert Review of Anti-infective Therapy 7, 999–1013.

Czekalski, N., Diez, E.G., Buergmann, H., 2014. Wastewater as a point source of antibiotic-resistance genes in the sediment of a freshwater lake. The ISME Journal 8, 1381–1390.

D'Costa, V.M., King, C.E., Kalan, L., Morar, M., Sung, W.W.L., Schwarz, C., Froese, D., Zazula, G., Calmels, F., Debruyne, R., Golding, G.B., Poinar, H.N., Wright, G.D., 2011. Antibiotic resistance is ancient. Nature 477, 457–461.

Davin-Regli, A., Bolla, J.M., James, C.E., Lavigne, J.P., Chevalier, J., et al., 2008. Membrane permeability and regulation of drug "influx and efflux" in enterobacterial pathogens. Current Drug Targets 9, 750–759.

DiazGranados, C.A., Zimmer, S.M., Klein, M., Jernigan, J.A., August 1, 2005. Comparison of mortality associated with vancomycin-resistant and vancomycin-susceptible enterococcal bloodstream infections: a meta-analysis. Clinical Infectious Diseases 41 (3), 327–333.

Du, L.F., Liu, W.K., 2012. Occurrence, fate, and ecotoxicity of antibiotics in agro-ecosystems. A review. Agronomy for Sustainable Development 32, 309–327.

Falkinham 3rd, J.O., Wall, T.E., Tanner, J.R., Tawaha, K., Alali, F.Q., Li, C., Oberlies, N.H., 2009. Proliferation of antibiotic-producing bacteria and concomitant antibiotic production as the basis for the antibiotic activity of Jordan's red soils. Applied and Environmental Microbiology 75, 2735–2741.

Fevre, C., Jbel, M., Passet, V., Weill, F.X., Grimont, P.A., Brisse, S., 2005. Six groups of the OXY β-lactamase evolved over millions of years in *Klebsiella oxytoca*. Antimicrobial Agents and Chemotherapy 49, 3453–3462.

Garau, G., Di Guilmi, A.M., Hall, B.G., 2005. Structure-based phylogeny of the metallo-beta-lactamases. Antimicrobial Agents and Chemotherapy 49, 2778–2784.

Hall, B.G., Barlow, M., 2004. Evolution of the serine beta-lactamases:past, present and future. Drug Resistance Updates 7, 111–123.

Kobayashi, T., Nonaka, L., Maruyama, F., Suzuki, S., 2007. Molecular evidence for the ancient origin of the ribosomal protection protein that mediates tetracycline resistance in bacteria. Journal of Molecular Evolution 65, 228–235.

Kumarasamy, K.K., Toleman, M.A., Walsh, T.R., Bagaria, J., Butt, F., Balakrishnan, R., Chaudhary, U., Doumith, M., Giske, C.G., Irfan, S., Krishnan, P., Kumar, A.V., Maharjan, S., Mushtaq, S., Noorie, T., Paterson, D.L., Pearson, A., Perry, C., Pike, R., Rao, B., Ray, U., Sarma, J.B., Sharma, M., Sheridan, E., Thirunarayan, M.A., Turton, J., Upadhyay, S., Warner, M., Welfare, W., Livermore, D.M., Woodford, N., 2010. Emergence of a new antibiotic resistance mechanism in India, Pakistan, and the UK: a molecular, biological, and epidemiological study. The Lancet Infectious Diseases 10, 597–602.

Lavilla Lerma, L., Benomar, N., Casado Muñoz, M.d.C., Gálvez, A., Abriouel, H., 2014. Antibiotic multiresistance analysis of mesophilic and psychrotrophic Pseudomonas spp. isolated from goat and lamb slaughterhouse surfaces throughout the meat production process. Applied and Environmental Microbiology 80, 6792–6806.

Lee, H.H., Molla, M.N., Cantor, C.R., Collins, J.J., 2010. Bacterial charity work leads to population-wide resistance. Nature 467, 82–85.

Li, B., Yang, Y., Ma, L., Ju, F., Guo, F., Tiedje, J.M., Zhang, T., 2015. Metagenomic and network analysis reveal wide distribution and co-occurrence of environmental antibiotic resistance genes. The ISME Journal 9, 2490–2502.

Liu, M., Zhang, Y., Yang, M., Tian, Z., Ren, L., Zhang, S., 2012. Abundance and distribution of tetracycline resistance genes and mobile elements in an oxytetracyclinen production wastewater treatment system. Environmental Science & Technology 46, 7551–7557.

Lyandres, O., 2012. Keeping Great Lakes Water Safe: Priorities for Protecting Against Emerging Chemical Pollutants. Alliance for the Great Lakes. https://www.csuohio.edu/urban/sites/csuohio.edu.urban/files/Drinking_Water_Great_Lakes.pdf.

Martinez, J.L., Coque, T.M., Baquero, F., 2015. What is a resistance gene? Ranking risk in resistomes. Nature Reviews Microbiology 13, 116–123.

McConnell, R.L., Abel, D.C. (Eds.), 2013. Environmental Geology Today. Jones and Bartlett Publishers, Sudbury, United States.

Nelson, M.L., Dinardo, A., Hochberg, J., Armelagos, G.J., 2010. Brief communication: mass spectroscopic characterization of tetracycline in the skeletal remains of an ancient population from Sudanese Nubia 350–550 CE. American Journal of Physical Anthropology 143, 151–154.

Nikaido, H., 2003. Molecular basis of bacterial outer membrane permeability revisited. Microbiology and Molecular Biology Reviews 67, 593–656.

Pages, J.M., James, C.E., Winterhalter, M., 2008. The porin and the permeating antibiotic: a selective diffusion barrier in Gram-negative bacteria. Nature Reviews Microbiology 6, 893–903.

Pruden, A., Pei, R., Storteboom, H., Carlson, K.H., 2006. Antibiotic resistance genes as emerging contaminants: studies in northern Colorado. Environmental Science & Technology 40, 7445–7450.

Reuther, R., 2009. Lake and river sediment monitoring. In: Environmental Monitoring Encyclopedia of Life Support Systems (EOLSS), vol. 2, pp. 120–147.

Rodriguez-Mozaz, S., Chamorro, S., Marti, E., Huerta, B., Gros, M., Sanchez-Melsio, A., Borrego, C.M., Barcelo, D., Balcazar, J.L., 2015. Occurrence of antibiotics and antibioticresistance genes in hospital and urban wastewaters and their impact on the receiving river. Water Research 69, 234–242.

Rollo, I.M., Williamson, J., Plackett, R.L., 1952. Acquired resistance to penicillin and to neoarsphenamine in Spirochaeta recurrentis. British Journal of Pharmacology and Chemotherapy 7, 33–41.
Rustam, I., Aminov, 2010. A brief history of the antibiotic era: lessons learned and challenges for the future. Frontiers in Microbiology | Antimicrobials, Resistance and Chemotherapy 1. https://doi.org/10.3389/fmicb.2010.00134. Article 134 | 2.
Song, J.S., Jeon, J.H., Lee, J.H., Jeong, S.H., Jeong, B.C., Kim, S.J., Lee, J.H., Lee, S.H., 2005. Molecular characterization of TEM-type betalactamases identified in cold-seep sediments of Edison Seamount (south of Lihir Island, Papua New Guinea). J.Microbiol. 43, 172–178.
Sredkova, M., 1998. Resistance to penicillins, aminoglycosides and vancomycin of clinical strains of enterococci. Ibidem 397–400.
Storteboom, H., Arabi, M., Davis, J.G., Crimi, B., Pruden, A., 2010. Tracking ntibioticresistance genes in the south Platte River Basin using molecular signatures of urban,agricultural, and ristine sources. Environmental Science & Technology 44, 7397–7404.
Surinder, K., 2012. Text book of Microbiology, first ed., ISBN 978-93-5025-510-0.
Sydnor, E.R., Perl, T.M., 2011. Hospital epidemiology and infection control in acute-care settings. Clinical Microbiology Reviews 24 (1), 141–173 [PubMed: 21233510].
Van Boeckel, T.P., Brower, C., Gilbert, M., Grenfell, B.T., Levin, S.A., Robinson, T.P., Teillant, A., Laxminarayan, R., 2015. Global trends in antimicrobial use in food animals. Proceedings of the National Academy of Sciences of the United States of America 112, 5649–5654.
Wright, G.D., Poinar, H., 2012. Antibiotic resistance is ancient: implications for drug discovery. Trends in Microbiology 20, 157–159.
Yang, Y., Xu, C., Cao, X., Lin, H., Wang, J., 2017. Antibiotic resistance genes in surface water of eutrophic urban lakes are related to heavy metals, antibiotics, lake morphology and anthropic impact. Ecotoxicology 26, 831–840.
Yin, Q., Yue, D.M., Peng, Y.K., Liu, Y., Xiao, L., 2013. Occurrence and distribution of antibiotic-resistant bacteria and transfer of resistance genes in Lake Taihu. Microbes and Environments 28, 479–486.
Yuyi, Y., Wenjuan, S., 2018. Antibiotics and antibiotic resistance genes in global lakes: a review and meta-analysis Environment. International 116, 60–73.
Zhang, W., Sturm, B.S.M., Knapp, C.W., Graham, D.W., 2009. Accumulation of tetracycline resistance genes in aquatic biofilms due to periodic waste loadings from swine lagoons. Environmental Science & Technology 43, 7643–7650.
Zhu, Y.G., Johnson, T.A., Su, J.Q., Qiao, M., Guo, G.X., Stedtfeld, R.D., Hashsham, A., Tiedje, J.M., 2013. Diverse and abundant antibiotic resistance genes in Chinese swine farms. Proceedings of the National Academy of Sciences 110, 3435–3440.

CHAPTER

Antibiotics and antimicrobial resistance mechanism of entry in the environment

8

B. Balabanova

Faculty of Agriculture, University "Goce Delčev", Štip, Republic of North Macedonia

8.1 General aspects of antibiotics use

8.1.1 Introduction to the term "antibiotic use"

The first true antimicrobial agents were introduced in 1935 (sulfonamides) and 1942 (penicillin). Since then, hundreds of new antimicrobial agents have been discovered and presented on the market (Kohanskiet al. 2010). Most antibiotics are prescribed to outpatient patients. Approximately 80% of the antibiotics in the outpatients are prescribed for the treatment of infections of the respiratory tract, for example, otitis media and sinusitis. Antibiotics are substances that can kill bacteria (bactericidal action) or prevent their growth (bacteriostatic action) (Kohanskiet al. 2010). They may be of natural origin, synthetically or semi-synthetically derived (Halling-Sørensen et al., 1998; Giger et al., 2003). Antibiotics are a good choice for treating bacterial infections, but do not affect viruses and fungi. Antibiotics were put into use in the early 1940s (Halling-Sørensen et al., 1998). Initially they were used only in hospitals to treat severe and life-threatening diseases such as tuberculosis, lung inflammation, and meningitis. Today, they are mostly used in primary health care, most of them when they are not needed or necessary, for mild infections, viral infections, and as preventive measures (WHO, 2002). These are conditions in which they do not act at all and can cause serious harm to the body. Today, it is almost impossible to find a person who has never used an antibiotic. They are also used for breeding animals on farms, as well as in agriculture and industry (Jemba 2002; Singer et al., 2003; Sarmah et al., 2006; Martinez 2009; Negreanu et al., 2012; Michael et al., 2013). Their frequent and widespread use leads to adverse consequences for people and the whole of society, as it leads to a weakening of immunity and the development of resistant bacterial species. At this point, it is a problem of global (world) scale in public health (Kohanskiet al. 2010).

Antibiotics are usually prescribed for upper respiratory tract infections (such as acute cough, colds, nasal secretions, sore throat, bronchitis) and middle ear infections, despite the evidence that antibiotics are not effective in most of these cases. Many people keep unused medicines (WHO, 2002). Keeping unused antibiotics to treat the next illness may seem like a good idea, but only the doctor can tell when there really is a need for antibiotics, and if so, which ones are best for the disease.

Antibiotics can cause allergic reactions in some people (WHO, 2002). Very common side effects of antibiotics may include diarrhea or constipation, nausea, and fungal infections. Antibiotics, especially

Antibiotics and Antimicrobial Resistance Genes in the Environment. https://doi.org/10.1016/B978-0-12-818882-8.00008-5

those with a wide range of effects, destroy both good and bad bacteria and thus indirectly reduce immunity (Igbinosa and Odjadjare, 2015). Only the intestines contain about 1.5 million good bacteria, without which we could not live (Igbinosa and Odjadjare, 2015). When we destroy a lot of them by taking an antibiotic, the pathway for the proliferation of pathogenic bacteria opens, leading to disorders in the digestive system, headache, fatigue, insomnia, stomach pain, and other issues (Ferber, 2003).

In addition, excessive reproduction of fungi and molds occurs. The fungi release toxins that affect the immune cells in the body, which further weakens immunity. Taking a large number of antibiotics without the control of a health care worker inflicts major damage to health in general.

8.1.2 Antibiotic resistance to bacteria

Antibiotics kill most bacteria, but some bacteria manage to develop protection, and in that case antibiotics stop acting on them. The more frequently antibiotics are taken, the greater the danger of developing resistance. Today there are bacteria that are resistant to most known antibiotics, and infections with such bacteria endanger health and life. The main reason for the occurrence of bacterial resistance is excessive and unconscious use of antibiotics, their taking when not needed, discontinuation of antibiotic therapy before the treatment is completed, taking inadequate (smaller) doses than necessary, and disregarding the correct intervals during taking the therapy.

Resistance to bacteria is contagious—new generations of bacteria inherit resistant genes. People can transfer their resistant bacteria to each other, or, in other words, even if you personally did not use antibiotics you can infect infectious bacteria from other people (Halling-Sørensen et al., 1998). Over time, bacteria can become resistant to many types of antibiotics, and so there are super-resistant bacteria to almost all antibiotics. The consequences of this phenomenon are an increase in the rate of illness, increased mortality from some infectious diseases, and an increase in the cost of treatment. Due to this phenomenon, new and more expensive antibiotics are used, but with their improper use, the bacteria again become resistant to the antibiotics. Modern medicine is trying desperately to win the war on infections (Schlüsener and Bester, 2006; Muller et al., 2007).

Sometimes it is not possible to completely prevent the development of resistant bacteria, but we can slow it down with the proper use of antibiotics, which means following the doctor's recommendations:

- Antibiotics are not drugs for treating flu and colds,
- Bacterial and not viral infections are treated with an antibiotic,
- Antibiotics are not a remedy for reducing body temperature,
- The irrational use of antibiotics reduces their effectiveness,
- Antibiotics should be taken only on the recommendation of a doctor and at a specified time and the dose should be consumed whole,
- Improper use of the antibiotic creates resistant bacteria.

8.1.3 Correct use of antibiotics

Although initially antibiotics were produced from natural sources such as fungi, intensive development and production of a large number of synthetic variants soon developed (Schlüsener and Bester, 2006; Muller et al., 2007). Thanks to this, today we have a large number of different antibiotics. By their actions they can be:

1. bactericidal – those that destroy bacterial cells;
2. bacteriostatic – those that stop the growth and development of bacteria, after which the organism itself removes the remaining bacterial cells.

For most infectious diseases, both types are equally effective, but in severe infections and in persons with impaired immunity, the bactericidal antibiotic is better. By the breadth of the action, we distinguish antibiotics:

1. With a wide spectrum of action – which act on several types of bacteria;
2. With a narrower spectrum of action – affecting a small number of bacterial species.

Ideally, prior to prescribing antibiotics, laboratory tests should be performed to identify the bacteria and thus allow the physician to prescribe an effective antibiotic. Such searches last a day or two so they cannot help with the initial selection of antibiotics since it is essential to give the medicine as soon as possible.

Even if the bacterium is recognized and tested in the laboratory by its sensitivity to particular antibiotics, the choice is not simple. Sensitivity in laboratory conditions is not always the same as the one in the infected person (WHO, 2002; Schlüsener and Bester, 2006; Muller et al., 2007). The nature and severity of the disease, the side effects of the drug, possible allergies, and the cost of the drug must be taken into account for the selection of the medicinal product. Because of this, antibiotics are often prescribed empirically on the basis of symptoms of the disease and simple blood tests. Temperature, redness, pain, swelling, and difficulty in swallowing are symptoms that indicate possible bacterial infection. In severe infections, antibiotics are given by injection, intravenous or intramuscular. Some infections, due to the hardness or nature of the causative agents, require treatment with a combination of several antibiotics (Zuccato et al., 2010).

In addition to treating bacterial infections, antibiotics are also taken for prevention. It is justified to use antibiotics for preventive purposes in a small number of conditions, such as surgery and spleen removal. The action of antibiotics and their concentration in the organism is specific for each person (WHO, 2002; Schlüsener and Bester, 2006; Muller et al., 2007). Therefore, the antibiotic must be taken at appropriate time intervals. An antibiotic is typically taken for a number of days, the whole package, to completely eradicate the infection. It often happens that a person takes several antibiotics tablets and then when they begin to feel better, they stop taking the medicine (WHO, 2002; Schlüsener and Bester, 2006; Muller et al., 2007). In such cases, the infection has decreased, but it has not disappeared, and the antibiotic is exposed to bacteria that have remained and can create resistance to it. In this case, the infection returns, usually in a stronger form that will be more difficult to treat (Ferber, 2003). Attention should also be paid to taking antibiotics with respect to the meals and shelf life (antibiotic syrups for children are prepared immediately prior to dispensing in a pharmacy and have a short shelf life). They must be shaken before use, and some must be kept in the refrigerator. All information on proper administration and storage of antibiotics should be obtained when the drug is given to the pharmacy.

If, after completion of the treatment, the medicine remains in the package, it should be disposed of in pharmaceutical waste in the pharmacy, and not kept in the home store. Such stocks are often used by their own estimates for infections that are not usually bacterial in origin, such as viral inflammation of the upper respiratory tract, which reduces the effectiveness of antibiotics and increases bacterial resistance (D'Costa et al., 2011). Antibiotics must not be given to another person who has not been prescribed the medication due to possible hypersensitivity reactions (for example, life-threatening allergy to penicillin antibiotics).

8.2 General pathways of introduction of antibiotics in environment

Antibiotics are designed to fight bacterial infections. This means they will not work against viral-like illnesses such as colds or grips, and against fungal infections. Although some people require these medicines when they are sick, any good doctor would refuse to prescribe them in such cases. When it comes to viruses or grips, it is best to allow your body to choose only viruses. Antibacterial soaps are not particularly useful when it comes to hand disinfecting, but are definitely bad for the environment. They contain components that cannot be filtered, and then through the sewage they return to the environment and become part of the water cycle (Christian et al., 2003; Giger et al., 2003; Gulkowska et al., 2008; Fatta-Kassinos et al., 2010). This can have a big impact because all other organisms are exposed to their constituents (Gulkowska et al., 2008). One of the entry routes of antibiotics into the environment is untreated animal waste, left in the open for degradation. An unsustainable 80% of all antibiotics are prescribed for animals living in factories (Hirsch et al., 1999; Christian et al., 2003; Giger et al., 2003; Gulkowska et al., 2008; Fatta-Kassinos et al., 2010). The bad conditions in which they live are stressful for animals, which makes them more at risk of disease. This is why they receive high doses of antibiotics as a precautionary measure (Montforts, 1999). This unnecessary practice does nothing but increase the bacterial resistance to the drugs. Recent research has shown that a vast majority of people believe that resistance to antibiotics occurs because their body becomes resistant to the effects of medications, and that is the main reason why they decide to shorten their therapy (Hashmi et al., 2017). But that is a complete untruth. If you do not take the recommended dose, you will leave bacteria in your body and allow them to better adapt the next time the body is attacked (Witte 1998; Tolls 2001; Xu et al., 2007). Resistance to antibiotics refers to bacteria, which over time adapt and become stronger. Any animal waste can be used in further farming on the farm or industry, while inadequate handling of residues or cessation of the livestock pollutes the environment.

Animal waste from a farm can be very dangerous, especially when it comes to soil contamination, because it can contain hormones, pesticides, as well as antibiotics (Witte, 1998). There are two types of waste, liquid and solid. Solid is smaller and less dangerous to the environment because it stays on the ground and it stops there. Liquid wastes on modern farms, such as urine or liquids from washing or disinfection, are collected in the canals and can be very dangerous if exposed to watercourses, which our agricultural producers are unfortunately often doing. Animal waste includes animal body parts, bodies, bones, and all animal products not intended for human consumption (Negreanu et al., 2012). Because of this, not only individual farms, but also the industry and agricultural combines can be large environmental polluters, and according to waste management strategies most of the waste comes from the processing industry and from agriculture. Waste should be classified as a hazardous waste group that has to be adequately removed. Meat processing industry waste contains useful and useless parts, which creates the need for a control process for waste treatment (Negreanu et al., 2012). Bones can be smeared in bone meal, then some waste parts can be used to obtain soap. Unfortunately, a large amount of this clonal waste ends in the environment. When it comes to animal bodies, there is a major environmental problem (Ghosh and LaPara, 2007; Dolliver et al., 2008). Some municipalities have an organized animal-hygiene service (old or dead animals, stray dogs, animals that have died on farms),

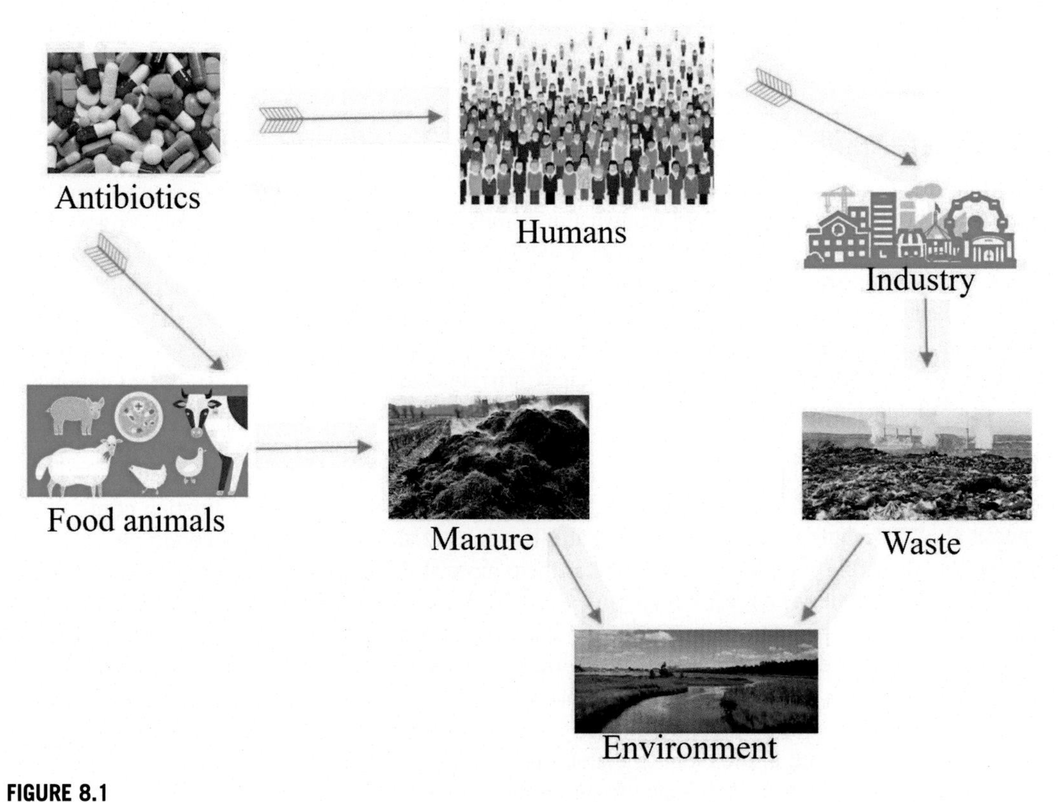

FIGURE 8.1

General pathways of introduction of antibiotics in environment.

so the operators in their professional vehicles carry the carcasses safely to the incineration sites. Another solution is the construction of temporary public tombs for animal bodies (Larsson and Fick, 2009; Liu et al., 2014; Negreanu et al., 2012). As zoo-hygiene does not work or does not exist in all municipalities, some farmers or owners of large livestock farms bury dead animals and carcasses in their backyards, or throw them into a dump or in watercourses that directly threaten the environment (Larsson and Fick, 2009; Liu et al., 2014; Negreanu et al., 2012).

8.2.1 Water/soil environment antibiotics exposure

Water dissolves industrial antibiotics that are bound to environmental matrices. Binding to soil particles (and sediments) delays its biodegradation and explains long-term permanence of the drugs in the environment (Li et al., 2004; Li and Zhang, 2011). Of course, soil particles also remove antibiotics from water, so that a kind of water–soil pharmacokinetics might be considered. Antimicrobial agents are retained in soil by its association with soil chemicals. For instance, Elliot soil humic acids produce complexation of antibiotics (Zhang et al., 2009; Cytryn, 2013). Interestingly, heavy metals (as methylmercury) also associate with humic acids, so that in the water film associated with soil organic particles several antimicrobial effects might be simultaneously present. Indeed it appears that in the presence of humic substances, in both dissolved and mineral-bound forms, environmental mobility of antibiotics might increase (Demanèche et al., 2008). Aluminum and iron oxides might alter these interactions by changing surface charge. For instance, sorption to such oxides results in different types of ciprofloxacin-surface complexes, probably changing the reactivity of fluoroquinolones in the soil–water interphase (Dolliver and Gupta, 2008; Knapp et al., 2010). It is to be noted that general alterations in water or in soil (as pH changes, or ionic strength) might alter these antibiotic–soil–water interactions, producing different levels of antibiotic release (dissolution) from soil particles. In a study, half-lives in soil have been estimated at 20–30 days for erythromycin or oleandomycin (Christian et al., 2003; Schlüsener and Bester, 2006; Negreanu et al., 2012).

8.2.2 Water/sludge environment antibiotics exposure

Antimicrobial agents as sulfonamides, macrolides, trimethoprim, cephalosporin, or fluoroquinolones can be found at potentially active concentrations in activated sludge treatment, and the antibiotic load during the year correlates with the variation in annual consumption data, being higher in the winter (Giger et al., 2003). The wastewater concentration of antimicrobials depends on the sludge–wastewater partition coefficient, but with fluoroquinolones, field experiments of sludge application to agricultural land confirmed long persistence of these compounds, but with limited mobility into the subsoil (Göbel et al., 2004; Schlüsener and Bester, 2006; Sukul and Spiteller, 2007). In compost toilets, amoxicillin decay is negligible, even in the presence of beta-lactamase-producing bacteria. Hydrophobic antibiotics, such as tetracycline or ciprofloxacin, were detected in all sludge samples from two Oslo city hospitals, but not in the collected influent samples, suggesting binding to effluent particles (Baquero et al., 2008). Similarly, fluoroquinolones were consistently found in hospital effluents (Zervos et al., 2003). The extensive use of antibiotics in human medicine, animal farming, and agriculture leads to antibiotic contamination of manure, which can be used as fertilizer. Leaching tests indicate that in general less than 1% of fluoroquinolones in the sludge reached the aqueous phase, which might indicate a relatively reduced mobility when sludge is used to fertilize soil. Nevertheless, that does not exclude localized biological effects on particulate material. Indeed high concentrations of fluoroquinolones were found in secondary sludge (sorption). Macrolides were frequently resistant to the processes carried out in sewage treatment plants in South China, and even higher concentrations were found in the final effluents than in the raw sewage (Xu et al., 2007; Watkinson et al., 2007).

8.3 Antimicrobial resistance

Microbiological resistance to antibiotics (antimicrobial resistance, AMR) is today a serious threat to public health and has taken on increasing importance in the world (Martinez, 2008). An increase in the number of multiresistant bacteria associated with the lack of new antibiotics threatens global health. Some patients are left without therapeutic options as some bacteria are resistant to all antibiotics (Martinez et al., 2007). Even more so, "old" antibiotics, and sometimes "newer ones" have been gradually removed from the market because they are not economically viable, although they are still effective. The issue of antimicrobial resistance is a major challenge and exceptional importance is attached to the challenges of maintaining public health (Hashmi et al., 2017).

Bacteria have the ability to adapt to the environment and to achieve antibiotic resistance in widespread use. A large number of discoveries of new antibiotics have been optimistic that the problem of resistance will be solved, but today it is known that the possibilities of action on the bacterial cell have already been greatly exploited. It is difficult to find new antibiotics that will completely destroy bacterial cells (Levy, 2002). By prescribing antibiotics for the treatment of viral etiology, incorrectly taking antibiotics, abusing the home stocks, and improperly storing antibiotic waste, a constant increase in resistance, many of the antibiotics that we have at our disposal are worthless. On the other hand, the pharmaceutical industry is increasingly working to research new antibiotics even though these are costly and long-lasting processes (Larsson and Fick, 2009). If we want to preserve antibiotics as effective drugs, it is necessary to act in both directions, and encourage industry to develop new antibiotics and rationalize the consumption of existing ones.

The rationalization possibilities are numerous, given that in practice, most antibiotics are spent in the treatment of viral etiology. Hospitals need to pay attention to and control hospital infections because once formed, super-resistant bacteria can easily spread in hospital environments.

In order to know the exact answer to the question of what antibiotic resistance is, it is necessary to emphasize that antibiotics are valuable drugs that prevent the growth of bacteria, and are used exclusively for the treatment of bacterial infections. Frequent and uncritical use of antibiotics results in bacteria developing resistance, i.e., they circumvent the ways in which antibiotics work on them and in that they are all more successful. Unfortunately, antibiotics are often taken unnecessarily, e.g., for treating viral respiratory infections whose initial symptoms overlap with similar but not proven bacterial infection at elevated temperatures or for the prevention of bacterial infections in pediatric practices (WHO, 2002).

8.3.1 Bacterial resistance mechanism and control

Bacteria have been present in the Earth for over 3.8 million years. Bacteria can multiply every 20 minutes in optimal conditions, and have excellent abilities to adapt to any changes in their surroundings. Some bacteria are known that they can tolerate temperatures of several hundred degrees or to survive hydrostatic pressure at thousands of feet below the water surface. Thanks to their DNA repair mechanism, some bacteria are resistant even to radiation. Their closest surroundings are subjected to nonparallel selection pressure. The use of antimicrobial agents favors naturally occurring bacteria drug-resistance. Sensitive bacteria die, and the most resistant survive, and it's possible that a person has already changed the composition of bacteria or relapse the frequencies of the strains in their own bacterial flora. Because the composition of human normal bacterial flora has not been fully studied so far, it is impossible to assess the changes that have been made and their consequences on human health (D'Costa et al., 2011).

So far, hundreds of different genes that are responsible for the bacterial resistance have been discovered. Bacteria are natural phenomena, capable to collect, compile new genes of resistance. Pneumococcal penicillin resistance genes, for example, are a compilation of other bacteria from the oral flora. The most worrying feature of bacteria is multiresistance, that is, the ability to survive more antimicrobial drugs at the same time. More resistant bacteria, which are of clinical significance, have multiple resistance (WHO, 2014).

Bacteria are able to collect resistance-coded genes which can be transferred from one bacteria to another. Furthermore, resistance can be encoded by mutations in chromosomal genes. Resistance is a problem in everyday clinical practice. Because the use of antimicrobial agents is necessary, resistant bacteria will remain a permanent problem. Even the health care system recommends to reduce the use of antibiotics. It seems that over time their uses grow, resulting in further complication of the problem of resistance. The new antimicrobial agents do not solve the problem. Although there is a constant development of new drugs, they give only a short-term solution to the problem. The development of a new drug takes at least 5–12 years (WHO, 2014). Several developed mechanisms can be applied for reducing bacterial resistance:

- Constant efforts to establish a precise diagnosis using laboratory tests and radiological investigations in accordance with the recommendations.
- Change of antibiotics only when necessary. Compulsory adherence to recommendations for different indications, only if there is a valid reason for doing so.
- Careful monitoring of the patient if a decision is made not to start with antibiotic therapy.
- Strict hand hygiene. Liquids based on alcohol are more effective than ordinary soaps in the reduction of contamination of the hands.

The extent of hygiene can be affected by the spread of resistant bacteria. In hospitals, the hand hygiene of staff and patients is the most important factor in the spread of bacteria. In many countries, the optimum climate is the reason for maintaining a large flora, which in turn leads to the problem of resistance. In the future, hygiene should be improved in all countries, and especially in outpatient clinics. For example, day care centers are important vectors in the spread of infections in children.

8.3.2 Critical issues

AMR is not limited to a geographical area or a particular country. In 2008, the European Center for Disease Prevention and Control estimated that across Europe 25,000 deaths and 2.5 million days of extended hospital treatment at an annual level were caused by multiresistant bacterial infections, which resulted in a cost of 1.5 billion euros. According to the Organization for European Economic Co-operation (OECD), an estimated 700,000 deaths worldwide could be caused by AMR every year. Despite steps taken, including efforts by the World Health Organization (WHO), the number of victims (mortality, morbidity) is steadily increasing, and prospects are getting weaker; According to the literature and epidemiological research, if nothing is done, resistance to antibiotics could by 2050 become the leading cause of mortality worldwide. Due to the great need to prevent infections and the rational use of antibiotics, the need to find new therapeutic options, preserve the efficacy of existing antibiotics, and curb the spread of resistance, many countries and international organizations have made ambitious plans over the past several years (Halling-Sørensen et al., 1998; WHO, 2014).

The various national, European, and international initiatives that have emerged over the last decade are a reflection of the common commitment to actively tackle this problem. It is essential that all participants join forces to avoid unnecessary duplication of efforts and to ensure better connectivity within the world's anti-AMR fight movement. Therefore, after considering the AMR problem in the world, the WHO, in cooperation with FAO and OIE, has developed a Global Plan of Action (GAP). GAP sets five major goals and highlights the One Health approach as a comprehensive approach to human health related to the animal world and the environment (WHO 2014).

Many states have committed themselves to developing and implementing a national strategy in line with GAP by mid-2017. The European Union has recently adopted ambitious Council conclusions on the next steps to combat AMR, which include the establishment of the One Health network among member states. The European Council and its agencies, as well as other international bodies (OECD, WHO Europe, ECDC, OIE, FAO) are in the best position to support countries in implementing national strategies to raise awareness of AMR and develop policies to control AMR in human and veterinary health sectors (WHO, 2014).

8.3.3 Risk assessment

The academic community and the pharmaceutical industry are in the best position to develop new medicines and alternative therapies, and state and international agencies should make efforts to strengthen surveillance and develop indicators to monitor antibiotic and resistance consumption in humans and animals. The obligation of health care professionals is to apply AMR guidelines and monitor their application. Accordingly, controlling resistance and maintaining the efficacy of antibiotics is a duty of all and can only be achieved through joint efforts with the aim of developing new antimicrobial drugs.

Although AMRs and healthcare-associated infections (HCAIs) are often considered separately, the association between them is well established and the same committees of many organizations face both of these problems. Prevention of infection and control strategies should go hand in hand with:

- Rational use of antibiotics,
- Appropriate means to control and monitor,
- Diagnostic tests to decide on appropriate therapy.

Strengthening public health initiatives to address this public health problem requires a unique approach at the European level, taking into account local features and existing initiatives. Although there are significant differences in AMR epidemiology and the organization of infection control in European countries, the basic principles of AMR control and HCAI prevention are the same. However, in order to overcome state differences, different organizations approach prevention and control of infection and control over the use of antibiotics, therefore many relevant stakeholders in each country need to be involved in solving problems.

In addressing the problem of antimicrobial resistance, it is perhaps the most important in all age groups to promote knowledge of the potential harm to the unjustified use of antibiotics and AMR in the case of citizens of all ages: from pre-school age to adults. Only such a comprehensive approach throughout the calendar year can prevent unwanted effects of overconsumption of antibiotics and preserve these valuable medicines to treat infections according to the targeted medical indication.

Pan-resistant strains of bacteria occur throughout the world, including gram-negative carbapenem-resistant bacteria, such as *Escherichia coli* and *Klebsiella pneumoniae*, as well as *Pseudomonas aeruginosa* and *Acinetobacter baumanii*, which have emerged in recent times since there are not many new species antibiotics.

8.4 Conclusions

The increased use of antibiotics and taking them when there is no need to reduces the power to act when the body really needs them. This leads to the development of antibiotic resistance to bacteria.

Resistance is a normal mechanism for protecting bacteria from the action of the antibiotic. Over time, resistance is increasing, which means there is a reduction in the number of effective antibiotics. Antibiotic resistance is a problem that affects everyone—policy makers, producers, doctors, and consumers. Awareness of the effects of inadequate and nonrational use of antibiotics globally is high. But in countries where the use of antibiotics is not regulated, and they can be taken without prescription, awareness is still very low.

References

Baquero, F., Martínez, J.L., Cantón, R., 2008. Antibiotics and antibiotic resistance in water environments. Current Opinion in Biotechnology 19 (3), 260–265.

Christian, T., Schneider, R.J., Färber, H.A., Skutlarek, D., Meyer, M.T., Goldbach, H.E., 2003. Determination of antibiotics residues in manure, soil, and surface waters. Acta Hydrochimica et Hydrobiologica 31, 34–44.

Cytryn, E., 2013. The soil resistome: the anthropogenic, the native, and the unknown. Soil Biology and Biochemistry 63, 18–23.

D'Costa, V.M., King, C.E., Kalan, L., Morar, M., Sung, W.W., Schwarz, C., Froese, D., Zazula, G., Calmels, F., Debruyne, R., Golding, G.B., 2011. Antibiotic resistance is ancient. Nature 477 (7365), 457–461.

Demanèche, S., Sanguin, H., Potè, J., Navarro, E., Bernillon, D., Mavingui, P., Wildi, W., Vogel, T.M., Simonet, P., 2008. Antibiotic-resistant soil bacteria in transgenic plant fields. Proceedings of the National Academy of Sciences of the United States of America 105, 3957–3962.

Dolliver, H., Gupta, S., 2008. Antibiotic losses in leaching and surface runoff from manure-amended agricultural land. Journal of Environmental Quality 37 (3), 1227–1237.

Dolliver, H., Gupta, S., Noll, S., 2008. Antibiotic degradation during manure composting. Journal of Environmental Quality 37 (3), 1245–1253.

Fatta-Kassinos, D., Hapeshi, E., Achilleos, A., Meric, S., Gros, M., Petrovic, M., Barcelo, D., 2010. Existence of pharmaceutical compounds in tertiary treated urban wastewater that is utilized for reuse applications. Water Resources Management 25, 1183–1193.

Ferber, D., 2003. Antibiotic resistance. WHO advises kicking the livestock antibiotic habit. Science 301, 1027.

Ghosh, S., LaPara, T.M., 2007. The effects of subtherapeutic antibiotic use in farm animals on the proliferation and persistence of antibiotic resistance among soil bacteria. The ISME Journal 1, 191–203.

Giger, W., Alder, A.C., Golet, E.M., Kohler, H.-P.E., McArdell, C.S., Molnar, E., Siegrist, H., Suter, M.J.-F., 2003. Occurrence and fate of antibiotics as trace contaminants in wastewaters sewage sludge and surface waters. Chimia 57, 485–491.

Gulkowska, A., Leung, H.W., So, M.K., Taniyasu, S., Yamashita, N., Yeung, L.W., Richardson, B.J., Lei, A.P., Giesy, J.P., Lam, P.K., 2008. Removal of antibiotics from wastewater by sewage treatment facilities in Hong Kong and Shenzhen, China. Water Research 42, 395–403.

Göbel, A., McArdell, C.S., Suter, M.J.F., Giger, W., 2004. Trace determination of macrolide and sulfonamide antimicrobials a human sulfonamide metabolite and trimethoprim in wastewater using liquid chromatography coupled to electrospray tandem mass spectrometry. Analytical Chemistry 76, 4756–4764.

Halling-Sørensen, B., Nielsen, S.N., Lanzky, P.F., Ingerslev, F., Lützhøft, H.H., Jørgensen, S.E., 1998. Occurrence, fate and effects of pharmaceutical substances in the environment-A review. Chemosphere 36 (2), 357–393.

Hashmi, M.Z., Strezov, V., Varma, A., 2017. Antibiotics and Antibiotic Resistance Genes in Soils, Soil Biology Edition, vol. 51. Springer International Publishing. https://doi.org/10.1007/978-3-319-66260-2_18.

Hirsch, R., Ternes, T., Haberer, K., Kratz, K.L., 1999. Occurrence of antibiotics in the aquatic environment. The Science of the Total Environment 225 (1), 109–118.

Igbinosa, E.O., Odjadjare, E.E., 2015. Antibiotics and Antibiotic resistance determinants: an undesired element in the environment. The battle against microbial pathogens: basic science. Technological Advances and Educational Programs 2, 858–866.

Jemba, P.K., 2002. The potential impact of veterinary and human therapeutic agents in manure and biosolids on plants grown on arable land: a review. Agriculture, Ecosystems & Environment 93 (1), 267–278.

Knapp, C.W., Dolfing, J., Ehlert, P.A.I., Graham, D.W., 2010. Evidence of increasing antibiotic resistance gene abundances in archived soils since 1940. Environmental Science & Technology 44, 580–587.

Kohanski, M.A., Dwyer, D.J., Collins, J.J., 2010. How antibiotics kill bacteria: from targets to networks. Nature Reviews Microbiology 8 (6), 423–435.

Larsson, D.J., Fick, J., 2009. Transparency throughout the production chain—a way to reduce pollution from the manufacturing of pharmaceuticals? Regulatory Toxicology and Pharmacology 53 (3), 161–163.

Levy, S.B., 2002. Factors impacting on the problem of antibiotic resistance. Journal of Antimicrobial Chemotherapy 49 (1), 25–30.

Li, B., Zhang, T., 2011. Mass flows and removal of antibiotics in two municipal wastewater treatment plants. Chemosphere 83, 1284–1289.

Li, S., Li, X., Wang, D., 2004. Membrane (RO-UF) filtration for antibiotic wastewater treatment and recovery of antibiotics. Separation and Purification Technology 34, 109–114.

Liu, Y., Zhang, J., Gao, B., Feng, S., 2014. Combined effects of two antibiotic contaminants on *Microcystis aeruginosa*. Journal of Hazardous Materials 279, 148–155.

Martinez, J.L., 2008. Antibiotics and antibiotic resistance genes in natural environments. Science 321, 365–367.

Martinez, J.L., 2009. Environmental pollution by antibiotics and by antibiotic resistance determinants. Environmental Pollution 157, 2893–2902.

Martinez, J.L., Baquero, F., Andersson, D.I., 2007. Predicting antibiotic resistance. Nature Reviews Microbiology 5, 958–965.

Michael, I., Rizzo, L., McArdell, C.S., Manaia, C.M., Merlin, C., Schwartz, T., Dagot, C., Fatta-Kassinos, D., 2013. Urban wastewater treatment plants as hotspots for the release of antibiotics in the environment: a review. Water Research 47 (3), 957–995.

Montforts, M.H.H.M., 1999. Environmental risk assessment for veterinary medicinal products Part 1. RIVM report 601300 001. In: Other than GMO-Containing and Immunological Products. National Institute for Public Health and Environment, Bilthoven, The Netherlands.

Muller, A., Coenen, S., Monnet, D.L., Goossens, H., 2007. European Surveillance of antimicrobial consumption (ESAC): outpatient antibiotic use in Europe, 1998–2005. Euro Surveillance 12. E071011-071011.

Negreanu, Y., Pasternak, Z., Jurkevitch, E., Cytryn, E., 2012. Impact of treated wastewater irrigation on antibiotic resistance in agricultural soils. Environmental Science & Technology 46 (9), 4800–4808.

Sarmah, A.K., Meyer, M.T., Boxall, A.B.A., 2006. A global perspective onthe use, sales, exposure pathways, occurrence, fate and effects of veterinary antibiotics (VAs) in the environment. Chemosphere 65, 725–759.

Schlüsener, M.P., Bester, K., 2006. Persistence of antibiotics such as macrolides, tiamulin and salinomycin in soil. Environmental Pollution 143 (3), 565–571.

Singer, R.S., Finch, R., Wegener, H.C., Bywater, R., Walters, J., Lipsitch, M., 2003. Antibiotic resistance – the interplay between antibiotic use in animals and human beings. The Lancet Infectious Diseases 3, 47–51.

Sukul, P., Spiteller, M., 2007. Fluoroquinolone antibiotics in the environment. Reviews of Environmental Contamination & Toxicology 191, 131–162.

Tolls, J., 2001. Sorption of veterinary pharmaceuticals-a review. Environmental Science & Technology 35, 3397–3406.

Watkinson, A.J., Murby, E.J., Costanzo, S.D., 2007. Removal of antibiotics in conventional and advanced wastewater treatment: implications for environmental discharge and wastewater recycling. Water Research 41, 4164–4176.

WHO, 2002. Use of Antimicrobials outside Human Medicine and Resultant Antimicrobial Resistance in Humans. Fact Sheet N°, vol. 268. World Health Organization, Geneva.
WHO, 2014. Antibiotic Resistance: Global Report on Surveillance 2014. World Health Organization, Geneva.
Witte, W., 1998. Medical consequences of antibiotic use in agriculture. Science 279, 996–997.
Xu, W., Zhang, G., Li, X., Zou, S., Li, P., Hu, Z., Li, J., 2007. Occurrence and elimination of antibiotics at four sewage treatment plants in the Pearl River Delta (PRD), South China. Water Research 41 (19), 4526–4534.
Zervos, M.J., Hershberger, E., Nicolau, D.P., Ritchie, D.J., Blackner, L.K., Coyle, E.A., Donnelly, A.J., Eckel, S.F., Eng, R.H., Hiltz, A., Kuyumjian, A.G., 2003. Relationship between fluoroquinolone use and changes in susceptibility to fluoroquinolones of selected pathogens in 10 United States teaching hospitals, 1991–2000. Clinical Infectious Diseases 37 (12), 1643–1648.
Zhang, Y., Marrs, C.F., Simon, C., Xi, C., 2009. Wastewater treatment contributes to selective increase of antibiotic resistance among Acinetobacter spp. The Science of the Total Environment 407, 3702–3706.
Zuccato, E., Castiglioni, S., Bagnati, R., Melis, M., Fanelli, R., 2010. Source, occurrence and fate of antibiotics in the Italian aquatic environment. Journal of Hazardous Materials 179, 1042–1048.

CHAPTER

Antibiotics, AMRs, and ARGs: fate in the environment

9

Muhammad Arshad[1], Rabeea Zafar[1,2]

[1]*Institute of Environmental Sciences and Engineering (IESE), School of Civil and Environmental Engineering (SCEE), National University of Sciences and Technology (NUST), Islamabad, Pakistan;* [2]*Department of Environmental Design, Health & Nutritional Sciences, Faculty of Sciences, Allama Iqbal Open University, Islamabad, Pakistan*

9.1 Introduction

The widespread use of antibiotics as therapeutics in animals and humans is well known. They are also used as feed-enhancing substances and growth promoters in animals (Zhu et al., 2013). Because of partial metabolism and absorption, they are released into the environment either unchanged or as metabolites (Chen et al., 2017). Due to incomplete removal of these antibiotics or their metabolites by wastewater treatment plants, they are commonly found in water bodies (Fig 9.1). Thus, antibiotics are continuously accumulating in domestic, hospital, and pharmaceutical wastes (Liu et al., 2018). Increasing release of antibiotics in lakes and rivers results in resistant strains, which ultimately reduces the therapeutic properties of antibiotics, and they become ineffective or the bacteria become resistant to the antibiotics.

9.2 Estimation of risk of developing antibiotic resistance

Antibiotic concentrations in aquatic products are dependent on: (a) living habits; (b) trophic level; and (c) properties of aquatic products. The exposure intensity of antibiotics follows the order of carnivorous > omnivorous > herbivores. In addition to this, the bioaccumulation factors of antibiotics have a major contribution in the fate of different antibiotics (Liu et al., 2018).

Environmental risk assessment guidelines for pharmaceutical products were established based upon the risk quotient (RQ) method, the ratio of measured maximum environmental concentration (ng/L) (MEC) and predicted no-effect concentration (ng/L) (PNEC), which is defined as "risk quotient" (Nie at al., 2015), and which is expressed in the below equation:

$$RQ = MEC/PNEC$$

PNEC value is dependent upon the toxicity data of various antibiotics (chronic and acute) (USFDA, 2010; EMEA, 2006; Hernando et al., 2006; Vryzas et al., 2011). The classification of environmental risks is done in four levels as per the calculated/experimental RQ values (Table 9.1):

Antibiotics and Antimicrobial Resistance Genes in the Environment. https://doi.org/10.1016/B978-0-12-818882-8.00009-7

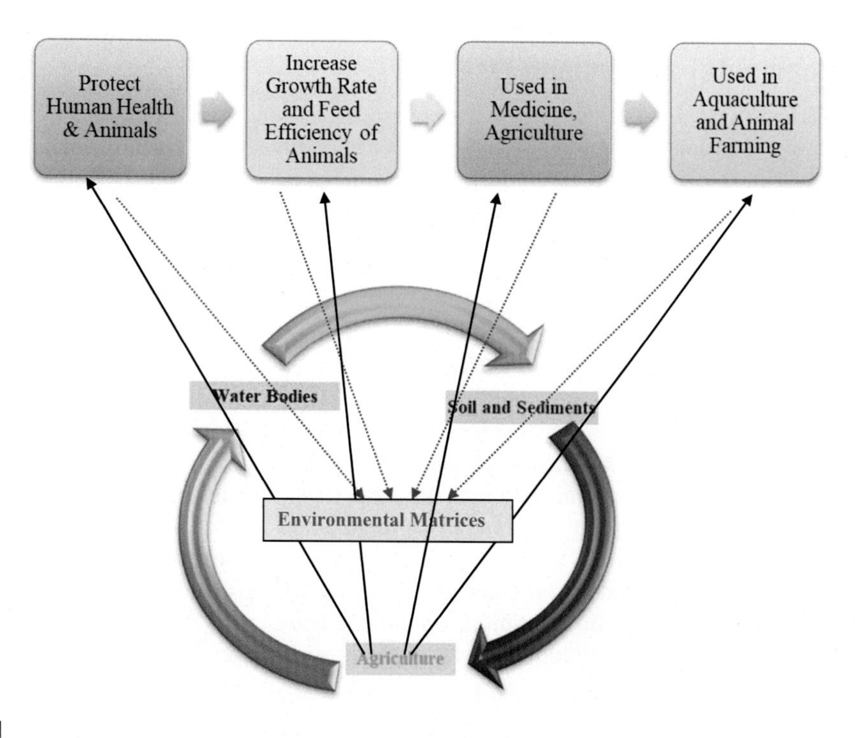

FIGURE 9.1

Use of antibiotics and their transport in different environmental matrices (As metabolites/unchanged through waste and back to food chain through wastewater irrigated agricultural fields).

Table 9.1 Environmental risk levels based upon RQ values.

RQ values	Environmental risk
Less than 0.01	No risk
Between 0.01–0.1	Low risk
Between 0.1–1	Medium risk
Greater than 1	High risk

Antibiotics always exist as mixtures, and though their amount/concentrations may be far less than maximum regulatory limits, toxic contaminants like pesticides, heavy metals, PAHs/PCBs, and other classes of pharmaceutical compounds can have synergistic effects on antibiotics. Therefore, the presence of antibiotics in different environmental situations cannot be ignored (Carvalho and Santos, 2016; Shi et al., 2016).

9.3 Environmental and human risk

For the calculations of RQs, the values of PNEC were used from Bengtsson-Palme and Larsson (2016). These values were based on the Minimal Inhibitory Concentrations (MICs) of all isolates (MIC 1%) that are included in the European Committee for Antimicrobial Susceptibility Testing (EUCAST) database of MIC. In this, the lowest value of MIC is considered. In case of minimal selective concentrations (MSCs) that should be less than MICs, then the lowest MIC is divided by 10, while calculating this, it is based upon the ratio of MIC/MSC. If the calculated value of RQ is above the risk limit, it is understood as the point where potential ecological risk may occur that can be validated through experimental values (Table 9.2). The environmental matrices contain a variety of bacterial strains that are in continuous influx and are affected by various contributing factors like nutrients, pH, light, and temperature (Hanna et al., 2018) (See Fig. 9.2).

The interaction between bacterial agent and a bacterium cell would facilitate the growth of resistant bacteria, which is explained by stimulus-response process. This supports the zero tolerance of antibiotics' presence in various environmental compartments. Even minute or trace levels of antibiotics can initiate resistant gene expression of bacteria (Gullberg et al., 2014).

Keeping in mind the above facts, food and drinking water will be major concern areas as human survival is dependent on them, and the presence of antibiotics can cause the growth of resistant strains and development of resistance against antibiotics in human body.

The continuous addition of antibiotics to water bodies and waste causes resistant bacteria and spread of resistance through antibiotic resistance genes; these are found in ground/drinking water and

Table 9.2 Risk characterization and hazard identification in antibiotic exposure.

Sr#	Risk	Estimation method	Risk levels	References
1	Ecological risk for antibiotic resistance	RQ = MEC/PNEC[1]	Low risk between 0.01 and 0.1 Moderate risk between 0.1 and 1 High risk >1	EC, 2003; Bengtsson-Palme and Larsson, 2016; Gullberg et al. (2011); Hanna et al., 2018
2	Potential human risk through drinking water	$PNEC_{DW} = 1000 \times ADI \times BW \times AT / IngR_{DW} \times EF \times ED$[2] $RQ = MEC/PNEC_{DW}$		WHO (2011)
3	Potential human risk through food/ vegetable consumption	$EDI = C_{food} \times IR_{food} \times \beta_{g/cup} \times \beta_{ww/dw})/BW$[3] $RQ = EDI/ADI$	$HQ > 0.1$[4]	USEPA 1996; WHO 1997 & 2011; USFDA 2010

[1]*RQ = Risk Quotient; MEC = Measured Environmental Concentrations; PNEC = Predicted No-Effect Concentration.*
[2]*PNECDW = Predicted No-Effect Concentration in Drinking Water; 1000 = Conversion Factor; ADI = Acceptable Daily Intake; BW=Body Weight (kg/person); AT = Average Exposure Time; IngDW = Ingestion rate of drinking water (L/person/day); EF = Exposure Frequency (days/year); ED = Exposure Duration (years).*
[3]*EDI = Estimated Daily Intake;* C_{food} *= Mean Concentration of Antibiotics in food (Dry Weight basis); IRfood = Equivalent of food/day;* $\beta_{g/cup}$ *= mass of food;* $\beta_{ww/dw}$ *= mean wet-to-dry conversion factor used by USEPA (for plant tissues in development of soil screening values); BW = body weight (kg/person).*
[4]*Indicates single exposure pathway assuming that humans are exposed through other pathways as well.*

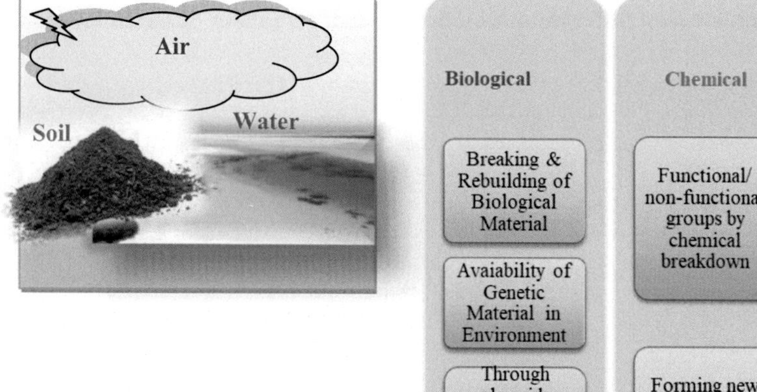

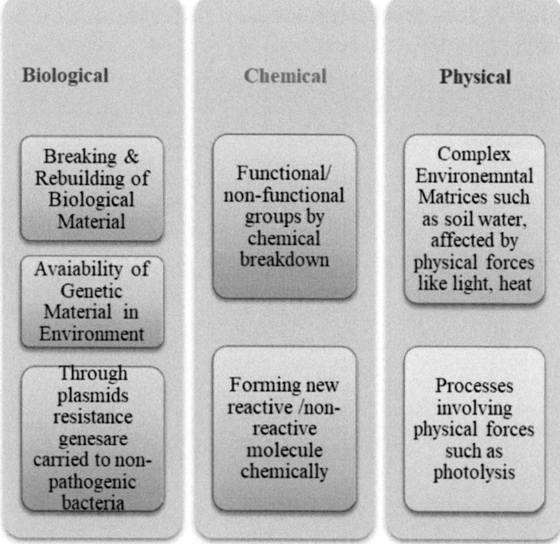

FIGURE 9.2

Interactive and independent processes between various environmental compartments support the movement of resistance bacteria supported by biological, chemical, and physical factors.

wastewater confirmed by numerous studies conducted globally. Table 9.3 shows the studies conducted on various resistant strains of bacteria and the possible sources.

9.4 Fate of antibiotics in soil

Antibiotics enter different matrices through use of manure as fertilizer, direct excretion of grazing animals, wastewater discharge that results in its presence in surface water, sediments, groundwater, and treated wastewater (Pan and Chu, 2016). The presence of most commonly used antibiotics in high concentrations, like sulfamethazine, chloramphenicol, tetracycline, erythromycin, and fluoroquinolones, affects the soil system adversely (Yan et al., 2013). They are present as charged, zwitterions, or neutral because they are ionizable under different conditions, thus having different properties that govern their persistence in soil as well as adsorption and degradation. The fate and behavior of some of the commonly used antibiotics in soil are summarized below:

This explains that antibiotics' adsorption in soil is governed by:

(1) Physicochemical properties of antibiotics and soil; (2) Degradation of antibiotics under aerobic or anaerobic conditions are apparently affected by soil microbial activities, its initial concentration, and oxygen status; (3) higher adsorption rate and slower degradation of antibiotic confirms its persistence in the surface soil and toxicity and risk to organisms in topsoil; and, (4) the weak adsorption and slow degradation of antibiotic indicates that it can penetrate toward gravity (Fig 9.3), resulting in ground water contamination (Bengtsson-Palme, 2018).

Table 9.3 Research conducted on different antibiotic-resistant strains in different environmental situations

Resistant strains	Source	Results	Reference
Pseudomonas aeruginosa	Drinking water and patients' feces	Resistant patterns and serotypes isolated	Shehabi et al., (2005)
Escherichia coli	Drinking water wells	Human acquisition of antibiotic resistance	Sun et al. (2017)
P. aeruginosa; Shigella **sp.**	Vegetable samples	Even low concentrations can develop resistance against antibiotics and are an indirect source of risk Concentrations of different antibiotics in plant tissues can help in developing resistant strains of coliform in the human gut; can trigger the horizontal gene transfer in the gut flora of bacteria	Gullberg et al. (2011), 2014; Allydice-Francis and Brown, 2012; Mokhtari et al. (2012)
Extended *E. coli* producing β-lactamase and animal-linked methicillin-resistant *Staphylococcus aureus* (MRSAs)	Food production chain	Can transmit infections in humans and can be transferred from one bacteria to another	Zhu et al., 2017; Aerestrup et al., 2008

From the above table, it can be inferred that there is dire need of studies on human health issues due to direct exposure to drinking water and vegetables contaminated with antibiotics; also the potential risks and hazards need to be assessed to set the regulatory limits and human health guidelines (Hanna et al., 2018).

Antibiotics are cyclic compounds having piperazine units, benzene ring, sulfonamides, morphiline, quinolone rings, and hexahydropyrimidines. Due to their structure they have meta-stable properties and form metabolites, hydroxylated forms, and conjugates in human and animal bodies, and are released into the environment. The fate of these active metabolites or unchanged compounds needs to be understood so that proper measures can be taken to safeguard the environment and human health (Banin et al., 2017). There are basically two main processes responsible for the fate and behavior of antibiotics in environmental matrices, as depicted in Fig. 9.4.

9.5 Fate of antibiotics in wastewater

Wastewaters are the major reservoirs for antibiotics released through hospital, municipal, and pharmaceuticals waste. Wastewater treatment plants are responsible for decontamination so that the concentration of antibiotics is reduced in water bodies and soil, once the treated wastewater is reused (Li et al., 2013). Due to persistent, recalcitrant and hydrophobic nature of antibiotics, it is very difficult

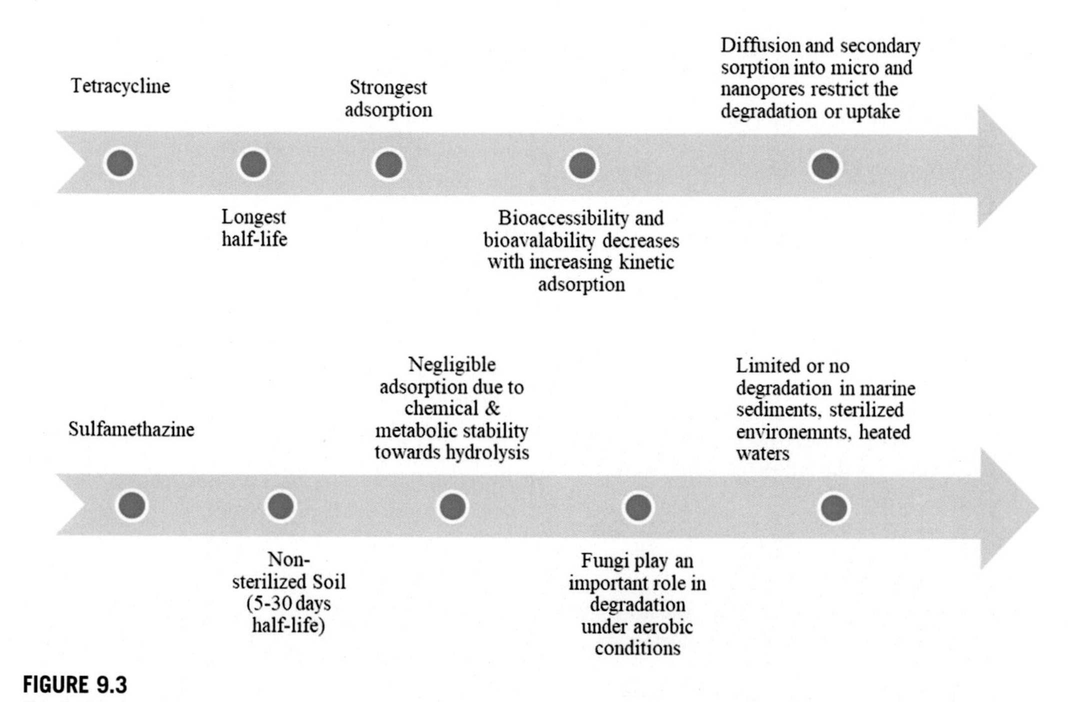

FIGURE 9.3

Fate and behavior of two commonly used antibiotics in soil (Jechalke et al., 2014; Garcia-Galan et al., 2011).

to remove them completely through wastewater treatment; also the presence of antibiotic resistant strains poses a challenge difficult to accomplish (Manzetti and Ghisi, 2014). Some studies showed that antibiotic levels are reduced through chlorination as compared to UV treatment, however, due to complexities in structures of antibiotics it is difficult to set one approach as a remedy. Hence, a combination of techniques, keeping in mind the properties of antibiotics, are to be devised for removal of low to medium to high concentrations of antibiotics in wastewaters (Manzetti and Stenersen, 2010).

The most effective and efficient method of antibiotics removal is a two-step sludge activation, which require days of processing. The process of degradation catalyzed by nitrifying bacteria significantly degraded antibiotics like trimethoprim, but antibiotics like ciprofloxacin and tetracycline were degraded very slowly. The decontamination technique must be a blend of bacterial cultures (nitrifying and oxidizing) and repetition by several steps. However, some new techniques for removing antibiotics efficiently from wastewater include deploying of Actinobacteria, Verrucomicrobia, Nitrospira, α-Proteobacteria and β-Proteobacteria species in three-step nitrification (Jiang et al., 2010; Manzetti and Ghisi, 2014).

Still the issue is not completely resolved as the low concentrations of antibiotics cannot be removed from a wastewater treatment plant and percolate in ground water reaching the drinking water. The transmission of antibiotic compounds/metabolites from sources to biota and back to the environment is through partial biotransformation and bioaccumulation and, finally, gradual deposition in soil and ground waters. Rate of biotransformation of antibiotic compounds is dependent on: (1) the nature of the antibiotic, its chemical structure and properties such as persistency; (2) pH levels of ground water

Catabolic fate

Detoxification System cytochrom P-450

Encompass 60 emebers of membrane-bound oxidase enzymeencoded genes

some of the 60 members act on toxic chemicals by catalytic conversion into solubilized forms

Excreted throeugh body fluids, sweat, urine

The xenobiotic selective group of P-450 is expressed in the 7th chromosome at the q22 cluster

Genetic cluster of this family, has the highest mutation rate; representing a genomic adaptability to different toxicants based upon time, exposure and evoution

Detoxification is also dependent upon Glutathione S-Transferase enzyme through conjugation with thiol group, tixic subsatnaces are solubilized and exreted through bile duct and urinary tract

Metabolic Fate

Depends on functional groups, chemical properties and reactive atoms in the structure

Results in deifferent metabolites of the same antibitoic

Hydroxylation of the central amine group, conjugation to glucuronic acid to the amide group, acetylation of the amine group were metabolic processes for antibiotics in humans

Some antibitoic structures are resistant to detoxification attempts frrm organisms and are thus released into environment unchanged

Antibitoic metaboites are of major concern and are most frequently released into environment causing pollution and wastewater are importnat support medium for the flourishing of resistant bacterial strains and transfer of resistant genes

FIGURE 9.4

Processes responsible for the release of antibiotic either as parent compound or as active metabolites (Stiborova et al., 2012; Palma et al., 2013; Eum et al., 2013; Manzetti and Stenersen, 2010).

and soils; (3) temperature, microbial population, and humidity in soil; and (4) physical properties of soil. These factors play a pivotal role in determining the half-life and rate of diffusion of antibiotics in environmental matrices (Karkman et al., 2018).

9.6 Fate of antibiotics in plants

Fate, behavior, and accumulation of antibiotics in a soil plant system are affected by physical, biological, and chemical processes occurring there (Herklotz et al., 2010). The uptake and translocation of antibiotics in plants is affected by physicochemical properties such as ionization, hydrophobicity, water solubility, and octanol-water partition coefficient (Holling et al., 2012; Jones-Lepp et al., 2010).

Uptake of antibiotics through roots helps in elimination of organic substances from soil. The elimination can be through:

1. Biotic processes such as photolysis (photodegradation); biodegradation by fungi or bacteria
2. Nonbiotic processes such as hydrolysis, sorption, oxidation-reduction, and photolysis.

Some of the above-mentioned processes occur in soil and some in wastewater treatment plants. Amongst them, biodegradation, photodegradation, and sorption are the most important processes occurring in soil matrix. In sorption, the soil phase or sorbent is the sediment or soil (Kinney et al., 2006). The organic matter facilitates the sorption process. Generally, soil acts as a filter and can minimize the uptake and accumulation of pollutants from wastewater in plant tissues.

9.7 Uptake of antibiotics by plants and translocation into tissues

Plant roots have different layers performing different functions. These are: (A) epidermis; (B) cortex; (C) endodermis; (D) Casparian strip (made up of cell wall, cell membrane, vacuole and nucleus); and (E) xylem and phloem.

Nutrients are absorbed through roots and transported to the vascular tissues. This transportation is facilitated in three ways:

i. Apoplastic (moving between cells along cell walls).
ii. Symplastic (moving through cells via plasmodesmata), and
iii. Transmembrane (moving through cells via cell membranes).

Antibiotics are available for uptake through epidermis in the rhizosphere region. Then they pass through the cortex, reach the vascular tissues, and through xylem are transported to the above-ground tissues (Carter et al., 2014). Antibiotics can affect ion channels and enzyme systems of plants, resulting in blocking the transport of essential nutrients required by plants (Jelic et al., 2011; Prosser et al., 2014).

Multiple factors that may affect the root concentration factors for antibiotics are:

i. Plant species, properties of soil, humidity, temperature, exposure time, and chemical concentration
ii. Accumulation in roots is expressed as root concentration factor, which is ratio of the concentration in roots to that in the exposure medium
iii. Accumulation in plant tissues is estimated as bioconcentration factor, which is the ratio of antibiotic concentration in the plant tissue to the concentration in the growth medium.
iv. Translocation of antibiotics in the plant tissues is expressed as translocation factor, which is a ratio of concentration in leaves to that in roots (Al-Farsi et al., 2017; Guasch et al., 2012).

The uptake, translocation, and accumulation of antibiotics by plants is dependent on three conditions: (A) antibiotics are basic in nature; (B) antibiotics are acidic in nature; or, (C) antibiotics are neutral in nature (Table 9.4).

Uptake by plants results in human consumption of contaminated fruit and vegetables, which can cause resistance to antimicrobial activity in humans, thus causing infections that are hard to treat and ultimately causing increased morbidity and mortality (Schimdt et al., 2008; Shenker et al., 2011).

Table 9.4 Nature of antibiotics and uptake by plants (Guasch et al., 2012).

Nature of antibiotics	Breakdown potential	Uptake by plant
Acidic	➢ Dissociation resulting in forming undissociated acids ➢ Release of anions	➢ ifficult to take up by plant because of negative potential and/or repulsion at cell membrane
Basic	➢ Dissociates in nutrient solution and water ➢ Release of cations and formation of neutral molecules	Uptake by: ➢ attraction of cation due to negatively charged biomembrane ➢ on trapping resulting in accumulation into vacuole ➢ igh partition in water so sorption on roots
Neutral		Hydrophobicity facilitates roots sorption and uptake from soil Pass through biomembrane reducing the bioaccumulation in roots due to molecular dissociation

9.8 Fate of AMR/ARBs and ARGs

In recent years, from 2000 to 2010, there was drastic increase in the consumption of antibiotics for human health, which is estimated to be 36% globally (Van Boeckel et al., 2014). These antibiotics are entering into surface and groundwater through wastewater treatment plants, municipal waste, surface runoff, agricultural fields irrigated with treated/untreated wastewater, or through animal manure or sewage sludge used as fertilizers in agriculture (Davies, 2012; Kümmerer, 2009). Antibiotics work through different mechanisms (Le Page et al., 2017), which include:

a. Inhibition of cell envelope synthesis
b. Inhibition of protein synthesis
c. Inhibition of nucleic acid (DNA/RNA) synthesis

In response to antibiotics, bacteria use two major genetic strategies to adapt to the antibiotics' "attack":

(i) mutations in gene(s) often associated with the mechanism of action of the compound
(ii) acquisition of foreign DNA coding for resistance determinants through horizontal gene transfer as depicted below.

Resistance was initially observed in health care settings, but it also emerged in the community, as a potential threat that was accepted worldwide for various species of microorganisms (Taylor et al., 2011). In addition, some bacteria, also known as "multidrug resistant strains" or "superbugs," became resistant to multiple antibiotics (Alanis, 2005; Ashbolt et al., 2013).

Indigenous bacterial population and communities are affected by the release of antimicrobial agents as these bacteria play the major role in biogeochemical cycles and the release of antimicrobial agents will disturb the ecosystem (Proia et al., 2018). These altered ecosystems support the spread of antibiotic resistant bacteria and increase the persistence of these bacteria. The water bodies receiving the waste containing antibiotics and having antibiotic-resistant bacterial strains provide a medium for the transfer of antibiotic resistance (mobile genetic elements) through horizontal gene transfer (Taylor et al., 2011).

9.9 Factors responsible for the fate of ARB and ARGs

Antibiotic-resistant bacteria and resistant genes have the ability to move from one environmental compartment to another, and they move in any direction (Robinson et al., 2016). To assess the importance of various sources of antibiotics, to understand antibiotic resistant bacteria in the environment it is necessary to determine their abundance, characteristics, natural variability, and their mobilization/transfer frequencies in different types of environments (Zhu et al., 2017; Larsson et al., 2018).

Key factors responsible for the fate of ARB and ARGs are: abiotic factor:pH, temperature, humidity, electrical conductivity, biological oxygen demand, chemical oxygen demand, total suspended solids, dissolved oxygen, total nitrogen, carbon, phosphorous, hydraulic retention time, sludge retention time, wastewater flow, dilution factor, antibiotic residues, presence of other contaminants such as metals, pesticides, etc.

Characteristics of bacteria: chemoorganoheterotrophs, neutrophilic, mesophilic, facultative or aerobic, composition of bacterial community, quantification of ARBs and ARGs (Manaia et al., 2018; Larssona et al., 2018).

9.10 Antibiotic resistant genes

Presence and flow of ARGs in the environment is a serious concern and emerging as the most critical to handle this problem. Transfer of ARGs is not restricted. It can happen within nonpathogenic species, within different pathogenic species, between pathogenic and nonpathogenic species, and among genetically distant species. This potential makes it more problematic for the environment as well as spread to no-target species, ultimately contaminating food chain. Water bodies can be an ideal medium for the accumulation and propagation of ARGs as they are directly under the anthropogenic influences (Rodriguez-Mozaz et al., 2015).

Conjugation is the major process involved in antibiotic resistance transfer. In a report in the literature, 80.8% of all tested strains were considered to transfer antibiotic resistance mainly through conjugation (Yan et al., 2013). Similar studies are prevalent on sediments of lakes as compared other environmental media. An interesting fact about the abundances of ARGs is that there are regional influences. As compared to underdeveloped and developing economies, there are higher levels of ARG exposure in developed ones. High population densities also contribute toward higher exposure of ARGs. Other factors including multidrug resistance can also be linked to the sources of resistant isolates, dosages, frequencies, types of antibiotics, etc.

Bacteria acquire resistance against antibiotics through horizontal gene transfer or gene mutation. In this process changes in bacterial membrane occur that actually prevent the antibiotics from entering into bacterial cells. Antibiotics are broken down through enzymes or through employing the efflux pumps; this helps in removal of antibiotic completely or its concentration is reduced to safe limits.

There are two forms of ARGs, iARGs and eARGs, which are intercellular ARGs and extracellular ARGs, respectively (Fig 9.5).

iARGs: Exist as intracellular DNA (iDNA) and resistance bacteria are spread or transferred through conjugation and/or transduction due to bacterial phages infection.

eARGs: Exist as extracellular DNA (eDNA) and facilitate the transfer and propagation of resistant bacterial strains through transformation (Liu et al., 2018; Dong et al., 2019).

The three mechanisms of horizontal gene transfer are mainly responsible for the spread of resistant bacterial strains in environmental matrices (Fig 9.6). Studies have shown that eDNA is more persistent and can live up to months and years in soil and sediments (Dong et al., 2019). The main reservoirs for ARGs are sediments, sludge from wastewater treatment plants, hospital and pharmaceutical wastewater treatment plants, and animal manure. Antibiotics can be completely removed through wastewater treatment plants, but the DNA of resistant bacterial strains would survive due to extremely high density of these strains in environmental matrices acting as their reservoirs of proliferation. This would ultimately cause the threat to human lives causing disease outbreaks with multidrug resistant bacteria taking a toll on health and lives of humans (Nnadozie et al., 2017).

9.11 Facilitation in spread of resistant genes through integrons

Integrons are a well-researched genetic transfer element, which provide a platform for assemblage of genes that in turn express and capture gene cassettes. Gene cassettes actually encode the determinants for antibiotic resistance. Integrons are divided into:

a. Mobile integrons (disseminated between bacteria)
b. Chromosomal integrons (stationary in bacteria)

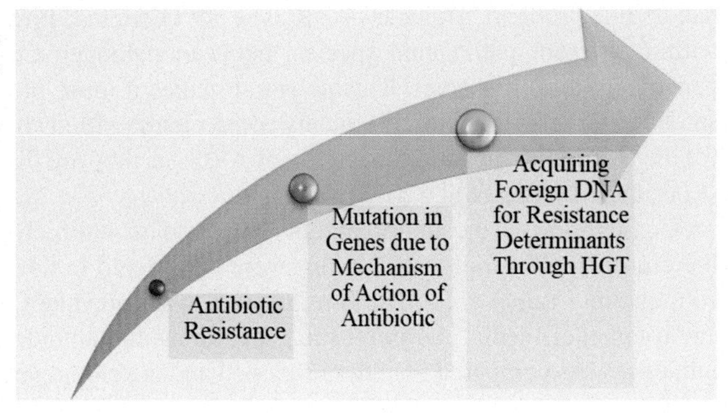

FIGURE 9.5

Mechanism of bacterial resistance in response to antibiotics.

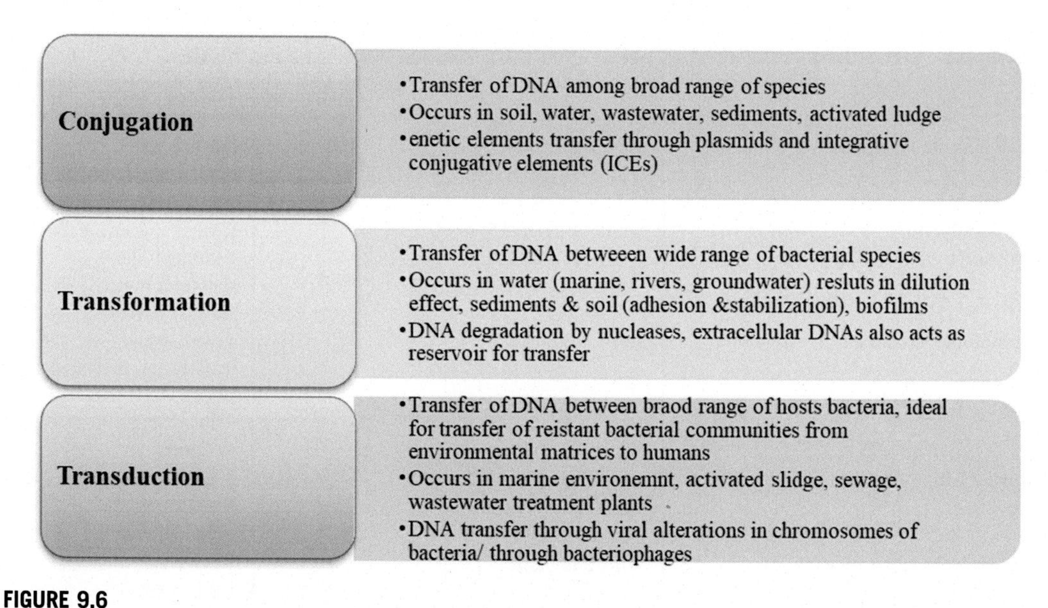

FIGURE 9.6

Mechanism of transfer of antibiotic resistant genes (ARGs) through horizontal gene transfer (HGT).

Mobile integron movement is facilitated by mobile genetic elements such as plasmids; they are also able to change their position/location in the host cell. Mobile integrons have generally limited cassettes in their cassette arrays, and due to encoding the resistance function and various phenotypic characteristics they provide the advantage of being adaptive to their host bacteria. The potential of integrons to acquire new gene cassettes results in comprehending that the total of all gene cassettes represents a metagenome that can be easily accessed by integrons. Thus, integrons facilitate the dissemination of antibiotic resistance through various environmental compartments.

Integrons have three main features: an integron-integrase gene (intI), a recombination site (attI), and a promoter (PC).

These help in capturing and expressing the external genes as a part of gene cassettes. These exogenous genes are again recombined into the attI site using intI's integrase activity and finally expressed from the promoter. This facilitates minimal disturbance to the resident/existing genome.

A proxy gene can be used as an indirect method of resistance detection and a marker for pollutants such as antibiotics, heavy metals, and disinfectants. Class 1 integron-integrase gene, intI1, can be used as a marker for pollution due to the following traits:

1. It indicates the presence of genes related to antibiotic resistance, presence of pesticides and other disinfectants and heavy metal contamination;
2. It has the capability to penetrate into a wide range of pathogenic and nonpathogenic bacteria of humans and other domestic animals;
3. As a result of environmental pressures, the presence of intI can change as they are located on mobile genetic elements and can transfer from one bacterium to another. Different proxy genes representing various classes of antibiotics are depicted in Table 9.5.

Table 9.5 ARGs dissemination and proxy genes for various classes of antibiotics.

Class of antibiotics	Proxy genes	Main reservoirs
Sulfonamide resistance	sul I and sul II	Surface water, rivers, marine environment, wastewater
Trimethoprim resistance	dfr, drfA 1, dfrA5, dfrA 6,dfrA 12, dfrA17	Rivers, wastewater treatment plants, slaughter houses, surface waters
Quinolone resistance	qnrA and qnrB	Rivers, lakes, wastewater, activated sludge, soils irrigated with wastewater, wetlands
Tetracycline resistance	tetA, tetB, tetC, tetD, tetE, tetG, tetL, tetM, tetO, tetQ, tetS, tetW	Surface waters, rivers, wastewater, soil-activated sludge
Vancomycin resistance	vanA and vanB	Wastewater, livestock meat, marine environment, drinking water, poultry
Macrolide resistance	ermA, ermB, ermC, ermF, ermT, ermX	Milk, poultry, surface water, wastewater, animal manure

9.12 Conclusions

Anthropogenic release of antibiotics to the environment is critical and multiple issues have been highlighted linked with the presence of antibiotics. The fate of antibiotics, ARBs, and ARGs has been explored to some extent, and further work is required to assess the role of different contributing factors and exposure through multiple pathways. Ecological risk needs to be determined so that regulatory standards and guidelines can be developed to determine the safe limits of all the administered antibiotics. From various studies and research conducted, gaps are identified that need to be filled in order to control the spread of ARBs and ARGs in the environment. These gaps are: role of antibiotic-resistant bacteria and their contribution in various environmental compartments, contribution of different sources of antibiotics. To safeguard the health of humans from spread of resistant strains in the food chain and effective mitigation of antibiotic resistance through social, behavioral, economic, and technological interventions are current important issues for investigation. Further research needs to be done so that the concentrations at which antibiotic resistance occurs can be investigated. Information and data on these levels will help to prevent the dissemination of ARGs through various environmental matrices resulting in development of multi drug resistant strains of bacteria that will create havoc to human life in the future.

References

Aarestrup, F.M., Wegener, H.C., Collignon, P., 2008. Resistance in bacteria of the food chain: epidemiology and control strategies. Expert Review of Anti-Infective Therapy 6, 733–750.

Alanis, A.J., 2005. Resistance to antibiotics: are we in the post-antibiotic era? Archives of Medical Research 36, 697–705.

Al-Farsi, R.S., Ahmed, M., Al-Busaidi, A., Choudri, B.S., 2017. Translocation of pharmaceuticals and personal care products (PPCPs) into plant tissues: A review. Emerging Contaminants 3, 132–137.

Allydice-Francis, K., Brown, P.D., 2012. Diversity of antimicrobial resistance and virulence determinants in *Pseudomonas aeruginosa* associated with fresh vegetables. The Internet Journal of Microbiology, 426241. https://doi.org/10.1155/2012/426241.

Ashbolt, N.J., Amézquita, A., Backhaus, T., Borriello, P., Brandt, K.K., Collignon, P., Coors, A., Finley, R., Gaze, W.H., Heberer, T., Lawrence, J.R., Larsson, D.G., McEwen, S.A., Ryan, J.J., Schönfeld, J., Silley, P., Snape, J.R., Van den Eede, C., Topp, E., 2013. Human health risk assessment (HHRA) for environmental development and transfer of antibiotic resistance. Environmental Health Perspectives 121, 993–1001.

Banin, E., Hughes, D., Kuipers, P.O., 2017. Editorial: bacterial pathogens, antibiotics and antibiotic resistance. FEMS Microbiology Reviews 41, 450–452.

Bengtsson-Palme, J., 2018. The diversity of uncharacterized antibiotic resistance genes can be predicted from known gene variants—but not always. Microbiome 6, 125. https://doi.org/10.1186/s40168-018-0508-2.

Bengtsson-Palme, J., Larsson, D.G.J., 2016. Concentrations of antibiotics predicted to select for resistant bacteria: proposed limits for environmental regulation. Environment International 86, 140–149.

Carter, L.J., Harris, E., Williams, M., Ryan, J.J., Kookana, R.S., Boxall, A.B., 2014. Fate and uptake of pharmaceuticals in soileplant systems. Journal of Agricultural and Food Chemistry 62, 816–825.

Carvalho, I.T., Santos, L., 2016. Antibiotics in the aquatic environments: a review of the European scenario. Environment International 94, 736–757.

Chen, Y., Zhou, J.L., Cheng, L., Zheng, Y.Y., Xu, J., 2017. Sediment and salinity effects on the bioaccumulation of sulfamethoxazole in zebrafish (*Danio rerio*). Chemosphere 180, 467–475.

Davies, I.A., 2012. Effects of Antibiotics on Aquatic Microbes. Environment Department: University of York, UK.

Dong, P., Wang, H., Fang, T., Wang, Y., Ye, Q., 2019. Assessment of extracellular antibiotic resistance genes (eARGs) in typical environmental samples and the transforming ability of eARG. Environment International 125, 90–96.

EC (European Commission), 2003. In: Commission, E. (Ed.), European Commission Technical Guidance Document in Support of Commission Directive 93//67/EEC on Risk Assessment for New Notified Substances and Commission Regulation (EC) No. 1488/94 on Risk Assessment for Existing Substance, Part II, pp. 100–103.

Eum, K.D., Wang, F.T., Schwartz, J., Hersh, C.P., Kelsey, K., Wright, R.O., Spiro, A., Sparrow, D., Hu, H., Weisskopf, M.G., 2013. Modifying roles of glutathione S-transferase polymorphisms on the association between cumulative lead exposure and cognitive function. Neurotoxicology 39, 65–71.

European Medicines Agency (EMEA), 2006. Guideline on the Environmental Risk Assessment of Medicinal Products for Human Use EMEA/CHMP/SWP/4447/00 Corr 2*, pp. 1–12. June.

Garcia-Galan, M.J., Rodriguez-Rodriguez, C.E., Vicent, T., Caminal, G., Diaz-Cruz, M.S., Barcelo, D., 2011. Biodegradation of sulfamethazine by *Trametes versicolor*: removal from sewage sludge and identification of intermediate products by UPLC-QqTOF-MS. The Science of the Total Environment 409, 5505–5512.

Guasch, H., Ginebreda, A., Geiszinger, A., 2012. Emerging and Priority Pollutants in Rivers: Bringing Science into River Management Plans, vol. 19. Springer Science & Business Media.

Gullberg, E., Albrecht, L.M., Karlsson, C., Sandegren, L., Andersson, D.I., 2014. Selection of a multidrug resistance plasmid by sublethal levels of antibiotics and heavy metals. mBio 5. https://doi.org/10.1128/mBio.01918-14, 01918-14.

Gullberg, E., Cao, S., Berg, O.G., Ilbäck, C., Sandegren, L., Hughes, D., Andersson, D.I., 2011. Selection of resistant bacteria at very low antibiotic concentrations. PLoS Pathogens 7, e1002158. https://doi.org/10.1371/journal.ppat.1002158.

Hanna, N., Sun, P., Sun, Q., Li, X., Yang, X., Ji, X., Zou, H., Ottoson, J., Nilsson, L.E., Berglund, B., Dyar, O.J., Tamhankar, A.J., Stålsby Lundborg, C., 2018. Presence of antibiotic residues in various environmental compartments of Shandong province in eastern China: its potential for resistance development and ecological and human risk. Environment International 114, 131–142.

Herklotz, P.A., Gurung, P., Heuvel, B.V., Kinney, C.A., 2010. Uptake of human pharmaceuticals by plants grown under hydroponic conditions. Chemosphere 78, 1416–1421.

Hernando, M.D., Mezcua, M., Fernandez-Alba, A.R., Barcelo, D., 2006. Environmental risk assessment of pharmaceutical residues in wastewater effluents, surface waters and sediments. Talanta 69, 334–342.

Holling, C.S., Bailey, J.L., Heuvel, B.V., Kinney, C.A., 2012. Uptake of human pharmaceuticals and personal care products by cabbage (*Brassica campestris*) from fortified and biosolids-amended soils. Journal of Environmental Monitoring 14, 3029–3036.

Jechalke, S., Heuer, H., Siemens, J., Amelung, W., Smalla, K., 2014. Fate and effects of veterinary antibiotics in soil. Trends in Microbiology 22, 536–545.

Jelic, A., Gros, M., Ginebreda, A., Cespedes-Sánchez, R., Ventura, F., Petrovic, M., Barcelo, D., 2011. Occurrence, partition and removal of pharmaceuticals in sewage water and sludge during wastewater treatment. Water Research 45, 1165–1176.

Jiang, M., Wang, L., Ji, R., 2010. Biotic and abiotic degradation of four cephalosporin antibiotics in a lake surface water and sediment. Chemosphere 80, 1399–1405.

Jones-Lepp, T.L., Sanchez, C.A., Moy, T., Kazemi, R., 2010. Method development and application to determine potential plant uptake of antibiotics and other drugs in irrigated crop production systems. Journal of Agricultural and Food Chemistry 58, 11568–11573.

Karkman, A., Do, T.T., Walsh, F., Virta, M.P.J., 2018. Antibiotic-resistance genes in wastewater. Trends in Microbiology 26, 220–228.

Kinney, C.A., Furlong, E.T., Zaugg, S.D., Burkhardt, M.R., Werner, S.L., Cahill, J.D., Jorgensen, G.R., 2006. Survey of organic wastewater contaminants in biosolids destined for land application. Environmental Science and Technology 40, 7207–7215.

Kümmerer, K., 2009. Antibiotics in the aquatic environment – a review – part I. Chemosphere 75, 417–434.

Larsson, D.G.J., Andremont, A., Bengtsson-Palme, J., Brandt, K.K., de Roda Husman, A.M., Fagerstedt, P., Fick, J., Flach, C.F., Gaze, W.H., Kuroda, M., Kvint, K., Laxminarayan, R., Manaia, C.M., Nielsen, K.M., Plant, L., Ploy, M.C., Segovia, C., Simonet, P., Smalla, K., Snape, J., Topp, E., van Hengel, A.J., Verner-Jeffreys, D.W., Virta, M.P.J., Wellington, E.M., Wernersson, A.S., 2018. Critical knowledge gaps and research needs related to the environmental dimensions of antibiotic resistance. Environment International 117, 132–138.

Le Page, G., Gunnarsson, L., Snape, J., Tyler, C.R., 2017. Integrating human and environmental health in antibiotic risk assessment: a critical analysis of protection goals, species sensitivity and antimicrobial resistance. Environment International 109, 155–169.

Li, K., Ji, F., Liu, Y., Tong, Z., Zhan, X., Hu, Z., 2013. Adsorption removal of tetracycline from aqueous solution by anaerobic granular sludge: equilibrium and kinetic studies. Water Science and Technology 67, 1490–1496.

Liu, S.S., Qu, H.M., Yang, D., Hu, H., Liu, W.L., Qiu, Z.G., Hou, A.M., Guo, J., Li, J.W., Shen, Z.Q., Jin, M., 2018. Chlorine disinfection increases both intracellular and extracellular antibiotic resistance genes in a full-scale wastewater treatment plant. Water Research 136, 131–136.

Manaia, C.M., Rocha, J., Scaccia, N., Marano, R., Radu, E., Biancullo, F., Cerqueira, F., Fortunato, G., Iakovides, I.C., Zammit, I., Kampouris, I., Vaz-Moreira, I., Nunes, O.C., 2018. Antibiotic resistance in wastewater treatment plants: tackling the black box. Environment International 115, 312–324.

Manzetti, S., Ghisi, R., 2014. The environmental release and fate of antibiotics. Marine Pollution Bulletin 79, 7–15.

Manzetti, S., Stenersen, J.H.V., 2010. A critical view of the environmental condition of the Sognefjord. Marine Pollution Bulletin 60, 2167–2174.

Mokhtari, W., Nsaibia, S., Majouri, D., Ben Hassen, A., Gharbi, A., Aouni, M., 2012. Detection and characterization of Shigella species isolated from food and human stool samples in Nabeul, Tunisia, by molecular methods and culture techniques. Journal of Applied Microbiology 113, 209–222.

Nie, M., Yan, C., Dong, W., Liu, M., Zhou, J., Yang, Y., 2015. Occurrence, distribution and risk assessment of estrogens in surface water, suspended particulate matter, and sediments of the Yangtze Estuary. Chemosphere 127, 109–116.

Nnadozie, C.F., Kumari, S., Bux, F., 2017. Status of pathogens, antibiotic resistance genes and antibiotic residues in wastewater treatment systems. Reviews in Environmental Science and Biotechnology 16, 491–515.

Palma, B.B., Silva E Sousa, M., Urban, P., Rueff, J., Kranendonk, M., 2013. Functional characterization of eight human CYP1A2 variants: the role of cytochrome b5. Pharmacogenetics and Genomics 23, 41–52.

Pan, M., Chu, L.M., 2016. Adsorption and degradation of five selected antibiotics in agricultural soil. The Science of the Total Environment 545–546, 48–56.

Proia, L., Anzil, A., Subirats, J., Borrego, C., Farrè, M., Llorca, M., Balcázar, J.L., Servais, P., 2018. Antibiotic resistance along an urban river impacted by treated wastewaters. The Science of the Total Environment 628–629, 453–466.

Prosser, R.S., Lissemore, L., Topp, E., Sibley, P.K., 2014. Bioaccumulation of triclosan and triclocarban in plants grown in soils amended with municipal dewatered biosolids. Environmental Toxicology & Chemistry 33, 975–984.

Robinson, T.P., Bu, D.P., Carrique-Mas, J., Fevre, E.M., Gilbert, M., Grace, D., Hay, S.I., Jiwakanon, J., Kakkar, M., Kariuki, S., Laxminarayan, R., Lubroth, J., Magnusson, U., Thi Ngoc, P., Van Boeckel, T.P., Woolhouse, M.E., 2016. Antibiotic resistance is the quintessential One Health issue. Transactions of the Royal Society of Tropical Medicine & Hygiene 110, 377–380.

Rodriguez-Mozaz, S., Chamorro, S., Marti, E., Huerta, B., Gros, M., Sànchez-Melsió, A., Borrego, M.C., Barceló, D., Balcázar, L.J., 2015. Occurrence of antibiotics and antibiotic resistance genes in hospital and urban wastewaters and their impact on the receiving river. Water Research 69 (1), 234–242.

Schmidt, B., Ebert, J., Lamshöft, M., Thiede, B., Schumacher-Buffel, R., Ji, R., Corvini, P.F., Schäffer, A., 2008. Fate in soil of 14C-sulfadiazine residues contained in the manure of young pigs treated with a veterinary antibiotic. Journal of Environmental Science and Health, Part B 43, 8–20.

Shehabi, A.A., Masoud, H., Maslamani, F.A., 2005. Common antimicrobial resistance patterns, biotypes and serotypes found among *Pseudomonas aeruginosa* isolates from patient's stools and drinking water sources in Jordan. Journal of Chemotherapy 17, 179–183.

Shenker, M., Harush, D., Ben-Ari, J., Chefetz, B., 2011. Uptake of carbamazepine by cucumber plants: a case study related to irrigation with reclaimed wastewater. Chemosphere 82, 905–910.

Shi, J., Zheng, G.J.S., Wong, M.H., Liang, H., Li, Y., Wu, Y., Li, P., Liu, W., 2016. Health risks of polycyclic aromatic hydrocarbons via fish consumption in Haimen bay (China), downstream of an e-waste recycling site (Guiyu). Environmental Research 147, 233–240.

Stiborova, M., Cechova, T., Borek-Dohalska, L., Moserova, M., Frei, E., Schmeiser, H.H., Paca, J., Arlt, V.M., 2012. Activation and detoxification metabolism of urban air pollutants 2-nitrobenzanthrone and carcinogenic 3-nitrobenzanthrone by rat and mouse hepatic microsomes. Neuroendocrinology Letters 33–3, 8–15.

Sun, P., Bi, Z., Nilsson, M., Zheng, B., Berglund, B., StålsbyLundborg, C., Börjesson, S., Li, X., Chen, B., Yin, H., Nilsson, L.E., 2017. Occurrence of blaKPC-2, blaCTX-M, and mcr-1 in Enterobacteriaceae from well water in Rural China. Antimicrobial Agents and Chemotherapy 61. https://doi.org/10.1128/AAC.02569-16 e02569-16.

Taylor, N.G.H., Verner-Jeffreys, D.W., Baker-Austin, C., 2011. Aquatic systems: maintaining, mixing and mobilising antimicrobial resistance? Trends in Ecology & Evolution 26, 278–284.

USEPA, 1996. Supplemental Guidance for Developing Soil Screening Levels for Superfund Sites, Background Discussion for Soil–Plant–Human Exposure Pathway. United States Environmental Protection Agency, Washington, DC, US.

USFDA, 2010. In: Administration, U.S.F.a.D. (Ed.), Chapter 1—Food and Drug Administration, Department of Health and Human Services, Sub-chapter E—Animal Drugs, Feeds, and Related Products, Part 556 — Tolerances for Residues of New Animal Drugs in Food—Sub-Part B—specific Tolerance for Residues of New Animal Drugs — Section 556.286 — Flunixin. United States Food and Drug Administration, Washington, DC, USA, 21 CF 556286.

Van Boeckel, T.P., Gandra, S., Ashok, A., Caudron, Q., Grenfell, B.T., Levin, S.A., Laxminarayan, R., 2014. Global antibiotic consumption 2000 to 2010: an analysis of national pharmaceutical sales data. The Lancet Infectious Diseases 14, 742–750.

Vryzas, Z., Alexoudis, C., Vassiliou, G., Galanis, K., Papadopoulou-Mourkidou, E., 2011. Determination and aquatic risk assessment of pesticide residues in riparian drainage canals in Northeastern Greece. Ecotoxicology and Environmental Safety 74, 174–181.

WHO, 1997. Guidelines for Predicting Dietary Intake of Pesticide Residues. WHO/FSF/FOS/97.7.

WHO, 2011. Guidelines for Drinking Water Quality, fourth ed. World Health Organization, Geneva. 2001. WHO.

Yan, C.X., Yang, Y., Zhou, J.L., Liu, M., Nie, M.H., Shi, H., Gu, L., 2013. Antibiotics in the surface water of the Yangtze Estuary: occurrence, distribution and risk assessment. Environmental Pollution 175, 22–29.

Zhu, S., Chen, H., Li, J., 2013. Sources, distribution and potential risks of pharmaceuticals and personal care products in Qingshan Lake basin, Eastern China. Ecotoxicology and Environmental Safety 6, 154–159.

Zhu, Y.G., Zhao, Y., Li, B., Huang, C.L., Zhang, S.Y., Yu, S., Chen, Y.S., Zhang, T., Gillings, M.R., Su, J.Q., 2017. Continental-scale pollution of estuaries with antibiotic resistance genes. Nat. Microbiol. 2, 16270. https://doi.org/10.1038/nmicrobiol.2016.270.

CHAPTER

On the edge of a precipice: the global emergence and dissemination of plasmid-borne *mcr* genes that confer resistance to colistin, a last-resort antibiotic

10

Jouman Hassan[1], Lara El-Gemayel[2], Isam Bashour[2], Issmat I. Kassem[1,3]

[1]*Department of Nutrition and Food Sciences, Faculty of Agricultural and Food Sciences, American University of Beirut (AUB), Beirut, Lebanon;* [2]*Department of Agriculture, Faculty of Agricultural and Food Sciences, AUB, Beirut, Lebanon;* [3]*Center for Food Safety, Department of Food Science and Technology, University of Georgia, GA, United States*

10.1 A brief history of colistin

Colistin (polymixin E) is an antibiotic that belongs to the polymyxins class. Colistin is composed of a cyclic decapeptide linked to a fatty acid chain and is active against many gram-negative bacteria (Kempf et al., 2016). Due to its low renal clearance and toxicity, colistin use as a treatment for bacterial infections in humans was abandoned in the 1970s (Kempf et al., 2013). However, with the rise in complicated and resistant infections, colistin was reintroduced as a last-resort drug to mainly treat patients infected with multidrug-resistant gram-negative bacteria. Until 2015, resistance to colistin was mainly thought to be due to chromosol mutations that are transmissible vertically via bacterial progeny (Wang et al., 2018a). However, during a surveillance project on antimicrobial resistance in China, a colistin-resistant *Escherichia coli* strain (SHP45) was isolated from a pig (Liu et al., 2016). This strain harbored the first described plasmid-borne colistin-resistance gene, which was dubbed *mcr-1* (mobile colistin resistance). This gene was found to be transmissible laterally between *Enterobacteriaceae*. *mcr-1* encodes a phosphoethanolamine transferase enzyme, which catalyzes the addition of phosphoethanolamine to lipid A, hence reducing the affinity of colistin to its target lipopolysaccharide. It has been argued that *mcr-1* is less frequently encountered in isolates of human origins in comparison to those of food-animal origins. Consequently, this suggested that *mcr-1*-mediated colistin resistance could have emerged in animals and eventually spread to humans. The lateral mobilization of colistin resistance is regarded as a significant public health hazard, because it can result in limiting available treatment options for important infections (Carattoli, 2013). The latter was further confirmed by the global emergence of the plasmid-borne colistin resistance genes in a variety of hosts and niches (Wang et al., 2018a; Kempf et al., 2016). On a global level, laterally mobile colistin resistance sums up the rising threat of antimicrobial resistance and serves as a call for action to tackle this peremptory challenge that faces humanity.

Antibiotics and Antimicrobial Resistance Genes in the Environment. https://doi.org/10.1016/B978-0-12-818882-8.00010-3

10.2 Colistin use in animal farming practices

Colistin has been historically used in animal farming practices, mainly for treating intestinal infections with gram-negative bacteria (Catry et al., 2015). However, colistin has been also used for other purposes such as growth promotion, metaphylaxis, and prophylaxis (Apostolakos and Piccirillo, 2018). In a survey conducted by Kassem et al. (2019), 12 different veterinary drugs that contained colistin were identified. These drugs were mainly promoted for poultry and covered a wide range of animal diseases. Additionally, Kempf et al. (2016) reported that colistin has been used as a growth promoter in India, Vietnam, China, Japan, and many other countries or administered alone or in combination topically, orally, by injection, or through the intramammary route for therapeutic purposes.

Colistin applications extended to treat both aquatic and terrestrial animal species. For example, colistin was deployed in (1) treating gastrointestinal infections that were triggered by noninvasive *Enterobacteriaceae* in cattle, small ruminants, poultry, and pigs, (2) managing enteric infections initiated caused by *Salmonella* and *E. coli* (i.e., postweaning diarrhea in piglets), and (3) metaphylaxis in cattle and pig production (Apostolakos and Piccirillo, 2018). Notably, in pig production, polymyxins have been preferred for the management of diarrhea (Kempf et al., 2016), whereas colistin (given orally) was chiefly used to treat mild colibacillosis in poultry (Apostolakos and Piccirillo, 2018).

In animal farming practices, such as metaphylaxis, colistin-containing drugs were primarily administered via the oral route in feed or drinking water. This implied that large amounts of these drugs were used. For example, it is estimated that 495 tons of polymyxins were orally given to swine and poultry in Europe in 2013 (Apostolakos and Piccirillo, 2018). Notably, 93%–95% of the farms in France used polypeptide antibiotics (including colistin) orally. Additionally, 90% of the farms deployed colistin in the postweaning period, while 48% used it during the maternity period in sows (Chauvin, 2010). Therefore, it is not surprising that colistin was the fifth most sold group of antimicrobials administered in agriculture in Europe in 2010 (Paterson and Harris, 2016). Globally, China used to be the highest colistin (~17.5 million tons) producer and consumer (Liu et al., 2016).

In 2012, the World Health Organization categorized colistin as a critically important medicine for humans (Paterson and Harris, 2016) because of its impact on multidrug-resistant gram-negative bacilli. For this reason, there have been several calls for limiting and banning the use of polymyxins in agriculture and animal husbandry. The latter was also spurred by the emergence of plasmid-mediated colistin resistance (*mcr*) in bacteria of animal origin (Liu et al., 2016). For example, in China, the percentage of *mcr*-positive isolates in pigs was 14.4% in 2012, and it increased to 20.9% in 2014. In retail meats, the percentage of *mcr*-positive isolates increased from 4.9% to 28% between 2011 and 2014. Notably, the increase in pork was from 6.3% to 22.3% during the same period (Liu et al., 2016).

In an effort to curtail the farming-associated dissemination of *mcr*, many countries such as China and Brazil have moved to ban and/or restrict the use of colistin in agriculture (Holmes et al., 2018). However, it is feared that these restrictions, although useful and needed, might not revert the tide of the dissemination of *mcr* (Wang et al., 2018a). The latter is likely related to the robustness of the plasmids that carry *mcr* and the fitness costs associated with the acquisition of these genetic elements.

10.3 Emergence of mobile colistin resistance on the global stage

Until 2015, when Liu et al. (2016) discovered the new plasmid-mediated colistin-resistance mechanism (*mcr-1*), resistance was thought to be mediated by chromosomal mutations (Coppi et al., 2018) and that the lateral transfer of resistance among bacteria was limited (Delgado-Blas et al., 2016). Today, *mcr-1* has been detected in several countries on five continents (Wang et al., 2018a). Furthermore, researchers have identified several other *mcr* genes (*mcr-2* to *9*) and additional variants in many countries (Xavier et al., 2016b; Carattoli et al., 2017). For example, *mcr-2* was identified in *E. coli* isolates from cattle and pigs in Belgium. *mcr-3* was identified in *E. coli* isolated from pigs in Malaysia, in *Salmonella* Typhimurium in the United States, and in *Klebsiella pneumoniae* in Thailand. *mcr-3* was also discovered in Europe in *E. coli* and *S.* Typhimurium from food-producing animals as well as humans. *mcr-5* is thought to have originated in *Cupriavidus gilardii*, while most recently, *mcr-9* was detected in *Salmonella enterica* isolated from a human patient in the United States (Quiroga et al., 2019). Here, the discussion will focus mainly on *mcr-1*, because it is the most studied and reported of all of these genes.

In addition to association with agriculture, several studies demonstrated that *mcr-1* can also occur in a wide variety of environments and niches such as recreational waters at public urban beaches in Brazil (Zhang et al., 2016), fecal samples from otherwise healthy individuals (von Wintersdorff et al., 2016; Wang et al., 2017), and the Haihe River in China (Yang et al., 2017a). Part of the success of *mcr-1* was attributed to its occurrence in different genetic backgrounds. *mcr-1* was identified on a wide variety of plasmid types, including IncI2, IncHI2, and IncX. The plasmids carry other important antimicrobial-resistance genes such as those encoding extended-spectrum β-lactamases (ESBL) and carbapenemases (Wang et al., 2018a). However, despite evidence of the *mcr-1* global distribution and success, little is known on its origin, acquisition, emergence, and spread (Wang et al., 2018a). Below, the global distribution of *mcr* will be highlighted and the discussion will shed light on regions of the world that previously did not receive a lot of attention.

10.3.1 Americas

Several studies have documented *mcr-1* in the Americas (Table 10.1). In the United States, colistin has not been widely marketed for use in animals (Matamoros et al., 2017). To evaluate the presence of *mcr-1* in *Enterobacteriaceae*, a survey of 2,003 cecal samples from cattle, swine, turkey, and chicken from slaughterhouses in the United States was performed (Meinersmann et al., 2017). As a result, two *E. coli* were isolated from swine and were shown to harbor *mcr-1* on IncI2 plasmids. In Eastern Canada, Pilote et al. (2019) showed that *mcr-1* was found in 6 out of 10 swine confinement facilities. That was equivalent to 60% of the visited pig facilities. In Latin America, isolates of *E. coli* harboring *mcr-1* and belonging to different sequence types have been identified in several countries such as Venezuela, Argentina, Ecuador, and Brazil (Merida-Vieyra et al., 2019; Quiroga et al., 2019) (Table 10.1). Furthermore, 49% of *E. coli* isolates from broilers were resistant to colistin in Argentina, and 19.5% of 41 chicken meat samples were positive for *mcr-1* in Brazil (Vinueza-Burgos et al., 2019). In Ecuador, 3.4% of the isolates (6 out of 176) carried *mcr-1* (Vinueza-Burgos et al., 2019). The latter was the first report on the occurrence of *mcr-1* in Ecuador. However, *mcr-2*, *mcr-3*, and *mcr-4* and other determinants haven't been identified in the region yet. This can be attributed to a lack of extensive screening and monitoring for these genes, and/or perhaps the genes did not spread to these countries yet.

Table 10.1 A compilation of studies reporting colistin resistance and *mcr-1* in the Americas.

Species	Origin	Year	Country	*mcr* Gene	Reference
Escherichia coli	Chicken and pig	2012–13	Brazil	*mcr-1*	Fernandes et al. (2016)
E. coli	Human (blood, urine, abscess, abdomen, bone)	2016	Argentina	*mcr-1*	Rapoport et al. (2016)
E. coli	Healthy poultry	2013–16	Argentina	*mcr-1*	Quiroga et al. (2019)
E. coli	Food (chicken meat)	2016	Brazil	*mcr-1*	Dominguez et al. (2017)
E. coli	Human (urine, blood, bone, peritoneal fluid)	2016	Brazil	*mcr-1*	Quiroga et al. (2019)
E. coli	Human (peritoneal fluid)	2016	Ecuador	*mcr-1*	Ortega-Paredes et al. (2016)
Klebsiella pneumoniae	Human	2016	Brazil	*mcr-1*	Aires et al. (2017)
Salmonella Typhimurium	Human (fecal matter, vaginal discharge, urine, blood and abscess)	2013–16	Colombia	*mcr-1*	Saavedra et al. (2017)
E. coli	Water from public beaches	2016	Brazil	*mcr-1*	Zhang et al. (2016)
E. coli	Swine fecal sample	2017	United States	*mcr-1*	Meinersmann et al. (2017)
E. coli	Swine	2019	Eastern Canada	*mcr-1*	Pilote et al. (2019)
K. pneumoniae	Human	2015	Argentina	*mcr-1*	Quiroga et al. (2019)
E. coli	Human	2017	Mexico	*mcr-1*	Merida-Vieyra et al. (2019)
E. coli	Pig stool	2015	Mexico	*mcr-1*	Merida-Vieyra et al. (2019)
E. coli	Broiler chicken farms	2013–14	Ecuador	*mcr-1*	Vinueza-Burgos et al. (2019)
E. coli	Human and swine	2015	Venezuela	*mcr-1*	Delgado-Blas et al. (2016)

Merida-Vieyra et al. (2019) reported the first *E. coli* (EC-PAG-733) isolate holding the *mcr-1* in Mexico from a fecal sample of a child with cancer. The gene was located on an IncI2 plasmid. Additionally, in Mexico, screening of pig stool samples identified an *E. coli* isolate (ST44) that harbored *mcr-1*.

In a study conducted in Colombia (Saavedra et al., 2017), 5,887 *E. coli* isolates were screened and 513 were found resistant to polymyxins. Two of these isolates harbored *mcr-1* on the chromosome, while the other *mcr-1*-positive isolates carried the gene on plasmids. Consequently, this further suggested that *mcr-1* might be mainly circulating in Colombia via transferable plasmids among colistin-resistant *Enterobacteriaceae*.

In Venezuela, Delgado-Blas et al. (2016) detected the first *mcr-1* in the country in human and swine feces. Among the 93 samples of *E. coli*, two (BB1290 and BB1291) were collected from a 43-year-old man and a pig, respectively. Analysis demonstrated a 100% similarity to *mcr-1* on an IncI2 plasmid. Notably, in addition to *mcr-1*, the human isolate carried several genes that encode resistance to aminoglycosides, beta-lactams, macrolides, sulfonamides, tetracycline, trimethoprim, and fluoroquinolones. However, the plasmids mediated resistance to colistin only. Although *mcr-1* was identified from the patient's feces, it was reported that the patient had no contact with animals, did not travel, and wasn't treated with colistin. Therefore, the source remains unknown, which highlights the need for further investigation.

In addition to using colistin against *Enterobacteriaceae*, the drug has been increasingly administered to treat infections with *Acinetobacter baumannii* and *Pseudomonas aeruginosa*. In Latin America, colistin resistance observed in these species was estimated at 1.5% of all the isolates tested.

10.3.2 Middle East and North Africa (MENA) Region

Despite many challenges in infrastructure and antimicrobial stewardship, and including several countries with political instability and civil wars, the MENA region remains among the least studied and investigated in terms of the spread of *mcr* and colistin resistance. However, the wide dissemination *mcr-1* to different parts of the world and its detection in humans, animals, and the environment, it became a necessity to study and monitor the emergence and prevalence of this mobile genetic marker in MENA countries. This is crucial to curb the global spread of resistance, partially due to the fact that the area hosts the largest religious gatherings in the world that are attended by people from all nations. Therefore, the possibilities of these people to be exposed and spread resistance to their countries should be considered.

10.3.2.1 Algeria

The mobile colistin resistance gene (*mcr-1*) was first detected on poultry chicken farms in Algeria (Arcilla et al., 2016). This genetic marker was in an *E. coli* isolated from chicken fecal matter. Another study in 2018 identified the relatively high dissemination of *mcr-1* in different poultry farms across Algeria, where it was detected in 20.6% of the chicken fecal samples (Chabou et al., 2018). These studies indicated the presence and emergence of this genetic marker in *E. coli* across many poultry farms in Algeria. Although *mcr-1* was detected in poultry farms, no further studies were conducted to assess the burden and the spread of this marker in the farm environment and beyond. Poultry farms can easily contaminate nearby water sources and soils via the use of farm waste as fertilizers in sustainable agriculture practices. Also, no studies were conducted on other food-producing animals to assess the dissemination of *mcr-1* to other farm-animals in Algeria. Therefore, there is a significant gap in understanding the prevalence of *mcr* in agriculture and food-producing animals in Algeria.

Yanat et al. (2016) reported that the mobile colistin resistance gene was also discovered in two human clinical cases in Algerian hospitals. The first case was associated with an *E. coli* (ST405) isolate harboring *mcr-1*, and it was retrieved from a sperm sample of a 29-year-old male. Further analysis showed that *mcr-1* was located on an incompatibility plasmid that was highly transmissible to other naïve *E. coli* as confirmed by conjugation studies. The high transferability frequency of this plasmid highlights the potential risk of the fast dissemination of the genetic marker to other bacteria in Algeria. The *E. coli* isolate also harbored other genes that encoded resistance to cephalosporins. The second case was a retrospective study conducted in 2016, where six colistin-resistant *E. coli* isolates were further tested for *mcr*. One of the six colistin-resistant isolates was found positive for *mcr-1*. This *E. coli* was isolated from a urine sample in 2011 from an 18-year-old patient with no travel history outside Algeria (Berrazeg et al., 2016). Detecting *mcr-1* in a sample from 2011 shows that this genetic marker was found in the Algerian community a long time ago, but due to the lack of robust surveillance and screening, this marker was not readily identified.

In 2018, a study conducted on wild Barbary Macaques in Algeria showed that one out of the 86 fecal samples collected was *mcr-1* positive (Bachiri et al., 2018). This gene was in an *E. coli* ST405, which was also resistant to different classes of antibiotics, including beta-lactams, aminoglycosides, and fluoroquinolones. The detection of *mcr-1* in *E. coli* from wildlife in Algeria exacerbates the problem, because wildlife plays a role in zoonosis and is a source to further disseminate resistance markers to new and pristine environments through fecal contamination. Drali et al. (2018) showed that *mcr-1* was also detected in two out of the 256 isolates collected from seawater in western and eastern Algeria. The strains harboring the gene were identified as *E. coli* belonging to different ST types, ST23 and ST115, indicating that *mcr-1* was found in different backgrounds in Algeria. These two beaches were contaminated from hospitals, and agricultural and industrial wastes, and swimming was prohibited. *mcr-1* in water sources is also highly problematic, because water is an ideal vehicle of transmission and spread of these genes locally and to nearby countries.

Taken together, *mcr-1* was detected in poultry, humans, wildlife, and environmental samples in Algeria, highlighting potential public health risks. Other *mcr* were not screened for and the source of *mcr* in Algeria remains under investigation, indicating the need for further studies.

10.3.2.2 Lebanon

In Lebanon, colistin resistance was briefly studied in clinical settings and was found to be associated with chromosomal mutations that are not readily transmissible between species. Recently, a few studies have been conducted in an effort to determine the spread of *mcr* in Lebanon, a country that lacks baseline data on antimicrobial resistance in nonclinical settings. Three papers published in 2017 and 2018 highlight that Lebanon might potentially have a high prevalence of *mcr-1*, especially in the agricultural sector. Dandachi et al. (2018a) assessed the prevalence of ESBL and colistin resistance in swine farms in Lebanon. One hundred and fourteen fecal samples were collected from pig farms in Lebanon and 111 *E. coli* were subsequently isolated. The majority (98%) of the *E. coli* isolated were identified as ESBL and 20.7% were positive for *mcr-1*. Furthermore, two studies reported the detection of *E. coli* harboring *mcr-1* in poultry (Hmede and Kassem, 2018; Dandachi et al., 2018b). In one study, the majority (98%) of the colistin-resistant *E. coli* tested were positive for *mcr-1*; the other 2% were negative for *mcr 1* to *3* (Hmede and Kassem, 2018). All of the *E. coli* isolates were also multidrug-resistant (MDR). The numbers reported in the literature on mobile colistin resistance in Lebanon are very high in poultry. The latter suggested that the high levels of *mcr-1* in Lebanon might be mainly due to weak regulations and enforcement on the proper use of colistin in poultry farming. Although colistin is a last resort antibiotic, colistin and colistin-based medications were readily available in veterinary stores and are extensively used in animal production as reported by a study conducted in 2018 in Lebanon (Kassem et al., 2019). In, it appears that there is a high reliance on colistin to treat complicated human infections (Kassem et al., 2019). Furthermore, in recent studies, *mcr-1*-positive

E. coli were detected in a high proportion of irrigation water samples and in the environment (drinking water, well water, and sewage) of Syrian War refugee camps that are hosted in Lebanon (Hmede et al., 2019; Alhaj Sulaiman and Kassem, 2019a,b). The lack of data on *mcr* in humans does not imply an absence of the genes in the Lebanese population, because the high levels detected in animal production and environment might be an indicator of the exposure of the population to this genetic element. Therefore, more data and studies are needed to investigate mobile colistin resistance, especially the detection and screening of *mcr* and its variants in the population and in agricultural settings.

10.3.2.3 Egypt

In Egypt, 241 clinical samples of gram-negative bacteria were collected from a hospital in Cairo in 2015 and screened for the presence of *mcr-1*. One sample that was collected from the sputum of a patient in the intensive care unit with no history of traveling abroad tested positive. The isolated bacterium was identified as *E. coli* and was resistant to colistin as well as ciprofloxacin, kanamycin, tetracycline, ampicillin, cefotaxime, and nalidixic acid (Elnahriry et al., 2016). This *E. coli* belonged to ST1011, which was also associated with avian fecal *E. coli* in Egypt. The latter suggested a possible zoonotic transfer of the *mcr-1*-positive strain from animals to humans. The detection of *mcr* in a human isolate in Egypt implied the potential spread of this mobile genetic element in the community, which will further jeopardize and limit treatment options. Notably, this study was the first report on *mcr-1* in the African continent (Elnahriry et al., 2016). However, no further studies were done to assess the burden and source of *mcr-1* in the population or hospitalized patients and other *mcr* were not tested.

Following the detection of *mcr-1* in humans, the gene was identified in a cow sample in 2016. In that study, 38 isolates were tested and only one was identified as *mcr-1* positive. MLST typing showed that the *mcr-1*-positive isolate was *E. coli* ST10 that was also resistant to other antimicrobial agents like aminoglycosides, quinolones, chloramphenicol, ampicillin, amoxicillin, and tetracycline. Notably, the cow suffered from subclinical mastitis (Khalifa et al., 2016).

In 2018, 576 fecal samples were collected from broiler chickens from several farms along the Nile delta in Egypt. The majority (87.5%) of the isolated bacteria were identified as *E. coli*, out of which 12.5% were ESBL-producing, 1.8% were ESBL-producing and carbapenems resistant, and 7.9% were phenotypically and genetically resistant to colistin, harboring *mcr-1* (Moawad et al., 2018). The latter was the first detection of *mcr-1* in poultry in Egypt. The presence of these farms in areas near the Nile delta poses a risk of contamination of water. Therefore, more efforts and studies should be done to test for environmental contamination as well as the burden of the use of colistin in the agricultural sector in Egypt.

In 2018, a study detected the presence of *mcr-1* in raw milk products in Egypt (Hammad et al., 2019). Specifically, 200 karisha cheeses prepared from raw milk were collected. The data showed that *mcr-1* was present in four out of the 200 cheese samples. *E. coli* strains harbored the *mcr-1* and belonged to three different phylogenetic groups, which highlights that *mcr-1* occurs in diverse genetic backgrounds in Egypt.

10.3.2.4 Tunisia, Morocco, and Sudan

In 2016, a study was conducted in Tunisia to screen for the presence of *mcr-1* in poultry. Fifty-two samples were collected from three farms. All the *E. coli* isolates detected were ESBL-producing and carried bla_{CTX-M}. Some *E. coli* also harbored *mcr-1* (Grami et al., 2016). This finding was the first to assess the presence of *mcr-1* in ESBL producing *E. coli* in Tunisia. Moreover, this highlighted the prevalence of *mcr-1* in poultry farms in Tunisia. Another study confirmed the previous findings and detected *mcr-1* in poultry, where 4% of the cefotaxime resistant *E. coli* also harbored *mcr-1* (Maamar et al., 2018).

In Morocco, mcr-1 was detected in E. coli isolated from fecal samples of septicemic broilers chickens (Rahmatallah et al., 2018). Only three isolates tested positive for mcr-1. Notably, a study was conducted in 2016 to determine the presence of *mcr-1* in pilgrims in Mecca (Saudi Arabia). One hundred eighty-two fecal samples were tested in 2013 and 2014. Two of the *mcr-1*-positive isolates were detected from Moroccans (Leangapichart et al., 2016). The origin of this gene was not identified and whether it was contracted during the Hajj is still unknown. Since Hajj is usually crowded; this setting can act as a reservoir and a hub for the dissemination and spread of resistance genes. The two *mcr-1*-positive individuals can also transmit this mobile gene to their home countries and communities.

In Sudan, only one study was conducted to detect the presence of *mcr-1* in a clinical setting. In 2017, 50 *Enterobacteriaceae* isolates were collected from a hospital in Khartoum and screened for the presence of *mcr-1*. The majority of the isolates were identified as *E. coli* (76%). Seven *E. coli* isolates (14%) were positive for *mcr-1*. This was the first and only report on *mcr-1* in Sudan.

Taken together, more extensive studies are needed to determine the prevalence of *mcr* in food-producing animals and the environment in Tunisia, Morocco, and Sudan (Altayb et al., 2018).

10.3.2.5 The Gulf Region

In Saudi Arabia, only two papers report on *mcr* in humans, identifying *mcr-1* and *mcr-5*, respectively. The latter is the first time *mcr-5* was reported in the MENA region. *mcr-5* was detected in *E. coli* with no further clarification on the origin and characteristics of this gene (Redhwan et al., 2019). This highlighted the need to expand screening in the region, because the absence of *mcr-1* cannot imply the lack of mobile elements conferring colistin resistance. The *mcr-1* positive isolate was an *E. coli* from a blood sample collected in 2012 (Sonnevend et al., 2016). This *E. coli* isolate was a multidrug-resistant strain that also harbored bla_{NDM-1}, indicating that it was both colistin and carbapenem resistant. Therefore, this strain was resistant to antibiotics that are commonly used to treat MDR infections. Also, the plasmid harboring *mcr-1* was classified as IncHI2, which was previously detected in animals, probably highlighting a possible route of transmission to humans.

In Qatar, out of the 17 two fecal samples collected from two poultry farms, 90 isolates were identified as *E. coli* and 14 (15.5%) were *mcr-1* positive, while two were also ESBL producers (Eltai et al., 2018). Also, *mcr-1* was detected in pooled *E. coli* isolates that originated from humans in Qatar (Redhwan et al., 2019). Therefore, *mcr-1* was detected in both animals and humans in Qatar.

In Oman, 22 colistin-resistant *E. coli* isolates from a clinical setting were screened for *mcr-1* and *mcr-2* (Mohsin et al., 2018). A single *E. coli* harbored *mcr-1*, and no isolates were positive for *mcr-2*. Further analysis and sequencing of the plasmid harboring *mcr-1* did not detect other resistant genes. MLST sequencing showed that the *E. coli* belonged to ST10 that has been detected worldwide and is known to spread ESBL and quinolone resistance.

In the United Arab Emirates, only one study reported *mcr-1* in a blood sample form a hospitalized patient in 2013. The *mcr-1* was detected in *E. coli* and was carried on an IncI2 plasmid. This *E. coli* harbored other resistant genes like bla_{CTX-M} (Sonnevend et al., 2016). The same study reported two other *E. coli* isolates that were *mcr-1* positive. These isolates were retrieved from a blood sample in 2012 and a wound infection in 2015, respectively.

In Kuwait, no published studies on the emergence of *mcr* or its variants are reported in the literature.

10.3.2.6 Jordan, Turkey, Iran, Iraq, Syria, Palestine, Libya, and Yemen

In Jordan, only one study investigated the occurrence of *mcr* in human isolates. In this study, multiplex PCR was used to screen for *mcr* in more than 1,000 gram-negative human isolates that included *Klebsiella* and *E. coli* from Jordan, Egypt and the Gulf region. Pooled *E. coli* and *Klebsiella* from Jordan were positive for *mcr-1*, which indicated that the gene was already found in humans in Jordan (Redhwan et al., 2019). Further studies need to be conducted for full analysis of the burden of *mcr* in humans and agricultural settings and to test the source of these genes in Jordan.

In Turkey, a recent study targeted poultry meat and retail poultry meat (n = 80) in two different provinces to assess the presence of *mcr-1* in *E. coli*. The gene was detected in four *E. coli* strains that belonged to different sequence types. Also, these strains harbored other resistance genes. This was the first detection of *mcr-1* in poultry meat in Turkey (Kurekci et al., 2018).

No data were found on the detection of *mcr*-related colistin-resistance in Iran and the studies linked colistin resistance to chromosomal mutations. No data are available on *mcr* in Iraq, Syria, Palestine, Libya, and Yemen.

Taken together, studies on *mcr* in the MENA region are still very limited (in number and scope) and are mainly restricted to screening for *mcr-1*. *mcr* was only detected in environmental samples in Algeria and in Lebanon. Therefore, emphasis should be placed on studying environmental samples to assess if transmission is occurring in the animal farm—environment—human continuum. With the increased emergence of multidrug-resistant gram-negative bacterial pathogens, treatment options are becoming more restricted. Also, with increased emergence of carbapenem-resistance genes, the need for colistin is more urgent. This is especially true for countries in the MENA that have been affected by war and have been stressed by war-induced immigration and political unrest. Furthermore, many MENA countries suffer from lack of resources and infrastructure, while antimicrobial stewardship is still underdeveloped (Fig. 10.1).

10.3.3 Europe

After the detection of *mcr-1* in southern China (Liu et al., 2016), European countries started to conduct surveillance and retrospective studies to screen for the presence of this gene and other variants. Poirel et al. (2017) speculated that the mobile colistin-resistant genes have emerged and spread due to the excessive use of colistin in animal treatment, prophylaxis, and growth promotion in Asian countries. However, while some countries in Europe have banned the use of colistin as a growth promoter, countries like Spain and Italy still use it for other purposes in animal production, necessitating studying local factors and the role of different European countries in the emergence and spread of these genes.

10.3.3.1 Belgium

In Belgium, Xavier et al. (2016b) conducted a surveillance study on 105 colistin-resistant *E. coli* samples from 53 porcine and 52 bovine animals suffering from diarrheal diseases in 2011—12. The results demonstrated that 12.5% (n = 13) were positive for *mcr-1*, out of which 11.5% were bovine

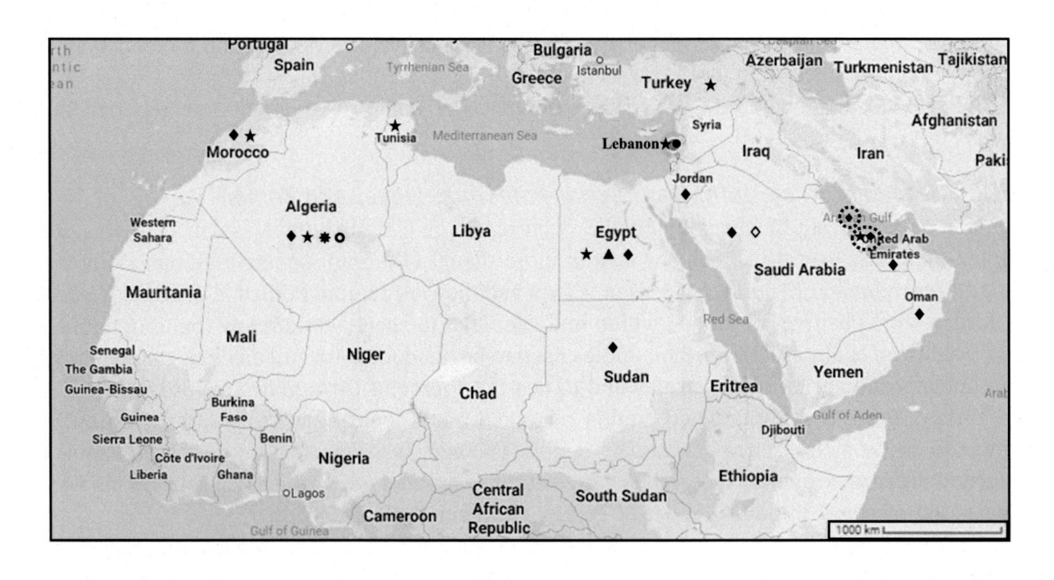

◆ Human (*mcr-1*)
◇ *mcr-5*
▲ Cow
★ Chicken
O Macaque
✱ Sea water
● Swine
Bahrain
Qatar

FIGURE 10.1

Mapping of *mcr* genes detected in the MENA region (Lebanon, Oman, Yemen, UAE, Saudi Arabi, Iraq, Qatar, Kuwait, Bahrain, Jordan, Syria, Palestine, Egypt, Djibouti, Sudan, Libya, Tunisia, Algeria, Mauritania, and Western Sahara) and their sources. Different shapes on the map indicate different sources of *mcr*. This is the first attempt to analyze and map these genes in the MENA region.

(n = 6) and 13.2% were porcine (n = 7) samples. Furthermore, the *mcr-1* negative isolates were tested for other *mcr* genes; *mcr-2* was detected in these samples, and it showed 76.2% homology to *mcr-1* and was harbored on an IncX4 plasmid. This *mcr* was also shown to be transferable to colistin-susceptible *E. coli* through conjugation studies, and it resulted in higher MIC values in the trans-conjugants. *mcr-2* was detected in 20.1% of the porcine isolates. The relatively high prevalence of *mcr-2* was thought to be due to the higher transferability frequencies of the IncX4 plasmid harboring *mcr-2* in comparison to the IncFII harboring *mcr-1* (Xavier et al., 2016a, 2016b). A study conducted in 2017 detected another new *mcr* in *E. coli* collected from pigs suffering from postweaning diarrhea. Specifically, samples from Belgian pigs were taken from Belgian pigs and were tested for *mcr 1* to *4*. Fifteen isolates were identified as colistin-resistant and 13.3% harbored *mcr-4*. Therefore, these reports identify the prevalence of *mcr-1*, *mcr-2*, and *mcr-4* in *E. coli* from pigs and cows in Belgium (Carattoli et al., 2017). Further studies should be conducted to investigate if these mobile elements are found in humans and/or environmental samples.

10.3.3.2 Poland

In Poland, *mcr-1* was detected in a female patient who was hospitalized due to a urinary tract infection and pneumonia. *mcr-1* was found in an *E. coli* isolate that was also resistant to different antibiotics, including cephalosporins, penicillin, amoxicillin, clavulanic acid, tetracyclines, and aztreonam. The MLST sequencing showed that the *E. coli* strain belonged to ST617 (Izdebski et al., 2016). This finding is the only report on *mcr-1* in Poland.

10.3.3.3 Sweden

In Sweden, *mcr-1* was detected in *E. coli* from human fecal samples. Two *E. coli* isolates were shown to be positive for *mcr-1*; however, the individuals' histories showed that they had previously traveled to Asia. Therefore, the gene might have been acquired outside of Sweden (Skov and Monnet, 2016).

10.3.3.4 Finland

In Finland, colistin use is limited to pathogens that are resistant to carbapenems (Gröndahl-Yli-Hannuksela et al., 2018). After the detection of *mcr-1* in different parts of the world, Finland conducted a prospective study to screen for *mcr-1* in clinical human fecal samples collected from healthy volunteers. One isolate from a 26-year-old volunteer was shown to be positive for *mcr-1* (Gröndahl-Yli-Hannuksela et al., 2018). The isolate was identified as a non-ESBL-producing *E. coli*. This highlighted that *mcr-1* can also be detected in countries with limited use of colistin.

10.3.3.5 Norway

In Norway, *mcr-1* was detected in a nonpathogenic *E. coli* identified through a retrospective study. This *E. coli* isolate was from the fecal matter of a patient who previously traveled to India and suffered from travelers' diarrhea (Solheim et al., 2016). Also, another study by Jørgensen et al. (2017) reported the presence of *mcr-1* in environmental samples. Seawater samples tested positive for *E. coli* ST10 harboring *mcr-1*. This study highlighted that colistin-resistant genes were also found in environmental samples in Norway. The source of this gene has not been identified; however, it is speculated that it might be due to fecal contamination from ship toilets, migratory birds, or nearby farms. This detection is of immense importance, because water plays a significant role in the transfer of resistant markers to different niches and environments.

10.3.3.6 Denmark

In Denmark, *mcr-1* harboring *E. coli* were isolated from a patient with a bloodstream infection as well as from five retail chicken samples. The *mcr-1*-positive *E. coli* isolated from the human also carried beta-lactamase genes, $bla_{\text{CTX-M}}$ and *ampC*. The *E. coli* isolated from the chicken samples harbored either $bla_{\text{SHV-12}}$ *or ampC* or both. MLST sequencing showed that one of the chicken *mcr-1*-positive *E. coli* isolates belonged to ST131 (Hasman et al., 2015). This sequence type is usually associated with human urinary tract and bloodstream infections and is rarely detected in chickens. The detection of the *mcr-1* in this *E. coli* strain raises further concerns on the transmission of the mobile elements between hosts and the cognate evolution of resistance in agriculturally and clinically important strains.

In 2017, a study was conducted on ESBL/*ampC E. coli* isolates to check for the cooccurrence of *mcr* and beta-lactamase genes. One sample was identified from a hospitalized patient with pyelonephritis. The *E. coli* isolated was *mcr-3* positive, and this was the first detected case of *mcr-3* outside of Asia (Roer et al., 2017). The *mcr-3* positive patient had a previous travel history to Thailand; however, the origin of the gene has not been identified. The *E. coli* belonged to ST131, a sequence type that was previously detected in Danish patients.

In Denmark, the *mcr* genes were also detected in *Salmonella.* In a study conducted in 2016, *mcr-1* was identified in four *Salmonella* Typhimurium isolates from humans. One of the *mcr-1* positive *Salmonella* carriers had a previous travel history to Thailand (Torpdahl et al., 2017). Litrup et al. (2017) also reported *mcr* genes in *Salmonella* isolates from Denmark. In that study, 10 isolates were found to be *mcr* positive, out of which one harbored both *mcr-1* and *mcr-3*, while the other isolates were positive for *mcr-3.* Eight of the 10 *mcr-3*-positive isolates carried also ESBL genes. Notably, these isolates were retrieved from people who previously traveled to Asia. Therefore, most of the *mcr-3* positive isolates were associated with patients that had visited Asia; perhaps indicating that the gene might have been acquired during travel to Asia. If true, this further suggests an important role of travel and human movement in the transfer and the dissemination of these resistance genes to different parts of the world; even to countries where low prevalence of *mcr* is expected. This also highlights the difficulty of controlling this problem due to continuous human, animal, and food movement.

10.3.3.7 Spain

In 2016, *mcr-1* was detected in 15 *E. coli* isolates collected from human clinical samples in Spanish hospitals (Prim et al., 2016). A recent report identified five multidrug resistant (MDR) *mcr-1*-positive *E. coli* isolates in patients suffering from adult acute myeloid leukemia (AML). This is highly problematic, because leukemic patients have low immunity, and they are highly susceptible to pathogens. Therefore, the presence of multidrug-resistant *E. coli* harboring *mcr-1* might limit treatment options and colistin effectiveness in these individuals (Lalaoui et al., 2019).

Reports have confirmed the presence of *mcr-1* in sewage water from wastewater treatment plants in Spain. A large number (33.3%) of the tested isolates was *mcr-1* positive. Of these isolates, nine were *E. coli* and one was *K. pneumonia.* Both species harbored $bla_{\text{CTX-M-55}}$ and $bla_{\text{TEM-1}}$ as well. The isolates were resistant to different classes of antibiotics, including quinolones, aminoglycosides, beta-lactams, and third-generation cephalosporins (Ovejero et al., 2017). The detection of *mcr-1* in environmental samples, like sewage water, is highly problematic, because this plays a significant role in the dissemination of pathogens and genetic elements that may encode resistance.

mcr-1 and other *mcr* genes also reported in animal production in Spain. A study on colistin resistance in commensal bacteria from healthy animals in slaughterhouses revealed that 23 *E. coli* isolates from pig samples harbored *mcr-1* (El Garch et al., 2018). Another study reported the presence of *mcr-1* in both *E. coli* and *Salmonella* isolates from different animal farms across Spain. Specifically, nine *mcr-1*-positive isolates were detected; out of which five were *E. coli* isolated from turkey feces (n = 1) and swine feces (n = 4), while four *Salmonella* isolates were retrieved from swine lymph nodes (Quesada et al., 2016). Other *mcr* were detected in pigs that suffered from postweaning diarrhea in Spain; notably nine isolates harbored *mcr-4* (Carattoli et al., 2017). Additionally, enteropathogenic *E. coli* isolated from post-weaning swine were collected between 2006 and 2017 and screened for colistin-resistance and for *mcr- 1* to *5*. The study showed that 140 isolates were *mcr* positive; with a high prevalence of *mcr-4* (n=102 isolates), while 37 isolates were *mcr-1*-positive and 5 isolates harbored *mcr-5*. The latter was the first discovery of *mcr-5* in Spain. Notably, different *mcr* genes; for example, *mcr-1/mcr-4*, *mcr-1/mcr-5* and *mcr-4/mcr-5* were shown to co-occur in the same isolates, respectively (García et al., 2018). Furthermore, screening of young calves in Spain resulted in the isolation of 6 *E. coli* that were found to be positive for *mcr-1* and one of the isolates was positive for both *mcr-1* and *mcr-3* (Hernández et al., 2017).

Taken together, data suggest that Spain is one of the European countries with relatively wide detection of *mcr* genes in humans, animals, and the environment. Also, different *mcr* genes (*mcr-1, mcr-3, mcr-4, mcr-5*) were detected and were associated with disparate bacterial species, including *E. coli, K. pneumonia*, and *Salmonella*. Perhaps, these observations could collectively be associated with Spain's overall consumption of colistin as well as the 14-fold increase in the use of this antibiotic between 2007 and 2014 (Curcio et al., 2017; Prim et al., 2016).

10.3.3.8 Italy

Italy and Spain's consumption of colistin is estimated to be 20 mg/kg in biomass, which is a marked difference in comparison to northern European countries (Curcio et al., 2017). Given that post-weaning diarrhea in pigs is mainly treated with colistin in Italy, a retrospective study was conducted to screen for colistin resistance and *mcr-1* in *E. coli* isolated from pigs. The majority (72.5%) of the *E. coli* were positive for *mcr-1*; while the other isolates (27.5 %) were negative for *mcr-1* and *mcr-2*. Another *mcr* gene was detected in a pig isolate in Italy in 2013. This isolate was identified to be a colistin-resistant *Salmonella enterica* serovar Typhimurium that harbored *mcr-4* (Carattoli et al., 2017). Therefore, like Spain, the relatively high prevalence of *mcr-1* in pigs in Italy might be associated with the use of colistin in animal production in this country (Curcio et al., 2017).

In the clinical setting, an *mcr-1* variant, *mcr-1.2*, was identified in an Italian hospital (Di Pilato et al., 2016). In this study, carbapenem-resistant *K. pneumonia* was tested for the presence of mobile colistin resistance. Notably, the carbapenem-resistant *K. pneumonia* was isolated from a leukemic child. Although the strain tested positive for *mcr-1*, the gene differed from the already known *mcr-1* by a single amino acid and was designated as *mcr-1.2*. This was the first description of a carbapenem resistant *K. pneumonia* that harbored *mcr-1.2*. Furthermore, the strain was MDR and belonged to ST 515, which is known to be problematic. This discovery was important, because it highlighted the presence of an MDR strain that was resistant to critically important antibiotics (carbapenems and colistin) in an immunocompromised individual (Di Pilato et al., 2016). Another study reported the presence of the *mcr-1* variant *(mcr-1.2)* in an *E. coli* isolated from human blood samples (Simoni et al., 2018).

10.3.3.9 Germany

In Germany, 577 isolates were collected from human, animal and environmental samples and screened for *mcr-1*. Four *E. coli* isolates were found to be *mcr-1* positive; out of which 3 were retrieved from swine samples, while the fourth isolate was detected in a human sample. The latter *E. coli* isolate also harbored the carbapenem resistance gene, bla_{KPC-2} (Falgenhauer et al., 2016). In another study, 10,609 *E. coli* were collected from different farms, slaughterhouses, and retail meat products between 2010 and 2015 and screened for the presence of *mcr-1*. Five hundred and five isolates were phenotypically resistant to colistin; out of which 402 harbored *mcr-1*. The highest prevalence of *mcr-1* was observed in turkeys (11.8%), broiler chickens (6.7%), and chicken meat (4.3%) (Irrgang et al., 2016). In contrast, relatively lower prevalence of *mcr-1* was noted in veal calves (2.4%), while only 3 *mcr-1* positive isolates were detected in layer chickens, and no *mcr-1* was found in cattle samples. Interestingly, *mcr-1* was detected in Germany in a sample from 2010, indicating that the gene was already present but was undetected until screening efforts were prompted by the discovery of *mcr-1* in China. Therefore, a relatively high prevalence of *mcr-1* in turkeys and broiler chickens were detected in Germany. However, trends of *mcr-1* incidences were not homogeneous; perhaps due to the 15.8% decrease in colistin use in Germany between 2011 and 2014.

10.3.3.10 Netherlands

In the Netherlands, relatively low use of colistin has been reported in both human medicine and animal farming applications. Polymyxins accounted for 0.4% of the antibiotics that were administered to broilers in 2014 (Kluytmans-van den Bergh et al., 2016). A similar percentage was observed in primary care and hospital settings, where polymyxins accounted for 0.1% and approximately 0.3% of all systemic antimicrobials, respectively (Kluytmans-van den Bergh et al., 2016). Subsequently, a relatively low prevalence of *mcr-1* was also reported in the Netherlands. Notably, a retrospective screening of *mcr-1* in a collection of 2,471 *Enterobacteriaceae* isolates from retail chicken meat, hospitalized patients, clinical cultures, and healthcare associated outbreaks was conducted in 2015. Only 3 ESBL-producing *E. coli* isolates from retail chicken meat were found to harbor *mcr-1*. One of these isolates was collected in 2009, while the other two in 2014. Kluytmans-van den Bergh et al., 2016 also reported that *mcr-1* was not identified in the human isolates tested. However, a study screened ESBL-producing and colistin-resistant *E. coli* that were collected between 2012-2013 from 9 travelers 1–2 weeks after their return to the Netherlands (Arcilla et al., 2016). *mcr-1* was found in 6 of the aforementioned *E. coli* isolates. These travelers had visited Thailand, Vietnam, Cambodia, Laos, Tunisia, Peru, Bolivia, Colombia, and China. Given that the ESBL-producing *E. coli* were absent in the tested individuals before travel, it was assumed that these *mcr-1* containing isolates might have been acquired during travel (Arcilla et al., 2016). The relatively low prevalence of *mcr-1* in the Netherlands was confirmed in other studies. For example, screening of *E. coli* and *Salmonella* isolates collected from food animals at slaughter between 2002 and 2014 showed the presence of *mcr-1* in 5 *E. coli* out of the 1,547 isolates tested (El Garch et al., 2018). Another study reported the detection of *mcr-1* in 1% of *Salmonella* from poultry meat (Veldman et al., 2016), while 0.3% of fecal samples from healthy animals were *mcr-1* positive (El Garch et al., 2018).

10.3.3.11 Switzerland

In Switzerland, *mcr-1* was reported in two *E. coli* associated with 2 different cases of bacteremia (Nordmann et al., 2016a). Both strains were found to harbor the beta-lactamase genes bla_{TEM-1} and bla_{TEM-52}, respectively. Subsequently, a retrospective study conducted on 257 non-duplicated isolates from human blood cultures showed that *mcr-1* was absent in these bacteria, which included important species such as *E. coli*, *K, pneumoniae*, and *Salmonella* serovar enteritidis (Nordmann et al., 2016b). However, *mcr-1*-positive *E. coli* (belonging to ST10 and ST5) were identified in fecal samples collected in 2015 from two Swiss travelers returning from India and one HIV positive individual (Donà et al.,

2017). Furthermore, screening of 2,049 non-duplicate isolates from human urine samples identified *mcr-1* in an *E. coli* strain that belonged to sequence type ST428, which was previously associated with infections in broiler breeders (Liassine et al., 2016). Interestingly, the first case of *mcr-1*-positive *Salmonella enterica* subsp. *enterica* isolated from the blood of a 77 years-old male patient was detected in Switzerland in 2017 (Carroll et al., 2018). *mcr-1* was also reported in six colistin-resistant *E. coli* isolated from retail chicken meat, which were imported from Germany (Donà et al., 2017). Furthermore, the gene was detected in 1 and 2 ESBL-producing *Enterobacteriaceae* isolated from the river Birs and from vegetables imported from Thailand and Vietnam, respectively (Zurfuh et al., 2016). These finding are of notable importance, because they highlight 1) potential environmental (river water) contamination/reservoirs and 2) the role of food imports in the transmission of the gene to countries with antimicrobial stewardship policies. Therefore, the role of travel and trade in the global transmission of colistin-resistant bacteria and *mcr* must not be overlooked.

10.3.3.12 France

Mobile colistin resistance was reported in humans, animals, and food products in France (Baron et al., 2016). These genetic determinants of colistin resistance also occurred in different bacterial species, including *K. pneumonia, E. coli, Enterobacter cloacae*, and *Salmonella* spp. Relatively high levels of colistin-resistance and *mcr-1* were detected in ESBL-producing *E. coli* isolated from veal calves suffering from diarrheal disease. Of the 517 isolates tested, 106 (21%) were *mcr-1* positive (Haenni et al., 2016). Notably, *mcr-1* was detected in an *E. coli* that dated back to 2005. Furthermore, evaluation of colistin resistance in 1,696 commensal *E. coli* isolates collected from French livestock between 2007 and 2014 identified *mcr-1* in pigs (0.5% prevalence), broiler chickens (1.8%) and turkeys (5.9%) (Perrin-Guyomard et al., 2016). In food products, *mcr-1* was detected in 4 *Salmonella* that were isolated from sausages, retail chicken products and ready-to-eat pies (Arcilla et al., 2016). *mcr-1* was identified in two colistin-resistant *K. pneumoniae* isolated from humans in France (1 from Marseille and 1 from Angers) (Rolain et al., 2016). Additionally, a study on the fecal carriage of colistin-resistant bacteria in a university hospital in western France reported that acquired resistance was low (1.4%) (Saly et al., 2017). However, an *mcr-1*-positive ESBL-producing *K. pneumonia* that exhibited high resistance to colistin was isolated from a patient with fungal meningitis (Caspar et al., 2017). Furthermore, *mcr-1* was detected in an ESBL-producing *E. cloacae* that was isolated from a 50-year-old Algerian women hospitalized in France. Given that *mcr-1*-positive *E. cloacae* has only been previously described in Asia, the gene and the bacterium might have been acquired outside of France. Regardless, this further suggested that people's movement (travel) can potentially contribute to the dissemination of these genetic markers (Baron et al., 2017).

10.3.3.13 Portugal

In Portugal, colistin is used in food animal production. This might possibly contribute to the relatively high levels of colistin resistance especially in pigs. *mcr-1* was reported in *Salmonella* spp. isolated from pigs, pork products, and human samples (Campos et al., 2016). In another study, *mcr-1* was detected in 98 isolates (*E. coli* and *K. pneumoniae*) that were retrieved from 100 pigs on two farms (Kieffer et al., 2017). Interestingly, six weeks before the study, colistin was used in feed for metaphylaxis on these two pig farms (Kieffer et al., 2017). A nationwide study was conducted in Portugal to screen for *mcr-1* in 1,840 *Enterobacteriaceae* isolated from food-producing animals (bovine, swine and poultry), meat (bovine and swine), meat products and animal feed. *mcr-1* was found in 8% of the *E. coli* and 0.47% of the *Salmonella enterica* isolates (Clemente et al., 2019). Notably, 45.7% of the *mcr-1*-positive *E. coli* were either ESBL or plasmid-mediated AmpC beta-lactamase co-producers. Furthermore, *mcr-1* was also detected in *E. coli* from turkeys (27% of the isolates), swine (10.1%), swine meat (5.1%) and broilers (2%) (Clemente et al., 2019). A notable finding was the detection of *mcr-1* in carbapenemase-producing *K. pneumoniae* in hospitalized patients in Portugal (Mendes et al., 2018). Specifically, 24 carbapenemase-producing and *mcr-1*-positive *K. pneumoniae* isolates were retrieved from 16 hospitalized patients (out of 5,361 patients screened for carbapenemase producers). Of these isolates, 17 were retrieved from the patients' gastrointestinal tract (colonizers), while the rest of the isolates were found in different samples, including urine, blood, and other biological fluids (Mendes et al., 2018). Given that colistin is administered for the treatment of carbapenem-resistant bacteria, the occurrence of *mcr-1* in these aforementioned isolates is of paramount importance.

10.3.3.14 United Kingdom

In the United Kingdom (UK), a study conducted on human and food-samples detected *mcr-1*-positive *Salmonella* and *E. coli* (Doumith et al., 2016). Specifically, retrospective analysis of the genomes of ~24,000 *Enterobacteriaceae* resulted in the identification of *mcr-1* in 15 *Salmonella* and *E. coli* isolates. Most of the isolates (10 *Salmonella enterica* and 3 *E. coli*) were associated with human patients. Interestingly, six *mcr-1*-positve *Salmonella* isolates were from patients who had recently visited Asia. Additionally, two *mcr-1*-positive *S.* Paratyphi B var Java were detected in poultry meat imported from the European Union (EU) (Doumith et al., 2016). Anjum et al. (2016) reported the detection of *mcr-1* in two isolates (*E. coli* and a *S.* Typhimurium var Copenhagen) associated with a pig farm in the UK. However, the *mcr-1* in *E. coli* was not transmissible to other bacteria, while the gene associated with *Salmonella* could be transferred via conjugation but at a low frequency.

10.3.4 Africa: South Africa

Although some countries in the African continent were discussed in the section on the MENA region, it should be noted that limited data are available on colistin-resistance and *mcr* in other countries in Africa. Two manuscripts reported the detection of *mcr-1* in both animals (chickens) and humans in South Africa. In a study, 7 clonally unrelated *mcr-1*-positive *E. coli* were isolated from hospitalized and community patients in South Africa. Notably, one of these isolates also harbored *floR* (encodes resistance to florfenicol, an antibiotic used in animal production), which suggested zoonotic potential and origins of this *E. coli* strain (Poirel et al., 2016). Furthermore, it was reported that colistin resistance in South Africa increased from 4.5% in 2008 to 13.6% in 2015 (Perreten et al., 2016). Interestingly, 19 Avian Pathogenic *E. coli* (APEC) isolated from broiler chicken air sacs in 2015 were *mcr-1* positive (Perreten et al., 2016). The latter has two main significant ramifications, 1) *mcr* might emerge in novel zoonotic strains, and 2) the genes will spread to strains that might affect animal health and production.

10.3.5 Asia

10.3.5.1 China

As mentioned previously, laterally transmissible colistin resistance and the mobile colistin resistance gene, *mcr-1*, were first reported by Liu et al (2016) in China. *mcr-1* was identified in an *E. coli* strain (SHP45) isolated from a pig in Southern China (Liu et al., 2016). Following this discovery, an extensive screening for *mcr-1* in pigs (804 isolates), in raw meat (523 isolates), and, retrospectively, in clinical isolates (902 *E. coli* and 420 *K. pneumoniae*) was conducted. Subsequently, *mcr-1* was detected in *E. coli* in ~ 15%, 21%, and 1.5% of the raw meat, animal (pigs), and clinical samples, respectively (Liu et al., 2016). Furthermore, *mcr-1*-positive *K. pneumoniae* was retrieved from 0.7% of the clinical isolates. Interestingly, epidemiological studies in China detected *mcr-1* in an *E. coli* isolate that dated back to 1980. Furthermore, retrospective studies showed that *mcr-1* increased from 5.2% in 2009 to 30% in 2014 in *E. coli* isolated from chicken samples (Sun et al., 2018). Subsequently, the occurrence of *mcr-1* on animal farms was further documented in various studies. For example, a study conducted in 2017 reported *mcr-1* in 5.11% of *E. coli* associated with chickens from 13 provinces (Yang et al., 2017b). Also, a notably high prevalence of *mcr-1*-positive *E. coli* was observed in fecal matter of pigs on different farms across China. Specifically, 76.2% of the fecal samples harbored *mcr-1* (Tong et al., 2018).

It appears that the prevalence of *mcr-1* is still low in the Chinese clinical settings. A retrospective cross-sectional analysis of isolates collected at two hospitals from 2007 to 2015, identified *mcr-1* in different gram-negative bacteria. Specifically, the gene was detected in 1% of the *E. coli* isolates, less than 1% of *K. pneumonia*, less than 1% of *E. cloacae*, and 1% of *Enterobacter aerogenes* (Wang et al., 2017). Furthermore, the first *Salmonella* isolates that were *mcr-1*-positive were reported in China in 2012. A retrospective study detected 37 *mcr-1*-positive *Salmonella* Typhimurium in human samples. Some of the isolates were also ESBL-producing or MDR, while 24 isolates were resistant to the drugs of choice for treating infections with *Salmonella*; mainly quinolones and cephalosporins (Lu et al., 2019). *mcr-1* was also detected in multidrug-resistant *Shigella sonnei* isolates from diarrheal fecal samples during a retrospective study that analyzed 1,650 historical strains collected from patients with acute diarrhea or dysentery between 2003 and 2015 (Ma et al., 2018). Notably, *mcr-1* was detected in hospital sewage samples, where the gene was carried on the chromosome or plasmids of different bacterial species, including *K. pneumoniae, Kluyvera* spp. and *E. coli* strains (Zhao and Zong, 2016). *mcr-1* in sewage poses further significant risks, because sewage systems are hubs for antimicrobial resistant genes and might serve as an important niche for the emergence of novel colistin-resistant and MDR strains of clinical importance. *mcr-1* was not only detected in human and food animal samples, but it was also reported in vegetables and companion animals in China. Specifically, two *Raoultella ornithinolytica*, and seven *E. coli* strains recovered from lettuce and tomato samples were found to be *mcr-1* positive (Luo et al., 2017). Furthermore, *mcr-1* was detected in 8.7% of *Enterobacteriaceae* isolated from cats and dogs in Beijing (Lei et al., 2017), while *mcr-1 and* bla_{NDM-5} were reported in an *E. coli* isolated from a rectal swab of a pet cat that suffered from fever and diarrhea. This *E. coli* strain was resistant to all antibiotics tested, including cephalosporins and carbapenems (Sun et al., 2016). Other *mcr* have also been reported in China. For example, a study conducted on a pig farm showed the presence of *mcr-3*-positive *E. coli* isolated from a pig and the farm environment (flies and soils) and 80% of these isolates were found to co-carry *mcr-1* (Wang et al., 2019). Recently, *K. pneumoniae* isolated form livestock in China showed high resistance to colistin but was negative to all the *mcr* genes tested (*1* to *7*). However, whole genome sequencing showed that the strain carried a novel *mcr*, *mcr-8*. This strain was isolated from a pig sample and also carried $bla\text{-}_{NDM}$ (Wang et al., 2018b). Due to the wide distribution of *mcr* and their emergence in a variety of bacterial species, hosts and niches in China, the Chinese government banned the use of colistin in animal feeds and restricted it to human therapy (Walsh and Wu, 2016).

10.3.5.2 Vietnam

In Vietnam, a study was conducted to screen for *mcr-1* in fecal samples from humans and backyard chickens. It was shown that *mcr-1* was detected in 59.4% and 20.6% of chicken and human fecal samples, respectively. Furthermore, *mcr-1* was found in 12.8% and 4% of the *E. coli* isolated from chickens and chicken farmers, respectively (Trung et al., 2017). *mcr-1* was also reported in hospital-acquired *E. coli* isolates; however, it appeared that the gene was carried chromosomally (Tada et al., 2017). In another study that analyzed local foods (meat and seafood products), 97% of the colistin-resistant *E. coli* isolates harbored *mcr-1*, while 3% were *mcr-3*-positive. Both genes were found to occur in either ESBL-producing or AmpC-producing *E. coli*, and they were confirmed to be transmissible by conjugation experiments (Yamaguchi et al., 2018).

10.3.5.3 **India, Thailand, Laos and Malaysia**

In India, three *mcr-1*-positive *E. coli* were identified in Indian foods, including meat, fish and raw vegetables (Ghafur et al., 2019). Also, in a study conducted in a clinical setting, *mcr-1* was reported in four *K. pneumonia*; however, the gene appeared to be chromosomal and not on a plasmid (Singh et al., 2018).

In Thailand, a nationwide study identified *mcr* in different bacterial strains. Specifically, four *mcr-3*-positive isolates were retrieved from human samples; out of which three were *E. coli* (from blood and abscess samples) and one was *K. pneumonia* (from wound samples). Furthermore, *mcr-1* was detected in *E. coli* isolated from wound samples (Wise et al., 2018). The gene has also been shown to occur in *E. coli* isolated from asymptomatic people in Thailand (Olaitan et al., 2016). Mobile colistin resistance was also detected in flies in Thailand. These flies were shown to carry *mcr-1* and *mcr-3* (Fukuda et al., 2018).

In Laos, *mcr-1*-positive *E. coli* were identified in pigs and asymptomatic humans (Olaitan et al., 2016). Notably, one of these isolates was reported to have transferred from a pig to a farmer, which further emphasizes zoonosis as a route of transmission for *mcr.*

In Malaysia, a study conducted on more than 900 bacterial isolates from the environment, humans and animals reported six *mcr-1*-positive *E. coli* from different samples, including humans, retail chicken, animal feed, pond water and pig fecal matter. This indicated the occurrence of *mcr-1* in different niches in Malaysia (Yu et al., 2016).

10.3.5.4 Japan, Korea, and Taiwan

In Japan. *E. coli* from food-producing animals were screened for the presence of *mcr-1* and *mcr-2*. It was shown that *mcr-1* was carried by 39 isolates from cattle, pigs, and broilers chickens, while *mcr-2* was not detected (Kawanishi et al., 2017). In the clinical settings, *mcr-1* was detected in *E. coli* that belonged to ST5702, which is suspected to have originated in livestock and was transmitted to the patient. Furthermore, *mcr-1* was found in four *E. coli* isolates (ST782 and ST456) as well as one *K. pneumonia* (ST1296) from five patients in Okinawa, Japan (Tada et al., 2018).

In South Korea, *mcr-1* was reported in *E. coli* from a hospitalized patient. Furthermore, *mcr-1* and *mcr-3* were detected in multidrug resistant *E. coli* from healthy pigs (Belaynehe et al., 2018). Additionally, 11 *mcr-1*-positive *E. coli* were detected in chicken fecal samples and chicken carcasses as well as a diseased pig (Lim et al., 2016). Notably, Yoon et al., (2018) presented evidence that suggested the possible transfer of *mcr-1*-positive *E. coli* isolates from chickens to humans in South Korea (Yoon et al., 2018). The emergence of mobile colistin resistance in South Korea was speculated to be associated with the increased annual consumption of the antibiotic in food-animal production (Lim et al., 2016).

In Taiwan, *mcr-1* and the *bla*$_{NDM-9}$ were detected in a patient with a urinary tract infection. These genes were carried in an *E. coli* strain that was also resistant to all other antibiotics tested (Lin et al., 2019). This posed a significant concern because of the possibility of limiting treatment options if these strains and genes became highly prevalent.

10.3.6 Australia

In comparison to neighboring countries where colistin is extensively applied in agriculture, the use of this antibiotic in healthcare and agriculture in Australia is limited (Ellem et al. 2017). Furthermore, polymyxins, including colistin, are not in the top 20 antimicrobial drugs prescribed in Australia. ar. However, *mcr-1* was identified in two colistin-resistant *E. coli* isolated from patients in New South Wales. The isolates were retrieved from urine samples of two hospitalized elderly women, respectively. The plasmids carrying the *mcr-1* were identical to previously identified *mcr-1*-carrying plasmid from the Middle East and Asia. Ellem et al. (2017) argued that although the two patients did not travel outside Australia, the low prevalence of *mcr-1* and colistin resistance in *Enterobacteriaceae* suggested that the gene might have not established in Australia yet.

10.4 Stepping away from the precipice: Conclusions and recommendations

Colistin, a last-resort antibiotic, is used to treat complicated infections with MDR gram-negative bacteria. Hence, the emergence of mobile colistin resistance will jeopardize treatment options and limit the efficacy of colistin, which is more problematic given (1) the weak initiatives and incentives to discover new antibiotics, (2) the absence of accessible antibiotic-independent interventions to control some infectious diseases in humans, and (3) dissemination of the mobile colistin-resistant genes (*mcr*) worldwide. Although the topic has received global attention, more data and studies are still required, especially in low- and middle-income countries with challenges in resources and infrastructure. This is also true for many countries in the MENA region, Africa, and Asia. Additionally, in countries with debilitated infrastructure, the study of *mcr*-mediated colistin resistance should expand to include humans, agriculture, and the environment. This is important, because these countries can become a source of transmission of the genes to other countries. Therefore, there should be an orchestrated global effort to monitor and control *mcr* reservoirs in low- and middle-income countries.

It is perhaps also prudent to move toward a global ban of the use of colistin in controversial agricultural applications such as growth promotion. This is because animal farming practices can inadvertently spread *mcr* to humans and the environment. However, although the emergence of *mcr* might have been driven arguably by agricultural practices, the medical sector also has a part to play via adopting prudent measures for the use of antibiotics in therapy. The problem of *mcr* is best investigated under the lens of the One Health approach, because antibiotic resistance genes are easily transmissible and can spread widely to affect many niches. It should not be forgotten that the dissemination of these genetic markers is not restricted by national borders. This is clear in the understudied but significant role of travel and trade (food and animals) in introducing *mcr* to new environments and countries (even those with established antimicrobial stewardship policies). Other routes of transmission also need to be assessed and these include migratory movement of wild animals. For example, a recent study detected the presence of *mcr-1* in *E. coli* isolated from herring gulls which migrate across Europe in the winter; potentially contributing to the dissemination of the genes to different European countries (Ruzauskas and Vaskeviciute, 2016). Therefore, *mcr* and colistin resistance exemplify a challenge that concerns the global community and will likely require multinational and transboundary resources and expertise to tackle it. Inaction is a dangerous alternative that threatens the treatment of human and animal infectious diseases and might herald an era where simple infections are once again life threatening.

References

Aires, C.A.M., da Conceição-Neto, O.C., E Oliveira, T.R.T., Dias, C.F., Montezzi, L.F., Picão, R.C., Albano, R.M., Asensi, M.D., Carvalho-Assef, A.P.D.A., 2017. Emergence of the plasmid-mediated mcr-1 gene in clinical KPC-2-producing *Klebsiella pneumoniae* sequence type 392 in Brazil. Antimicrobial Agents and Chemotherapy 61 e00317-17.

Altayb, H.N., Siddig, M.A., EL Amin, N.M., Maowia, A.I.H., Mukhtar, M., 2018. Molecular characterization of CTX-M ESBLs among pathogenic enterobacteriaceae isolated from different regions in Sudan. Global Advanced Research Journal of Microbiology (GARJM) 7, 040–047.

Anjum, M.F., Duggett, N.A., AbuOun, M., Randall, L., Nunez-Garcia, J., Ellis, R.J., Rogers, J., Horton, R., Brena, C., Williamson, S., Martelli, F., Davies, R., Teale, C., 2016. Colistinresistance in Salmonella and Escherichia coli isolates from a pig farm in GreatBritain. J Antimicrob Chemother 71 (8), 2306–2313. https://doi.org/10.1093/jac/dkw149.

Apostolakos, I., Piccirillo, A., 2018. A review on the current situation and challenges of colistin resistance in poultry production. Avian Pathology 47, 546–558.

Arcilla, M.S., VAN Hattem, J.M., Matamoros, S., Melles, D.C., Penders, J., DE Jong, M.D., Schultsz, C., 2016. Dissemination of the mcr-1 colistin resistance gene. The Lancet Infectious Diseases 16, 147–149.

Bachiri, T., Lalaoui, R., Bakour, S., Allouache, M., Belkebla, N., Rolain, J.M., Touati, A., 2018. First report of the plasmid-mediated colistin resistance gene mcr-1 in *Escherichia coli* ST405 isolated from wildlife in Bejaia, Algeria. Microbial Drug Resistance 24, 890–895.

Baron, S., Hadjadj, L., Rolain, J.M., Olaitan, A.O., 2016. Molecular mechanisms of polymyxinresistance: knowns and unknowns. Int J Antimicrob Agents 48 (6), 583–591. https://doi.org/10.1016/j.ijantimicag.2016.06.023.

Baron, S., Bardet, L., Dubourg, G., Fichaux, M., Rolain, J.-M., 2017. mcr-1 plasmid-mediated colistin resistance gene detection in an *Enterobacter cloacae* clinical isolate in France. Journal of Global Antimicrobial Resistance 10, 35.

Belaynehe, K.M., Shin, S.W., Park, K.Y., Jang, J.Y., Won, H.G., Yoon, I.J., Yoo, H.S., 2019. GeneticAnalysis of p17S-208 Plasmid Encoding the Colistin Resistance mcr-3 Gene inEscherichia coli Isolated from Swine in South Korea. Microb Drug Resist 25 (3), 457–461. https://doi.org/10.1089/mdr.2018.0132. Epub 2018 Oct 31. PubMed PMID:30379604.

Berrazeg, M., Hadjadj, L., Ayad, A., Drissi, M., Rolain, J.-M., 2016. First detected human case in Algeria of mcr-1 plasmid-mediated colistin resistance in a 2011 *Escherichia coli* isolate. Antimicrobial Agents and Chemotherapy 60, 6996–6997.

Campos, J., Cristino, L., Peixe, L., Antunes, P., 2016. MCR-1 in multidrug-resistant and copper-tolerant clinically relevant Salmonella 1, 4,[5], 12: i:-and S. Rissen clones in Portugal, 2011 to 2015. Euro Surveillance 21, 30270.

Carattoli, A., 2013. Plasmids and the spread of resistance. International Journal of Medical Microbiology 303, 298–304.

Carattoli, A., Villa, L., Feudi, C., Curcio, L., Orsini, S., Luppi, A., Pezzotti, G., Magistrali, C.F., 2017. Novel plasmid-mediated colistin resistance mcr-4 gene in Salmonella and *Escherichia coli*, Italy 2013, Spain and Belgium, 2015 to 2016. Euro Surveillance 22.

Carroll, L.M., Zurfluh, K., Jang, H., Gopinath, G., Nüesch-Inderbinen, M., Poirel, L., Nordmann, P., Stephan, R., Guldimann, C., 2018. First report of an mcr-1-harboring Salmonella enterica subsp. enterica serotype 4,5,12:i:- strain isolated from blood of a patient in Switzerland, 2018 Nov Int J Antimicrob Agents 52 (5), 740–741.

Caspar, Y., Maillet, M., Pavese, P., Francony, G., Brion, J.P., Mallaret, M.R., Bonnet, R., Robin, F., Beyrouthy, R., Maurin, M., 2017 May. mcr-1 Colistin Resistance in ESBL-ProducingKlebsiella pneumoniae, France. Emerg Infect Dis 23 (5), 874–876. https://doi.org/10.3201/eid2305.161942.

Catry, B., Cavaleri, M., Baptiste, K., Grave, K., Grein, K., Holm, A., Jukes, H., Liebana, E., Navas, A.L., Mackay, D., 2015. Use of colistin-containing products within the European Union and European Economic Area (EU/EEA): development of resistance in animals and possible impact on human and animal health. International Journal of Antimicrobial Agents 46, 297–306.

Chabou, S., Leulmi, H., Rolain, J.-M., 2018. Emergence of mcr-1-mediated colistin resistance in *Escherichia coli* isolates from poultry in Algeria. Journal of Global Antimicrobial Resistance 16, 115.

Chauvin, C., 2010. Etude des acquisitions de médicaments vétérinaires contenant des antibiotiques dans un échantillon d'élevages porcins naisseurs-engraisseurs année 2008 et comparaison 2008/2005. Anses 33.

Clemente, L., Manageiro, V., Correia, I., Amaro, A., Albuquerque, T., Themudo, P., Ferreira, E., Caniça, M., 2019. Revealing mcr-1-positive ESBL-producing *Escherichia coli* strains among Enterobacteriaceae from food-producing animals (bovine, swine and poultry) and meat (bovine and swine), Portugal, 2010–2015. International Journal of Food Microbiology 296, 37–42.

Coppi, M., Cannatelli, A., Antonelli, A., Baccani, I., DI Pilato, V., Sennati, S., Giani, T., Rossolini, G.M., 2018. A simple phenotypic method for screening of MCR-1-mediated colistin resistance. Clinical Microbiology and Infection 24, 201. e1–201. e3.

Curcio, L., Luppi, A., Bonilauri, P., Gherpelli, Y., Pezzotti, G., Pesciaroli, M., Magistrali, C.F., 2017. Detection of the colistin resistance gene mcr-1 in pathogenic *Escherichia coli* from pigs affected by post-weaning diarrhoea in Italy. Journal of Global Antimicrobial Resistance 10, 80–83.

Dandachi, I., Fayad, E., EL-Bazzal, B., Daoud, Z., Rolain, J.-M., 2018a. Prevalence of extended-Spectrum Beta-lactamase-producing gram-negative bacilli and emergence of mcr-1 Colistin resistance gene in Lebanese swine farms. Microbial Drug Resistance 25, 233–240.

Dandachi, I., Leangapichart, T., Daoud, Z., Rolain, J.-M., 2018b. First detection of mcr-1 plasmid-mediated colistin-resistant *Escherichia coli* in Lebanese poultry. Journal of Global Antimicrobial Resistance 12, 137.

Delgado-Blas, J.F., Ovejero, C.M., Abadia-Patiño, L., Gonzalez-Zorn, B., 2016. Coexistence of mcr-1 and blaNDM-1 in *Escherichia coli* from Venezuela. Antimicrobial Agents and Chemotherapy 60, 6356–6358.

DI Pilato, V., Arena, F., Tascini, C., Cannatelli, A., DE Angelis, L.H., Fortunato, S., Giani, T., Menichetti, F., Rossolini, G.M., 2016. mcr-1.2, a new mcr variant carried on a transferable plasmid from a colistin-resistant KPC carbapenemase-producing *Klebsiella pneumoniae* strain of sequence type 512. Antimicrobial Agents and Chemotherapy 60, 5612–5615.

Dominguez, J.E., Espinosa, R.A.F., Redondo, L.M., Cejas, D., Gutkind, G.O., Chacana, P.A., DI Conza, J.A., Fernández-Miyakawa, M.E., 2017. Plasmid-mediated colistin resistance in *Escherichia coli* recovered from healthy poultry. Revista Argentina de Microbiología 49, 297–298.

Donà, V., Bernasconi, O.J., Pires, J., Collaud, A., Overesch, G., Ramette, A., Perreten, V., Endimiani, A., 2017. Heterogeneous genetic location of mcr-1 in colistin-resistant *Escherichia coli* isolates from humans and retail chicken meat in Switzerland: emergence of mcr-1-carrying IncK2 plasmids. Antimicrobial Agents and Chemotherapy 61 e01245-17.

Doumith, M., Godbole, G., Ashton, P., Larkin, L., Dallman, T., Day, M., Day, M., Muller-Pebody, B., Ellington, M.J., DE Pinna, E., 2016. Detection of the plasmid-mediated mcr-1 gene conferring colistin resistance in human and food isolates of *Salmonella enterica* and *Escherichia coli* in England and Wales. Journal of Antimicrobial Chemotherapy 71, 2300–2305.

Drali, R., Berrazeg, M., Zidouni, L.L., Hamitouche, F., Abbas, A.A., Deriet, A., Mouffok, F., 2018. Emergence of mcr-1 plasmid-mediated colistin-resistant *Escherichia coli* isolates from seawater. The Science of the Total Environment 642, 90–94.

EL Garch, F., DE Jong, A., Bertrand, X., Hocquet, D., Sauget, M., 2018. mcr-1-like detection in commensal *Escherichia coli* and Salmonella spp. from food-producing animals at slaughter in Europe. Veterinary Microbiology 213, 42–46.

Ellem, J.A., Ginn, A.N., Chen, S.C.-A., Ferguson, J., Partridge, S.R., Iredell, J.R., 2017. Locally acquired mcr-1 in *Escherichia coli*, Australia, 2011 and 2013. Emerging Infectious Diseases 23, 1160.

Elnahriry, S.S., Khalifa, H.O., Soliman, A.M., Ahmed, A.M., Hussein, A.M., Shimamoto, T., Shimamoto, T., 2016. Emergence of plasmid-mediated colistin resistance gene mcr-1 in a clinical *Escherichia coli* isolate from Egypt. Antimicrobial Agents and Chemotherapy 60, 3249–3250.

Eltai, N.O., Abdfarag, E.A., AL-Romaihi, H., Wehedy, E., Mahmoud, M.H., Alawad, O.K., AL-Hajri, M.M., Thani, A.A.A., Yassine, H.M., 2018. Antibiotic resistance profile of commensal *Escherichia coli* isolated from broiler chickens in Qatar. Journal of Food Protection 81, 302–307.

Falgenhauer, L., Waezsada, S.-E., Yao, Y., Imirzalioglu, C., Käsbohrer, A., Roesler, U., Michael, G.B., Schwarz, S., Werner, G., Kreienbrock, L., 2016. Colistin resistance gene mcr-1 in extended-spectrum β-lactamase-producing and carbapenemase-producing gram-negative bacteria in Germany. The Lancet Infectious Diseases 16, 282–283.

Fernandes, M., Moura, Q., Sartori, L., Silva, K., Cunha, M., Esposito, F., Lopes, R., Otutumi, L., Goncalves, D., Dropa, M., Matte, M.H., Monte, D., Landgraf, M., Francisco Gr, Bueno MF, de Oliveira Garcia D, Knobl T, Moreno Am, Lincopan N, 2016. Silent dissemination of colistin-resistant *Escherichia coli* in South America 90 could contribute to the global spread of the mcr-1 gene. Euro Surveillance 21, 91.

Fukuda, A., Usui, M., Okubo, T., Tagaki, C., Sukpanyatham, N., Tamura, Y., 2018. Co-harboring of cephalosporin (bla)/colistin (mcr) resistance genes among Enterobacteriaceae from flies in Thailand. FEMS Microbiology Letters 365, fny178.

García, V., GARCÍA-Meniño, I., Mora, A., Flament-Simon, S.C., Díaz-Jiménez, D., Blanco, J.E., Alonso, M.P., Blanco, J., 2018. Co-occurrence of mcr-1, mcr-4 and mcr-5 genes in multidrug-resistant ST10 Enterotoxigenic and Shiga toxin-producing *Escherichia coli* in Spain (2006-2017). International Journal of Antimicrobial Agents 52, 104–108.

Ghafur, A., Shankar, C., Gnanasoundari, P., Venkatesan, M., Mani, D., Thirunarayanan, M., Veeraraghavan, B., 2019. Detection of chromosomal and plasmid-mediated mechanisms of colistin resistance in *Escherichia coli* and *Klebsiella pneumoniae* from Indian food samples. Journal of Global Antimicrobial Resistance 16, 48–52.

Grami, R., Mansour, W., Mehri, W., Bouallègue, O., Boujaâfar, N., Madec, J.-Y., Haenni, M., 2016. Impact of food animal trade on the spread of mcr-1-mediated colistin resistance, Tunisia, July 2015. Euro Surveillance 21, 30144.

Gröndahl-Yli-Hannuksela, K., Lönnqvist, E., Kallonen, T., Lindholm, L., Jalava, J., Rantakokko-Jalava, K., Vuopio, J., 2018. The first human report of mobile colistin resistance gene, mcr-1, in Finland. Apmis 126, 413–417.

Haenni, M., Poirel, L., Kieffer, N., Châtre, P., Saras, E., Métayer, V., Dumoulin, R., Nordmann, P., Madec, J.-Y., 2016. Co-occurrence of extended spectrum β lactamase and MCR-1 encoding genes on plasmids. The Lancet Infectious Diseases 16, 281–282.

Hammad, A.M., Hoffmann, M., Gonzalez-Escalona, N., Abbas, N.H., Yao, K., Koenig, S., Allué-Guardia, A., Eppinger, M., 2019 Sep. Genomic features of colistin resistant Escherichia coli ST69 strain harboring mcr-1 on IncHI2 plasmid from raw milk cheese in Egypt. Infect Genet Evol 73, 126–131. https://doi.org/10.1016/j.meegid.2019.04.021.

Hasman, H., Hammerum, A.M., Hansen, F., Hendriksen, R.S., Olesen, B., Agersø, Y., Zankari, E., Leekitcharoenphon, P., Stegger, M., Kaas, R.S., 2015. Detection of mcr-1 encoding plasmid-mediated colistin-resistant *Escherichia coli* isolates from human bloodstream infection and imported chicken meat, Denmark 2015. Euro Surveillance 20, 1–5.

Hernández, M., Iglesias, M.R., Rodríguez-Lázaro, D., Gallardo, A., Quijada, N.M., MIGUELA-Villoldo, P., Campos, M.J., Píriz, S., López-Orozco, G., DE Frutos, C., 2017. Co-occurrence of colistin-resistance genes mcr-1 and mcr-3 among multidrug-resistant *Escherichia coli* isolated from cattle, Spain, September 2015. Euro Surveillance 22.

Hmede, Z., Kassem, I.I., 2018. The colistin resistance gene, mcr-1, is prevalent in commensal *E. coli* isolated from Lebanese pre-harvest poultry. Antimicrobial Agents and Chemotherapy, AAC 01304–01318.

Hmede, Z., Sulaiman, A.A.A., Jaafar, H., Kassem II, 2019 Jul. Emergence of plasmid-borne colistin resistance gene mcr-1 in multidrug-resistant Escherichia coli isolated from irrigation water in Lebanon. Int J Antimicrob Agents 54 (1), 102–104. https://doi.org/10.1016/j.ijantimicag.2019.05.005. Epub 2019 May 7. PubMed PMID: 31071466.

Holmes, A., Holmes, M., Gottlieb, T., Price, L.B., Sundsfjord, A., 2018 Jan 29. End non-essential useof antimicrobials in livestock. BMJ 360, k259. https://doi.org/10.1136/bmj.k259.

Kluytmans-van den Bergh, M., Huizinga, P., Bonten, M., Bos, M., De bruyne, K., Friedrich, A., Rossen, J., Savelkoul, P., Kluytmans, J., 2016. Presence of mcr-1-positive Enterobacteriaceae in retail chicken meat but not in humans in the Netherlands since 2009. Eurosurveillance 21.

Irrgang, A., Roschanski, N., Tenhagen, B.-A., Grobbel, M., Skladnikiewicz-Ziemer, T., Thomas, K., Roesler, U., Kaesbohrer, A., 2016. Prevalence of mcr-1 in *E. coli* from livestock and food in Germany, 2010–2015. PLoS One 11, e0159863.

Izdebski, R., Baraniak, A., Bojarska, K., Urbanowicz, P., Fiett, J., Pomorska-Wesołowska, M., Hryniewicz, W., Gniadkowski, M., Żabicka, D., 2016. Mobile MCR-1-associated resistance to colistin in Poland. Journal of Antimicrobial Chemotherapy 71, 2331–2333.

Jørgensen, S.B., Søraas, A., Arnesen, L.S., Leegaard, T., Sundsfjord, A., Jenum, P.A., 2017. First environmental sample containing plasmid-mediated colistin-resistant ESBL-producing *Escherichia coli* detected in Norway. Apmis 125, 822–825.

Kassem II, Hijazi, M.A., Saab, R., 2019 Mar. On a collision course: The availability and use of colistin-containing drugs in human therapeutics and food-animal farming in Lebanon. J Glob Antimicrob Resist 16, 162–164. https://doi.org/10.1016/j.jgar.2019.01.019.

Kawanishi, M., Abo, H., Ozawa, M., Uchiyama, M., Shirakawa, T., Suzuki, S., Shima, A., Yamashita, A., Sekizuka, T., Kato, K., 2017. Prevalence of colistin resistance gene mcr-1 and absence of mcr-2 in *Escherichia coli* isolated from healthy food-producing animals in Japan. Antimicrobial Agents and Chemotherapy 61 e02057-16.

Kempf, I., Fleury, M.A., Drider, D., Bruneau, M., Sanders, P., Chauvin, C., Madec, J.-Y., Jouy, E., 2013. What do we know about resistance to colistin in Enterobacteriaceae in avian and pig production in Europe? International Journal of Antimicrobial Agents 42, 379–383.

Kempf, I., Jouy, E., Chauvin, C., 2016. Colistin use and colistin resistance in bacteria from animals. International Journal of Antimicrobial Agents 48, 598–606.

Khalifa, H.O., Ahmed, A.M., Oreiby, A.F., Eid, A.M., Shimamoto, T., Shimamoto, T., 2016. Characterisation of the plasmid-mediated colistin resistance gene mcr-1 in *Escherichia coli* isolated from animals in Egypt. International Journal of Antimicrobial Agents 47, 413.

Kieffer, N., Aires-De-Sousa, M., Nordmann, P., Poirel, L., 2017. High rate of MCR-1–producing *Escherichia coli* and *Klebsiella pneumoniae* among pigs, Portugal. Emerging Infectious Diseases 23, 2023.

Kurekci, C., Aydin, M., Nalbantoglu, O.U., Gundogdu, A., 2018. First report of *Escherichia coli* carrying the mobile colistin resistance gene mcr-1 in Turkey. Journal of global antimicrobial resistance 15, 169.

Lalaoui, R., Djukovic, A., Bakour, S., Sanz, J., Gonzalez-Barbera, E.M., Salavert, M., López-Hontangas, J.L., Sanz, M.A., Xavier, K.B., Kuster, B., Debrauwer, L., Ubeda, C., Rolain, J.M., 2019 Aug. Detection of plasmid-mediated colistin resistance, mcr-1 gene, in Escherichia coli isolated from high-risk patients with acute leukemia in Spain. J Infect Chemother 25 (8), 605–609. https://doi.org/10.1016/j.jiac.2019.03.007.

Leangapichart, T., Gautret, P., Brouqui, P., Mimish, Z., Raoult, D., Rolain, J.-M., 2016. Acquisition of mcr-1 plasmid-mediated colistin resistance in *Escherichia coli* and *Klebsiella pneumoniae* during Hajj 2013 and 2014. Antimicrobial Agents and Chemotherapy 60, 6998–6999.

Lei, L., Wang, Y., Schwarz, S., Walsh, T.R., Ou, Y., Wu, Y., Li, M., Shen, Z., 2017 Apr. mcr-1 inEnterobacteriaceae from Companion Animals, Beijing, China, 2012–2016. EmergInfect Dis 23 (4), 710–711. https://doi.org/10.3201/eid2304.161732.

Liassine, N., Assouvie, L., Descombes, M.C., Tendon, V.D., Kieffer, N., Poirel, L., Nordmann, P., 2016. Very low prevalence of MCR-1/MCR-2 plasmid-mediated colistin resistance in urinary tract Enterobacteriaceae in Switzerland. Int J Infect Dis. Oct 51, 4–5. https://doi.org/10.1016/j.ijid.2016.08.008.

Lim, S.K., Kang, H.Y., Lee, K., Moon, D.C., Lee, H.S., Jung, S.C., 2016 Oct 21. First Detection of the mcr-1 Gene in Escherichia coli Isolated from Livestock between 2013 and 2015 in South Korea. Antimicrob Agents Chemother 60 (11), 6991–6993. https://doi.org/10.1128/AAC.01472-16.

Lin, Y.-C., Kuroda, M., Suzuki, S., Mu, J.-J., 2019. Emergence of an *Escherichia coli* strain co-harbouring mcr-1 and blaNDM-9 from a urinary tract infection in Taiwan. Journal of Global Antimicrobial Resistance 16, 286–290.

Litrup, E., Kiil, K., Hammerum, A.M., Roer, L., Nielsen, E.M., Torpdahl, M., 2017. Plasmid-borne colistin resistance gene mcr-3 in Salmonella isolates from human infections, Denmark, 2009–17. Euro Surveillance 22.

Liu, Y.-Y., Wang, Y., Walsh, T.R., Yi, L.-X., Zhang, R., Spencer, J., Doi, Y., Tian, G., Dong, B., Huang, X., 2016. Emergence of plasmid-mediated colistin resistance mechanism MCR-1 in animals and human beings in China: a microbiological and molecular biological study. The Lancet Infectious Diseases 16, 161–168.

Lu, X., Zeng, M., Xu, J., Zhou, H., Gu, B., Li, Z., Jin, H., Wang, X., Zhang, W., Hu, Y., Xiao, W., Zhu, B., Xu, X., Kan, B., 2019 Apr. Epidemiologic and genomic insights on mcr-1-harbouringSalmonella from diarrhoeal outpatients in Shanghai, China, 2006–2016. EBioMedicine 42, 133–144. https://doi.org/10.1016/j.ebiom.2019.03.006.

Luo, J., Yao, X., Lv, L., Doi, Y., Huang, X., Huang, S., Liu, J.-H., 2017. Emergence of mcr-1 in Raoultella ornithinolytica and *Escherichia coli* isolates from retail vegetables in China. Antimicrobial Agents and Chemotherapy 61 e01139-17.

Ma, Q., Huang, Y., Wang, J., Xu, X., Hawkey, J., Yang, C., Liang, B., Hu, X., Wu, F., Yang, X., 2018. Multidrug-resistant Shigella sonnei carrying the plasmid-mediated mcr-1 gene in China. International Journal of Antimicrobial Agents 52, 14–21.

Maamar, E., Alonso, C.A., Hamzaoui, Z., Dakhli, N., Abbassi, M.S., Ferjani, S., Saidani, M., Boubaker, I.B.-B., Torres, C., 2018. Emergence of plasmid-mediated colistin-resistance in CMY-2-producing *Escherichia coli* of lineage ST2197 in a Tunisian poultry farm. International Journal of Food Microbiology 269, 60–63.

Matamoros, S., VAN Hattem, J.M., Arcilla, M.S., Willemse, N., Melles, D.C., Penders, J., Vinh, T.N., Hoa, N.T., DE Jong, M.D., Schultsz, C., 2017. Global phylogenetic analysis of *Escherichia coli* and plasmids carrying the mcr-1 gene indicates bacterial diversity but plasmid restriction. Scientific Reports 7, 15364.

Meinersmann, R.J., Ladely, S.R., Plumblee, J.R., Cook, K.L., Thacker, E., 2017. Prevalence of mcr-1 in the cecal contents of food animals in the United States. Antimicrobial Agents and Chemotherapy 61 e02244-16.

Mendes, A.C., Novais, Â., Campos, J., Rodrigues, C., Santos, C., Antunes, P., Ramos, H., Peixe, L., 2018. mcr-1 in carbapenemase-producing *Klebsiella pneumoniae* with hospitalized patients, Portugal, 2016–2017. Emerging Infectious Diseases 24, 762.

Merida-Vieyra, J., De Colsa-Ranero, A., Arzate-Barbosa, P., Arias-De La Garza, E., Méndez-Tenorio, A., MURCIA-Garzón, J., AQUINO-Andrade, A., 2019. First clinical isolate of *Escherichia coli* harboring mcr-1 gene in Mexico. PLoS One 14, e0214648.

Moawad, A.A., Hotzel, H., Neubauer, H., Ehricht, R., Monecke, S., Tomaso, H., Hafez, H.M., Roesler, U., EL-Adawy, H., 2018. Antimicrobial resistance in Enterobacteriaceae from healthy broilers in Egypt: emergence of colistin-resistant and extended-spectrum β-lactamase-producing *Escherichia coli*. Gut Pathogens 10, 39.

Mohsin, J., Pál, T., Petersen, J.E., Darwish, D., Ghazawi, A., Ashraf, T., Sonnevend, A., 2018. Plasmid-mediated colistin resistance gene mcr-1 in an *Escherichia coli* ST10 bloodstream isolate in the Sultanate of Oman. Microbial Drug Resistance 24, 278–282.

Nordmann, P., Lienhard, R., Kieffer, N., Clerc, O., Poirel, L., 2016a. Plasmid-mediated colistin-resistant *Escherichia coli* in bacteremia in Switzerland. Clinical Infectious Diseases 62, 1322–1323.

Nordmann, P., Assouvie, L., Prod'Hom, G., Poirel, L., Greub, G., 2016b. Screening of plasmid-mediated MCR-1 colistin-resistance from bacteremia. Eur J Clin Microbiol Infect Dis. Nov 35 (11), 1891–1892.

Ortega-Paredes, D., Barba, P., Zurita, J., 2016. Colistin-resistant *Escherichia coli* clinical isolate harbouring the mcr-1 gene in Ecuador. Epidemiology and Infection 144, 2967–2970.

Ovejero, C., Delgado-Blas, J., Calero-Caceres, W., Muniesa, M., Gonzalez-Zorn, B., 2017. Spread of mcr-1-carrying Enterobacteriaceae in sewage water from Spain. Journal of Antimicrobial Chemotherapy 72, 1050–1053.

Paterson, D.L., Harris, P.N., 2016. Colistin resistance: a major breach in our last line of defence. The Lancet Infectious Diseases 16, 132–133.

Perreten, V., Strauss, C., Collaud, A., Gerber, D., 2016 Jun 20. Colistin Resistance Gene mcr-1 inAvian-Pathogenic Escherichia coli in South Africa. Antimicrob Agents Chemother 60 (7), 4414–4415. https://doi.org/10.1128/AAC.00548-16. Print 2016 Jul. PubMed PMID: 27161625.

Perrin-Guyomard, A., Bruneau, M., Houée, P., Deleurme, K., Legrandois, P., Poirier, C., Soumet, C., Sanders, P., 2016. Prevalence of mcr-1 in commensal *Escherichia coli* from French livestock, 2007 to 2014. Euro Surveillance 21, 1–3.

Pilote, J., Létourneau, V., Girard, M., Duchaine, C., 2019. Quantification of airborne dust, endotoxins, human pathogens and antibiotic and metal resistance genes in Eastern Canadian swine confinement buildings. Aerobiologia 1–14.

Poirel, L., Kieffer, N., Brink, A., Coetze, J., Jayol, A., Nordmann, P., 2016. Genetic features of MCR-1-producing colistin-resistant *Escherichia coli* isolates in South Africa. Antimicrobial Agents and Chemotherapy 60, 4394–4397.

Poirel, L., Jayol, A., Nordmann, P., 2017. Polymyxins: antibacterial activity, susceptibility testing, and resistance mechanisms encoded by plasmids or chromosomes. Clinical Microbiology Reviews 30, 557–596.

Prim, N., Rivera, A., Rodríguez-Navarro, J., Español, M., Turbau, M., Coll, P., Mirelis, B., 2016. Detection of mcr-1 colistin resistance gene in polyclonal *Escherichia coli* isolates in Barcelona, Spain, 2012 to 2015. Euro Surveillance 21, 30183.

Quesada, A., Ugarte-Ruiz, M., Iglesias, M.R., Porrero, M.C., Martínez, R., Florez-Cuadrado, D., Campos, M.J., García, M., Píriz, S., Sáez, J.L., 2016. Detection of plasmid mediated colistin resistance (MCR-1) in *Escherichia coli* and *Salmonella enterica* isolated from poultry and swine in Spain. Research in Veterinary Science 105, 134–135.

Quiroga, C., Nastro, M., DI Conza, J., 2019. Current scenario of plasmid-mediated colistin resistance in Latin America. Revista Argentina de Microbiología 51, 93–100.

Rahmatallah, N., EL Rhaffouli, H., Laraqui, A., Sekhsokh, Y., Lahlou Amine, I., EL Houadfi, M., Fassi Fihri, O., Dec, 2018. Detection of Colistin Encoding Resistance Genes MCR-1 in Isolates Recovered from Broiler Chickens in Morocco. Saudi J. Pathol. Microbiol. Vol-3 (Iss-12), 520–521.

Rapoport, M., Faccone, D., Pasteran, F., Ceriana, P., Albornoz, E., Petroni, A., Corso, A., Group, M., 2016. First description of mcr-1-mediated colistin resistance in human infections caused by *Escherichia coli* in Latin America. Antimicrobial Agents and Chemotherapy 60, 4412–4413.

Redhwan, A., Choudhury, M., AL Harbi, B., Kutbi, A., Alfaresi, M., Aljindan, R., Balkhy, H., AL Johani, S., Ibrahim, E., Deshmukh, A., 2019. A snapshot about the mobile colistin resistance (mcr) in the Middle East and North Africa region. Journal of Infection and Public Health 12, 149–150.

Roer, L., Hansen, F., Stegger, M., Sönksen, U.W., Hasman, H., Hammerum, A.M., 2017. Novel mcr-3 variant, encoding mobile colistin resistance. In: An ST131 *Escherichia coli* Isolate from Bloodstream Infection, Denmark, 2014. Eurosurveillance, p. 22.

Rolain, J.-M., Kempf, M., Leangapichart, T., Chabou, S., Olaitan, A.O., LE Page, S., Morand, S., Raoult, D., 2016. Plasmid-mediated mcr-1 gene in colistin-resistant clinical isolates of *Klebsiella pneumoniae* in France and Laos. Antimicrobial Agents and Chemotherapy 60, 6994–6995.

Ruzauskas, M., Vaskeviciute, L., 2016. Detection of the mcr-1 gene in *Escherichia coli* prevalent in the migratory bird species *Larus argentatus*. Journal of Antimicrobial Chemotherapy 71, 2333–2334.

Saavedra, S.Y., Diaz, L., Wiesner, M., Correa, A., Arévalo, S.A., Reyes, J., Hidalgo, A.M., DE LA Cadena, E., Perenguez, M., Montaño, L.A., 2017. Genomic and molecular characterization of clinical isolates of Enterobacteriaceae harboring mcr-1 in Colombia, 2002 to 2016. Antimicrobial Agents and Chemotherapy 61 e00841-17.

Saly, M., Jayol, A., Poirel, L., Megraud, F., Nordmann, P., Dubois, V., 2017. Prevalence of faecal carriage of colistin-resistant Gram-negative rods in a university hospital in western France, 2016. Journal of Medical Microbiology 66, 842–843.

Simoni, S., Morroni, G., Brenciani, A., Vincenzi, C., Cirioni, O., Castelletti, S., Varaldo, P.E., Giovanetti, E., Mingoia, M., 2018. Spread of colistin resistance gene mcr-1 in Italy: characterization of the mcr-1.2 allelic variant in a colistin-resistant blood isolate of *Escherichia coli*. Diagnostic Microbiology and Infectious Disease 91, 66–68.

Singh, S., Pathak, A., Kumar, A., Rahman, M., Singh, A., Gonzalez-Zorn, B., Prasad, K.N., 2018. Emergence of chromosome-borne colistin resistance gene mcr-1 in clinical isolates of *Klebsiella pneumoniae* from India. Antimicrobial Agents and Chemotherapy 62 e01885-17.

Skov, R.L., Monnet, D.L., 2016. Plasmid-mediated colistin resistance (mcr-1 gene): three months later, the story unfolds. Euro Surveillance 21, 30155.

Solheim, M., Bohlin, J., Ulstad, C.R., Schau, S.J., Naseer, U., Dahle, U.R., Wester, A.L., 2016. Plasmid-mediated colistin-resistant *Escherichia coli* detected from 2014 in Norway. International Journal of Antimicrobial Agents 48, 227.

Sonnevend, A., Ghazawi, A., Alqahtani, M., Shibl, A., Jamal, W., Hashmey, R., Pal, T., 2016. Plasmid-mediated colistin resistance in *Escherichia coli* from the arabian peninsula. International Journal of Infectious Diseases 50, 85–90.

Sulaiman, A.A.A., Kassem II, 2019a. First report of the plasmid-borne colistinresistance gene (mcr-1) in Proteus mirabilis isolated from domestic and sewerwaters in Syrian refugee camps. Travel Med Infect Dis 101482. https://doi.org/10.1016/j.tmaid.2019.101482 [Epub ahead of print] PubMed PMID: 31521803.

Sulaiman, A.A.A., Kassem II, 2019b. First report on the detection of the plasmid-bornecolistin resistance gene mcr-1 in multi-drug resistant E. coli isolated from domestic and sewer waters in Syrian refugee camps in Lebanon. Travel Med InfectDis 30, 117–120. https://doi.org/10.1016/j.tmaid.2019.06.014. Epub 2019 Jun 28. PubMed PMID: 31260746.

Sun, J., Yang, R.-S., Zhang, Q., Feng, Y., Fang, L.-X., Xia, J., Li, L., Lv, X.-Y., Duan, J.-H., Liao, X.-P., 2016. Co-transfer of bla NDM-5 and mcr-1 by an IncX3−X4 hybrid plasmid in *Escherichia coli*. Nature microbiology 1, 16176.

Sun, J., Zhang, H., Liu, Y.-H., Feng, Y., 2018. Towards understanding MCR-like colistin resistance. Trends in Microbiology 26, 794−808.

Tada, T., Nhung, P.H., Shimada, K., Tsuchiya, M., Phuong, D.M., Anh, N.Q., Ohmagari, N., Kirikae, T., 2017. Emergence of colistin-resistant *Escherichia coli* clinical isolates harboring mcr-1 in Vietnam. International Journal of Infectious Diseases 63, 72−73.

Tada, T., Uechi, K., Nakasone, I., Nakamatsu, M., Satou, K., Hirano, T., Kirikae, T., Fujita, J., 2018. Emergence of IncX4 plasmids encoding mcr-1 in a clinical isolate of *Klebsiella pneumoniae* in Japan. International Journal of Infectious Diseases 75, 98−100.

Tong, H., Liu, J., Yao, X., Jia, H., Wei, J., Shao, D., Liu, K., Qiu, Y., Ma, Z., Li, B., 2018. High carriage rate of mcr-1 and antimicrobial resistance profiles of mcr-1-positive *Escherichia coli* isolates in swine faecal samples collected from eighteen provinces in China. Veterinary Microbiology 225, 53−57.

Torpdahl, M., Hasman, H., Litrup, E., Skov, R.L., Nielsen, E.M., Hammerum, A.M., 2017. Detection of mcr-1-encoding plasmid-mediated colistin-resistant Salmonella isolates from human infection in Denmark. International Journal of Antimicrobial Agents 2, 261−262.

Trung, N.V., Matamoros, S., Carrique-Mas, J.J., Nghia, N.H., Nhung, N.T., Chieu, T.T.B., Mai, H.H., VAN Rooijen, W., Campbell, J., Wagenaar, J.A., 2017. Zoonotic transmission of mcr-1 colistin resistance gene from small-scale poultry farms, Vietnam. Emerging Infectious Diseases 23, 529.

Veldman, K., Van Essen-Zandbergen, A., Rapallini, M., Wit, B., Heymans, R., Van Pelt, W., Mevius, D., 2016. Location of colistin resistance gene mcr-1 in Enterobacteriaceae from livestock and meat. Journal of Antimicrobial Chemotherapy 71, 2340−2342.

Vinueza-Burgos, C., Ortega-Paredes, D., Narváez, C., DE Zutter, L., Zurita, J., 2019. Characterization of cefotaxime resistant *Escherichia coli* isolated from broiler farms in Ecuador. PLoS One 14, e0207567.

Walsh, T.R., Wu, Y., 2016. China bans colistin as a feed additive for animals. The Lancet Infectious Diseases 16, 1102−1103.

Wang, Y., Tian, G.-B., Zhang, R., Shen, Y., Tyrrell, J.M., Huang, X., Zhou, H., Lei, L., Li, H.-Y., Doi, Y., 2017. Prevalence, risk factors, outcomes, and molecular epidemiology of mcr-1-positive Enterobacteriaceae in patients and healthy adults from China: an epidemiological and clinical study. The Lancet Infectious Diseases 17, 390−399.

Wang, R., Dorp, L., Shaw, L.P., Bradley, P., Wang, Q., Wang, X., Jin, L., Zhang, Q., Liu, Y., Rieux, A., 2018a. The global distribution and spread of the mobilized colistin resistance gene mcr-1. Nature Communications 9, 1179.

Wang, X., Wang, Y., Zhou, Y., Li, J., Yin, W., Wang, S., Zhang, S., Shen, J., Shen, Z., Wang, Y., 2018b. Emergence of a novel mobile colistin resistance gene, mcr-8, in NDM-producing *Klebsiella pneumoniae*. Emerging Microbes & Infections 7, 122.

Wang, Z., Fu, Y., Schwarz, S., Yin, W., Walsh, T.R., Zhou, Y., He, J., Jiang, H., Wang, Y., Wang, S., 2019. Genetic environment of colistin resistance genes mcr-1 and mcr-3 in *Escherichia coli* from one pig farm in China. Veterinary Microbiology 230, 56−61.

Von Wintersdorff, C.J., Wolffs, P.F., VAN Niekerk, J.M., Beuken, E., VAN Alphen, L.B., Stobberingh, E.E., Oude Lashof, A.M., Hoebe, C.J., Savelkoul, P.H., Penders, J., 2016. Detection of the plasmid-mediated colistin-resistance gene mcr-1 in faecal metagenomes of Dutch travellers. Journal of Antimicrobial Chemotherapy 71, 3416−3419.

Wise, M.G., Estabrook, M.A., Sahm, D.F., Stone, G.G., Kazmierczak, K.M., 2018. Prevalence of mcr-type genes among colistin-resistant Enterobacteriaceae collected in 2014-2016 as part of the INFORM global surveillance program. PLoS One 13, e0195281.

Xavier, B., Lammens, C., Butaye, P., Goossens, H., Malhotra-Kumar, S., 2016a. Complete sequence of an IncFII plasmid harbouring the colistin resistance gene mcr-1 isolated from Belgian pig farms. Journal of Antimicrobial Chemotherapy 71, 2342–2344.

Xavier, B.B., Lammens, C., Ruhal, R., Kumar-Singh, S., Butaye, P., Goossens, H., Malhotra-Kumar, S., 2016b. Identification of a novel plasmid-mediated colistin-resistance gene, mcr-2, in *Escherichia coli*, Belgium, June 2016. EuroSurveillance Monthly 21, 30280.

Yamaguchi, T., Kawahara, R., Harada, K., Teruya, S., Nakayama, T., Motooka, D., Nakamura, S., Nguyen, P.D., Kumeda, Y., Van Dang, C., 2018. The presence of colistin resistance gene mcr-1 and-3 in ESBL producing *Escherichia coli* isolated from food in Ho Chi Minh City, Vietnam. FEMS Microbiology Letters 365 fny100.

Yanat, B., Machuca, J., Yahia, R.D., Touati, A., Pascual, Á., Rodríguez-Martínez, J.-M., 2016. First report of the plasmid-mediated colistin resistance gene mcr-1 in a clinical *Escherichia coli* isolate in Algeria. International Journal of Antimicrobial Agents 48, 760.

Yang, D., Qiu, Z., Shen, Z., Zhao, H., Jin, M., Li, H., Liu, W., Li, J.-W., 2017a. The occurrence of the colistin resistance gene mcr-1 in the Haihe River (China). International Journal of Environmental Research and Public Health 14, 576.

Yang, Y.-Q., Li, Y.-X., Song, T., Yang, Y.-X., Jiang, W., Zhang, A.-Y., Guo, X.-Y., Liu, B.-H., Wang, Y.-X., Lei, C.-W., 2017b. Colistin resistance gene mcr-1 and its variant in *Escherichia coli* isolates from chickens in China. Antimicrobial Agents and Chemotherapy 61, e01204–e01216.

Yoon, E.-J., Hong, J.S., Yang, J.W., Lee, K.J., Lee, H., Jeong, S.H., 2018. Detection of mcr-1 plasmids in Enterobacteriaceae isolates from human specimens: comparison with those in *Escherichia coli* isolates from livestock in Korea. Annals of Laboratory Medicine 38, 555–562.

Yu, C.Y., Ang, G.Y., Chin, P.S., Ngeow, Y.F., Yin, W.-F., Chan, K.-G., 2016. Emergence of mcr-1-mediated colistin resistance in *Escherichia coli* in Malaysia. International Journal of Antimicrobial Agents 47, 504.

Zhang, X.-F., Doi, Y., Huang, X., Li, H.-Y., Zhong, L.-L., Zeng, K.-J., Zhang, Y.-F., Patil, S., Tian, G.-B., 2016. Possible transmission of mcr-1–harboring *Escherichia coli* between companion animals and human. Emerging Infectious Diseases 22, 1679.

Zhao, F., Zong, Z., 2016. Kluyvera ascorbata strain from hospital sewage carrying the mcr-1 colistin resistance gene. Antimicrobial Agents and Chemotherapy 60, 7498–7501.

Zurfuh, K., Poirel, L., Nordmann, P., Nüesch-Inderbinen, M., Hächler, H., Stephan, R., 2016. Occurrence of the plasmid-borne mcr-1 colistin resistance gene in extended-spectrum-β-lactamase-producing Enterobacteriaceae in river water and imported vegetable samples in Switzerland. Antimicrobial Agents and Chemotherapy 60, 2594–2595.

CHAPTER

Uptake mechanism of antibiotics in plants

11

Safdar Ali Mirza[1], Muhammad Afzaal[2], Sajida Begum[1], Taha Arooj[1], Muniza Almas[1], Sarfraz Ahmed[3], Muhammad Younus[3]

[1]*Department of Botany, GC University, Lahore, Punjab, Pakistan;* [2]*Sustainable Development Study Center (Env.Science) GC University, Lahore, Punjab, Pakistan;* [3]*Department of Basic Sciences, University of Veterinary and Animal Sciences, Narowal, Punjab, Pakistan*

11.1 Introduction

Antimicrobial contaminants and the effects they cause are creating alarming distress due to their wide practice and determination in the atmosphere. A vast variety of antibiotics and antimicrobial agents related to animals and humans have been observed in different habitats like in animal wastes, residues, metropolitan surface water, manufacturing waste water, drinking water, as well as in soil. The issues that could appear due to antibiotic effluence in agro-ecosystems include outbursts of multiresistant varieties of bacteria as well as dispersal of genes related to resistance against these agents in the atmosphere. Still, it is indeterminate that eco-friendly antimicrobial agents, varieties of bacteria and the genes related to antimicrobial resistance can enter into endophytic systems of plants directly, unprotected to contaminants. Gradually, because of the appearance of antibiotics in plants, the interest shifted to the investigations about total symbionts that have developed resistance to these antibiotics with time in plants. Later on, after the discovery of these genes in bacterial species, these were quantified and categorized in different classes, viz. *sul1*, *sul2*, *tet X* and $bla_{CTX\text{-}M}$ in which the first two resemble sulfamethoxazole related resistance while the later one codes for tetracycline and cephalexin, respectively. These genes were reported in symbionts and amplified with time due to which these genes started their development in symbiotic systems of food crops. Soon, these genes enter into the farming fields and the food crops become exposed to these genes. The magnitude of their effect on that particular crop depends upon the properties of the genes as well as species in which they were reported and on absorption capability as well as the habitat and climatic conditions. During the process of tarnishing, the antibiotics e.g., amoxicillin, tylosin, oxytetracycline, tetracycline, ciprofloxacin, ofloxacin, trimetharopin, sulfamethoxazole etc. are absorbed by the emerging root hairs of crops e.g., rice, wheat, lettuce, cucumber, potato, tomato, cabbage and other vegetable plants, and ultimately these antibiotics accumulate in those root hairs. After their accumulation, these antibiotics are transferred to the edible part of that crop, which is consumed by humans. The effect of these antibiotics on human health is not known to a greater extent and more investigations are required. A few of the known effects reported are sensitivities or allergic reactions caused by long-time exposure to particular antibiotic as

Antibiotics and Antimicrobial Resistance Genes in the Environment. https://doi.org/10.1016/B978-0-12-818882-8.00011-5

well as disturbance of the gastrointestinal tract (GIT). The pollutants containing antibiotics are increasing day by day and these have ultimately made their way to the crops consumed by humans and animals. Investigations are currently underway regarding the effects of antibiotics accumulation in plants as well as their determination level after accumulation. Moreover, possible hazards of antibiotics and the degree of their hazardous effects on human health are under consideration.

11.2 Genes related to antibiotic resistance developed in plant endosymbionts

Insufficient information is available about the probable impacts of antibiotic pressure for developing resistance and survival of these antibiotics over time. These include bacterial varieties that become resistant, the symbionts as well as the genes responsible for antibiotic development. There is a huge diversity of such symbionts that includes mutualistic and communalistic organisms that reside in plant tissues, including xylem, phloem, and in roots, as well as in the aerial parts. Wei et al. (2016) reported resistance in symbionts isolated from medicinal plants. Still it is under investigation that whether this resistance is confined to symbionts related to crop plants or the pollutants with antibiotic contamination are responsible for dissemination in other plants. The major concern these days is basically the effect of antibiotics on the developing plants and the degree of toxicity of these antibiotics to plants as well as the capability of that plant to hold these antimicrobial agents. Basically, the emergence of antibiotics in vegetables appeared when the concentration of antibiotics increased in soil.

11.3 Effects of antibiotic exposure on endosymbionts

Many antibiotics inhibit seed germination, plant growth and may interrupt growth regulation. Exposure to a certain antibiotic compounds may cause severe effects e.g., nitro-furan in spring onions may be metabolized it into hydrazine-related chemicals that may be carcinogenic and genotoxic substances were detected in edible parts of the plant. Moreover, these human-made antibiotics have severe effects on plants themselves as well as on edaphic characteristics. Human and animal health is basically associated with plants and soil and many researchers are focusing on the complex association of symbionts with their environment and the possible impacts of synthetic antibiotics on plant and human health. One of the experiments showed that antibiotics like tetracycline and amoxicillin are taken up by carrot and lettuce plants and accumulated resulting in hazardous effects. Hazardous effects on range and forms of nitrifying soil bacterial populations have been reported for oxytetracycline from dung of livestock animals. According to some of the latest research in Europe, it has been reported that a minor amount of antibiotics can exert highly damaging effects on plants. One such condition is reduced germination rate of seedlings as well as reduction in biomass. The degree of effects depends upon the species of plant being affected and is particularly noticeable in herbal plants, mainly from accumulation of sulfadiazine and penicillin in soil. The study suggested that unnecessary use of antibiotics can affect the frequency and variety of plants as well as the whole plant communities of the food crops. The nitrogen cycle is being affected by silver nanoparticles, which are found in many antibiotics and when these antibiotics are discharged in water bodies, they ultimately affect the phytoplankton that have a major role in the cycle so basically there are different factors like aquatic life, plants and soil organisms that are interlinked to the health of all other organisms of planet and these factors are being contaminated by antibiotic pollutants with hazardous effects. No doubt, antibiotics are important for survival and treatment of different infections and due to these antibiotics we have saved many lives, but their extensive use has severe effects on crops, animals as well as humans, so their use should be reduced (Zhang et al., 2017).

11.4 Types of antibiotics in soil

1. Pesticides
2. Polycyclic aromatic hydrocarbons
3. Perfluorinated compounds
4. Pharmaceutical compounds
5. Hygiene products
6. Nanoparticles made by genetic modifications

According to previous research, bioaccumulation is more complex and higher in roots of plants as compared to the upper parts. The mechanisms by which these compounds are absorbed are affected by the types of contaminants, the plant variety and the type of symbiont (Venkata et al., 2018).

11.4.1 New class of antibiotics

Soil is mostly defined as dead mud, but actually soil is the mixture of a rich variety of microbes that produce highly poisonous antibiotic compounds to other bacteria and might provide us with basic principles for exploration of new antibiotics in order to cope with the emerging multiresistant pathogens. Hover et al. (2018) reported that investigators found a new class of antibiotics from soil samples. For this purpose, 2000 samples were sent by scientists across the United States and the researchers found bacteria present in all samples. DNA of these bacterial species were examined and a new category of antibiotics called malacidins was discovered. Before application of it to humans, further research on the possible hazards is still required. This particular class showed antibiotic properties against many resistant varieties of pathogens as found by many lab experiments. The most remarkable effect was reported in case of *Staphylococcus aureus* that causes severe dermal infections in rats and is resistant to methicillin and often called a "superbug" Every year, about ten thousand lives are threatened by this superbug in the United States.

11.5 Consumption of antimicrobial agents from soil through animal dung

Antimicrobials are employed in food of livestock animals in order to improve their development. These agents are not digested completely in the GIT and are excreted out in the form of solid waste and urine. The solid waste, also known as dung, is a source of nutrients and is used commonly in agricultural practices worldwide as an organic source and a most trusted support to crops.

Application of dung to agricultural land is a common practice worldwide, including United States in this way the wastage of animals is also handled. The amount of dung applied to soil depends upon the condition of soil and need of crop. Basically the soil need and nutrient concentrations are not usually considered, so larger amounts are applied as compared to required amount. Researchers are focusing on antibiotics in dung-fertilized soil, their amount and effect on animals and humans. The widely used antibiotics are nicarbazin, penicillin, monensin, virginiamycin, tylosin, amprolium, sulfamethazine and a variety of tetracyclines including chlortetracycline and oxytetracycline. Little is known about the harmful effects of these antibiotics on surroundings, including their application to dung-fertilized soil and what happens as these reach the human body when they are consumed, such as from vegetables grown in the dung-fertilized soils (Kumar et al., 2005).

11.6 Mechanism of uptake of antimicrobial agents by plants

Plants as primary producers in the ecosystem in all the nutrient cycles provide a route for antibiotic dispersal due to their ability to absorb them. Investigators reported elevated amounts of antibiotic-resistant endophytic bacteria (AREBs) in dung-fertilized soils of agricultural lands. The most commonly emerging antibiotics in farms of China are sulfonamides, tetracyclines, and cephalosporins. Different reports appear from time to time and provide information about modifications of these antibiotics and appearance of resistance against these in soil and aquatic habitats. Park et al. (2016) tested different concentrations and combinations of these antibiotic-resistant endophytic bacteria in agricultural crops through vessel field experiments.

11.7 Animal manure, a source of antibiotics

Animal waste or dung has some quantities of antibiotics in it that are not degraded are transferred to fields during manure application experiments were conducted to estimate the possible effects of these antibiotics on plant development and growth and the mechanism by which they respond to antibiotic exposure. Phytotoxic levels of antibiotics were determined affecting the growth of plants at different stages including growth pattern at 50% MIC level. At minimum inhibitory concentration, antibiotics reduced the yield as these act as inhibitors that disturb the plant's metabolism. Different studies conclude that low concentrations of antibiotics are useful for proper growth of plants, but if concentrations are increased, they prove fatal for the plant for example, β-lactans, which is not found in plant cells, can affect the cell's liability while working over bacterial cell wall. Antibiotic uptake and accumulation in plants depends on plant species and the class of antibiotics being employed (Pan et al., 2014, Chowdhury et al., 2016).

11.8 Factors affecting uptake mechanisms of antibiotics in plants

Plants are affected by different properties of water including ionization potential, solubility of water, and different metabolites present in water. Sometimes the concentration of antibiotic in plants remains the same even if the concentration elevates in the soil. The possible reasons for this mechanism are:

1. Saturation level of these compounds in water
2. Antibiotic storage in those cells from which they can be easily removed
3. Accelerated release of antibiotics from plant during stress (Chowdhury et al., 2016).

Agricultural crops accumulating antibiotic residues are consumed by humans may develop resistance against these antibiotics and as a result, superbugs resistant to a wide range of antibiotics appear (Franklin et al., 2015).

11.9 Effect of antibiotic exposure on endosymbionts

Experiments designed to investigate the effects of antibiotic residues on endosymbionts in agricultural crops showed that these crops polluted by the residues of antibiotics may contribute to development of AREBs. Three major antibiotic classes contribute to formation of AREBs and the resistance genes in symbionts of plants, which were discussed earlier. One of the factors that contributes to their formation and growth is the natural symbiotic organisms in the soil that develops association with the nutrients in form of solution. Development of microbial populations and induction of high-level resistance is dependent on selective pressure of antibiotics; this phenomenon has been proved even in habitats with minor antibiotic concentrations (Heuer et al., 2011; Sandegren, 2014).

11.10 Role of antibiotic resistant endophytic bacteria in plant uptake

In the natural bacterial flora, the soil-dwelling and nitrifying bacteria are the sole source of symbiotic development and enter in plants through the conducting tissues, resulting in formation of antibiotic resistant endophytic bacteria in plants, which are persistent symbionts (Compant et al., 2010). Even minor residues of antibiotics accumulated in plants below MIC, may provide selection pressure for symbiotic organisms, which results in formation of AREBs and provide them selection advantage for their survival in contaminated habitats. Different genes for resistance are present in plants, e.g., *Sul1* and *Sul2* for sulfonamide and $bla_{\text{CTX-M}}$ for β-lactans resistance (Makowska et al., 2015). These genes are mostly linked with other genes for resistance e.g., *Sul1* gene is linked to *Tn21* integron and *Sul2* gene is often linked to *IncQ* and *qPCR* family of small plasmids. Diversity of resistance genes is found in antibiotic-polluted habitats that show their survival even under high selection pressure (Skold, 2000). In a study by Kang et. al., (2013), he showed the mechanism of antibiotic uptake by plants and development of multidrug-resistant pathogens in dung-fertilized soils on which agricultural crops or vegetables are sown (Yang et al., 2016). As the vegetables grow, the antibiotics are incorporated in symbionts and play a role in development of superbugs and they play a vital role in development of resistance against antibiotics when consumed by humans. Thus it is important to investigate the response of terrestrial agricultural crops, especially vegetables, to the soil antibiotics. There is need for further investigation to find factors for composition of AREBs under stress of different antibiotics in the agricultural fields. The result will serve as database for the possible risks associated with application of antibiotics and food security (Zhang et al., 2017).

References

Chowdhury, F., Langenkämper, G., Grote, M., 2016. Studies on uptake and distribution of antibiotics in red cabbage. J. Verbr. Lebensm. 11, 61–69.

Compant, S., Clément, C., Sessitsch, A., 2010. Plant growth-promoting bacteria in the rhizo- and endosphere of plants: their role, colonization, mechanisms involved and prospects for utilization. Soil Biology and Biochemistry 42, 669–678.

Franklin, A.M., Williams, C.F., Andrews, D.M., Woodward, E.E., Watson, J.E., 2015. Uptake of three antibiotics and an antiepileptic drug by wheat crops spray irrigated with wastewater treatment plant effluent. Journal of Environmental Quality 45, 546.

Heuer, H., Solehati, Q., Zimmerling, U., Kleineidam, K., Schloter, M., Müller, T., Focks, A., Thiele-Bruhn, S., Smalla, K., 2011. Accumulation of sulfonamide resistance genes in arable soils due to repeated application of manure containing sulfadiazine. Applied and Environmental Microbiology 77, 2527–2530.

Hover, B.M., Kim, S.H., Katz, M., Charlop-Powers, Z., Owen, J.G., Ternei, M.A., Maniko, J., Estrela, A.B., Molina, H., Park, S., 2018. Culture-independent discovery of the malacidins as calcium-dependent antibiotics with activity against multidrug-resistant gram-positive pathogens. Nature Microbiology 3, 415–422.

Kang, D.H., Gupta, S., Rosen, C., Fritz, V., Singh, A., Chander, Y., Murray, H., Rohwer, C., 2013. Antibiotic uptake by vegetable crops from manure-applied soils. Journal of Agricultural and Food Chemistry 61, 9992–10001.

Kumar, K., Gupta, S.C., Baidoo, S.K., Chander, Y., Rosen, C.J., 2005. Antibiotic uptake by plants from soil fertilized with animal manure. Journal of Environmental Quality 34, 2082–2085.

Makowska, N., Koczura, R., Mokracka, J., 2015. Class 1 integrase, sulfonamide and tetracycline resistance genes in wastewater treatment plant and surface water. Chemosphere 144, 1665–1673.

Pan, M., Wong, C.K.C., Chu, L.M., 2014. Distribution of antibiotics in wastewater-irrigated soils and their accumulation in vegetable crops in the Pearl River Delta, Southern China. Journal of Agricultural and Food Chemistry 62, 11062–11069.

Park, S.B., Kim, S.J., Kim, S.C., 2016. Evaluating plant uptake of veterinary antibiotics with hydroponic method. Korean Journal of Soil Science and Fertilizer 49, 242–250.

Sandegren, L., 2014. Selection of antibiotic resistance at very low antibiotic concentrations. Upsala Journal of Medical Sciences 119, 103–107.

Sköld, O., 2000. Sulfonamide resistance: mechanisms and trends. Drug Resistance Updates 3, 155–160.

Venkata, L., Pullagurala, R., Rawat, S., Adisa Jose, I.O., Torresdey, L.G., 2018. Plant uptake and translocation of contaminants of emerging concern in soil. The Science of the Total Environment 1585–1596.

Wei, R.C., Gorge, F., Zhang, L.L., Hou, X., Cao, Y.N., Gong, L., Chen, M., Wang, R., Bao, E., 2016. Occurrence of 13 veterinary drugs in animal manure-amended soils in Eastern China. Chemosphere 144, 2377–2383. PubMed.

Yang, Q.X., Zhang, H., Guo, Y.H., Tian, T.T., 2016. Influence of chicken manure fertilization on antibiotic-resistant bacteria in soil and the endophytic bacteria of pakchoi. International Journal of Environmental Research and Public Health 13, 662.

Zhang, H., Xunan, L., Qingxiang, Y., Deng, R., Linqian, B., 2017. Plant growth, antibiotic uptake and prevalence of antibiotic resistance in an endophytic system of pakchoi under antibiotic exposure. International Journal of Environmental Research and Public Health 14 (11), 1336.

CHAPTER

Modeling the spread of antibiotics and AMR/ARGs in soil

12

Srujana Kathi

Guest Faculty, Department of Ecology and Environmental Sciences, Pondicherry University, Puducherry, India

12.1 Introduction

Soil is a natural pool of antibiotics and antibiotic resistance genes (ARGs) (Xie et al., 2018). When antibiotics are applied to soil along with manure, they are necessarily going to increase the abundance and transfer of resistance genes. Antibiotics are found in groundwater in areas with rigorous livestock farming (Kivits et al., 2018). ARGs in soil affect the structure and functioning of soil microbial communities. The dissipation of antimicrobial resistance (AMR)/ARGs gets improved in the rhizosphere. Uninterrupted discharge of antibiotics from appropriated practices into soil water protracted bioavailability (Jechalke et al., 2014). Landfills are expected to be significant reservoirs of antibiotics and ARGs as they obtain unexploited and undesirable antibiotics and ARGs in municipal solid waste. ARGs may decay with age of garbage (Song et al., 2016). Beach soil is a pool of ARGs and salinity is a major aspect shaping the dispersal of soil resistome (Zhang et al., 2019).

Actual management of antibiotic jeopardies in the environment requires a consideration of the factors accountable for the appearance, diffusion, and conservation of antibiotic resistance (Greenfield et al., 2018). Mathematical modeling is now an essential part in refining our thoughts on antibiotic resistance (Opatowski et al., 2011). Birkegård et al. (2018) stated that modeling the biological procedures automatically delivers a vision into AMR incidence, transmission, and tenacity, integrating the vagueness and inconsistency of the system using stochastic modeling. Widespread sensitivity analysis and model validation should be done in forthcoming days by means of information that can contribute to model advance, parameterization, and authentication. Mathematical modeling allows us to forecast the blowout of resistance and to a small degree to regulate its undercurrents (Arepeva et al., 2015). Discerning pressure applied by the extensive use of antibacterial drugs is hastening the growth of resilient bacterial populations. Numerous vigorous models were projected to investigate the issue of bacterial resistance in relation to antibacterial substances and their usage in human and animal populations (Ramaswamy et al., 2018). Models enhance pragmatic experimentation extra proficient, less cost by evading moral impasses. Inside-host models have been based on: (1) appearance of antibiotic resistance, (2) microbial suitability and assortment for antibiotic-resistant strains associated with antibiotic-sensitive strains, and (3) antibiotic tolerance. Population-level models have been utilized to appreciate and envisage the effects of treatment

Antibiotics and Antimicrobial Resistance Genes in the Environment. https://doi.org/10.1016/B978-0-12-818882-8.00012-7

protocols like antibiotic cycling (Spicknall et al., 2013). Incidence, destiny, and ecotoxicity of antibiotics in agro-ecosystems could be addressed by source control plus reducing use and lowering environmental discharge through pretreatments of urban wastes and manures is a feasible way to lessen undesirable influences of antibiotics in agro-ecosystems (Du and Liu, 2012).

12.2 Fate and degradation of antibiotics in soil

Sewage sludge is usually designated a noteworthy pool of AMR/ARGs that could arrive in agricultural systems as fertilizer after composting. Soil kinds and the inconsistency of sludge composts could have prejudiced the destiny of ARB subsequent to the land submission of sludge composts. Fluctuations in the microbial community contributed supplements to the dynamics of ARGs in red and black soil as well as other factors, including coselection from heavy metals, horizontal gene transfer, biomass, and environmental factors (Zhang et al., 2018). Antibiotics are bioactive materials applied for living organisms as growth advocates and medicines for the conduct of various diseases. They get released into the soil environment as animal manure and biosolid adjustments in agricultural fields (Pan and Chu, 2017). The chemical nature regulates the tenacity of the antibiotic in soil (Holmes et al., 2016). Soil-metal concentrations must be determined when shaping risks due to AR in the environment and the propagation of resistance (Knapp et al., 2017). Biodegradation results in the production of different metabolites with minimized toxicity (Akram et al., 2017). Chemical conditions such as moisture content and nitrate concentrations contained by the refuse can be correlated to antibiotics and ARGs, signifying environmental factors' influence on the circulation of antibiotics and ARGs in landfill matrix (Song et al., 2016). The determination of ARGS in the soil is contingent on physicochemical properties of the remainder, physiognomies of the soil, and climatic aspects like rainfall, temperature, and humidity. Antibiotics disturb microbial biodiversity of soil microorganisms by altering their enzymatic activity and capacity to metabolize diverse carbon sources, as well as by fluctuating the overall microbial biomass and the comparative profusion of gram-positive bacteria, gram-negative bacteria, and fungi amongst microbial communities (Cycon et al., 2019). Some ARB and ARGs persist during mesophilic anaerobic digestion. Thermophilic treatments are highly operative at reducing ARB and ARGs (Youngquist et al., 2016). Usage of compost product posed elevated risk of resistance selection in soil ecosystem (Zhang et al., 2019). Horizontal gene transfer shows a vital role in the spread of ARGs from manures (Xie et al., 2018). Surface transport of antibiotics through runoff was accredited to deferred infiltration of water into the soil as surface sealing through manure and particle-bound transport (Akram et al., 2017). Depending on the antibiotic species and soil properties, residues can be transferred to groundwater and surface water through leaching and runoff and can potentially be absorbed by plants (Kim et al., 2011). Diverse plant organs and tissues have the capacity to uptake and stockpile the antibiotics, typically in roots, cotyledons, and cotyledon petioles parts (Akram et al., 2017). Antibiotics in soils assort from transport processes, sorption, and leakage processes (Kivits et al., 2018). Sorption characteristics of sulfadimethoxine, sulfamethoxazole, and sulfamethazine are contingent on pH (Park and Huwe, 2016). Establishment and degradation of all detected metabolites was dependent on soil type (Koba et al., 2017). Airborne particulate matter derived from feed yards enabled scattering of numerous veterinary antibiotics, as well as microbial communities comprising ARG (McEachran et al., 2015).

12.3 Modeling of antibiotic resistance genes in soil

Mathematical modeling is an invaluable means for decoding the mechanisms of AMR expansion and blowout, and can aid us to examine and suggest novel control strategies (Birkegård et al., 2018). Building an antibiotic-resistance model that suitably designates struggle between antibiotic-sensitive and resistant strains for a specific agent requires understanding whether inside-host strain coexistence ensues, replacement infection is probable, and strain conversion is possible (Spicknall et al., 2013). Mathematical models for spread of AMR in bacterial populations have effectively discovered the equilibrium between the fitness benefit to hosts of resistance in contradiction to the cost to hosts of plasmid carriage (Baker et al., 2016). We simulate how the population of resistant bacteria fluctuates over realistic timescales and deliberate how variations in the parameter values may modify these time courses. Through parameter variation and sensitivity analysis, we are able to draw conclusions about the importance of the model parameters, which could possibly be used in documentation of control measures to limit emerging AMR (Baker et al., 2016). Environmental antibiotic pollution has unidentified effects on resistance gene levels. Selective pressure can be appraised in surface waters, sediments, and wastewaters. Gene abundance might be correlated with antibiotic pressure by means of linear mixed models. Furthermore, antibiotic pressure and matrix should be considered in resistance risk assessment (Duarte et al., 2019). Septian et al., (2019) explored single- and bisolute sorptions of sulfadiazine (SDZ) and ciprofloxacin (CIP) onto montmorillonite and kaolinite at pH values of 5 and 8. Freundlich and Langmuir models were used and fit the experimental data well for single-solute sorption. Competitive sorption models such as Sheindorf–Rebhun–Sheintuch, Murali–Aylmore, and the modified-extended Langmuir model (MELM) were used to model bisolute systems. The MELM delivered the best prediction with SDZ sorption onto montmorillonite at pH 8 and CIP onto kaolinite at pH 5 and 8 in SDZ/CIP system happening synergistically, whereas others followed antagonistically. Mathematical modeling can be related for time series analysis. A mathematical model was developed that could well clarify the time course of *sul* gene profusion by considering the cost of *sul* genes, horizontal gene transfer, and selection of the resistant populations in the manifestation of sulfadiazine (Heuer et al., 2008).

Risk analysis of the European Union region for 12 antimicrobials using a spatial assessment was accomplished. An estimation of antibiotic release was made, exposure was assessed, consequences were modeled based on soil use, and risk was projected by combining release, exposure, and consequences using spatial multicriteria decision analysis. A final risk value for soil vulnerability was calculated for each antibiotic studied and represented in chloropleth maps (ArcGIS 9.3). Getis-Ord Gi statistic was used to identify clusters of areas at high risk for antibiotic soil contamination. They reported that enrofloxacin was the highest-risk antibiotic in the European Union, followed by tetracyclines, tylosin, and sulfadiazine. The highest risk values were found in Belgium, Ireland, Netherlands, Switzerland, Denmark, Germany, and the United Kingdom. This methodology can be applied successfully for evaluating the contamination potential of antibiotics over large areas with limited input data (De la Torre et al., 2012). Improved annotation of antibiotic resistance determinants analyses was enabled by Resfams, a new curated database of protein families and associated highly precise and accurate profile hidden Markov models, established for antibiotic resistance function and organized by ontology. Environmental and human-associated microbial communities harbor discrete resistance genes, signifying that antibiotic resistance functions are constrained by ecology (Gibson et al., 2015).

Nets-Within-Nets formalism can be used to professionally model antibiotic resistance population dynamics at both the individual, population and at the single microbiota level. This approach assimilates heterogeneous data in the same model, highly useful when generating computational models for complex biological systems. Simulations permit to overtly take into account timing and stochastic events (Bardini et al., 2018). Williams et al. (2018) deliberated models and a conforming lab that explores how a population of bacteria can advance antibiotic resistance, with stress on dismissing common delusions surrounding the mechanism of antibiotic resistance.

Concentration of the antibiotics in soils can be projected using algorithms from the Committee for Medicinal Products for Veterinary Use. These algorithms use information on typical treatment doses and durations for products containing the antibiotic, average masses of the treated animal, animal housing characteristics, manure/slurry characteristics, and restrictions on fertilizer submission to soils (William-Nguyen et al., 2016). Models based on stochastic differential equations are also used in studies of mutation and conjugation. Mutation and conjugation are imperative mechanisms for the expansion of resistance (Philipsen et al., 2010). A combinatorial method that uses an amalgamation of both continuous and categorical attribute modeling might be suitable in a digital soil mapping work flow (Malone et al., 2017). A modified form of the Collins-Selleck disinfection kinetic model was applied to study thermophilic anaerobic digestion for eliminating ARGs from residual municipal wastewater solids. This model permits the direct comparison of different operating conditions on anaerobic digestion performance in mitigating the measure of ARGs in wastewater solids and could be used to design full-scale anaerobic digesters (Burch et al., 2016). The validated models are highly useful tools to rationalize the approaches for monitoring and prediction of ARGs (Wu et al., 2019). Glushchenko et al. (2019) highlighted the agent-based VERA model, which shows simulating the spread of pathogen with interpretation of possible horizontal transfer of resistance determinants from commensal microbiota community. Exploring model behavior, they presented a number of nonlinear dependencies, together with the exponential nature of dependence of total of the diseased on average resistance of a pathogen. It is crucial to build an accurate model of the population dynamics of bacteria within warehoused manures and slurries and to contain such intricacies in order to advance effective control measures (Baker et al., 2016).

12.4 Concluding remarks

Worldwide antibiotic resistance is an increasing challenge on account of escalating release of anthropogenic pollutants. Accessible research has focused heavily on human health effects of antibiotic resistance genes. Analysis of antibiotics in agro-ecosystems is limited by several matrix inferences. Knowledge gaps between assessed and predicted ARG concentrations should be addressed sufficiently for offering acceptable models to comprehend the incidence, fate, and dissemination of antibiotics and antibiotic resistance genes in the soil.

References

Akram, R., Amin, A., Hashmi, M.Z., Wahid, A., Mubeen, M., Hammad, H.M., Fahad, S., Nasim, W., 2017. Fate of antibiotics in soil. In: Antibiotics and Antibiotics Resistance Genes in Soils. Springer, Cham, pp. 207–220.

Arepeva, M., Kolbin, A., Kurylev, A., Balykina, J., Sidorenko, S., 2015. What should be considered if you decide to build your own mathematical model for predicting the development of bacterial resistance? Recommendations based on a systematic review of the literature. Frontiers in Microbiology 6, 352.

Baker, M., Hobman, J.L., Dodd, C.E., Ramsden, S.J., Stekel, D.J., 2016. Mathematical modelling of antimicrobial resistance in agricultural waste highlights importance of gene transfer rate. FEMS Microbiology Ecology 92 (4).

Bardini, R., Di Carlo, S., Politano, G., Benso, A., 2018. Modeling antibiotic resistance in the microbiota using multi-level Petri Nets. BMC Systems Biology 12 (6), 108.

Birkegård, A.C., Halasa, T., Toft, N., Folkesson, A., Græsbøll, K., 2018. Send more data: a systematic review of mathematical models of antimicrobial resistance. Antimicrobial Resistance and Infection Control 7 (1), 117.

Burch, T.R., Sadowsky, M.J., LaPara, T.M., 2016. Modeling the fate of antibiotic resistance genes and class 1 integrons during thermophilic anaerobic digestion of municipal wastewater solids. Applied Microbiology and Biotechnology 100 (3), 1437–1444.

Cycoń, M., Mrozik, A., Piotrowska-Seget, Z., 2019. Antibiotics in the soil environment—degradation and their impact on microbial activity and diversity. Frontiers in Microbiology 10.

De la Torre, A., Iglesias, I., Carballo, M., Ramírez, P., Muñoz, M.J., 2012. An approach for mapping the vulnerability of European Union soils to antibiotic contamination. The Science of the Total Environment 414, 672–679.

Du, L., Liu, W., 2012. Occurrence, fate, and ecotoxicity of antibiotics in agro-ecosystems. A review. Agronomy for Sustainable Development 32 (2), 309–327.

Duarte, D.J., Oldenkamp, R., Ragas, A.M., 2019. Modelling environmental antibiotic-resistance gene abundance: a meta-analysis. The Science of the Total Environment 659, 335–341.

Gibson, M.K., Forsberg, K.J., Dantas, G., 2015. Improved annotation of antibiotic resistance determinants reveals microbial resistomes cluster by ecology. The ISME Journal 9 (1), 207.

Glushchenko, O.E., Prianichnikov, N.A., Olekhnovich, E.I., Manolov, A.I., Tyakht, A.V., Starikova, E.V.,, Ilina, E.I., 2019. VERA: agent-based modeling transmission of antibiotic resistance between human pathogens and gut microbiota. Bioinformatics 35 (19), 3803–3811.

Greenfield, B.K., Shaked, S., Marrs, C.F., Nelson, P., Raxter, I., Xi, C.,., Jolliet, O., 2018. Modeling the emergence of antibiotic resistance in the environment: an analytical solution for the minimum selection concentration. Antimicrobial Agents and Chemotherapy 62 (3) e01686-17.

Heuer, H., Focks, A., Lamshöft, M., Smalla, K., Matthies, M., Spiteller, M., 2008. Fate of sulfadiazine administered to pigs and its quantitative effect on the dynamics of bacterial resistance genes in manure and manured soil. Soil Biology and Biochemistry 40 (7), 1892–1900.

Holmes, A.H., Moore, L.S., Sundsfjord, A., Steinbakk, M., Regmi, S., Karkey, A.,., Piddock, L.J., 2016. Understanding the mechanisms and drivers of antimicrobial resistance. The Lancet 387 (10014), 176–187.

Jechalke, S., Heuer, H., Siemens, J., Amelung, W., Smalla, K., 2014. Fate and effects of veterinary antibiotics in soil. Trends in Microbiology 22 (9), 536–545.

Kim, K.R., Owens, G., Kwon, S.I., So, K.H., Lee, D.B., Ok, Y.S., 2011. Occurrence and environmental fate of veterinary antibiotics in the terrestrial environment. Water, Air, & Soil Pollution 214 (1–4), 163–174.

Kivits, T., Broers, H.P., Beeltje, H., van Vliet, M., Griffioen, J., 2018. Presence and fate of veterinary antibiotics in age-dated groundwater in areas with intensive livestock farming. Environmental Pollution 241, 988–998.

Knapp, C.W., Callan, A.C., Aitken, B., Shearn, R., Koenders, A., Hinwood, A., 2017. Relationship between antibiotic resistance genes and metals in residential soil samples from Western Australia. Environmental Science and Pollution Research 24 (3), 2484–2494.

Koba, O., Golovko, O., Kodešová, R., Fér, M., Grabic, R., 2017. Antibiotics degradation in soil: a case of clindamycin, trimethoprim, sulfamethoxazole and their transformation products. Environmental Pollution 220, 1251–1263.

Malone, B.P., Minasny, B., McBratney, A.B., 2017. Combining continuous and categorical modeling: digital soil mapping of soil horizons and their depths. In: Using R for Digital Soil Mapping. Springer, Cham, pp. 231–244.

McEachran, A.D., Blackwell, B.R., Hanson, J.D., Wooten, K.J., Mayer, G.D., Cox, S.B., Smith, P.N., 2015. Antibiotics, bacteria, and antibiotic resistance genes: aerial transport from cattle feed yards via particulate matter. Environmental Health Perspectives 123 (4), 337–343.

Opatowski, L., Guillemot, D., Boelle, P.Y., Temime, L., 2011. Contribution of mathematical modeling to the fight against bacterial antibiotic resistance. Current Opinion in Infectious Diseases 24 (3), 279–287.

Pan, M., Chu, L.M., 2017. Fate of antibiotics in soil and their uptake by edible crops. The Science of the Total Environment 599, 500–512.

Park, J.Y., Huwe, B., 2016. Effect of pH and soil structure on transport of sulfonamide antibiotics in agricultural soils. Environmental Pollution 213, 561–570.

Philipsen, K.R., Christiansen, L.E., Madsen, H., 2010. Nonlinear Stochastic Modelling of Antimicrobial Resistance in Bacterial Populations.

Ramaswamy, V.K., Vargiu, A.V., Malloci, G., Dreier, J., Ruggerone, P., 2018. Molecular determinants of the promiscuity of MexB and MexY multidrug transporters of Pseudomonas aeruginosa. Frontiers in microbiology 9, 1144.

Septian, A., Oh, S., Shin, W.S., 2019. Sorption of antibiotics onto montmorillonite and kaolinite: competition modelling. Environmental technology 40 (22), 2940–2953.

Song, L., Li, L., Yang, S., Lan, J., He, H., McElmurry, S.P., Zhao, Y., 2016. Sulfamethoxazole, tetracycline and oxytetracycline and related antibiotic resistance genes in a large-scale landfill, China. Science of the Total Environment 551, 9–15.

Spicknall, I.H., Foxman, B., Marrs, C.F., Eisenberg, J.N., 2013. A modeling framework for the evolution and spread of antibiotic resistance: literature review and model categorization. American Journal of Epidemiology 178 (4), 508–520.

Williams, M.A., Friedrichsen, P.J., Sadler, T.D., Brown, P.J., 2018. Modeling the emergence of antibiotic resistance in bacterial populations. The American Biology Teacher 80 (3), 214–220.

Williams-Nguyen, J., Sallach, J.B., Bartelt-Hunt, S., Boxall, A.B., Durso, L.M., McLain, J.E., Zilles, J.L., 2016. Antibiotics and antibiotic resistance in agroecosystems: state of the science. Journal of Environmental Quality 45 (2), 394–406.

Wu, D., Wang, B.H., Xie, B., 2019. Validated predictive modelling of sulfonamide and beta-lactam resistance genes in landfill leachates. Journal of environmental management 241, 123–130.

Xie, W.Y., Shen, Q., Zhao, F.J., 2018. Antibiotics and antibiotic resistance from animal manures to soil: a review. European Journal of Soil Science 69 (1), 181–195.

Youngquist, C.P., Mitchell, S.M., Cogger, C.G., 2016. Fate of antibiotics and antibiotic resistance during digestion and composting: a Review. Journal of Environmental Quality 45 (2), 537–545.

Zhang, J., Sui, Q., Tong, J., Zhong, H., Wang, Y., Chen, M., Wei, Y., 2018. Soil types influence the fate of antibiotic-resistant bacteria and antibiotic resistance genes following the land application of sludge composts. Environment International 118, 34–43.

Zhang, M., He, L.Y., Liu, Y.S., Zhao, J.L., Liu, W.R., Zhang, J.N., Ying, G.G., 2019a. Fate of veterinary antibiotics during animal manure composting. The Science of the Total Environment 650, 1363–1370.

Zhang, Y.J., Hu, H.W., Yan, H., Wang, J.T., Lam, S.K., Chen, Q.L.,., He, J.Z., 2019b. Salinity as a predominant factor modulating the distribution patterns of antibiotic resistance genes in ocean and river beach soils. The Science of the Total Environment 668, 193–203.

Further reading

Ramsay, D.E., Invik, J., Checkley, S.L., Gow, S.P., Osgood, N.D., Waldner, C.L., 2018. Application of dynamic modelling techniques to the problem of antibacterial use and resistance: a scoping review. Epidemiology and Infection 146 (16), 2014–2027.

CHAPTER

Metagenomics and methods development for the determination of antibiotics and AMR/ARGS 13

Surojeet Das, Aashna Srivastava, Sunil Kumar
Faculty of Biotechnology, Institute of Bio-Sciences and Technology, Shri Ramswaroop Memorial University, Barabanki, Uttar Pradesh, India

13.1 Introduction

Developments in genomic technologies have tendered an innovative slant for the risk assessment and monitoring of environmental health. High-throughput sequencing of the entire microbial fauna provides a better glimpse of the community and practical composition, instead of a conservative analysis which is more genes- and species-specific (Tyson and Hugenholtz, 2008). The new techniques thus adopted are more dependent on culture-independent approaches and could access genomic information from bacteria that are not culturable (Schleifer et al., 1995). A huge quantity of genomic data can be generated in a very short time by these technologies as they are more laboratory-intensive and less dependent on labor (Glenn, 2011). The direct extraction, sequencing, and analysis of DNA from a colony of microorganisms, popularly called shotgun metagenomics (Handelsman, 2004), is one high-throughput approach which is in sync with next-generation sequencing and has potential usage for maintaining a close watch on environmental public health.

The relevance of metagenomics in environmental health function is yet to be completely understood; however, this method has been utilized to trace fecal contamination in watersheds through community composition profiling (Sercu et al., 2010), identification of sewage contamination indicators (Peccia and Bibby, 2013; Andreishcheva et al., 2010), and to detect pathogens in wastewater (Ye and Zhang, 2011). However, elucidation of the huge amount of data produced throws successive challenges for public health decision makers. Keeping in view the importance of genomic signal in relation to risk, in order to reach a decision to define the genomic-level response along with identification of the financial implications of using these methods over the more transitional methods will be imperative to driving the metagenomic data into public health decision making.

Antibiotic resistance is a worldwide phenomenon and is an increasing cause of morbidity and mortality (Eisenstein et al., 2011). It is when bacteria evolve under discerning pressure that resistance occurs. This, in turn, results in resistance to the antibiotics that are used to treat their infection. Although most of the investigations related to antibiotic resistance have concentrated on pathogenic bacteria in clinical settings, antibiotic resistance determinants (ARDs) and antibiotic resistance have

Antibiotics and Antimicrobial Resistance Genes in the Environment. https://doi.org/10.1016/B978-0-12-818882-8.00013-9

been discovered to be pervasive in environmental bacteria (Wright, 2010); additionally, the major share of the resistance genes discovered in pathogenic bacteria have progressed or are obtained from environmental microbial communities (Martinez, 2009). ARDs here relate to the genomic factors connected to the presence and spreading of antibiotic resistance genes (ARGs), including mobile genetic elements (MGEs) like plasmids, phages, metal resistance genes (MRGs), and transposable elements (TEs), all of which have been depicted to coselect for ARGs (Wright, 2007). A large quantity of ARDs has been found in the antibiotic resistomes of natural environments including marine, soil, freshwater, and wastewater ecosystems (Wang et al., 2010; Fang et al., 2009). In several instances, these genes have displayed resistance to selected antibiotics (Edwards and Schmieder, 2012). Selective pressure supporting these genes may be the reason for the presence of resistance genes in the environment, including overuse of antibiotics and their misuse during clinical treatment and in aquaculture and agricultural applications alongside metal pollution. ARDs are finally transmitted into coastal systems and watersheds via animal waste, urban/agricultural runoff, and above all sewage, thus forming environmental reservoirs of ARDs. Humans can be open to these elements through various forms of food, including agricultural crops, seafood, and livestock; activities like swimming; drinking of contaminated drinking water; and direct contact with organisms laden with bacteria that are resistant to antibiotics (Boxall et al., 2013).

Infrequent and incomplete monitoring for antibiotic resistance in the marine environment has been the trend as it has majorly focused on determining antibiotic levels in various water media (Gagnon et al., 2009). Moreover, public health surveillance or management of the water quality decision structure has not accepted environmental monitoring of antibiotic resistance, probably because of a lack of data and due to risk and risk metrics. Alternatively, global efforts for surveillance have been carried out by the National Antimicrobial Resistance Monitoring System for Enteric Bacteria (Centers for Disease Control and Prevention, 2014) and the European Antimicrobial Resistance Surveillance Network (European Centre for Disease Prevention and Control, 2014) have concentrated on the pervasiveness of antibiotic usage and resistance isolated in clinical and public health settings (Sigauque et al., 2011). There is an emergent need for the identification and control of usually uncharacterized environmental ARD reservoirs due to the global extent of antibiotic resistance, and due to the surfacing of multidrug bacterial pressure, along with mounting reports of its widespread presence in the environment.

13.2 Antimicrobial analysis by the metagenomic method

It has been observed that most metagenomic studies primarily follow similar steps. The first step is to single out decontaminated DNA from specimens gathered from the environment using DNA extraction kits. These kits are particularly designed for extraction of inhibitor-free DNA from difficult-to-culture organisms, like soil and water, which otherwise cannot be achieved using other nonmetagenomic extraction kits, as usually humic and fulvic acids are not filtered out by these kits. Extraction in such kits is followed by either direct NGS or elimination with endonucleases for subsequent cloning and transformation into bacterial hosts. These transformants are then used to create libraries which are subsequently extracted out and then sequenced. Assessment of the function of genes within the clones can be evaluated through their expression in the transformant (Meyer et al., 2012).

13.3 Advances in metagenomic analysis for evaluating antimicrobial resistance

13.3.1 Human microbiome analysis

Bengtsson-Palme and team in 2015 used shotgun metagenomic sequencing in their samples. The samples were fecal specimens which were taken before and after exchange programs. On observation it was discovered that ARGs were increased by a great quantity. There was a notable increase of 7.7-fold in trimethoprim, a 2.6-fold increase in sulfonamide, and an increase of 2.6-fold in β-lactams. This increase was observed despite the fact that no antibiotics were administered during that period. However, visible variation in encoding genes for resistance, particularly to popularly used antibiotics such as β-lactams, aminoglycosides, and tetracyclines (Kristiansson et al., 2015), was observed. It was evident from the study that traveling to different environments may affect the ARG profile of the microbiome and in turn this may also result in resistant microbes causing a possible infection. However, the sequencing could not detect the low-abundance genes or taxa as culture identified the ESBL-encoding genes in Enterobacteriaceae (Pal et al., 2015; Reimer et al., 2017). The world is becoming increasingly globalized and hence it is of extreme interest to investigate further the consequences of travel on the distribution of ARGs, especially in the wake of this discovery.

In order to explore the outcome of antibiotics on the microbiome studies were conducted on the growth of intestinal ARGs in two individuals who were fit and healthy. These subjects were not exposed to quinolone antibiotics and this was done by administration of ciprofloxacin for a 6-day course of treatment (Weidenmaier et al., 2015). Two subjects experienced varied effect of antibiotics on ARG groups, in particular the class D β-lactamases. Increased intestinal ARGs were also observed in the subjects during the course of antibiotic administration. The study revealed that the ARG composition returned to their original form in about 4 weeks after the treatment in both the subjects, of course in different measures. A fixed-and-random effects model of computing selection pressure was used to compute the segment of ARGs of a particular antibiotic per dose. This when taken up in clinical practices can be utilized to understand the consequences of therapeutic regimens, especially on the intestinal microbiome. This led to clinical application of microbiomics to typify the human microbiome in order to administer a therapeutic intervention with nominal dysbiosis, which was hence established. However, this is only possible when there is proper investigation of the effect of antibiotics on the ARG pool. The sample size of two people was too restrictive to generalize the findings as there was no guarantee of reproducibility. It must be reiterated that it was due to antibiotic administration that there was a shift in the composition of the intestinal resistome. Hence, it cannot be suggested that ciprofloxacin resistance is mediated by D β-lactamases.

There was further deliberation that by administering cefprozil to healthy subjects it was shown that there was a dysbiotic impact of antibiotics influenced by the human gut microbiome. This was done by analyzing the stool after treatment and 3 months thereafter (Ouameur et al., 2016). Most subjects experienced an increase in *Lachnoclostridium bolteae* after exposure to antibiotics, with a subgroup of subjects already having *Enterobacter cloacae*. Lower initial microbiome diversity was associated with this effect. The genes which had an affect (increased) by antibiotic exposure were beta-lactamase (blaCepA), arr2 (rifampicin), and mef(G), despite the fact that exposure to antibiotics by the subjects continued to be individually explicit.

13.3.2 Pathogenomics analysis

It was stated that the diagnostics of the pathogen relies on recognition of previously acknowledged etiological agents (Patrick et al., 2013; Osei Sekyere, 2018). Regardless of the fact that there are series of tests available, for example, microscopy, immunoassays, and culture-based investigations, the etiologies of most of the samples sent to laboratories continue to be undiagnosed, probably because the etiologies may happen to be new or perhaps are yet to be targeted. This occurs in 60% of encephalitis and 40% of gastroenteritis cases (Allred et al., 2008; Keir et al., 2011).

A part solution to the above-stated limitation is provided by metagenomics, which is primarily culture independent and also pathogen agnostic. The generated sequence can utilize data in order to forecast virulence genes and resistance. Metagenomics for detection of pathogens without any previous knowledge has been exploited (Li et al., 2016a,b). Diarrhea in stool samples was investigated by the team which showed a method for the identification of tetracycline and β-lactamase ARGs as the most ubiquitous of the present ARGs. The pathogens which were mixed up with the infection contained anellovirus, parechovirus, *Clostridium perfringens*, norovirus, sapovirus, and *Clostridium difficile* (Elward et al., 2016). In a very similar study metataxonomic and metagenomic approaches were compared to techniques for appropriate culture in scientific pathology and thereby a conclusion was drawn stating that metagenomics is apt as a clinical diagnostic tool for patients who are facing pneumonia linked with a ventilator (Castro-Nallar et al., 2016). However, metagenomics is not the answer for samples which are misdiagnosed or undiagnosed, which calls for research and development on better pathogen diagnostics.

In recent studies it was reported that enterotoxigenic *E. coli* (ETEC), which induces diarrhea in healthy subjects, led to a radical alteration in the host's composition of the *E. coli* microbiome. Commensal *E. coli* was swapped with ETEC until the antibiotics were administered. Resistance of *E. coli* commensals to β-lactams and ciprofloxacin allowed the latter to reinhabit the gut 6–17 h post antibiotics administration. There was no resistance between the commensals and ETEC was reported, which was quite noteworthy. The capability of ETEC to substitute commensal *E. coli* proving it to cause diarrhea reassures us of the role pathogens play in dysbiosis (Rasko et al., 2018).

13.3.3 Soil microbiome

The cause of the risk of antibiotic resistance is the contribution of microbes which produce antibiotics found in the top soil and other aspects of the environment over the years (Wright and Perry, 2013). Intervention by humans such as the use of antibiotics in agricultural (antibiotics used in agriculture) has shown the way for an augmentation in selection pressure, which subsequently can lead to an impact on the environmental ecosystem, diversity, and distribution of the metaresistome. Substantial written examples signifying the movement of ecological ARGs into human pathogens, which conveys the point that origination of experimental resistance, probably has been from the surroundings (Pehrsson et al., 2013). In fact, there are several strong evidences to support the conclusion that environmental resistomes and genes code resistance to β-lactams.

In 2016, investigations were made into the occurrence and profusion of linezolid and floefenicol ARGs in soils next to pig corrals. High-percentage occurrence of florfenicol ARGs was established next to areas where there was heavy use of florfenicol. Widespread use of florfenicol in farm animals and the extent of soils contaminated with waste produced by pigs had a huge possibility of diffusion of florfenicol ARGs (Zhao et al., 2016). This probable diffusion of ARG makes it perturbing as ARG can proliferate within the surroundings and subsequently to people, causing health hazards. It is imperative that the use of veterinary antibiotics is very limited, thereby reducing AMR.

Discovery of traces of ARGs in a comparatively clear environment signifies AMR as a pervasive innate process which may take place without assistance from anthropogenic aggravation, although their foremost purpose may not be to intercede resistance. The primary natural function of ARGs in bacteria is still to be established; however, their utilization as fortification against competition from antibiotic producers has been acknowledged. This becomes more evident from a study conducted on various ARGs, primarily linked with fluoroquinolone, efflux pumps, sulfonamide resistance, and vancomycin in clear Arctic wetland (McLain et al., 2017). However, there is negligible proof that AMR determinants apart from conferring resistance to xenobiotics are also involved in vital cell processes inclusive of biosynthetic pathway regulation, detoxification, homeostasis, virulence, or survival and growth (Ballard et al., 2007). It is important to note that not all ARGs which occur naturally are a threat to human health, in fact, the menace they present largely depends on whether they are carrying commensals or pathogens.

To help predict and minimize the occurrence and evaluation of AMR it is important that there is a better understanding of the ecological AMR function. In fact, there is huge potential to understand the impact of human habitats and ecological systems on the development and extent of damage to nature. With the increasing traces of ARGs, even in immaculate environments, it has become imperative to use metatranscriptomic analysis to determine the degree to which they are expressed, whether moderately or entirely, and also to determine more tasks in these environments besides potential resistance (Narasimhan et al., 2016).

13.3.4 Marine environment analysis

Discovery of ARGs was made in marine environments rich in both cultivable and uncultivable organisms like rivers and oceans (Chen et al., 2013). The existence, quantity, and dissemination of ARGs may vary in different surroundings and will hugely depend on the presence or absence of anthropogenic activities. The team in their analysis carried out metagenomic profiling on areas which had human impacts, such as Pearl River Estuary in the ocean beds of the South China Sea situated south of China. Macrolide and polypeptide ARGs were the most commonly identified ARGs in the South China Sea with efflux pumps as the primary mechanism. However, there were traces of fluoroquinolone, aminoglycoside and, sulfonamide ARGs discovered in the Pearl River Creek, which draws a parallel with the commonly used antibiotics. The pristine environment witnessed a lower diversity in both genotype and resistance mechanisms than that heavily impacted by human activities. The study done by this team presented a more in-depth description of the impact of urbanization on the microbial world, such as freshwater ecosystems in this particular study, which is in stark contrast to other studies which primarily focused on only certain aspects of urbanization, such as microbial density, chemical pollution, and nutrient modification (Nasu et al., 1997). A study examining the impact of untreated (natural), partly, or completely treated sewage on the environment resistome along with bacterial community found in a river situated in India identified that ARGs like blaVIM, blaOXA-48, blaNDM, blaKPC, and blaIMP-type carbapenemases, along with m and cr-1 genes and tet(X) that correspondingly provide resistance to colistin, carbapenems, and tigecycline which were ultimately alternative antibiotics (Larsson et al., 2014; Earl et al., 2017). Proper sewage system and management is a big challenge in developing countries and inadequate treatment and sewage management can be part of the cause of AMR by dissemination of antibiotic-resistant bacteria. There may be waterborne infection due to ineffective sewage treatment, which may lead to a bigger issue of increased antibiotic use. A study revealed a unique mobile β-lactamase which hydrolyzes carbapenems. The research also brought to light seven supposedly unique ARGs, which comprise six β-lactamases and one amikacin resistance gene. There are several studies similar to this one which use metagenomics to study the aquatic environment and which showed the path to the discovery of bacitracin, sulfonamide, β-lactam, tetracycline, chloramphenicol, macrolide, and glycopeptide ARGs (Faustman et al., 2012; Li et al., 2016a,b).

13.3.5 Wastewater treatment effluents analysis

A prominent source of various types of bacteria and ARGs is wastewater treatment plants linked with human pathogens. Liquid and solid waste from various sources such as hospital, industries, and communities is collected by wastewater treatment plants (WWTPs), making them a valuable source of resident pathogens. Due to the changing microbial population, activated and digested sludge, along with in-flow can aid the proliferation of metal ARGs and ARGs through HGT. Six plasmid DNAs from two municipal WWTPs which had tetracycline and quinolone in maximum quantities were reported. This culture-independent metagenomic approach supplied more data in a reduced time span and at a lower cost, which is a better option than the traditional plasmid analysis methods (Li et al., 2015a,b).

As per several other studies in this field, WWTPs are an important sources of putative novel plasmids, which includes tetracycline resistance genes and environmental ARGs (Verner-Jeffreys et al., 2015), along with sulfonamide resistance genes (Tang et al., 2016) and β-lactam resistance genes (Staley et al., 2015).

A WWTP was challenged with an antibiotic blend of azithromycin, trimethoprim, norfloxacin, and sulfamethoxazole and then a study was carried out on the effects of antibiotics on the bacterial community (Gonzalez-Martinez et al., 2018). The results were interpreted for sul123 (sulfamethoxazole) and ermF, carA, and msrA (erythromycin). Mutations in gyrA and grlB indicated resistance to norfloxacin.

Furthermore, the effluents containing ARG, such as from places like municipal hospitals and dairy farms, leave an impact on the receiving environment, i.e., a river reservoir, which was analyzed by matching up the gene profusion for both the source and receiving environment (Baker-Austin et al., 2017). It is due to the high quantity of β-lactam resistance gene transcripts present that hospital antibiotic usage is prolonged. The contribution of high ARG levels was primarily from effluents in the receiving aquatic environments. Antibiotic usage at effluent sources was associated with noteworthy ARG expressions, which was an indication that an increase in ARG expression and distribution was directly proportional to antibiotic pollution.

A first of its kind study done by Rowe and team is of great interest as it coalesces metatranscriptomics and metagenomics, and also makes an effort to correlate ARG manifestations in the environment with antibiotic selection pressure. All previous and parallel studies have concentrated on the anthropogenic impact on the resistance in receiving waters. It is important to note here that despite the study linking the overexpressions of ARGs to antibiotic usage, there are probably other factors playing a role as well, such as the temperature of sewage and metabolic activity of the used samples.

Therefore, there is definite scope for additional studies to reinforce the correlation between ARG expression and the use of antibiotics at effluent sources (Baker-Austin et al., 2017).

13.3.6 Drinking water analysis

A research discovered that the process of chlorination contributes to an increase in ampC, aphA2, ermA, tetG, blaTEM-1, tetA, and ermB, and there is a considerable reduction in sull genes in water used for drinking. The study also confirmed that chlorination may contribute to the concentration of various ARGs, along with MGEs (mobilome) (Li et al., 2013). Most of the bacteria present, post the chlorination process were resilient to cephalothin, chloramphenicol, and trimethoprim. Additionally, another study revealed that remaining chlorine from water treated with chlorine resulted in bacterial community alteration, such as bacA, bacitracin resistance gene, and manyfold ARGs, and it was primarily transmitted by chlorine-resistant *Acidovorax* and *Pseudomonas* (Li et al., 2015a,b). Although chlorination is extensively used to purify water for the purpose of drinking, the exercise also holds good for bacteria that are resistant and ARGs. More thorough study and further research is necessary before conclusions can be drawn for recommendations.

13.3.7 Analysis of veterinary and agricultural sources

There is an extensive use of antibiotics in development, advancement, metaphylaxis, therapeutics, and prophylaxis (Osei Sekyere and Adu, 2015). China uses approximately 21,000 of the total of 97,000 metric tons of antibiotics produced annually in animal husbandry. Hence, there is an observation of an increase in the quantity of resistant bacteria residing in animal gut, amplified by the detail that almost half of the dispensed antibiotics are not soaked up in the animal gut but are discarded in the feces, which in turn exposes the environment to subtherapeutic levels of antibiotics and contributes further to AMR (Jiang et al., 2016).

A study was carried out to evaluate the diversity of the tetracycline mobilome with a swine compost sample. This led to the discovery of two new tetracycline ARGs (TRGs), namely tet(W/N/W), encoding mosaic ribosomal protection, and tet(59), encoding a tetracycline efflux pump, which were discovered along with 17 other specific TRGs. The finding of novel TRGs after years of research and efforts points to our currently limited information on the livestock metaresistome and AMR.

Until the current surfacing of the mer-1 gene, which depicted transferability from veterinary to human medicine, the influence of antibiotic usage in animal husbandry on human diseases was greatly debated (Sekyere, 2016).

The impact of composting on the microbes found on the transcriptional reaction of ARGs and in manure has been researched by recounting alterations in the resistome process due to composting. There was a specific observation made in relation to the aggregated expression of these ARGs by making a comparison between metatranscriptomic and metagenomic data for the varying centers of population of microbial post composting. Composting particularly abridged the manifestation point of TRGs, tetS-tetM-tetO-tetW, with a negligible effect on fluoroquinolone and sulfonamide resistance gene expression. Despite the fact that the microbial population altered throughout the process, it is important to note that the core resistome persisted. Again, the process reduced ARG-bearing pathogens of clinical importance, RNA viruses and bacteriophages (Wang et al., 2017).

13.4 Conclusions

NGS metagenomics, despite being an extremely important area of microbiology, still has challenges. Virome assays for example involve complicated sample and nucleic acid work-ups, although NGS of all DNA is possible in a given sample. A huge quantity of taxonomically vague sequences have been discarded. Strain resolved binning can end up being difficult and taxa which are low in number may be challenging to identify. It is difficult to study the genetic environments of detected ARGs and the phylogeny of species that possess these functions (Martinez, 2009). Depending on the type, microbes of interest and depth of sequencing microbes can be costlier, especially when it is combined with the requisite of superior technical know-how (Reimer et al., 2017). Acquiring high-quality DNA is another challenge as there is a great possibility that it may be tainted by environmental objects coextracted with them, such as fulvic and humic acids. High-performance extraction kits can however minimize the challenge to an extent, though the performance is majorly influenced by the physiochemical nature of the environment (Gomes et al., 2013).

Large-scale application of metatranscriptomics is restricted by challenges despite having extensive promise. The highest composed RNA is from ribosomal RNA, a huge quantity of which impacts the dilution of mRNA, which is the foremost objective of metatranscriptomics (Aguiar-Pulido et al., 2016). Furthermore, differentiating between host and microbial RNA can be difficult, although there are commercial fortification kits available. Third, mRNA which is the integral part of the sample before sequencing, is highly unstable and finally, reference databases are fewer as far as coverage is concerned (Narasimhan et al., 2016).

References

Allred, A.F., Kirkwood, C.D., Wang, D., Klein, E.J., Tarr, P.I., Finkbeiner, S.R., 2008. Metagenomic analysis of human diarrhea: viral detection and discovery. PLoS Pathogens 4, e1000011.

Andreishcheva, E.N., Sogin, M.L., McLellan, S.L., Huse, S.M., Mueller-Spitz, S.R., 2010. Diversity and population structure of sewage-derived microorganisms in wastewater treatment plant influent. Environmental Microbiology 12, 378–392.

Baker-Austin, C., Micallef, C., Maskell, D.J., Verner-Jeffreys, D.W., Ryan, J.J., Pearce, G.P., Rowe, W.P., 2017. Overexpression of antibiotic resistance genes in hospital effluents over time. Journal of Antimicrobial Chemotherapy 72, 1617–1623.

Ballard, J.D., Groh, J.L., Krumholz, L.R., Luo, Q., 2007. Genes that enhance the ecological fitness of *Shewanella oneidensis* MR-1 in sediments reveal the value of antibiotic resistance. Applied and Environmental Microbiology 73, 492–498.

Boxall, A.B., Feil, E.J., Wellington, E.M., Cross, P., Hawkey, P.M., Gaze, W.H., et al., 2013. The role of the natural environment in the emergence of antibiotic resistance in gram-negative bacteria. The Lancet Infectious Diseases 13, 155–165.

Castro-Nallar, E., Hoffman, E.P., Toma, I., Pérez-Losada, M., Hilton, S.K., McCaffrey, T.A., et al., 2016. Metataxonomic and metagenomic approaches vs. culture-based techniques for clinical pathology. Frontiers in Microbiology 17.

Centers for Disease Control and Prevention, 2014. National Antimicrobial Resistance Monitoring System for Enteric Bacteria (NARMS) Homepage. Available: http://www.cdc.gov/narms/.

Chen, B., Yu, K., Zhang, T., Li, X., Liang, X., Yang, Y., 2013. Metagenomic profiles of antibiotic resistance genes (ARGs) between human impacted estuary and deep ocean sediments. Environmental Science and Technology 47, 12753–12760.

Earl, A.M., Ernst, C.M., Cerqueira, G.C., Dekker, J.P., Feldgarden, M., Grad, Y.H., et al., 2017. Multi-institute analysis of carbapenem resistance reveals remarkable diversity, unexplained mechanisms, and limited clonal outbreaks. Proceedings of the National Academy of Sciences 114, 1135–1140.

Edwards, R., Schmieder, R., 2012. Insights into antibiotic resistance through metagenomic approaches. Future Microbiology 7, 73–89.

Eisenstein, B., Dantas, G., Davies, J., Bush, K., Courvalin, P., Huovinen, P., et al., 2011. Tackling antibiotic resistance. Nature Reviews Microbiology 9, 894–896.

Elward, A., Haslam, D.B., Mihindukulasuriya, K.A., Wylie, K.M., El Feghaly, R.E., Zhou, Y., et al., 2016. Metagenomic approach for identification of the pathogens associated with diarrhea in stool specimens. Journal of Clinical Microbiology 54, 368–375.

European Centre for Disease Prevention and Control, 2014. European Antimicrobial Resistance Surveillance Network (EARS-Net) Homepage. Available: http://www.ecdc.europa.eu/en/activities/surveillance/EARS-Net.

Fang, H.H., Zhang, T., Zhang, X.X., 2009. Antibiotic resistance genes in water environment. Applied Microbiology and Biotechnology 82, 397–414.

Faustman, E.M., Port, J.A., Wallace, J.C., Griffith, W.C., 2012. Metagenomic profiling of microbial composition and antibiotic resistance determinants in Puget Sound. PLoS One 7, e48000.

Gagnon, C., Francçois, M., Segura, P.A., Sauve, S., 2009. Review of the occurrence of anti-infectives in contaminated wastewaters and natural and drinking waters. Environmental Health Perspectives 117, 675–684.

Glenn, T.C., 2011. Field guide to next-generation DNA sequencers. Molecular Ecology Resources 11, 759–769.

Gomes, E.S., Lemos, E.G.d.M., Schuch, V., 2013. Biotechnology of polyketides: new breath of life for the novel antibiotic genetic pathways discovery through metagenomics. Brazilian Journal of Microbiology 44, 1007–1034.

Gonzalez-Martinez, A., Margareto, A., Rodriguez-Sanchez, A., Pesciaroli, C., Barcelo, D., Diaz-Cruz, S., Vahala, R., 2018. Linking the effect of antibiotics on partial-nitritation biofilters: performance, microbial communities and microbial activities. Frontiers in Microbiology 19, 354.

Handelsman, J., 2004. Metagenomics: application of genomics to uncultured microorganisms. Microbiology and Molecular Biology Reviews 68, 669–685.

Jiang, H., Zhao, Q., Wang, S., Du, X.D., Wang, Y., Wang, Z., et al., 2016. Prevalence and abundance of florfenicol and linezolid resistance genes in soils adjacent to swine feedlots. Scientific Reports 6, 32192.

Keir, G., Ambrose, H., Clewley, J., Granerod, J., Davies, N., Cunningham, R., et al., 2011. Diagnostic strategy used to establish etiologies of encephalitis in a prospective cohort of patients in England. Journal of Clinical Microbiology 49, 3576–3583.

Kristiansson, E., Palmgren, H., Bengtsson-Palme, J., Angelin, M., Huss, M., Kjellqvist, S., et al., 2015. The human gut microbiome as a transporter of antibiotic resistance genes between continents. Antimicrobial Agents and Chemotherapy 59, 6551–6560.

Larsson, D.J., Graham, D.W., Davies, J., Snape, J., Collignon, P., 2014. Underappreciated role of regionally poor water quality on globally increasing antibiotic resistance. Environmental Science and Technology. ACS Publications.

Li, A., Shi, P., Cheng, S., Jia, S., Zhang, T., Zhang, X.X., 2013. Metagenomic insights into chlorination effects on microbial antibiotic resistance in drinking water. Water Research 47, 111–120.

Li, A.D., Li, L.G., Zhang, T., 2015a. Exploring antibiotic resistance genes and metal resistance genes in plasmid metagenomes from wastewater treatment plants. Frontiers in Microbiology 6, 1025.

Li, B., Guo, F., Qiu, J.-W., Li, X., Deng, Y., Yang, Y., et al., 2016a. Impacts of human activities on distribution of sulfate-reducing prokaryotes and antibiotic resistance genes in marine coastal sediments of Hong Kong. FEMS Microbiology and Ecology 192, fiw128.

Li, B., Shi, P., Hu, Q., Jia, S., Zhang, T., Zhang, X.X., 2015b. Bacterial community shift drives antibiotic resistance promotion during drinking water chlorination. Environmental Science and Technology 149, 12271–12279.

Li, B., Xiao, K.-Q., Ma, L., Bao, P., Zhang, T., Zhou, X., Zhu, Y.G., 2016b. Metagenomic profiles of antibiotic resistance genes in paddy soils from South China. FEMS Microbiology and Ecology 192, fiw023.

Martinez, J.L., 2009. The role of natural environments in the evolution of resistance traits in pathogenic bacteria. Proceedings of the Royal Society B Biological Sciences 276, 2521–2530.

McLain, J.E., Diaz, K.S., Rich, V.I., 2017. Searching for antibiotic resistance genes in a pristine Arctic wetland. Journal of Contemporary Water Research & Education 160, 42–59.

Meyer, F., Gilbert, J., Thomas, T., 2012. Metagenomics – a guide from sampling to data analysis. Microbial Informatics and Experimentation 2, 3.

Narasimhan, G., Mathee, K., Cickovski, T., Aguiar-Pulido, V., Suarez-Ulloa, V., Huang, W., 2016. Metagenomics, metatranscriptomics, and metabolomics approaches for microbiome analysis: supplementary issue: bioinformatics methods and applications for big metagenomics data. Evolutionary Bioinformatics 12, S36436. EBO.

Nasu, M., YAMAGucHI, N., Kenzaka, T., 1997. Rapid in situ enumeration of physiologically active bacteria in river waters using fluorescent probes. Microbes and Environment 12, 1–8.

Osei Sekyere, J., 2018. Candida auris: a systematic review and meta-analysis of current updates on an emerging multidrug-resistant pathogen. Microbiology Open e00578.

Osei Sekyere, J., Adu, F., 2015. Prevalence of multidrug resistance among *Salmonella enterica* serovar Typhimurium isolated from pig faeces in Ashanti region, Ghana. International Journal of Antibiotics 2015.

Ouameur, A.A., Dridi, B., Raymond, F., Gingras, H., Déraspe, M., Iqbal, N., et al., 2016. The initial state of the human gut microbiome determines its reshaping by antibiotics. The ISME Journal 10, 707–720.

Pal, C., Larsson, D.G.J., Bengtsson-Palme, J., Eriksson, K.M., Thorell, K., Hartmann, M., Nilsson, R.H., 2015. METAXA2: improved identification and taxonomic classification of small and large subunit rRNA in metagenomic data. Molecular Ecology and Resources 15, 1403–1414.

Patrick, D.M., Gardy, J.L., Tang, P., Miller, R.R., Montoya, V., 2013. Metagenomics for pathogen detection in public health. Genome Medicine 5, 81.

Peccia, J., Bibby, K., 2013. Identification of viral pathogen diversity in sewage sludge by metagenome analysis. Environmental Science and Technology 47, 1945–1951.

Pehrsson, E.C., Dantas, G., Forsberg, K.J., Gibson, M.K., Ahmadi, S., 2013. Novel resistance functions uncovered using functional metagenomic investigations of resistance reservoirs. Frontiers in Microbiology 4.

Rasko, D.A., Michalski, J.M., Zanetti, L., Tennant, S.M., Richter, T.K., Chen, W.H., 2018. Responses of the human gut *Escherichia coli* population to pathogen and antibiotic disturbances. mSystems 3.

Reimer, A., Pagotto, F., Forbes, J.D., Ronholm, J., Knox, N.C., 2017. Metagenomics: the next culture-independent game changer. Frontiers in Microbiology 18, 1069.

Schleifer, K.H., Amann, R.I., Ludwig, W., 1995. Phylogenetic identification and in-situ detection of individual microbial-cells without cultivation. Microbiological Reviews 59, 143–169.

Sekyere, J.O., 2016. Current state of resistance to antibiotics of last-resort in South Africa: a review from a public health perspective. Frontiers in Public Health 4.

Sercu, B., Wu, C.H., Brodie, E.L., Wong, J., Van de Werfhorst, L.C., DeSantis, T.Z., et al., 2010. Characterization of coastal urban watershed bacterial communities leads to alternative community-based indicators. PLoS One 5, e11285.

Sigauque, B., Grundmann, H., Klugman, K.P., Ramon-Pardo, P., Walsh, T., Khan, W., et al., 2011. A framework for global surveillance of antibiotic resistance. Drug Resist Updates 14, 79–87.

Staley, C., Cotner, J.B., Phillips, J., Sadowsky, M.J., Wang, P., Gould, T.J., 2015. High-throughput functional screening reveals low frequency of antibiotic resistance genes in DNA recovered from the Upper Mississippi River. Journal of Water and Health 13, 693–703.

Tang, J., Huang, K., Ye, L., He, X., Zhang, X.-X., Bu, Y., et al., 2016. Metagenomic analysis of bacterial community composition and antibiotic resistance genes in a wastewater treatment plant and its receiving surface water. Ecotoxicology and Environmental Safety 132, 260–269.

Tyson, G.W., Hugenholtz, P., 2008. Microbiology: metagenomics. Nature 455, 481–483.

Verner-Jeffreys, D., Baker, K.S., Rowe, W., et al., 2015. Search engine for antimicrobial resistance: a cloud compatible pipeline and web interface for rapidly detecting antimicrobial resistance genes directly from sequence data. PLoS One 10, e0133492.

Wang, C., Dong, D., Strong, P., Wu, W., Zhu, W., Qin, Y., Ma, Z., 2017. Microbial phylogeny determines transcriptional response of resistome to dynamic composting processes. Microbiome 5, 103.

Wang, H.H., Allen, H.K., Davies, J., Donato, J., Handelsman, J., Cloud-Hansen, K.A., 2010. Call of the wild: antibiotic resistance genes in natural environments. Nature Reviews Microbiology 8, 251–259.

Weidenmaier, C., Huson, D.H., Autenrieth, I.B., El-Hadidi, M., Schütz, M., Willmann, M., Peter, S., 2015. Antibiotic selection pressure determination through sequence-based metagenomics. Antimicrobial Agents and Chemotherapy 59, 7335–7345.

Wright, G.D., 2007. The antibiotic resistome: the nexus of chemical and genetic diversity. Nature Reviews Microbiology 5, 175–186.

Wright, G.D., 2010. Antibiotic resistance in the environment: a link to the clinic? Current Opinion in Microbiology 13, 589–594.

Wright, G.D., Perry, J.A., 2013. The antibiotic resistance "mobilome": searching for the link between environment and clinic. Frontiers in Microbiology 14.

Ye, L., Zhang, T., 2011. Pathogenic bacteria in sewage treatment plants as revealed by 454 pyrosequencing. Environmental Science and Technology 45, 7173–7179.

Zhao, F., Ma, H., Huang, K., Tao, W., Zhang, X.X., Wang, Z., et al., 2016. High levels of antibiotic resistance genes and their correlations with bacterial community and mobile genetic elements in pharmaceutical wastewater treatment bioreactors. PLoS One 11, e0156854.

CHAPTER

Global trends in ARGs measured by HT-qPCR platforms

14

Hassan Waseem[1,2], Hamza Saleem ur Rehman[2], Jafar Ali[3], Muhammad Javed Iqbal[2,4], Muhammad Ishtiaq Ali[1]

[1]*Environmental Microbiology Laboratory, Department of Microbiology, Quaid-i-Azam University, Islamabad, Pakistan;* [2]*Department of Biotechnology, University of Sialkot, Sialkot, Pakistan;* [3]*Key Laboratory of Environmental Nanotechnology and Health Effects, Research Center for Eco-Environmental Sciences, Chinese Academy of Sciences, Beijing, China;* [4]*Department of Biochemistry and Biotechnology, University of Gujrat, Gujrat, Pakistan*

14.1 Introduction

Antimicrobial resistance (AMR) is the ability of microorganisms to resist the actions of antimicrobial agents. It is an emerging public health issue worldwide (Zurfluh et al., 2017). The abuse and inappropriate disposal of antibiotics have caused widespread dissemination of antibiotics in several environments (Waseem et al., 2017a). This has resulted in an incommensurate increase in antimicrobial-resistant bacteria, which, in turn, is influencing the treatment of infections in both humans and animals worldwide (Klein et al., 2018; Waseem et al., 2019b). AMR burdens the health care systems by taking longer treatment time, which in turn, can have severe economic impacts as well (Marston et al., 2016).

There is a diverse array of screening methods available that involve both phenotypic and genetic characterization of AMR present in individual bacterial strains and different metagenomic communities (Dhawde et al., 2018; Munir et al., 2017; Williams et al., 2017). Some of them are routinely used in diagnostic laboratories, while others are still confined to research laboratories and are in different stages of development (Georgios et al., 2014). However, testing of bacteria directly against antimicrobials, to routinely assess the activity of drugs, is the most pragmatic way of screening AMR in clinical settings.

Some of the most common phenotypic tests used in routine laboratory practices for AMR detection include dilution and disk diffusion tests. These are the two basic methods that have been developed by the Clinical and Laboratory Standards Institute (CLSI) (CLSI, 2019). Dilution tests analyze bacterial growth in agar or broth containing antimicrobials in a series of dilutions. The lowest concentration of antimicrobial agent inhibiting the observable growth of a microbe is called minimum inhibitory concentration (MIC) value. Whereas in the other method (disk diffusion), different concentrations are created by using multiple filter paper disks, each containing a single concentration of drug. Size of the resultant growth inhibition zone determines the quantitative susceptibility of the microbe to the particular drug under testing. Phenotypic resistance in nonsusceptible bacteria is often associated with the changes in the genetic makeup of bacteria. Such changes can either be intrinsic, in which spontaneous mutations change the genetic composition, or acquired where antimicrobial-resistance genes

Antibiotics and Antimicrobial Resistance Genes in the Environment. https://doi.org/10.1016/B978-0-12-818882-8.00014-0

(ARGs) are acquired from the environmental or other resistant bacteria. Phenotypic resistance determination methods are advantageous in certain situations, i.e., when resistance to the same antimicrobial agent is caused by different mechanisms. But these methods are time-consuming and laborious and, in some instances, may take up to several days in rendering the results because of slow bacterial growth (Ota et al., 2019; Samhan et al., 2017). The major problem associated with phenotypic methods is their inability to evaluate the risks conferred by the acquisition or transmission of the repertoire of resistance mechanisms. Another associated problem with phenotypic methods is their inability to detect resistance in viable but nonculturable bacteria (Ramamurthy et al., 2014). Even in the case of culturable bacteria, phenotypic methods cannot take into account the silent or unexpressed ARGs. Silent ARGs are a source of risk because they cannot only be transmitted across other bacteria via horizontal gene transfer (HGT) but may be expressed in the same organisms under different environmental conditions.

Molecular methods, including qPCR, can detect ARGs promptly with reliability. With the exponential growth and development in sequencing technologies and bioinformatics databases, the number of identified ARGs conferring resistance against hundreds of antibiotics has reached to thousands (Waseem et al., 2017b). Consequently, the need for high-throughput screens for the surveillance of these ARGs in numerous samples was greatly felt by the research community. Additionally, the development of new ARGs consistently by different bacterial species has also prompted the researchers to continuously add, modify, and replace the targets in the high-throughput screen. As a result, flexible high-throughput qPCR systems for monitoring gene expressions in wide range of samples hit the market.

HT-qPCR is one of the most advanced versions of qPCR technology. It provides high throughput, efficient sample use, and cost-effectiveness along with the already established strongholds of conventional real-time qPCR such as high sensitivity and specificity. We know that AMR needs to be monitored in different environments and populations for the purpose of general surveillance and analysis of the efficacy of different strategic interventions and treatment technologies to halt the spread of AMR dissemination. Several countries have already initiated monitoring programs at national levels for controlling antimicrobial resistance in the clinical and environmental settings (Dejsirilert et al., 2009; Karp et al., 2017; Turnidge, 2017). Some are also in the process of evolving a comprehensive monitoring program (Saleem et al., 2018). Such programs are aiming to decipher the implications of AMR prevalence and the emergence of antibiotic-resistant bacteria.

A comprehensive data analysis by our group has identified 62 relevant studies where different HT-qPCR systems have been used for analyzing, identifying, and profiling ARGs in an array of samples collected from different parts of the world (Fig. 14.1). A total of four types of HT-qPCR platforms were used in all the reported studies, but SmartChip Real-Time PCR (previously WaferGen; now TaKaRa) is the extensively employed high-throughput platform, among others. The reaction wells in SmartChip are 100 nL, and it can accommodate 5184 real-time PCR reactions in a single run. One significant advantage of using the SmartChip platform for profiling or validation of ARGs is that they offer two versions of the SmartChip, i.e., (1) predispensed chips where the PCR assays are already dispensed into the reaction wells and the user has to add only samples and PCR reagents; and (2) another, more flexible version, MyDesign chips, don't contain any PCR assays and SmartChip multisample nanodispenser can be used to add the samples and PCR assays into the chips.

We will not be discussing the specifications of other HT-qPCR platforms as that is beyond the scope of this chapter, but the relevant information about the high-throughput capacity, reaction volumes, and flexibility can be retrieved from our recently published review article (Waseem et al., 2019c). In many

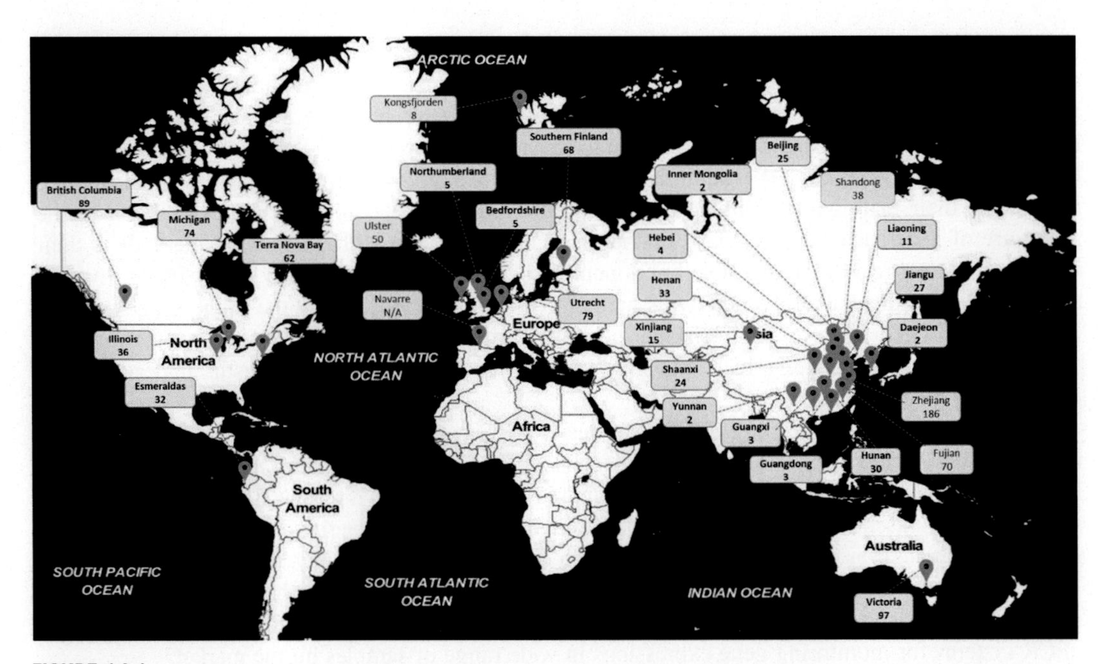

FIGURE 14.1

World map showing location and number of samples that have been analyzed on HT-qPCR technology.

of the initially published studies, high-throughput qPCR assays were performed using 296 targets comprising of ARGs, MGEs, and 16S rRNA gene (housekeeping gene) against 16 samples. Recently, the AMR panels have been evolved to contain 384 targets against four samples. The ease of flexibility and throughput of the Takara SmartChip system have been instrumental in making this system most popular among the scientific community exploring the antibiotic resistome.

Assessment and evaluation of AMR trends within and among strata is the primary constituent in any AMR-related epidemiological investigation. Such evaluations are very much essential for the identification of critical contributing factors and for designing comprehensive strategies for mitigating the spread of AMR. To assess AMR in the metagenome, DNA of the bacterial communities is first extracted from the collected samples (e.g., animal feces) and, after going through a quality and quantity check, run for amplification on HT-qPCR.

14.1.1 Use of HT-qPCR for measuring AMR in aquatic environments

Aquatic environments are very important and critical reservoirs of resistant bacteria and associated ARGs. Both resistant microbes and ARGs can be introduced into humans due to their direct contact with water. Anthropogenic activities are the leading cause of the presence of ARGs in aquatic environments, including wastewater treatment plants, rivers, and surface water, etc., and HT-qPCR technology has been extensively used by scientists to explore the intricate AMR dynamics in different aquatic environments (Table 14.1).

Table 14.1 ARG detection in aquatic samples utilizing HT-qPCR platforms.

Sr. No.	Sample type	No. of primers detected/No. Of primers in array	HT-qPCR platforms	References
1	Water from drinking water treatment plant and distribution system	188[a]/296	SmartChip Real-Time PCR System (Takara)	Xu et al. (2016)
2	River water	236/296	SmartChip Real-Time PCR System (Takara)	Zheng et al. (2017)
3	River water	212/296	SmartChip Real-Time PCR System (Takara)	Ouyang et al. (2015)
4	Urban WWTP samples	184/296	SmartChip Real-Time PCR System (Takara)	Karkman et al. (2016a)
5	Hospital sewage and wastewater treatment plant (WWTP) samples	67/84	BioMark Dynamic Array	Buelow et al. (2018)
6	Surface water and WWTP samples	190/296	SmartChip Real-Time PCR System (Takara)	Stedtfeld et al. (2016)
7	River water	36/384	SmartChip Real-Time PCR System (Takara)	Stedtfeld et al. (2017)
8	River water	124/296	SmartChip Real-Time PCR System (Takara)	Zhou et al. (2017)
9	Watershed samples	13/13	BioMark Dynamic Array	Uyaguari-Díaz et al. (2018)
10	Wastewater	66/296	SmartChip Real-Time PCR System (Takara)	Jong et al. (2018)
11	Groundwater near landfill	179/296	SmartChip Real-Time PCR System (Takara)	(Chen et al., 2017c)
12	Wastewater	168/296	SmartChip Real-Time PCR System (Takara)	Jiao et al. (2017)
13	Wastewater	125/296	SmartChip Real-Time PCR System (Takara)	Lin et al. (2016)
14	Pharmaceutical (antibiotic) contaminated wastewater	N/A/296	SmartChip Real-Time PCR System (Takara)	Tang et al. (2017)
15	Untreated hospital wastewater	178/296	Applied Biosystems ViiA 7 Real-Time PCR System (Wcgene)	(Wang et al., 2018c)
16	Water and sludge samples from WWTP	NA/296	SmartChip Real-Time PCR System (Takara)	Jiao et al. (2018)
17	Drinking water treatment plants samples	48/384	Applied Biosystems Open Array Platform	Chang et al. (2019)

Continued

Table 14.1 ARG detection in aquatic samples utilizing HT-qPCR platforms.—cont'd

Sr. No.	Sample type	No. of primers detected/No. Of primers in array	HT-qPCR platforms	References
18	Water samples (BAC treatment)	149[a]/296	SmartChip Real-Time PCR System (Takara)	(Zheng et al., 2018)
19	Sediment and water samples	75/296	SmartChip Real-Time PCR System (Takara)	Han et al. (2018)
20	Urban wastewater treatment plants (UWWTP) samples	254/384	SmartChip Real-Time PCR System (Takara)	Pärnänen et al. (2019)
21	Untreated municipal wastewater, treated municipal wastewater and drinking water samples.	36/47	BioMark Dynamic Array	Sandberg et al. (2018)
22	Estuary water samples	259/296	SmartChip Real-Time PCR System (Takara)	Zhu et al. (2017)
23	Storm drain outfall samples	35/47	BioMark Dynamic Array	Ahmed et al. (2018)
24	Fish farm sediments	71/296	SmartChip Real-Time PCR System (Takara)	Muziasari et al. (2016)

[a]*indicates the average of two or more values.*

14.1.2 Wastewater treatment plants

Wastewater treatment plants (WWTPs) are one of the most important receptors and sources of environmental AMR. The importance of surveillance of WWTPs to mitigate the dissemination of AMR is already evident (Waseem et al., 2018). The high-throughput data generated by HT-qPCR will be useful for global surveillances of AMR in wastewaters.

There are many reported studies where the AMR status of WWTPs is being investigated by employing HT-qPCR technology. For example, a trans-European AMR surveillance study has recently investigated the AMR status of the influent and effluent of the WWTPs distributed among seven European countries (Pärnänen et al., 2019). A total of 229 ARGs and 25 MGEs were detected and analyzed in a total of 168 collected samples during the study. Societal and environmental factors such as antibiotic consumption, size of the WWTP, and environmental temperatures were found to be critical for the perseverance and spread of AMR. Strong correlations in the patterns/trends between clinical and environmental AMR were also observed. Another study has also reported the use of HT-qPCR technology for deciphering the seasonal variations of AMR in a WWTP with tertiary treatment in Helsinki, Finland (Karkman et al., 2016b). A total of 296 primers targeting transposase (11 primer sets) and ARGs (285 primer sets) were used. All the transposase primers and 175 ARG primer sets were detected. The study concluded that raw sewage entering the WWTPs could be a rich source of ARGs and transposases. Additionally, it was speculated, based on the results, that a WWTP with a tertiary treatment system can effectively control AMR dissemination into the environment.

There are many instances where, apart from antibiotics, presence of chemicals or heavy metals can coselect bacterial antibiotic resistance (Lin et al., 2016; Pal et al., 2017). HT-qPCR technology has also been employed in attempts to decipher the intricate coselection mechanisms in wastewater treatment plants. In one such effort, the abundance and distribution of ARGs was investigated in different types of wastewaters (dyed and domestic) by using HT-qPCR and RNA sequencing technology (Jiao et al., 2017). The study revealed that the presence of dyeing chemicals in WWTPs could influence the HGT of ARGs. In another study, HT-qPCR technology was used to assess 39 ARGs, five metal resistance genes, and three integron genes in samples collected from nontreated and treated UWWTPs (Sandberg et al., 2018). The established HT-qPCR technology was successfully able to quantify antimicrobial resistance and other genes in environmental samples. In yet another study, the effect of chlorination on the resistance status of WWTPs was evaluated (Lin et al., 2016). A total of 296 primers (285 ARGs and 10 MGEs) were targeted in secondary effluents of a WWTP after chlorination. Surprisingly, the study concluded that, except for a small negligible percentage of ARGs (4.8%), all the detected ARGs were decreased after chlorination. Apart from conventional wastewater treatment plants, HT-qPCR technology has also been extensively utilized to explore the AMR in drinking water treatment systems and hospital wastewaters (Buelow et al., 2018; Tang et al., 2017; Wang et al., 2018c; Xu et al., 2016).

As WWTPs are scarce in developing countries, so the effects of other small-scale AMR treatment technologies have also been evaluated to gauge their impact on ARGs removal. We know that the primary purpose of bioreactors is managing wastes and producing energy and not the removal of AMR. Therefore, aerobic-denitrifying downflow hanging sponge bioreactors were cooptimized for total nitrogen and ARGs removal (Jong et al., 2018). HT-qPCR technology was used to assess the removal of ARGs and MGEs in bioreactors as a function of four different percent bypasses by volume, i.e., 0%, 10%, 20%, and 30%. A total of 296 primer sets were used during the study, out of which 59 ARGs and 7 MGEs were detected in all the assessed samples. All the systems were able to remove more than 90% of the ARGs, but the removal rates of total nitrogen and ARGs were varied in different bypasses. The cooptimal reductions of both the parameters were achieved at 20% bypass.

14.1.3 River waters

Although WWTPs are considered extremely important in maintaining water quality by removing different contaminants, they, reportedly, have limited influence on the occurrence or removal of ARGs from waters. As reported in many of the studies mentioned above, WWTPs are unable to remove the ARGs from effluent wastewater altogether. In fact, in many cases, the abundance of ARGs was found to be more in effluent than influent. The scientific community has already highlighted the fragmentary removal of ARGs from wastewaters is critically affecting the water bodies (Rodriguez-Mozaz et al., 2015).

Different research groups have employed HT-qPCR technology for the surveillance of river and other water bodies. For example, Zheng and coworkers have used this technology for characterizing the spatial distribution of ARGs in East Tioxi River, Zhejiang (Zheng et al., 2017). Similar investigations by using the same technology have also been employed on the water samples collected from other rivers of China (Ouyang et al., 2015; Zhou et al., 2017). In various instances, the HT-qPCR investigations of different river samples have been used for the validation of a new method. For example, Stedtfeld and coworkers have used HT-qPCR technology for the validation of an isothermal (LAMP) assay for the amplification of class 1 integrase gene, which is regarded as a proxy of anthropogenic pollution and total ARGs abundance (Stedtfeld et al., 2017b).

Such HT-qPCR based semiquantitative studies can provide broader insights into the aquatic dynamics of AMR more extensively. In the future, other WWTP designs and operational controls and parameters can be evaluated, and other water bodies in different parts of the world can be monitored by using the highly parallel qPCR array. Such monitoring would be essential to maximize the effectiveness of AMR containment strategies.

14.2 Use of HT-qPCR for measuring AMR in soil

Soil is known to be a sink and a significant reservoir of ARGs. Because of the vast diversity of factors influencing the soil resistome, researchers have employed the HT-qPCR technology, in different studies, for understanding the impact of various factors on different soil resistomes (Table 14.2).

Utilization of animal manure significantly increases the risks of AMR dissemination into the soil. One such recently been published study investigated the potentially varying impacts of different manure management practices, using the 384 HT-qPCR primer array, on the soil resistome and microbiome of corn and pasture fields (Chen et al., 2019). Similarly, in another study, AMR status of agricultural soils impacted with chemical fertilizers and animal manure was evaluated with the same array (Wang et al., 2018b). Like animal manure, the application of sewage sludge on agricultural land is another common irrigation practice. Sewage sludge has been reported as hot spots of AMR (Calero-Cáceres et al., 2014). The impact of the use of sewage sludge, in different doses and frequencies, on agricultural soil was evaluated by determining the presence of ARGs and MGEs utilizing a highly parallel nanofluidic HT-qPCR system. The study concluded that the application of sewage sludge in the agricultural soils had ultimately enhanced the abundance of ARGs and MGEs (Urra et al., 2019).

Salinity is known to modulate microbial composition (Tang et al., 2012) and thus can impact the related ARGs. A wide variety of ARGs was profiled using the HT-qPCR in soil samples collected from the ocean and river beaches to understand the effect of salinity on soil resistome. A total of 110 ARGs were detected in 61 soil samples (42 ocean beach; 19 river beach). The results of the high-throughput study revealed that salinity-related properties significantly influenced ARGs and salinity was found to be an important factor that shapes the dynamics of the soil resistome (Zhang et al., 2019).

Water conservation is a global challenge that can influence the existence of human life on this planet. Reclamation and reuse of effluent water, for nonpotable uses like irrigation, is an important water conservation strategy (Chen et al., 2015). Release of antibiotics residues, ARGs, and ARBs can contribute significantly to the dissemination of AMR in the environment. A high-throughput qPCR approach was used for the profiling of the ARGs in urban park soil samples. A total of 147 genes were detected, and the study concluded that reclaimed water irrigation could result in the enrichment of ARGs in soils (Wang et al., 2014).

Microcosms are artificially created small ecosystems that can imitate and predict the dynamics of natural ecosystems. Microcosms-based studies can also be employed to understand the ecological risk assessments (Rohr et al., 2016). The effects of different copper concentrations on ARGs and copper resistant genes have been investigated in laboratory microcosms (Kang et al., 2018). A 296 primer HT-qPCR array was used, and a total of 126 unique ARGs across all samples were detected. The study provided empirical evidence that copper exposure for a shorter period can cause stress to the bacterial community and influence the resistome of the soil.

Table 14.2 ARGs detection in soil samples utilizing HT-qPCR platforms.

Sr. No.	Sample type	No. of primers detected/No. Of primers in array	HT-qPCR platforms	References
1	Urban soil irrigated with reclaimed water	151/296	SmartChip Real-Time PCR System (Takara)	Wang et al. (2014)
2	Soil/manure of swine farms	149/313	Applied Biosystem Open Array Platform	Zhu et al. (2013)
3	Soil/manure of dairy cattle and swine farms	182/384	SmartChip Real-Time PCR System (Takara)	Muurinen et al. (2017)
4	Soil amended with sewage sludge and chicken manure	135/296	SmartChip Real-Time PCR System (Takara)	Chen et al. (2016)
5	Soil amended by manure and compost	144/296	SmartChip Real-Time PCR System (Takara)	Gou et al. (2018)
6	Agricultural soil	153/296	Bio-Rad CFX384™ Real-Time PCR Detection System	Hu et al. (2016)
7	Copper-contaminated agricultural soil	152/296	SmartChip Real-Time PCR System (Takara)	Kang et al. (2018)
8	Ocean and river soil samples	118/296	SmartChip Real-Time PCR System (Takara)	Zhang et al. (2019)
9	Nickel-contaminated agricultural soil	133/296	Bio-Rad CFX384™ Real-Time PCR Detection System	Hu et al. (2017)
10	Soil from Antarctica	73/384	SmartChip Real-Time PCR System (Takara)	Wang et al. (2016)
11	Soil containing sewage sludge	75/296	SmartChip Real-Time PCR System (Takara)	Xie et al. (2016)
12	Soil containing penicillin mycelial dreg (PMD)	37/55	SmartChip Real-Time PCR System (Takara)	(Wang et al., 2018a)
13	Soil samples (soil-plant systems)	36/48	SmartChip Real-Time PCR System (Takara)	Cui et al. (2018)
14	Manure-amended soil	240/296	SmartChip Real-Time PCR System (Takara)	Chen et al. (2017a)
15	Soil, rhizosphere, and phyllosphere	175/296	SmartChip Real-Time PCR System (Takara)	Chen et al. (2017b)
16	Soil affected by high anthropogenic activity	222/296	SmartChip Real-Time PCR System (Takara)	Xiang et al. (2018)
17	Soil amended with biogas slurry	86/320	Applied Biosystems ViiA 7 Real-time PCR System (Wcgene Biotechnology, Shanghai, China)	Pu et al. (2018)
18	Composite soil	80[a]/95	BioMark dynamic array	Urra et al. (2019)

Continued

Table 14.2 ARGs detection in soil samples utilizing HT-qPCR platforms.—cont'd

Sr. No.	Sample type	No. of primers detected/No. Of primers in array	HT-qPCR platforms	References
19	Metal-polluted urban soil	175/296	SmartChip Real-Time PCR System (Takara)	Zhao et al. (2019)
20	Manure-aamended soil	90/296	SmartChip Real-Time PCR System (Takara)	Zhou et al. (2019)
21	Cornfield and pasture soil	89/384	SmartChip Real-Time PCR System (Takara)	Chen et al. (2019)
22	Soil from dry land (peanut) and paddy (rice) fields	95/384	SmartChip Real-Time PCR System (Takara)	(Wang et al., 2018b)
23	High arctic soil	131/296	SmartChip Real-Time PCR System (Takara)	McCann et al. (2019)
24	Soil amended with manure and chemical fertilizers	101/296	SmartChip Real-Time PCR System (Takara)	Wolters et al. (2018)

[a]indicates the average of two or more values.

Similarly, other toxic metals such as As, Ni, Zn, etc., can also act as coselective agents and thus can contribute to the proliferation of ARGs in the environment. Soil samples from the Belfast metropolitan area, Ireland were investigated for the presence of ARGs and MGEs by employing HT-qPCR technology (Zhao et al., 2019). A total of unique 164 ARGs were detected across all the analyzed samples. Cooccurrence patterns were revealed between specific metals (As, Cd, Co, Cr, Cu. Hg, Ni, and Zn) and related ARGs. Such studies highlight the role of toxic metals in coselecting the ARGs in different soils and present a risk for the dissemination of ARGs in the environment.

Genes in pristine and noncontaminated soils plausibly represent ancestral genes that have been vertically transferred over time (Van Goethem et al., 2018). ARGs in environments can serve as a signature for the presence of archaic bacterial strains. Arctic and Antarctic regions are considered to be relatively remote and pristine environments. Soils from these regions are considered to be less or nonimpacted with the anthropogenic activities, and the resistome in such soils may comprise the ancestral gene diversity. For understanding the differences in dynamics between putative autochthonous and acquired allochthonous ARGs, it is necessary to compare the differentially impacted soils. To achieve this purpose, HT-qPCR was used to analyze the ARGs and MGEs among eight soil clusters in the Kongsfjorden region of Svalbard in the High Arctic (McCann et al., 2019). In a different but similar study, impact of soil characteristics on resistome of the soil at Terra Nova Bay was evaluated (Wang et al., 2016). A total of 62 soil samples were collected from the vicinity of old Gondwana Research Station and new Jang Bogo Research Station. Only 73 ARGs and MGEs in total were detected in all 62 samples when run against the 384 primer sets. Both the studies had identified that soil characteristics and anthropogenic activities could influence the diversity and abundance of resistant genes in the soil. Another study, investigating the spatial and temporal distribution of ARGs, had characterized and compared the ARGs in arable and pristine soils by using HT-qPCR (Xiang et al., 2018).

Characterization of such sites by using high-throughput technologies, including HT-qPCR, would be essential to understand the baseline of AMR and the subsequent establishment of benchmarks for tracking and comparing the AMR dissemination in other environmental matrices. These studies are also helpful in understanding the influence of anthropogenic activities on the distribution of ARGs in the environment.

14.3 Use of HT-qPCR for measuring AMR in gut microbiomes

The prevalence of ARGs in the gut microbiome has been considered as an increasing human health threat in recent years. The gut microbiota can serve as reservoir, accumulator, and place for transferability of ARGs across different species and environmental matrices (Table 14.3).

Animal manure has already been known as a reservoir of ARGs. Animal farms can influence the ARGs in the gut microbiome. Approximately 80% of the antibiotics produced in the United States end up being used in animal farms (Guglielmi, 2017). Due to regular and consistent antibiotic treatment on farm animals, their gut microbiome is accumulated and enriched with ARGs. Therefore, the use of HT-qPCR for decoding the AMR dynamics in the gut microbiome and associated matrices like animal farms is indispensable.

Looft and colleagues were the first ones to use HT-qPCR technology for investigating the ARGs in the gut microbiome of any animal (Looft et al., 2012). They compared two groups of farm pigs raised in highly controlled environments and examined the prevalence, abundance, and diversity of microbial functional genes and ARGs in both group of littermates, where one group received a diet containing performance-enhancing antibiotics and the other received the same diet but without the antibiotics. The phylogenetic, metagenomic, and quantitative PCR-based approaches indicated that the ARGs

Table 14.3 ARG detection in gut microbiomes utilizing HT-qPCR platforms.

Sr. No.	Sample type	No. of primers detected/No. Of primers in array	HT-qPCR platforms	References
1	Swine gut microbiome	57/85.5[a]	Applied Biosystem Open Array Platform	Looft et al. (2012)
2	Chicken gut microbiome	187/384	SmartChip Real-Time PCR System (Takara)	Guo et al. (2018)
3	Intestinal content of fish	28/384	SmartChip Real-Time PCR System (Takara)	Muziasari et al. (2017)
4	Human gut microbiome	46/384	BioMark Dynamic Array	Buelow et al. (2017)
5	Swine gut microbiome	154/296	SmartChip Real-Time PCR System (Takara)	Zhao et al. (2018)
6	Mice gut microbiome	65/384	SmartChip Real-Time PCR System (Takara)	(Stedtfeld et al., 2017a.)
7	Pig manure (microbiome)	191/296	SmartChip Real-Time PCR System (Takara)	Lu et al. (2017)

[a]*indicates the average of two or more values.*

were exponentially increased in the medicated swine microbiome despite a high background of resistance genes in nonmedicated swine. Similarly, HT-qPCR was used to investigate the ARGs in the manure of piglets and adult pigs on different diets. Concentrations of heavy metals and antibiotics were found to be positively correlated with the diversity of ARGs (Lu et al., 2017).

Chicken feces have also been known to play their role in the dissemination of ARGs into the environment (Waseem et al., 2019a). Studies were traditionally focused on investigating single microorganisms or a few genes in chicken gut microbiota and meat (Álvarez-Fernández et al., 2013; Kozak et al., 2009). Advances in HT-qPCR technology have pushed the researchers to use this technology for analyzing the AMR status in chicken feces. For example, a study was performed for investigating the resistomes in the fecal samples of chickens (Guo et al., 2018). Pools of fecal samples were collected from the production chickens and household chickens from a rural village in northern Ecuador. A qPCR array detected the high abundance of ARGs and MGEs in production chickens in comparison to household chickens (up to a 157-fold difference). Apart from agricultural animals such as cows, chicken, pigs, etc., the gut microbiomes of wild animals have also been investigated for the presence of antimicrobial resistance (Mo et al., 2018; Radhouani et al., 2014). Stedtfeld et al. (2018) during the upgradation of HT-qPCR primer array have found a significant number of ARGs detected in the DNA extracted from the fecal samples of tree shrew and bongo (Stedtfeld et al., 2018).

14.4 Conclusion

In summary, HT-qPCR technology has mostly been focused on evaluating AMR in different environmental matrices. Several studies have also investigated the ARGs in gut microbiomes of different animal species. There have been some reports of the use of this technology for the analysis of clinical or human samples, but the number of such studies is negligible. However, in the foreseeable future, this technology is expected to take its place in routine laboratory procedures as well. It is anticipated that with the technological improvements, HT-qPCR will find its way into many other environmental niches as well.

References

Ahmed, W., Zhang, Q., Lobos, A., Senkbeil, J., Sadowsky, M.J., Harwood, V.J., Saeidi, N., Marinoni, O., Ishii, S., 2018. Precipitation influences pathogenic bacteria and antibiotic resistance gene abundance in storm drain outfalls in coastal sub-tropical waters. Environment International 116, 308–318. https://doi.org/10.1016/j.envint.2018.04.005.

Álvarez-Fernández, E., Cancelo, A., Díaz-Vega, C., Capita, R., Alonso-Calleja, C., 2013. Antimicrobial resistance in *E. coli* isolates from conventionally and organically reared poultry: a comparison of agar disk diffusion and Sensi Test Gram-negative methods. Food Control 30, 227–234. https://doi.org/10.1016/J.FOODCONT.2012.06.005.

Buelow, E., Bayjanov, J.R., Majoor, E., Willems, R.J., Bonten, M.J., Schmitt, H., van Schaik, W., 2018. Limited influence of hospital wastewater on the microbiome and resistome of wastewater in a community sewerage system. FEMS Microbiology Ecology 94. https://doi.org/10.1093/femsec/fiy087.

Buelow, E., Bello González, T.D.J., Fuentes, S., de Steenhuijsen Piters, W.A.A., Lahti, L., Bayjanov, J.R., Majoor, E.A.M., Braat, J.C., van Mourik, M.S.M., Oostdijk, E.A.N., Willems, R.J.L., Bonten, M.J.M., van Passel, M.W.J., Smidt, H., van Schaik, W., 2017. Comparative gut microbiota and resistome profiling of intensive care patients receiving selective digestive tract decontamination and healthy subjects. Microbiome 5, 88. https://doi.org/10.1186/s40168-017-0309-z.

Calero-Cáceres, W., Melgarejo, A., Colomer-Lluch, M., Stoll, C., Lucena, F., Jofre, J., Muniesa, M., 2014. Sludge as a potential important source of antibiotic resistance genes in both the bacterial and bacteriophage fractions. Environmental Science and Technology 48, 7602–7611. https://doi.org/10.1021/es501851s.

Chang, F., Shen, S., Shi, P., Zhang, H., Ye, L., Zhou, Q., Pan, Y., Li, A., 2019. Antimicrobial resins with quaternary ammonium salts as a supplement to combat the antibiotic resistome in drinking water treatment plants. Chemosphere 221, 132–140. https://doi.org/10.1016/j.chemosphere.2019.01.047.

Chen, Q.L., An, X.L., Li, H., Zhu, Y.G., Su, J.Q., Cui, L., 2017a. Do manure-borne or indigenous soil microorganisms influence the spread of antibiotic resistance genes in manured soil? Soil Biology and Biochemistry 114, 229–237. https://doi.org/10.1016/J.SOILBIO.2017.07.022.

Chen, Q.L., An, X.L., Zhu, Y.G., Su, J.Q., Gillings, M.R., Ye, Z.L., Cui, L., 2017b. Application of struvite alters the antibiotic resistome in soil, rhizosphere, and phyllosphere. Environmental Science and Technology 51, 8149–8157. https://doi.org/10.1021/acs.est.7b01420.

Chen, Q.L., Li, H., Zhou, X.Y., Zhao, Y., Su, J.Q., Zhang, X., Huang, F.Y., 2017c. An underappreciated hotspot of antibiotic resistance: the groundwater near the municipal solid waste landfill. The Science of the Total Environment 609, 966–973. https://doi.org/10.1016/j.scitotenv.2017.07.164.

Chen, Q., An, X., Li, H., Su, J., Ma, Y., Zhu, Y.-G., 2016. Long-term field application of sewage sludge increases the abundance of antibiotic resistance genes in soil. Environment International 92–93, 1–10. https://doi.org/10.1016/j.envint.2016.03.026.

Chen, W., Bai, Y., Zhang, W., Lyu, S., Jiao, W., Chen, W., Bai, Y., Zhang, W., Lyu, S., Jiao, W., 2015. Perceptions of different stakeholders on reclaimed water reuse: the case of Beijing, China. Sustainability 7, 9696–9710. https://doi.org/10.3390/su7079696.

Chen, Z., Zhang, W., Yang, L., Stedtfeld, R.D., Peng, A., Gu, C., Boyd, S.A., Li, H., 2019. Antibiotic resistance genes and bacterial communities in cornfield and pasture soils receiving swine and dairy manures. Environmental Pollution 248, 947–957. https://doi.org/10.1016/j.envpol.2019.02.093.

CLSI Standards & Guidelines. https://clsi.org/standards/(Accessed 4 May, 2019).

Cui, E.P., Gao, F., Liu, Y., Fan, X.Y., Li, Z.Y., Du, Z.J., Hu, C., Neal, A.L., 2018. Amendment soil with biochar to control antibiotic resistance genes under unconventional water resources irrigation: proceed with caution. Environmental Pollution 240, 475–484. https://doi.org/10.1016/j.envpol.2018.04.143.

Dejsirilert, S., Suankratay, C., Trakulsomboon, S., Thongmali, O., Sawanpanyalert, P., Aswapokee, N., Tantisiriwat, W., 2009. National Antimicrobial Resistance Surveillance, Thailand (NARST) data among clinical isolates of *Pseudomonas aeruginosa* in Thailand from 2000 to 2005. Journal of the Medical Association of Thailand 92 (Suppl. 4). S68-75.

Dhawde, R., Macaden, R., Saranath, D., Nilgiriwala, K., Ghadge, A., Birdi, T., Dhawde, R., Macaden, R., Saranath, D., Nilgiriwala, K., Ghadge, A., Birdi, T., 2018. Antibiotic resistance characterization of environmental *E. coli* isolated from river Mula-Mutha, Pune district, India. International Journal of Environmental Research and Public Health 15, 1247. https://doi.org/10.3390/ijerph15061247.

Georgios, M., Egki, T., Effrosyni, S., 2014. Phenotypic and molecular methods for the detection of antibiotic resistance mechanisms in gram negative nosocomial pathogens. In: Trends in Infectious Diseases. https://doi.org/10.5772/57582. InTech.

Gou, M., Hu, H.W., Zhang, Y.J., Wang, J.T., Hayden, H., Tang, Y.Q., He, J.Z., 2018. Aerobic composting reduces antibiotic resistance genes in cattle manure and the resistome dissemination in agricultural soils. The Science of the Total Environment 612, 1300–1310. https://doi.org/10.1016/J.SCITOTENV.2017.09.028.

Guglielmi, G., 2017. Are antibiotics turning livestock into superbug factories? Science (-80). https://doi.org/10.1126/science.aaq0783.

Guo, X., Stedtfeld, R.D., Hedman, H., Eisenberg, J.N.S., Trueba, G., Yin, D., Tiedje, J.M., Zhang, L., 2018. Antibiotic resistome associated with small-scale poultry production in rural Ecuador. Environmental Science and Technology 52, 8165–8172. https://doi.org/10.1021/acs.est.8b01667.

Han, Y., Wang, J., Zhao, Z., Chen, J., Lu, H., Liu, G., 2018. Combined impact of fishmeal and tetracycline on resistomes in mariculture sediment. Environmental Pollution 242, 1711–1719. https://doi.org/10.1016/j.envpol.2018.07.101.

Hu, H.W., Wang, J.T., Li, J., Li, J.J., Ma, Y.B., Chen, D., He, J.Z., 2016. Field-based evidence for copper contamination induced changes of antibiotic resistance in agricultural soils. Environmental Microbiology 18, 3896–3909. https://doi.org/10.1111/1462-2920.13370.

Hu, H.W., Wang, J.T., Li, J., Shi, X.Z., Ma, Y.B., Chen, D., He, J.Z., 2017. Long-term nickel contamination increases the occurrence of antibiotic resistance genes in agricultural soils. Environmental Science and Technology 51, 790–800. https://doi.org/10.1021/acs.est.6b03383.

Jiao, Y.N., Chen, H., Gao, R.X., Zhu, Y.G., Rensing, C., 2017. Organic compounds stimulate horizontal transfer of antibiotic resistance genes in mixed wastewater treatment systems. Chemosphere 184, 53–61. https://doi.org/10.1016/j.chemosphere.2017.05.149.

Jiao, Y.N., Zhou, Z.C., Chen, T., Wei, Y.Y., Zheng, J., Gao, R.X., Chen, H., 2018. Biomarkers of antibiotic resistance genes during seasonal changes in wastewater treatment systems. Environmental Pollution 234, 79–87. https://doi.org/10.1016/J.ENVPOL.2017.11.048.

Jong, M.C., Su, J.Q., Bunce, J.T., Harwood, C.R., Snape, J.R., Zhu, Y.G., Graham, D.W., 2018. Co-optimization of sponge-core bioreactors for removing total nitrogen and antibiotic resistance genes from domestic wastewater. The Science of the Total Environment 634, 1417–1423. https://doi.org/10.1016/J.SCITOTENV.2018.04.044.

Kang, W., Zhang, Y.-J., Shi, X., He, J.-Z., Hu, H.-W., 2018. Short-term copper exposure as a selection pressure for antibiotic resistance and metal resistance in an agricultural soil. Environmental Science & Pollution Research 25, 29314–29324. https://doi.org/10.1007/s11356-018-2978-y.

Karkman, A., Johnson, T.A., Lyra, C., Stedtfeld, R.D., Tamminen, M., Tiedje, J.M., Virta, M., 2016a. High-throughput quantification of antibiotic resistance genes from an urban wastewater treatment plant. FEMS Microbiology Ecology 92. https://doi.org/10.1093/femsec/fiw014 fiw014.

Karkman, A., Johnson, T.A., Lyra, C., Stedtfeld, R.D., Tamminen, M., Tiedje, J.M., Virta, M., 2016b. High-throughput quantification of antibiotic resistance genes from an urban wastewater treatment plant. FEMS Microbiology Ecology 92. https://doi.org/10.1093/femsec/fiw014.

Karp, B.E., Tate, H., Plumblee, J.R., Dessai, U., Whichard, J.M., Thacker, E.L., Hale, K.R., Wilson, W., Friedman, C.R., Griffin, P.M., McDermott, P.F., 2017. National antimicrobial resistance monitoring system: two decades of advancing public health through integrated surveillance of antimicrobial resistance. Food-bourne Pathogens & Disease 14, 545–557. https://doi.org/10.1089/fpd.2017.2283.

Klein, E.Y., Van Boeckel, T.P., Martinez, E.M., Pant, S., Gandra, S., Levin, S.A., Goossens, H., Laxminarayan, R., 2018. Global increase and geographic convergence in antibiotic consumption between 2000 and 2015. Proceedings of the National Academy of Sciences of the United States of America 115, E3463–E3470. https://doi.org/10.1073/pnas.1717295115.

Kozak, G.K., Pearl, D.L., Parkman, J., Reid-Smith, R.J., Deckert, A., Boerlin, P., 2009. Distribution of sulfonamide resistance genes in *Escherichia coli* and Salmonella isolates from swine and chickens at abattoirs in Ontario and Québec, Canada. Applied and Environmental Microbiology 75, 5999–6001. https://doi.org/10.1128/AEM.02844-08.

Lin, W., Zhang, M., Zhang, S., Yu, X., 2016. Can chlorination co-select antibiotic-resistance genes? Chemosphere 156, 412–419. https://doi.org/10.1016/j.chemosphere.2016.04.139.

Looft, T., Johnson, T.A., Allen, H.K., Bayles, D.O., Alt, D.P., Stedtfeld, R.D., Sul, W.J., Stedtfeld, T.M., Chai, B., Cole, J.R., Hashsham, S.A., Tiedje, J.M., Stanton, T.B., 2012. In-feed antibiotic effects on the swine intestinal microbiome. Proceedings of the National Academy of Sciences of the United States of America 109, 1691–1696. https://doi.org/10.1073/pnas.1120238109.

Lu, X.M., Li, W.F., Li, C.B., 2017. Characterization and quantification of antibiotic resistance genes in manure of piglets and adult pigs fed on different diets. Environmental Pollution 229, 102–110. https://doi.org/10.1016/j.envpol.2017.05.080.

Marston, H.D., Dixon, D.M., Knisely, J.M., Palmore, T.N., Fauci, A.S., 2016. Antimicrobial resistance. Journal of the American Medical Association 316, 1193. https://doi.org/10.1001/jama.2016.11764.

McCann, C.M., Christgen, B., Roberts, J.A., Su, J.-Q., Arnold, K.E., Gray, N.D., Zhu, Y.-G., Graham, D.W., 2019. Understanding drivers of antibiotic resistance genes in High Arctic soil ecosystems. Environment International 125, 497–504. https://doi.org/10.1016/J.ENVINT.2019.01.034.

Mo, S.S., Urdahl, A.M., Madslien, K., Sunde, M., Nesse, L.L., Slettemeås, J.S., Norström, M., 2018. What does the fox say? Monitoring antimicrobial resistance in the environment using wild red foxes as an indicator. PLoS One 13. https://doi.org/10.1371/journal.pone.0198019 e0198019.

Munir, A., Waseem, H., Williams, M.R., Stedtfeld, R.D., Gulari, E., Tiedje, J.M., Hashsham, S.A., 2017. Modeling hybridization kinetics of gene probes in a DNA biochip using FEMLAB. Microarrays 6. https://doi.org/10.3390/microarrays6020009.

Muurinen, J., Stedtfeld, R., Karkman, A., Pärnänen, K., Tiedje, J., Virta, M., 2017. Influence of manure application on the environmental resistome under Finnish agricultural practice with restricted antibiotic use. Environmental Science and Technology 51, 5989–5999. https://doi.org/10.1021/acs.est.7b00551.

Muziasari, W.I., Pärnänen, K., Johnson, T.A., Lyra, C., Karkman, A., Stedtfeld, R.D., Tamminen, M., Tiedje, J.M., Virta, M., 2016. Aquaculture changes the profile of antibiotic resistance and mobile genetic element associated genes in Baltic Sea sediments. FEMS Microbiology Ecology 92. https://doi.org/10.1093/femsec/fiw052 fiw052.

Muziasari, W.I., Pitkänen, L.K., Sørum, H., Stedtfeld, R.D., Tiedje, J.M., Virta, M., 2017. The resistome of farmed fish feces contributes to the enrichment of antibiotic resistance genes in sediments below baltic sea fish farms. Frontiers in Microbiology 7, 2137. https://doi.org/10.3389/fmicb.2016.02137.

Ota, Y., Furuhashi, K., Nanba, T., Yamanaka, K., Ishikawa, J., Nagura, O., Hamada, E., Maekawa, M., 2019. A rapid and simple detection method for phenotypic antimicrobial resistance in *Escherichia coli* by loop-mediated isothermal amplification. Journal of Medical Microbiology 68, 169–177. https://doi.org/10.1099/jmm.0.000903.

Ouyang, W.Y., Huang, F.Y., Zhao, Y., Li, H., Su, J.Q., 2015. Increased levels of antibiotic resistance in urban stream of Jiulongjiang River, China. Applied Microbiology and Biotechnology 99, 5697–5707. https://doi.org/10.1007/s00253-015-6416-5.

Pal, C., Asiani, K., Arya, S., Rensing, C., Stekel, D.J., Larsson, D.G.J., Hobman, J.L., 2017. Metal resistance and its association with antibiotic resistance. In: Advances in Microbial Physiology, pp. 261–313. https://doi.org/10.1016/bs.ampbs.2017.02.001.

Pärnänen, K.M.M., Narciso-da-Rocha, C., Kneis, D., Berendonk, T.U., Cacace, D., Do, T.T., Elpers, C., Fatta-Kassinos, D., Henriques, I., Jaeger, T., Karkman, A., Martinez, J.L., Michael, S.G., Michael-Kordatou, I., O'Sullivan, K., Rodriguez-Mozaz, S., Schwartz, T., Sheng, H., Sørum, H., Stedtfeld, R.D., Tiedje, J.M., Giustina, S.V. Della, Walsh, F., Vaz-Moreira, I., Virta, M., Manaia, C.M., 2019. Antibiotic resistance in European wastewater treatment plants mirrors the pattern of clinical antibiotic resistance prevalence. Science Advances 5. https://doi.org/10.1126/sciadv.aau9124 eaau9124.

Pu, C., Liu, H., Ding, G., Sun, Y., Yu, X., Chen, J., Ren, J., Gong, X., 2018. Impact of direct application of biogas slurry and residue in fields: in situ analysis of antibiotic resistance genes from pig manure to fields. Journal of Hazardous Materials 344, 441–449. https://doi.org/10.1016/j.jhazmat.2017.10.031.

Radhouani, H., Silva, N., Poeta, P., Torres, C., Correia, S., Igrejas, G., 2014. Potential impact of antimicrobial resistance in wildlife, environment and human health. Frontiers in Microbiology 5, 23. https://doi.org/10.3389/fmicb.2014.00023.

Ramamurthy, T., Ghosh, A., Pazhani, G.P., Shinoda, S., 2014. Current perspectives on viable but non-culturable (VBNC) pathogenic bacteria. Frontiers in Public Health 2, 103. https://doi.org/10.3389/fpubh.2014.00103.

Rodriguez-Mozaz, S., Chamorro, S., Marti, E., Huerta, B., Gros, M., Sànchez-Melsió, A., Borrego, C.M., Barceló, D., Balcázar, J.L., 2015. Occurrence of antibiotics and antibiotic resistance genes in hospital and urban wastewaters and their impact on the receiving river. Water Research 69, 234–242. https://doi.org/10.1016/j.watres.2014.11.021.

Rohr, J.R., Salice, C.J., Nisbet, R.M., 2016. The pros and cons of ecological risk assessment based on data from different levels of biological organization. Critical Reviews in Toxicology 46, 756–784. https://doi.org/10.1080/10408444.2016.1190685.

Saleem, Z., Hassali, M.A., Hashmi, F.K., 2018. Pakistan's national action plan for antimicrobial resistance: translating ideas into reality. The Lancet Infectious Diseases 18, 1066–1067. https://doi.org/10.1016/S1473-3099(18)30516-4.

Samhan, F.A., Stedtfeld, T.M., Waseem, H., Williams, M.R., Stedtfeld, R.D., Hashsham, S.A., 2017. On-filter direct amplification of *Legionella pneumophila* for rapid assessment of its abundance and viability. Water Research 121. https://doi.org/10.1016/j.watres.2017.05.028.

Sandberg, K.D., Ishii, S., LaPara, T.M., 2018. A Microfluidic quantitative polymerase chain reaction method for the simultaneous analysis of dozens of antibiotic resistance and heavy metal resistance genes. Environmental Science & Technology Letters 5, 20–25. https://doi.org/10.1021/acs.estlett.7b00552.

Stedtfeld, R.D., Guo, X., Stedtfeld, T.M., Sheng, H., Williams, M.R., Hauschild, K., Gunturu, S., Tift, L., Wang, F., Howe, A., Chai, B., Yin, D., Cole, J.R., Tiedje, J.M., Hashsham, S.A., 2018. Primer set 2.0 for highly parallel qPCR array targeting antibiotic resistance genes and mobile genetic elements. FEMS Microbiology Ecology 94. https://doi.org/10.1093/femsec/fiy130.

Stedtfeld, R.D., Stedtfeld, T.M., Fader, K.A., Williams, M.R., Quensen, J., Zacharewski, T.R., Tiedje, J.M., Hashsham, S.A., 2017a. TCDD influences reservoir of antibiotic resistance genes in murine gut microbiome. FEMS Microbiology Ecology.

Stedtfeld, R.D., Stedtfeld, T.M., Waseem, H., Fitschen-Brown, M., Guo, X., Chai, B., Williams, M.R., Shook, T., Logan, A., Graham, A., Chae, J.C., Sul, W.J., VanHouten, J., Cole, J.R., Zylstra, G.J., Tiedje, J.M., Upham, B.L., Hashsham, S.A., 2017b. Isothermal assay targeting class 1 integrase gene for environmental surveillance of antibiotic resistance markers. Journal of Environmental Management 198, 213–220. https://doi.org/10.1016/j.jenvman.2017.04.079.

Stedtfeld, R.D., Williams, M.R., Fakher, U., Johnson, T.A., Stedtfeld, T.M., Wang, F., Khalife, W.T., Hughes, M., Etchebarne, B.E., Tiedje, J.M., Hashsham, S.A., 2016. Antimicrobial resistance dashboard application for mapping environmental occurrence and resistant pathogens. FEMS Microbiology Ecology 92. https://doi.org/10.1093/femsec/fiw020 fiw020.

Tang, M., Dou, X., Wang, C., Tian, Z., Yang, M., Zhang, Y., 2017. Abundance and distribution of antibiotic resistance genes in a full-scale anaerobic–aerobic system alternately treating ribostamycin, spiramycin and paromomycin production wastewater. Environmental Geochemistry and Health 39, 1595–1605. https://doi.org/10.1007/s10653-017-9987-5.

Tang, X., Xie, G., Shao, K., Bayartu, S., Chen, Y., Gao, G., 2012. Influence of salinity on the bacterial community composition in Lake Bosten, a large oligosaline lake in arid Northwestern China. Applied and Environmental Microbiology 78, 4748–4751. https://doi.org/10.1128/AEM.07806-11.

Turnidge, J., 2017. Antimicrobial use and resistance in Australia. Australian Prescriber 40, 2–3. https://doi.org/10.18773/austprescr.2017.007.

Urra, J., Alkorta, I., Mijangos, I., Epelde, L., Garbisu, C., 2019. Application of sewage sludge to agricultural soil increases the abundance of antibiotic resistance genes without altering the composition of prokaryotic communities. The Science of the Total Environment 647, 1410–1420. https://doi.org/10.1016/j.scitotenv.2018.08.092.

Uyaguari-Díaz, M.I., Croxen, M.A., Luo, Z., Cronin, K.I., Chan, M., Baticados, W.N., Nesbitt, M.J., Li, S., Miller, K.M., Dooley, D., Hsiao, W., Isaac-Renton, J.L., Tang, P., Prystajecky, N., 2018. Human activity determines the presence of integron-associated and antibiotic resistance genes in Southwestern British columbia. Frontiers in Microbiology 9, 852. https://doi.org/10.3389/fmicb.2018.00852.

Van Goethem, M.W., Pierneef, R., Bezuidt, O.K.I., Van De Peer, Y., Cowan, D.A., Makhalanyane, T.P., 2018. A reservoir of 'historical' antibiotic resistance genes in remote pristine Antarctic soils. Microbiome 6, 40. https://doi.org/10.1186/s40168-018-0424-5.

Wang, B., Li, G., Cai, C., Zhang, J., Liu, H., 2018a. Assessing the safety of thermally processed penicillin mycelial dreg following the soil application: organic matter's maturation and antibiotic resistance genes. The Science of the Total Environment 636, 1463–1469. https://doi.org/10.1016/j.scitotenv.2018.04.288.

Wang, F.H., Qiao, M., Su, J.Q., Chen, Z., Zhou, X., Zhu, Y.G., 2014. High throughput profiling of antibiotic resistance genes in urban park soils with reclaimed water irrigation. Environmental Science and Technology 48, 9079–9085. https://doi.org/10.1021/es502615e.

Wang, F., Stedtfeld, R.D., Kim, O.S., Chai, B., Yang, L., Stedtfeld, T.M., Hong, S.G., Kim, D., Lim, H.S., Hashsham, S.A., Tiedje, J.M., Sul, W.J., 2016. Influence of soil characteristics and proximity to antarctic research stations on abundance of antibiotic resistance genes in soils. Environmental Science and Technology 50, 12621–12629. https://doi.org/10.1021/acs.est.6b02863.

Wang, F., Xu, M., Stedtfeld, R.D., Sheng, H., Fan, J., Liu, M., Chai, B., Soares de Carvalho, T., Li, H., Li, Z., Hashsham, S.A., Tiedje, J.M., 2018b. Long-Term effect of different fertilization and cropping systems on the soil antibiotic resistome. Environmental Science and Technology 52, 13037–13046. https://doi.org/10.1021/acs.est.8b04330.

Wang, Q., Wang, P., Yang, Q., 2018c. Occurrence and diversity of antibiotic resistance in untreated hospital wastewater. The Science of the Total Environment 621, 990–999. https://doi.org/10.1016/j.scitotenv.2017.10.128.

Waseem, H., Williams, M.R., Stedtfeld, R.D., Hashsham, S., 2017a. Antimicrobial resistance in the environment. Water Environment Research 89, 921–941. https://doi.org/10.2175/106143017X15023776270179.

Waseem, H., Ali, J., Jamal, A., Ali, M.I., 2019a. Potential dissemination of antimicrobial resistance from small scale poultry slaughter houses in Pakistan. Applied Ecology and Environmental Research 17, 3049–3063. https://doi.org/10.15666/aeer/1702_30493063.

Waseem, H., Ali, J., Sarwar, F., Khan, A., Rehman, H.S.U., Choudri, M., Arif, N., Subhan, M., Saleem, A.R., Jamal, A., Ali, M.I., 2019b. Assessment of knowledge and attitude trends towards antimicrobial resistance (AMR) among the community members, pharmacists/pharmacy owners and physicians in district Sialkot, Pakistan. Antimicrobial Resistance and Infection Control 8, 67. https://doi.org/10.1186/s13756-019-0517-3.

Waseem, H., Jameel, S., Ali, J., Saleem Ur Rehman, H., Tauseef, I., Farooq, U., Jamal, A., Ali, M., 2019c. Contributions and challenges of high throughput qPCR for determining antimicrobial resistance in the environment: a critical review. Molecules 24, 163. https://doi.org/10.3390/molecules24010163.

Waseem, H., Williams, M.R., Jameel, S., Hashsham, S.A., 2018. Antimicrobial resistance in the environment. Water Environment Research 90, 865–884. https://doi.org/10.2175/106143018X15289915807056.

Waseem, H., Williams, M.R., Stedtfeld, T., Chai, B., Stedtfeld, R.D., Cole, J.R., Tiedje, J.M., Hashsham, S.A., 2017b. Virulence factor activity relationships (VFARs): a bioinformatics perspective. Environmental Science Process and Impacts 19. https://doi.org/10.1039/c6em00689b.

Williams, M.R., Stedtfeld, R.D., Waseem, H., Stedtfeld, T., Upham, B., Khalife, W., Etchebarne, B., Hughes, M., Tiedje, J.M., Hashsham, S.A., 2017. Implications of direct amplification for measuring antimicrobial resistance using point-of-care devices. Analytical Methods 9. https://doi.org/10.1039/c6ay03405e.

Wolters, B., Fornefeld, E., Jechalke, S., Su, J.Q., Zhu, Y.G., Sørensen, S.J., Smalla, K., Jacquiod, S., 2018. Soil amendment with sewage sludge affects soil prokaryotic community composition, mobilome and resistome. FEMS Microbiology Ecology 95. https://doi.org/10.1093/femsec/fiy193.

Xiang, Q., Chen, Q.L., Zhu, D., An, X.L., Yang, X.R., Su, J.Q., Qiao, M., Zhu, Y.G., 2018. Spatial and temporal distribution of antibiotic resistomes in a peri-urban area is associated significantly with anthropogenic activities. Environmental Pollution 235, 525–533. https://doi.org/10.1016/j.envpol.2017.12.119.

Xie, W.Y., McGrath, S.P., Su, J.Q., Hirsch, P.R., Clark, I.M., Shen, Q., Zhu, Y.G., Zhao, F.J., 2016. Long-Term impact of field applications of sewage sludge on soil antibiotic resistome. Environmental Science and Technology 50, 12602–12611. https://doi.org/10.1021/acs.est.6b02138.

Xu, L., Ouyang, W., Qian, Y., Su, C., Su, J., Chen, H., 2016. High-throughput profiling of antibiotic resistance genes in drinking water treatment plants and distribution systems. Environmental Pollution 213, 119–126. https://doi.org/10.1016/j.envpol.2016.02.013.

Zhang, Y.J., Hu, H.W., Yan, H., Wang, J.T., Lam, S.K., Chen, Q.L., Chen, D., He, J.Z., 2019. Salinity as a predominant factor modulating the distribution patterns of antibiotic resistance genes in ocean and river beach soils. The Science of the Total Environment 668, 193–203. https://doi.org/10.1016/j.scitotenv.2019.02.454.

Zhao, Y., Cocerva, T., Cox, S., Tardif, S., Su, J.-Q., Zhu, Y.-G., Brandt, K.K., 2019. Evidence for co-selection of antibiotic resistance genes and mobile genetic elements in metal polluted urban soils. The Science of the Total Environment 656, 512–520. https://doi.org/10.1016/J.SCITOTENV.2018.11.372.

Zhao, Y., Su, J.Q., An, X.L., Huang, F.Y., Rensing, C., Brandt, K.K., Zhu, Y.G., 2018. Feed additives shift gut microbiota and enrich antibiotic resistance in swine gut. The Science of the Total Environment 621, 1224–1232. https://doi.org/10.1016/j.scitotenv.2017.10.106.

Zheng, J., Chen, T., Chen, H., 2018. Antibiotic resistome promotion in drinking water during biological activated carbon treatment: is it influenced by quorum sensing? The Science of the Total Environment 612, 1–8. https://doi.org/10.1016/j.scitotenv.2017.08.072.

Zheng, J., Gao, R., Wei, Y., Chen, T., Fan, J., Zhou, Z., Makimilua, T.B., Jiao, Y., Chen, H., 2017. High-throughput profiling and analysis of antibiotic resistance genes in East Tiaoxi River, China. Environmental Pollution 230, 648–654. https://doi.org/10.1016/j.envpol.2017.07.025.

Zhou, X., Qiao, M., Su, J.-Q., Wang, Y., Cao, Z.-H., Cheng, W.-D., Zhu, Y.-G., 2019. Turning pig manure into biochar can effectively mitigate antibiotic resistance genes as organic fertilizer. The Science of the Total Environment 649, 902–908. https://doi.org/10.1016/J.SCITOTENV.2018.08.368.

Zhou, Z.-C., Zheng, J., Wei, Y.-Y., Chen, T., Dahlgren, R.A., Shang, X., Chen, H., 2017. Antibiotic resistance genes in an urban river as impacted by bacterial community and physicochemical parameters. Environmental Science & Pollution Research 24, 23753–23762. https://doi.org/10.1007/s11356-017-0032-0.

Zhu, Y.-G., Johnson, T.A., Su, J.-Q., Qiao, M., Guo, G.-X., Stedtfeld, R.D., Hashsham, S.A., Tiedje, J.M., 2013. Diverse and abundant antibiotic resistance genes in Chinese swine farms. Proceedings of the National Academy of Sciences of the United States of America 110, 3435–3440. https://doi.org/10.1073/pnas.1222743110.

Zhu, Y.-G., Zhao, Y., Li, B., Huang, C.-L., Zhang, S.-Y., Yu, S., Chen, Y.-S., Zhang, T., Gillings, M.R., Su, J.-Q., 2017. Continental-scale pollution of estuaries with antibiotic resistance genes. Nature Microbiology 2, 16270. https://doi.org/10.1038/nmicrobiol.2016.270.

Zurfluh, K., Bagutti, C., Brodmann, P., Alt, M., Schulze, J., Fanning, S., Stephan, R., Nüesch-Inderbinen, M., 2017. Wastewater is a reservoir for clinically relevant carbapenemase- and 16s rRNA methylase-producing Enterobacteriaceae. International Journal of Antimicrobial Agents 50, 436–440. https://doi.org/10.1016/j.ijantimicag.2017.04.017.

CHAPTER

15 Databases, multiplexed PCR, and next-generation sequencing technologies for tracking AMR genes in the environment

Kanwal Rehman[1], Komal Jabeen[1,2], Tahir Ali Chohan[3], Muhammad Sajid Hamid Akash[4]

[1]*Department of Pharmacy, University of Agriculture, Faisalabad, Pakistan;* [2]*Institute of Physiology and Pharmacology, University of Agriculture, Faisalabad, Pakistan;* [3]*Institute of Pharmaceutical Sciences, University of Veterinary and Animal Sciences, Lahore, Pakistan;* [4]*Department of Pharmaceutical Chemistry, Government College University Faisalabad, Faisalabad, Pakistan*

15.1 Introduction

The environment is progressively documented for the participation of antimicrobial resistance (AMR) as a worldwide burden. AMR is basically a resistance to the effects of medications that are successfully used to treat specific microbial infections. AMR has extended to all regions and countries including Pakistan where abuse and/or overuse of antimicrobials is a major causative factor to provoke the burden of infections due to the resistance of microbes such as viruses, bacteria, fungi, and parasites. Resistance that can arise through different mechanisms is primarily acquired resistance by one species from another, which is a developed resistance in certain kind of species and is also termed as natural resistance and genetic mutation (Nelson and Williams, 2014). In the 21st century on of the most consequential issues in the environment is antimicrobial resistance. The use, misuse, and excessive use of antibiotics, as well as extensive distribution of antibiotics in a variety of environmental fields and their inappropriate clearance, has resulted in an inconsistent growing of AMR that is disturbing the treatment of infections in humans globally (O'Neill). AMR affects a large area of the health care system, including many health sectors, ultimately having a very large impact on the whole world population. Various kinds of microbes exist on surfaces, in water, and in air that are opportunistic microbes/pathogens and unkind toward human health, immune system and make humans more susceptible to infections. Several opportunistic pathogens and microbes that have been isolated from the environment and associated with humans are *Pantoea conspicua*, *Staphylococcus haemolyticus*, *Staphylococcus hominis*, *S. aureus*, *Acinetobacter pittii*, and members of the family of Enterobacteriaceae; as well, Corynebacterium and Propionibacterium have also been identified by culture-dependent and other advanced techniques (Castro et al., 2004; Checinska et al., 2015; Ichijo et al., 2016; Liu et al., 2012). Mutation due to AMR occurs by both possession of mobile genetic elements and alteration in chromosomal genes (Fu et al., 2013; Zhang et al., 2011). Globally and

Antibiotics and Antimicrobial Resistance Genes in the Environment. https://doi.org/10.1016/B978-0-12-818882-8.00015-2

regionally the financial and social burdens that are related with the AMR are quantified with the addition of insufficient form of evidence (Organization, 2014). The mortality ratio worldwide due to AMR is predicated to be close to 700,000 annually, and the ratio is expected to increase in 2050 to more than 10 million per year. These gradually augmented ratios are causing a serious health threat globally resulting from the incorporation of environmental dispersal of antibiotic resistance genes (ARGs) and infection of antibiotic-resistance bacteria (ARB) (O'Neil, 2014; Resistance, 2016). A 2014 surveillance by WHO was conducted in various regions of the world focusing on antimicrobial-resistance patterns at country level. This study found shortcomings in date allocation about the resistance of antibiotics in food chain, gaps in surveillance, and the lack of consent on methodology, along with other issues (Organization, 2014). Nowadays, a variety of screening technologies are available that are involved in both genotyping and phenotyping characteristics of AMRs. In diagnostic laboratories routinely used methods are usage of AMR genes tracking and other research tools like multiplexed polymerase chain reaction (PCR) and next-generation sequencing (NGS), which are being used in various stages by specialists and academicians. Currently various kinds of PCRs, including real-time, standard, and multiplex PCRs, are used to detect the presence of AMR genes (Chamberlain, 1990). And the NGS technology has improved the potential of quickly characterizing the AMR genes carried by a bacterium. The use of PCR in diagnostic laboratories is limited due to high cost and inaccessibility of sufficient sample volumes required for analysis. To increase the diagnostic competence of PCR and reduce the associated shortcomings, an alternative methodology is multiplex PCR. In multiplex PCR more than one primer pair is used in a reaction mixture and more than one target sequence can be amplified. This technique is preferable due to effortless, time saving, and less consumption of test sample within the laboratory. Multiplexed PCR has been used successfully in various applications like analysis of genes and RNA detection, diagnosis of nucleic acid, polymorphism/mutation, and quantitative analysis. Moreover, multiplex PCR has also been used in the field of antimicrobial infection for identification of bacteria, fungi, viruses, and parasites (Chamberlain et al., 1988; Jin et al., 1996; Rithidech et al., 1997; Sherlock et al., 1998; Shuber et al., 1993; Zimmermann et al., 1996). Although a greater number of techniques are being used for tracking the phenotyping and genotyping characterization of AMR genes in the environment, there are certain limitations in each of these methods, however, multiplex PCR and NGS technologies provide the in-depth information and identification of novel mechanisms of antimicrobial resistance. Further, it is considered fundamental to monitor the microbial resistance present in microbes not only in clinical settings but also in human samples to obtain perceptions about the database baseline levels of AMR genes present in the environment. Hence, this chapter will include detailed discussion about these techniques that are significantly important to perform screening of both genotype and phenotype.

15.2 Databases of antimicrobial resistance genes in the environment

One of the alarming situations is antibiotic resistance, because in the clinical environment microbial antibiotic resistance range is increasing and paths for developing new antibiotics are decreasing day by day. The bacteria develop resistance through mutation and/or polymorphism of gene resistance, due to which it is difficult to treat microbial infection (Cooper and Shlaes, 2011). For characterization and identification of such mutated and/or polymorphic genes, manually created databases are used, such as the Antibiotic Resistance (AR) Genes Database, which is helpful in providing the data against AMR. This AR database gives complete knowledge of mechanism of action, resistance profile, protein database, ontology, and genetic sequence (Liu and Pop, 2008).

According to recent studies, it is estimated that in the United States, one out of six, or about 48 million people, suffered from food-borne diseases. About 128,000 out of them were hospitalized and 3000 of them died annually (Control and Prevention, 2011) (http://www.cdc.gov/foodborneburden). Various studies have been done to describe the particular database information such as MvirDB (http://mvirdb.llnl.gov/) that focus on the microbial virulence genes, primarily helpful in characterization and identification of microbial virulence databases and investigating another study that is antibiotic resistance gene Online (ARGO) at Lahey Clinic (http://www.lahey.org/Studies/). The feature of β-lactamase ontology was primary introduced by Comprehensive Antibiotic Resistance Databases (CARD; http://arpcard.mcmaster.ca/).

Penicillin was discovered by Alexander Fleming in 1928. It totally changed the treatment of bacterial infections. The use of antibiotics is increasing day by day and this also increases the number of antibiotic resistance microbes. Recent studies in United States showed that methicillin-resistant *Staphylococcus aureus* (MRSA) caused more deaths as compared to HIV (Bancroft, 2007). Nowadays, more challenges are arising in the form of extensively drug and multidrug-resistant tuberculosis (XDR-TB and MDR- TB, respectively) (Sekiguchi et al., 2007).

Antibiotic resistance is a major health concern globally and there is a need for international efforts to control its emergence. A large amount of data is present that gives information about the occurrence of pathogenic resistance in humans, and this data shows that different part of Europe are resistant to different classes of antibiotics (Goossens and Sprenger, 1998). Antibiotic resistance rates are higher in central and southern Europe as compared to northern countries of Europe. One of the main leading causes of antibiotic resistance is irrational usage of antibiotics (Bronzwaer et al., 2002).

The data regarding the antibiotic-resistance genes such as macrolide-lincosamide-streptogramin (MLS) and tetracycline can be found from database nomenclature of tetracycline and MLS (http://faculty.washington.edu/marilynr/). The institute that produces the Pasteur database (http://bigsdb.web.pasteur.fr/) delivers multilocus sequence type information for *Klebsiella pneumoniae*.

The worldwide health burden of AMR database is its impact on the effectiveness of antibiotic efficacy to cure the infections. Mortality rate exceeds up to 700,000 annually due to antibiotic resistance. Databases provide new approaches for tracking AMR genes globally. Abundance of AMR genes varies in different countries, e.g., the highest AMR was found in Brazil, South Asia, and African countries. The spectrum of resistance was decreased in Australia and New Zealand (Organization, 2014). A larger number of AMR genes encode resistance mainly toward β-lactams, tetracyclines, sulfonamides, and macrolides. Phenicol and sulfonamide resistance are most abundant in African and Asian populations, while the United States and European countries had a large proportion of gene resistance against macrolides (Roberts, 2008).

The new, best approach for tracking the microbial and pathogenic resistance in the environment is databases. The institute that is National Center for Biotechnology Information (NCBI) (http://www.ncbi.nlm.nih.gov/nuccore) primarily struggled to inaugurate the international databases, for evaluation and detection of epidemic pressure at the international level in the form of widespread network of sequencers that play a vital role as national private and public health care assets.

15.3 Techniques used for tracking the AMR genes in the environment

The detection of antimicrobial vulnerability of a clinical isolate is fundamental for most favorable antimicrobial therapy of infected individuals. Due to growing species of microorganisms the determination of resistance mechanism in microbes is needed and necessary (Fluit et al., 2000; Fluit et al., 1999). Testing is mandatory not only for management of therapy but also to examine the microbial resistant or the resistance of genes in the environment. For instance, in *Neisseria gonorrhoeae* the existence of β-lactamase is correlated with the resistance against the action of penicillin. The expansion of resistance genes is dependent upon the level/mode of action; gene resistance is not basically related to the failure of treatment but expression level of therapy may be low. Several standard protocols are used to predict the AMR genes in the environment and databases, multiplexed PCR and NDS are cost effective and highly predictive for tracking the resistant genes in the environment (Liu et al., 1992; Livermore, 1987; Milatovic and Braveny, 1987).

15.3.1 Multiplex polymerase chain reaction

Multiplex PCR is a cost effective and effortless technique. It is used to amplify numerous fragments of DNA simultaneously and can be employed for detection of large genetic mutations. The thermal cycler reaction in multiplex PCR can be carried out by many separated PCR reactions that are conducted in a single reaction through using temperature-mediated DNA polymerase and various primer pairs (Fig. 15.1).

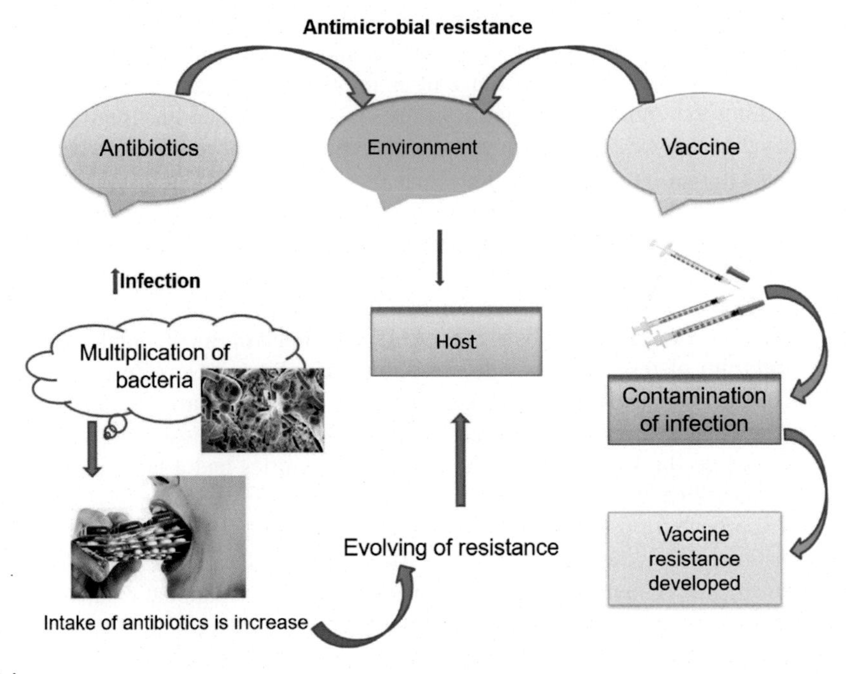

FIGURE 15.1

Schematic representation of process of multiplex PCR.

This multiple primer pair used in multiplex PCR reaction can work at the same annealing temperature during PCR reaction. Multiplex PCR is becoming a rapid and suitable screening technique in research and clinical laboratories. For the development of a well-organized multiple PCR reaction it is necessary to balance the concentration of PCR buffer, primers, deoxynucleotides (dNTPs), Taq polymerase, magnesium chloride, amount of DNA template, and cycling temperature.

Similarly, for very particular amplification of product in multiplex PCR the optimal combination of buffer concentration and annealing temperature is fundamental. The chances of spurious amplification are increased in multiplex PCR when the presence of primer pair is more than one due to the formation of primer dimers (Brownie et al., 1997). On the other hand, impaired annealing and extension rates may cause nonspecific components amplification, consumption of reaction components that may result in achieving undesired results. Thus to minimize nonspecific reactions, the optimization and validation of multiplex PCR reaction is necessary (Henegariu et al., 1997).

Conversely, the particular considerable parameters are primer designing, their length, targeted nucleic acid sequence, and their concentration (Mitsuhashi, 1996; Robertson and Walsh-Weller, 1998; Shuber et al., 1995; Wu et al., 1991; Kruger et al., 2017). Careful validation and evaluation throughout the procedure is essential, and detection of microbial agents is increased by multiplex PCR, because this technique is highly suitable for practical purpose to attain the multiple genes detection simultaneously (Markoulatos et al., 2002).

15.3.1.1 Application of multiplex PCR

15.3.1.1.1 Polymorphism/mutation

Application of multiplex PCR is done in many areas of polymorphism/mutation identification, quantitative, gene deletion, RNA detection, and nucleic acid analysis; in the field of infection this technique is valuable for identification of fungi, virus, parasites, and bacteria (Chamberlain et al., 1988; Chamberlain, 1990; Jin et al., 1996; Rithidech et al., 1997; Sherlock et al., 1998; Shuber et al., 1993; Zimmermann et al., 1996; Anjum et al., 2017, Strommenger et al., 2003).

15.3.1.1.2 Amplification of nucleic acid–based diagnosis and identification

Multiplex PCR technique is developed for amplification of nucleic acid–based diagnosis and identification of various microorganisms in a small volume of samples. for example *Escherichia coli*, *S. aureus*, *Staphylococcus warneri*, and *Haemophilus influenza* (Czilwik et al., 2015). During the last few years, the prevalence of MRSA has increased significantly globally and the infection due to MRSA such as bacteremia is responsible for increasing the mortality ratio associated with *S. aureus* infection. The multiplex PCR technique can play an important part in rapid identification of susceptible AMR genes that have impact on effect of severe infection caused by MRSA (Bergeron and Ouellette, 1998; Byl et al., 1999; Jean et al., 2004).The *mecA* gene that is coded for resistance for methicillin in MRSA has been detected by multiplex PCR (Behzadi et al., 2015).

15.3.1.1.3 Viral infections

Multiplex PCR is proving to be a dominant tool for identifying various infections that are caused by viruses such as encephalitis and meningitis. A variety of viruses such as cytomegalovirus, Epstein Barr virus, herpes simplex virus, poliovirus, varicella-zoster virus, adenovirus, arboviruses, paramyxoviruses, arenaviruses, and rabies are majorly related with neurological diseases. In neurological diseases, reliable, quick, and effective diagnosis is necessary to provide a rational source for therapy; a number of multiplex PCRs have been developed for accurate diagnosis of microbial-resistant genes (Casas et al., 1999; Casas et al., 1997; Cassinotti et al., 1996; Cassinotti and Siegl, 1998; Tenorio et al., 1993; Orr et al., 2018; Behzadi et al., 2015). In throat and nasopharyngeal infections, a nested multiplex RT-PCR that is used to detect parainfluenza virus (PIV) types 1, 2, and 3 by using three pairs of primers (Echevarría et al., 1998) has also been used. Similarly, *aacA-aphD* gene that is coded by bifunctional enzyme has been detected by multiplex PCR and is known to be a resistant gene against aminoglycosides such as kanamycin, gentamicin, amikacin, and tobramycin (Schmitz et al., 1999; Vanhoof, 1994; Das et al., 2015).

15.4 Next-generation sequencing

The Human Genome Project (HGP) in 1990 allowed the whole human genome sequencing. The Sanger sequencing, a golden standard technique, is useful for sequencing the DNA and was influential for HGP (Gibson and Muse, 2002).The platforms of NGS are improving the DNA and RNA sequencing competence in conditions of price, speed, and ease. In NGS millions of small fragments of DNA are sequenced in parallel direction. Frederick Sanger in 1977 developed a technique that is DNA sequencing technology that focuses on chain termination method, called Sanger sequencing. Another method of sequencing that is based on DNA chemical modification and cleavage subsequently at a specific base site was developed by Walter Gilbert. Sanger sequencing was adopted in the laboratory as a chief technology in a first generation and application of commercial sequencing, because of its lower radioactivity and high efficiencies. Sanger improved the NGS accuracy and rapidity, which was helpful in reducing manpower and cost. In addition to this, X-prize also played a role to accelerate the development and improvement of NGS. There are several kinds of sequencing techniques, namely, first, second, third, and fourth generation. The Sanger sequencing is a first-generation method, which uses fragment terminators in a form of dNTPs that allow separating the label fragments via gel electrophoresis. It became a golden standard in 1977. Nowadays, the most frequently used technique of NGS is second-generation sequencing that mainly consists of amplification, preparation, and sequencing steps. Millions of short fragments are sequenced at higher throughput in second-generation sequencing, due to which it has various clinical applications. Correspondingly, it has become a more preferable generation of sequencing then other sequencing methods. Similarly, the whole human genome can be sequenced within an hour at a lowest cost and effort in third-generation sequencing. The aim of conducting fourth-generation sequencing method is to directly analyze the cellular genome. The classification of NGS techniques has led to defeating the restrictions of conventional DNA sequencing and has initiated wide range of usage in molecular biology applications (Buermans and Den Dunnen, 2014). Several techniques are employed for sequencing, such as parallel signature sequencing, hybridization, and pyro sequencing (Bains and Smith, 1988; Brenner et al., 2000). These types of techniques are needed for achieving higher throughput, smaller fragments, improved sensitivity, and multiplexing, but in 2005 the pyro sequencing technique was to become first of the next-generation sequencers with an enlarged system of 454 by 454/Roche. The first commercially accessible high-yield sequencing platform used emulsion PCR of collected DNA fragments that affixed to microbeads. But nowadays new genomic sequencing FLX Titanium platform is being used, in which about one million beads are coated with DNA molecules, that is clonally amplified, followed by parallel pyrosequencing (Liu et al., 2012). On the other hand, Solexa/Illumina genome analyzer can increase the number/density of DNAs by generating DNA clonal cluster on glass surface by bridge amplification rather than on agarose beads. These improvements in Solexa/Illumina analyzer can overcome the trouble in quantifying the density/number of bases, while entire bases throughput is higher with improved computational method (Zerbino and Birney, 2008). Similarly, microarray is used in genome sequencing, the complete knowledge of query genome is necessary for arrays. It involves calculation of thousands of gene expressions in microorganisms, especially yeast, and helps transform the functional data of raw sequence of a genome (Shalon et al., 1996; Spellman et al., 1998). The major role of NGS in RNA and DNA sequencing for gene expression analysis of variants is identification of unknown genome determined and binding site of protein on DNA, along with tracking of various AMR genes in the environment (Figs. 15.2 and 15.3) (Pavlopoulos et al., 2013).

15.5 Conclusion

In this chapter, we have briefly discussed the multiplex PCR and NGS technique for tracking the AMR genes in the environment. Multiplex PCR optimizes results that can be achieved by using a number of optimal primer pairs in combination, and it has improved the efficiency of results. An increase in the number of microbial genes detection is highly efficient for practical purposes for attaining a concurrent

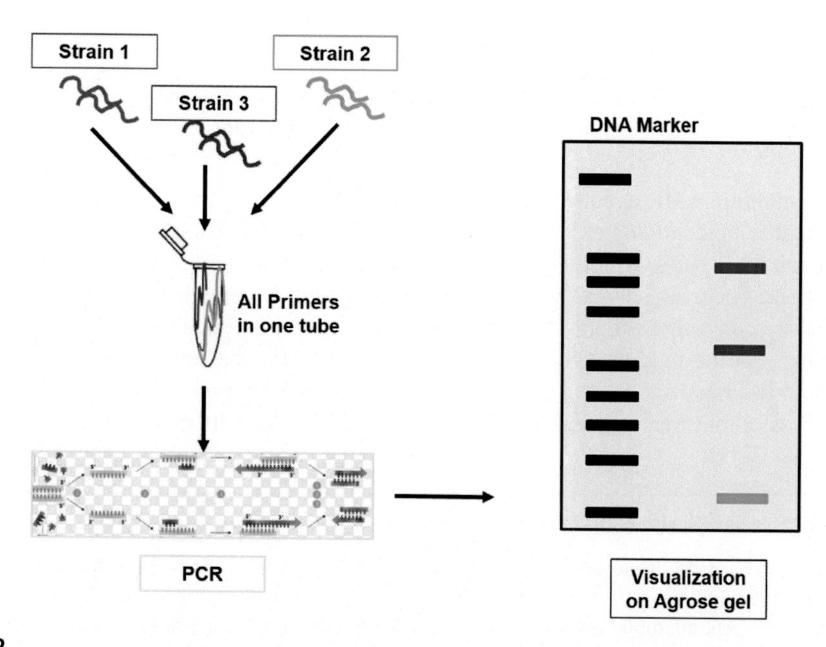

FIGURE 15.2

This represents next-generation sequencing of DNA library construction by using genomic DNA transformed into fragmented DNA and amplicons. Genome library is obtained with the aid of adapters and its ligation.

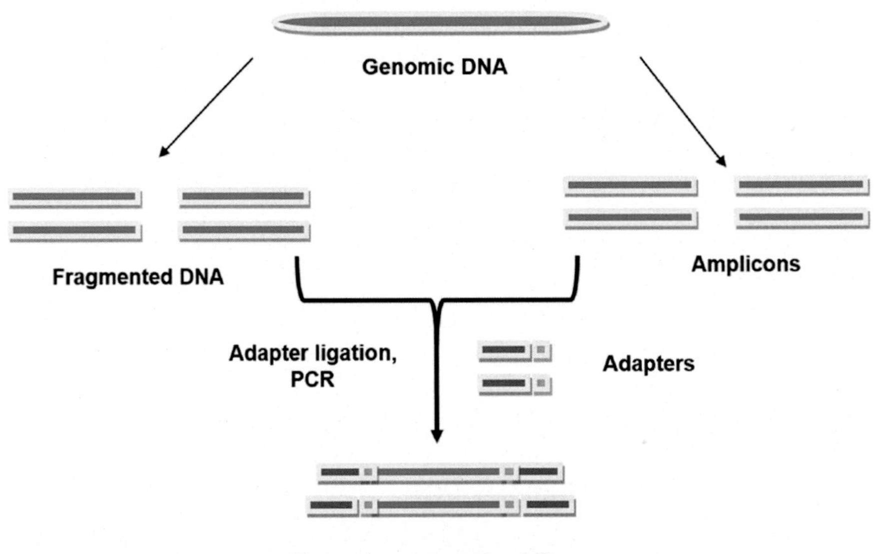

FIGURE 15.3

This represents next-generation sequencing of RNA library construction by using genomic RNA that can form a genomic library with the aid of adaptor ligation and by performing RT-PCR.

recognition of multiple AMR agents/genes that might be responsible for causing identical syndromes and/or comparable epidemiological features that can be done by using multiplex PCR. Another advanced technique is NGS, which considerably reduces the time for diagnosis and helps improve clinical treatment. There are several genotyping and phenotyping methods for instant multiplex PCR and NGS that are currently available to detect the antimicrobial resistant genes in the environment; these techniques can be used not only to examine or screen for the resistant microbes in clinical samples but also in animal and human subjects to obtain the baseline level of AMR in environment and measure the possibilities that can be taken to control and/or manage its rise in future.

References

Bains, W., Smith, G.C., 1988. A novel method for nucleic acid sequence determination. Journal of Theoretical Biology 135, 303–307.

Bancroft, E.A., 2007. Antimicrobial resistance: it's not just for hospitals. Jama 298 (15), 1803–1804.

Bergeron, M.G., Ouellette, M., 1998. Preventing antibiotic resistance through rapid genotypic identification of bacteria and of their antibiotic resistance genes in the clinical microbiology laboratory. Journal of Clinical Microbiology 36, 2169–2172.

Brenner, S., Williams, S.R., Vermaas, E.H., Storck, T., Moon, K., McCollum, C., Mao, J.-I., Luo, S., Kirchner, J.J., Eletr, S., 2000. In vitro cloning of complex mixtures of DNA on microbeads: physical separation of differentially expressed cDNAs. Proceedings of the National Academy of Sciences 97, 1665–1670.

Bronzwaer, S., Buchholz, U., Courvalin, P., Snell, J., Cornaglia, G., De Neeling, A., et al., 2002. Comparability of antimicrobial susceptibility test results from 22 European countries and Israel: an external quality assurance exercise of the European Antimicrobial Resistance Surveillance System (EARSS) in collaboration with the United Kingdom National External Quality Assurance Scheme (UK NEQAS). Journal of Antimicrobial Chemotherapy 50 (6), 953–964.

Brownie, J., Shawcross, S., Theaker, J., Whitcombe, D., Ferrie, R., Newton, C., Little, S., 1997. The elimination of primer-dimer accumulation in PCR. Nucleic Acids Research 25, 3235–3241.

Buermans, H., Den Dunnen, J., 2014. Next generation sequencing technology: advances and applications. Biochimica et Biophysica Acta – Molecular Basis of Disease 1842, 1932–1941.

Byl, B., Clevenbergh, P., Jacobs, F., Struelens, M.J., Zech, F., Kentos, A., Thys, J.-P., 1999. Impact of infectious diseases specialists and microbiological data on the appropriateness of antimicrobial therapy for bacteremia. Clinical Infectious Diseases 29, 60–66.

Casas, I., Pozo, F., Trallero, G., Echevarria, J., Tenorio, A., 1999. Viral diagnosis of neurological infection by RT multiplex PCR: a search for entero-and herpesviruses in a prospective study. Journal of Medical Virology 57, 145–151.

Casas, I., Tenorio, A., Echevarria, J., Klapper, P., Cleator, G., 1997. Detection of enteroviral RNA and specific DNA of herpesviruses by multiplex genome amplification. Journal of Virological Methods 66, 39–50.

Castro, V.A., Thrasher, A.N., Healy, M., Ott, C.M., Pierson, D.L., 2004. Microbial characterization during the early habitation of the International Space Station. Microbial Ecology 47 (2), 119–126.

Cassinotti, P., Mietz, H., Siegl, G., 1996. Suitability and clinical application of a multiplex nested PCR assay for the diagnosis of herpes simplex virus infections. Journal of Medical Virology 50, 75–81.

Cassinotti, P., Siegl, G., 1998. A nested-PCR assay for the simultaneous amplification of HSV-1, HSV-2, and HCMV genomes in patients with presumed herpetic CNS infections. Journal of Virological Methods 71, 105–114.

Chamberlain, J.S., Gibbs, R.A., Rainer, J.E., Nguyen, P.N., Thomas, C., 1988. Deletion screening of the Duchenne muscular dystrophy locus via multiplex DNA amplification. Nucleic Acids Research 16, 11141–11156.

Chamberlain, L., 1990. Multiplex PCR for the diagnosis of Duchenne muscular dystrophy. In: PCR Protocols: A Guide to Methods and Applications, pp. 272–281.

Checinska, A., Probst, A.J., Vaishampayan, P., White, J.R., Kumar, D., Stepanov, V.G., et al., 2015. Microbiomes of the dust particles collected from the international space station and spacecraft assembly facilities. Microbiome 3 (1), 50.

Control CfD, Prevention, 2011. National diabetes fact sheet: national estimates and general information on diabetes and prediabetes in the United States, 2011, 201 (1). US department of health and human services, centers for disease control and prevention, Atlanta, GA, pp. 2568–2569.

Cooper, M.A., Shlaes, D., 2011. Fix the antibiotics pipeline. Nature 472 (7341), 32.

Czilwik, G., Messinger, T., Strohmeier, O., Wadle, S., Von Stetten, F., Paust, N., Roth, G., Zengerle, R., Saarinen, P., Niittymäki, J., 2015. Rapid and fully automated bacterial pathogen detection on a centrifugal-microfluidic LabDisk using highly sensitive nested PCR with integrated sample preparation. Lab on a Chip 15, 3749–3759.

Echevarría, J.E., Erdman, D.D., Swierkosz, E.M., Holloway, B.P., Anderson, L.J., 1998. Simultaneous detection and identification of human parainfluenza viruses 1, 2, and 3 from clinical samples by multiplex PCR. Journal of Clinical Microbiology 36, 1388–1391.

Fluit, A.C., Jones, M.E., Schmitz, F.-J., Acar, J., Gupta, R., Verhoef, J., 2000. Antimicrobial resistance among urinary tract infection (UTI) isolates in Europe: results from the SENTRY Antimicrobial Surveillance Program 1997. Antonie van Leeuwenhoek 77, 147–152.

Fluit, A.C., Schmitz, F.J., Jones, M.E., Acar, J., Gupta, R., Verhoef, J., Group, S.P., 1999. Antimicrobial resistance among community-acquired pneumonia isolates in Europe: first results from the SENTRY Antimicrobial Surveillance Program 1997. International Journal of Infectious Diseases 3, 153–156.

Fu, Y., Zhang, W., Wang, H., Zhao, S., Chen, Y., Meng, F., Zhang, Y., Xu, H., Chen, X., Zhang, F., 2013. Specific patterns of gyr A mutations determine the resistance difference to ciprofloxacin and levofloxacin in *Klebsiella pneumoniae* and *Escherichia coli*. BMC Infectious Diseases 13, 8.

Gibson, G., Muse, S.V., 2002. A Primer of Genome Science. Sinauer Sunderland.

Goossens, H., Sprenger, M.J., 1998. Community acquired infections and bacterial resistance. Bmj 317 (7159), 654–657.

Henegariu, O., Heerema, N., Dlouhy, S., Vance, G., Vogt, P., 1997. Multiplex PCR: critical parameters and step-by-step protocol. Biotechniques 23, 504–511.

Ichijo, T., Yamaguchi, N., Tanigaki, F., Shirakawa, M., Nasu, M., 2016. Four-year bacterial monitoring in the International Space Station—Japanese Experiment Module "Kibo" with culture-independent approach. npj Microgravity 2, 16007.

Jin, L., Richards, A., Brown, D., 1996. Development of a dual target-PCR for detection and characterization of measles virus in clinical specimens. Molecular and Cellular Probes 10, 191–200.

Liu, L., Li, Y., Li, S., Hu, N., He, Y., Pong, R., Lin, D., Lu, L., Law, M., 2012. Comparison of next-generation sequencing systems. BioMed Research International.

Liu, P., Gur, D., Hall, L.M., Livermore, D., 1992. Survey of the prevalence of βlactamases amongst 1000 Gram-negative bacilli isolated consecutively at the Royal London Hospital. Journal of Antimicrobial Chemotherapy 30, 429–447.

Liu, B., Pop, M., 2008. ARDB—antibiotic resistance genes database. Nucleic Acids Research 37 (suppl_1), D443–D447.

Livermore, D., 1987. Clinical significance of beta-lactamase induction and stable derepression in gram-negative rods. European Journal of Clinical Microbiology 6, 439–445.

Markoulatos, P., Siafakas, N., Moncany, M., 2002. Multiplex polymerase chain reaction: a practical approach. Journal of Clinical Laboratory Analysis 16, 47–51.

Milatovic, D., Braveny, I., 1987. Development of resistance during antibiotic therapy. European Journal of Clinical Microbiology 6, 234–244.

Mitsuhashi, M., 1996. Technical report: Part 2. Basic requirements for designing optimal PCR primers. Journal of Clinical Laboratory Analysis 10, 285–293.

Nelson, K.E., Williams, C.M., 2014. Infectious Disease Epidemiology: Theory and Practice. Jones & Bartlett Publishers.

O'Neil, J., 2014. Review on Antimicrobial Resistance. Antimicrobial Resistance: Tackling a Crisis for the Health and Wealth of Nations.

O'Neill J. Review on antimicrobial resistance. Tackling a global health crisis: rapid diagnostics: stopping unnecessary use of antibiotics. Independent Review on AMR:1-36.

Organization, W.H., 2014. Antimicrobial Resistance Global Report on Surveillance: 2014 Summary. World Health Organization.

Pavlopoulos, G.A., Oulas, A., Iacucci, E., Sifrim, A., Moreau, Y., Schneider, R., Aerts, J., Iliopoulos, I., 2013. Unraveling genomic variation from next generation sequencing data. BioData Mining 6, 13.

Rithidech, K.N., Dunn, J.J., Gordon, C.R., 1997. Combining multiplex and touchdown PCR to screen murine microsatellite polymorphisms. Biotechniques 23, 36–44.

Resistance RoA, 2016. Tackling Drug-Resistant Infections Globally: Final Report and Recommendations. Review on Antimicrobial Resistance.

Roberts, M.C., 2008 May. Update on macrolide-lincosamide-streptogramin, ketolide, and oxazolidinone resistance genes. FEMS Microbiology Letters 282 (2), 147–159.

Robertson, J.M., Walsh-Weller, J., 1998. An introduction to PCR primer design and optimization of amplification reactions. In: Forensic DNA Profiling Protocols. Springer, pp. 121–154.

Schmitz, F.-J., Fluit, A.C., Gondolf, M., Beyrau, R., Lindenlauf, E., Verhoef, J., Heinz, H.-P., Jones, M.E., 1999. The prevalence of aminoglycoside resistance and corresponding resistance genes in clinical isolates of staphylococci from 19 European hospitals. Journal of Antimicrobial Chemotherapy 43, 253–259.

Sekiguchi, J.-I., Miyoshi-Akiyama, T., Augustynowicz-Kopeć, E., Zwolska, Z., Kirikae, F., Toyota, E., et al., 2007. Detection of multidrug resistance in Mycobacterium tuberculosis. Journal of Clinical microbiology 45 (1), 179–192.

Shalon, D., Smith, S.J., Brown, P.O., 1996. A DNA microarray system for analyzing complex DNA samples using two-color fluorescent probe hybridization. Genome Research 6, 639–645.

Sherlock, J., Cirigliano, V., Petrou, M., Tutschek, B., Adinolfi, M., 1998. Assessment of diagnostic quantitative fluorescent multiplex polymerase chain reaction assays performed on single cells. Annals of Human Genetics 62, 9–23.

Shuber, A.P., Grondin, V.J., Klinger, K.W., 1995. A simplified procedure for developing multiplex PCRs. Genome Research 5, 488–493.

Shuber, A.P., Skoletsky, J., Stern, R., Handelin, B.L., 1993. Efficient 12-mutation testing in the CFTR gene: a general model for complex mutation analysis. Human Molecular Genetics 2, 153–158.

Spellman, P.T., Sherlock, G., Zhang, M.Q., Iyer, V.R., Anders, K., Eisen, M.B., Brown, P.O., Botstein, D., Futcher, B., 1998. Comprehensive identification of cell cycle–regulated genes of the yeast *Saccharomyces cerevisiae* by microarray hybridization. Molecular Biology of the Cell 9, 3273–3297.

Tenorio, A., Echevarría, J.E., Casasa, I., Echevarría, J., Tabarés, E., 1993. Detection and typing of human herpesviruses by multiplex polymerase chain reaction. Journal of Virological Methods 44, 261–269.

Vanhoof, R., 1994. Hannecartpokorni & the Belgian Study Group of Hospital Infections: detection by polymerase chain reaction of genes encoding aminoglycoside-modifying enzymes in methicillin-resistant *Staphylococcus aureus* isolates of epidemic phage types. Journal of Medical Microbiology 41, 282–290.

Wu, D.Y., Ugozzoli, L., Pal, B.K., Qian, J., Wallace, R.B., 1991. The effect of temperature and oligonucleotide primer length on the specificity and efficiency of amplification by the polymerase chain reaction. DNA and Cell Biology 10, 233–238.

Zerbino, D.R., Birney, E., 2008. Velvet: algorithms for de novo short read assembly using de Bruijn graphs. Genome Research 18, 821–829.

Zhang, L., Kinkelaar, D., Huang, Y., Li, Y., Li, X., Wang, H.H., 2011. Acquired antibiotic resistance: are we born with it? Applied and Environmental Microbiology 77, 7134–7141.

Zimmermann, K., Schögl, D., Plaimauer, B., Mannhalter, J.W., 1996. Quantitative multiple competitive PCR of HIV-1 DNA in a single reaction tube. Biotechniques 21, 480–484.

CHAPTER

Toxicity of antibiotics

16

Kanwal Rehman[1], Saira Hafeez Kamran[2], Muhammad Sajid Hamid Akash[3]

[1]*Department of Pharmacy, University of Agriculture, Faisalabad, Pakistan;* [2]*Institute of Pharmacy, Gulab Devi Educational Complex, Pakistan;* [3]*Department of Pharmaceutical Chemistry, Government College University Faisalabad, Faisalabad, Pakistan*

16.1 Introduction

Antibiotics are the drugs that target the essential components necessary for the growth or nourishment of microbes. They have the capacity to inhibit the growth or kill the microorganisms. The clinically important microorganisms pathogenic to the human population are classified as bacteria, fungi, viruses, and protozoa (Rolain and Baquero, 2016). Since the discovery of natural, semi-synthetic, and synthetic compounds for the treatment of various bacterial infections, mankind has also been facing their toxicity and development of resistance. Allergic reactions, hepatotoxicity, nephrotoxicity, ototoxicity, and phototoxicity are the most common adverse events associated with the use of antibiotics (Goodman & Gilman, 2011).

16.2 History of antibiotic discovery

The discovery of antibiotics has contributed significantly to reduce the infectious disease burden in human population, one of the leading causes of morbidity and mortality. The beginning of the "antibiotic era" is usually associated with the novel work of Paul Ehrlich and Alexander Fleming. From 1904 to 1909, Paul Ehrlich synthesized and tested many compounds in his lab for the treatment of syphilis and toxoplasmosis, one of the major causes of mortality in that time. He found an organic arsenic compound, arsphenamine, also known as Compound 606 to be effective for the treatment of syphilis in humans and it was the clinical choice until the discovery of penicillin despite its serious side effects. One of the striking features of Compound 606 is that the exact mechanism of action is still not completely elucidated (Aminov, 2010). The accidental antibiotic discovery is associated with Alexander Fleming in 1928, who while studying a culture of *Staphylococcus aureus* observed a mold that contaminated his culture and inhibited the growth of bacteria in the culture. It has been observed in the history that the use of antibiotics by humans is actually very old. Traces of tetracyclines were found in human skeleton (350−550 CE) and isolated from the femoral midshafts of Roman period skeletons (Dakhleh Oasis, Egypt). The use of antimicrobials in alternative medicines has also been associated with qinghaosu used in traditional Chinese medicine for thousands of years in various illness. In 1972

Antibiotics and Antimicrobial Resistance Genes in the Environment. https://doi.org/10.1016/B978-0-12-818882-8.00016-4

artemisinins were extracted from qinghaosu by Professor Ty Youyu, who was awarded the Nobel Prize in Physiology or Medicine in 2015. Artemisinins are currently the most effective antimalarial therapy in different drug-resistant areas in the world (Aminov, 2010).

16.3 Toxicity testing

Toxicity testing plays a pivotal role in ensuring the safety from potentially dangerous substances. Every drug before being introduced in the market has to undergo extensive testing for pharmacodynamic and pharmacological properties. The importance of toxicological tests has increased due to history of therapeutic disasters. The therapeutic disaster of the sulfanilamide elixir for the treatment of various illnesses, introduced in United States in 1938 and in 1962, introduction of thalidomide for morning sickness in pregnant women changed the trends in the scientific community and hastened toxicity testing. The conventional methods for the assessment of toxic reactions of drugs are assessed by employing acute, subacute, and chronic toxicity testing during preclinical studies in animal models. However, these methods were highly criticized for heavy animal use by different animal welfare authorities across the globe. Different governmental agencies, scientific institutions, and the U.S. National Academy of Sciences called for rethinking methods that should be employed for evaluation of chemical safety (Krewski et al., 2010).

The dilemma for appropriate toxicological testing led to the development of alternative testing strategies (ATSs) employing in vitro and in silico methodologies to evaluate the changes in pharmacodynamic processes at cellular, biochemical, molecular, and tissue levels. High-throughput screening uses computational and advanced robotics for the testing of hundreds to thousands of compounds at a time. The toxicity testing of chemicals with poor data could be estimated with quantitative structure–activity relationship model approaches. The use of ATS in toxicological assessment of chemicals is promoted and encouraged in scientific research and development (Zaunbrecher et al., 2017).

16.4 Toxicity of antibacterial

16.4.1 Toxicity of beta-lactam antibiotics

The toxicity of β-lactam antibiotics is mainly manifested by hypersensitivity reactions, blood dyscrasias, nephrotoxicity, and neurotoxicity.

16.4.2 Hypersensitivity reactions

Immune-mediated drug hypersensitivity reactions are the most common untoward reactions associated with β-lactam antibiotics. These are adverse drug reactions that clinically resemble allergy immune response to a specific class of drugs depending on the pharmacodynamic and genetic factors of the host. β-lactam antibiotics are the most common drugs eliciting an immunological response. Most of the allergic reactions are mediated by immunoglobulin E or T cells. The first β-lactam antibiotic that induced the allergic reaction was benzylpenicillin, followed by amoxicillin, now being the most common drug to cause hypersensitivity reactions (Fig. 16.1) (Shepherd, 2003).

Penicillin and its breakdown products are major determinants of hypersensitivity reactions that act as haptens, binding with proteins after covalent reaction. Amongst the breakdown products, the most

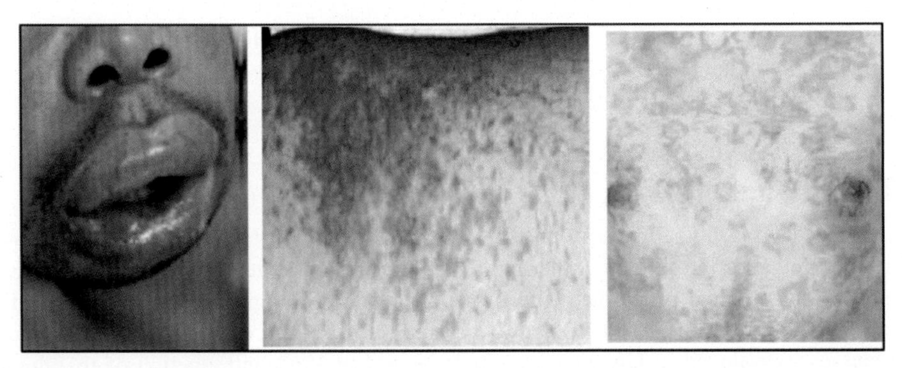

FIGURE 16.1

Various allergic reactions due to β-lactam antibiotics (*left* to *right*): Angioedema due to penicillin, maculopapular rash associated with flucloxacillin, urticaria associated with penicillin (Blumenthal et al., 2019; Kuruvilla and Khan, 2015).

abundantly found is major determinant moiety (MDM). This is a penicilloyl moiety formed when β-lactam ring is broken down. A large number of reactions occur due to MDM, however, some cases were reported where other breakdown products were majorly responsible. The rate of formation of antibodies against the major and minor breakdown products determines the severity of allergic reactions. A minor determinant mixture of penicilloyl-polylysine is used in a patch test to evaluate the kind of allergic reactions, which have been classified as immediate or nonimmediate, depending on the time interval between the application of patch on skin and onset of symptoms. The reactions that occur within the first hour of administration are immediate reactions, clinically manifested by urticaria, angioedema, bronchospasm, rhinitis, and anaphylactic shock. The penicillin skin patch test is an excellent diagnostic tool for the identification of immediate reactions. The nonimmediate reactions mainly include exfoliative dermatitis, pneumonitis, hepatitis, vasculitis, Stevens-Johnson syndrome and toxic-epidermal necrolysis (Fig. 16.2) (Demoly and Romano, 2005).

Hypersensitivity reactions to penicillins extend to other β-lactam antibiotics. Cross sensitivity is highest with cephalosporins and less with carbapenems. Angioedema and hypersensitivity reactions are most severe produced by the penicillins. Cross reactivity of such reactions with cephalosporins is very high, therefore administration of cephalosporins should be avoided in patients who have positive Coombs test. Cephalosporins are usually recommended and administered in patients having history of mild reactions to penicillin. Such patients are usually at low risk of development of allergic reactions, however, patients who had experienced recent severe immediate reactions to penicillin should avoid cephalosporins (Yılmaz and Özcengiz, 2017). Carbapenems are the most effective class of antibiotics in resistant cases. The common hypersensitivity reactions of carbapenems include rash, urticaria, and immediate hypersensitivity (El-Gamal et al., 2017). The incidence of hypersensitivity reactions with carbapenems is lower as compared to penicillin and cephalosporins (Table 16.1).

16.4.3 Other toxicities of β-lactam antibiotics

Penicillin possesses minimal direct toxicity. Blood dyscrasias are the second most common adverse effect of different penicillins, associated with defects in hemostasis, granulocytopenia, and bone

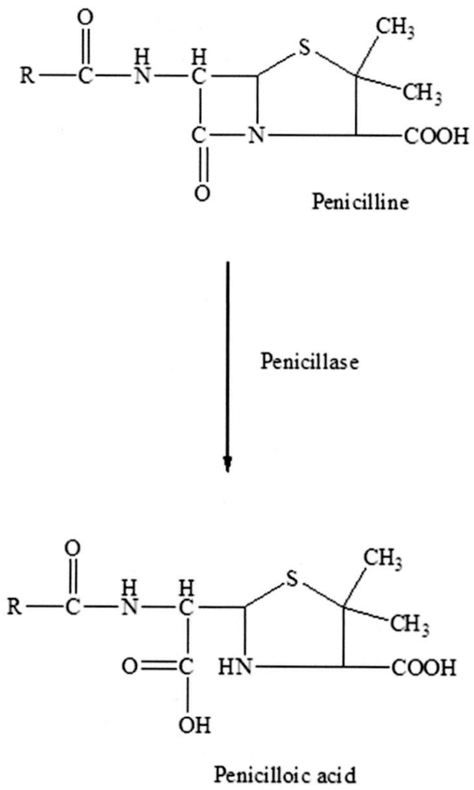

Penicilline

Penicillase

Penicilloic acid

FIGURE 16.2

Conversion of penicillin to penicilloic acid.

marrow depression. Cephalosporins have been reported to cause serious bleeding disorders and are associated with methythiotetrazole (MTT) group. MTT group is present in cefotetan, moxalactam, cefamandole, and cefaperazone. These drugs also cause intolerance to alcohol, a disulfiram-like reaction. Parenteral administration of penicillins may cause immediate reaction manifested by headache, dizziness, tinnitus, lethargy, confusion, twitching, hallucinations, and epileptic seizures. Gastrointestinal disturbances because of disturbances in *Clostridium difficile* may result in colitis. The therapy should be stopped or dose may be adjusted otherwise. Restoration of the microbial flora occurs within a few days after the therapy is discontinued. Penicillins are nephrotoxic particularly in patients having underlying kidney problems. Cephalosporins are also nephrotoxic. The nephrotoxicity is mostly dose related. High doses of cephalothin (8–12 g/day) have produced renal tubular necrosis. Concurrent administration of cephalosporins with aminoglycosides has produced synergistic effects and hastened damage to the kidneys. Carbapenems, especially imipenem, cause seizures in 1.5% of patients treated with these drugs. Central nervous system (CNS) toxicity is more common with carbapenems (Imani et al., 2017).

Table 16.1 Incidence of hypersensitivity reactions with different classes of β-lactam antibiotics.

Class	Types of reactions	Percentage of reactions
Penicillin	Skin rashes (maculopapular, urticarial, scarlatiniform, morbiliform, vesicular, and bullous) Angioedema Anaphylactic shock Vasculitis Serum sickness Exfoliative dermatitis Stevens-Johnson syndrome	0.7%–10%
Cephalosporins	Maculopapular rash Anaphylactic shock	1%
Carbapenems	Rashes	<1%

16.4.4 Toxicity of protein synthesis inhibitors

Different classes of protein synthesis inhibitors include aminoglycosides, macrolides, tetracyclines, and chloramphenicol.

16.4.4.1 Aminoglycosides

Aminoglycosides are natural and semi-synthetic derivatives of compounds isolated from the family of actinomycetes. Streptomycin was first isolated from *Streptomyces griseus*. All aminoglycosides contain an aminocyclitol or hexose ring (usually in the center) linked by a glycoside bond to two or more amino sugars. The most common clinically employed aminoglycosides include gentamicin, kanamycin, netilmicin, tobramycin, neomycin, paromomycin, and amikacin. These drugs have narrow therapeutic index, and almost all aminoglycosides produce reversible or irreversible ototoxicity (vestibular and cochlear) and nephrotoxicity. Since the introduction of aminoglycosides in clinical practice in 1944, these have been the most commonly used antibiotics for the treatment of gram-negative infections (Fig. 16.3). Aminoglycosides act by disrupting or prematurely terminating the formation of initiation complex of peptide formation. They bind to the 16S ribosomal RNA (12S in case of streptomycin) in the 30S ribosomal unit of the bacterial ribosome and thereby inhibit the synthesis of protein.

Long-term use of all aminoglycosides leads to the development of ototoxicity and nephrotoxicity. The ototoxicity of aminoglycoside may be reversible or irreversible depending on the dose of the drug administered. Two areas are highly affected by aminoglycoside toxicity, cochlear and vestibular. The administration of aminoglycoside therapy initially may lead to imbalance in the active ionic movement in the endolymph. This may lead to degeneration of the neurons and hair cells in the sensory perception areas in the cochlea. Once the sensory cells are destroyed, permanent loss of hearing may occur. Damaged cells in the base of the cochlea may lead to impairment or permanent loss of processing of high-frequency sounds and damage to the apex in the cochlea leads to loss of low-frequency sound processing. High doses and prolonged exposure of aminoglycosides can lead to the irreversible destruction of the cochlear and vestibular sensory cells, especially in infants. Neonatal cases of aminoglycoside antibiotic exposure in low- and middle-income countries is high due to low cost, high efficacy, and availability.

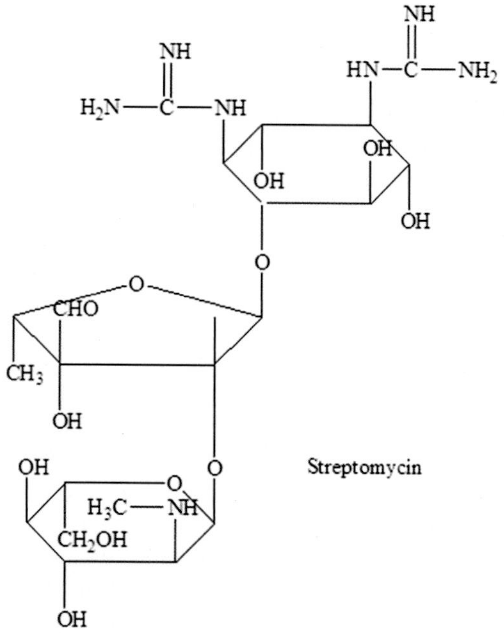

FIGURE 16.3

Chemical structure of streptomycin.

The incidence of ototoxicity induced by aminoglycosides in multidrug-resistant tuberculosis patients was estimated to be 22.9%. The use of aminoglycosides in tuberculosis patients should be considered as second-line therapy by clinicians (Sogebi et al., 2017). Genetic susceptibility to aminoglycosides has also been observed recently in a study conducted in China.

16.4.4.1.1 Nephrotoxicity

Aminoglycosides possess poor oral absorption and hence are administered intramuscularly or intravenously. Patients with severe gram-negative aerobic bacterial infections are intravenously treated with aminoglycosides. As these drugs are not metabolized by the liver, approximately 10% of the administered dose accumulates in the kidneys. Dose-dependent nephrotoxicity has been observed in 10%–25% of people therapeutically treated with aminoglycosides. The nephrotoxic effect of aminoglycosides is characterized by tubular necrosis, retention of fluid, and tubular obstruction. This results in increase in renal vascular resistance and decreased renal blood flow. Gentamicin has been reported to increase the mesangial cell contraction by increasing the cytosolic free calcium. The increased levels of calcium thereby cause activation of phospholipase A_2, thromboxane A_2, platelet activating factor, and prostaglandin A_2. The net effect is gentamicin-induced reduction of ultrafiltration coefficient and glomerular filtration rate. Another possible mechanism increasing tubular necrosis is mediation of reactive oxygen species (ROS) by aminoglycosides in renal tubules. ROS increases the susceptibility of renal failure (Martínez-Salgado et al., 2007).

16.4.5 Toxicity of tetracyclines

The most common side effects associated with tetracyclines are gastrointestinal (GI) disturbances manifested by GI distress and irritation; epigastric burning; abdominal pain leading to nausea, vomiting. and diarrhea; and esophagitis. Tetracycline tolerability can be enhanced by administering these drugs with food, however, administration with dairy products is contraindicated. These drugs should not be administered with dairy products or antacids as they can form chelates and would be unable to produce any therapeutic effect. This effect would cause serious toxic symptoms, including brown discoloration of teeth and bone retardation in infants and children. As tetracyclines are excreted into breast milk and placental fluid, hence they are contraindicated in lactating mothers and pregnant females. The deposition in the teeth and bones is associated with tetracycline-calcium orthophosphate complex. Tetracycline causes inflammatory reactions like esophagitis, pancreatitis, and ulcers. Because of the incomplete absorption of the tetracyclines from the gut, these drugs can cause disturbances in enteric flora. Suppression of the normal gut flora may cause overgrowth of resistant microorganisms. Occasionally these drugs may produce life-threatening pseudomembranous colitis due to *C. difficile*.

Hepatic and renal toxicity occur mostly in patients receiving high doses of tetracyclines. Hepatotoxicity has been reported in patients receiving tigecycline, minocycline, and doxycycline, however, these drugs have lesser renal side effects. Demeclocycline was reported to cause nephrogenic diabetes insipidus and this has been exploited for the treatment of syndrome of inappropriate secretion of antidiuretic hormone. Patients with underlying renal disease receiving tetracyclines were found to possess high levels of nitrogen-containing compounds (urea, uric acid, and creatinine). Individuals receiving tetracycline were reported to suffer from photosensitivity reactions when exposed to sunlight. Pigmentation of nails and onycholysis has also been reported, as well as blood disorders, particularly atypical lymphocytes, increase in leukocyte count, granulocyte toxicity, and thrombocytopenic purpura. Hypersensitivity reactions are rare with tetracycline. However, allergic responses like burning of eyes, cheilitis, glossitis, vaginitis, and anal or vulvular pruritus may occur and can persist for weeks after the treatment is stopped. Oral administration of tetracyclines can cause fever of varying degrees, eosinophilia, and anaphylactic shock or angioedema. Degraded tetracycline, if administered to patients, can cause Fanconi syndrome, characterized by polyuria, polydipsia, glycosuria, proteinuria, nausea and vomiting, and aminoaciduria.

16.4.6 Toxicity of macrolides

Macrolides have been considered to be one of the safest antibiotics since their introduction in 1952. Various members of this class are erythromycin, clarithromycin, azithromycin, tilmicosin, and tylosin. All members share large macrocyclic lactone ring. The most common side effect of these drugs is hepatotoxicity because the biliary route is the major route of excretion. Erythromycin estolate has been reported to cause cholestatic hepatitis, however, the stearate and ethylsuccinate salts have less incidence. This side effect has been reported in various clinical studies and begins 10–20 days after the antibiotic treatment. Other macrolides also have been reported to cause hepatic illness. In hepatocellular necrosis caused by macrolides, the SH groups of proteins bind to nitrosoalkanes that are produced from the transformation of macrolides. Symptoms of hepatic illness usually begin with gastrointestinal upset, including nausea, abdominal cramping followed by vomiting. This may cause acute cholecystitis followed by jaundice. Elevated transaminases, leukocytosis, fever, and eosinophilia have also been reported. Hepatotoxicity has also been observed with clarithromycin and azithromycin at high doses, although at a lower rate than with erythromycin (Lockwood et al., 2010). Azithromycin and clarithromycin have been reported to possess lower incidence of hepatotoxicity, whereas telithromycin was reported to possess severe hepatotoxic effects leading to liver transplantation and death (Brinker et al., 2009), (Dhiman and Chawla, 2005)

Erythromycin was first reported in 1991 to cause prolongation of QT interval. A 61-year-old female during fifth intravenous administration of erythromycin was reported with QT prolongation and torsade de pointes (Schoenenberger et al., 2009). Erythromycin has been reported to possess greatest risk of cardiotoxicity. Clarithromycin possesses QT interval prolongation but torsade de pointes has less incidence. The prolongation of QT interval is associated with ventricular arrhythmias. The number of cardiac toxicity cases reported with azithromycin are less frequent. The mechanism of cardiotoxicity is different as compared to clarithromycin and erythromycin (Guo et al., 2010). Clarithromycin and erythromycin block HERG potassium channels, which is proved to be one of the underlying mechanisms for cardiotoxic effects of these drugs (Stanat et al., 2003).

Different routes of administration of macrolides produce varied toxicities. Large oral doses of erythromycin produce epigastric distress, nausea, vomiting, and diarrhea, and intravenous administration also may cause similar symptoms. This side effect of erythromycin has been exploited to promote peristalsis and gastric emptying postoperatively and in patients with gastroparesis. Intravenous administration may also cause thrombophlebitis at the site of injection. The incidence of GI upset is less with the rest of the macrolides.

Macrolides cause various allergic reactions including Stevens-Johnson syndrome and toxic epidermal necrolysis. If a rash appears after any dose of macrolides, the drug should be immediately stopped and supportive care should be provided. Macrolides usually cause cutaneous hypersensitivity reactions. Intravenous administration of erythromycin has been associated with auditory impairment (Blumenthal et al., 2019).

16.4.7 Toxicity of chloramphenicol

With the advent of newer antibiotics and development of resistance to chloramphenicol, it is now rarely used in the clinical setup. Another limitation of chloramphenicol clinical use is severe blood dyscrasias produced by this drug. Chloramphenicol, introduced in 1947, provided excellent activity against many microorganisms and was largely employed for the treatment of typhoid fever, meningitis, and rickettsia infections. The most common and most important side effect of chloramphenicol is bone marrow depression leading to aplastic anemia, specifically when plasma concentration is more than 25 μg/mL. The hematopoietic system is affected by chloramphenicol by dose and an idiosyncratic response. The dose-related toxicities include decreased production of erythrocytes due to decreased incorporation of iron into heme, leukopenia, and thrombocytopenia. The idiosyncratic response is generally manifested by aplastic anemia and pancytopenia (Wareham and Wilson, 2002). Neonates administered chloramphenicol developed serious side effects, termed as gray baby syndrome. This syndrome developed due to underdeveloped hepatic metabolism in neonates required for the metabolism of chloramphenicol and also due to the inadequate renal excretion of chloramphenicol. The syndrome usually began after 4–8 days and manifested in the first 24 h by decreased appetite, vomiting, cyanosis, irregular respiration, and green stools. In the next 24 h, infants turned gray and became flaccid and hypothermic (Wareham and Wilson, 2002).

16.4.8 Toxicity of clindamycin and vancomycin

Clindamycin, a lincosamide introduced into clinical practice to treat skin and soft tissue infections and respiratory infections in patients allergic to β-lactam antibiotics, possesses higher incidence of development of lethal pseudomembranous colitis. Ten percent of cases were reported at various intervals and symptoms of colitis include fever, watery diarrhea, and leukocytosis. Immediate removal of drug with treatment with metronidazole and oral vancomycin is usually recommended. Other side effects include hypersensitivity reactions (Yılmaz and Özcengiz, 2017).

Vancomycin was introduced into clinical practice to mainly treat methicillin-resistant *S. aureus* infections. It is not the first-line agent owing to the side effects, which mainly include hypersensitivity reactions, phlebitis, hypotension and tachycardia, chills and fever. The hypersensitivity reactions may be Ig Ear non-IgE mediated. The non-IgE-mediated reactions have less hypotension and cardiac symptoms. Rapid intravenous infusion may produce extreme flushing, which is called as red man syndrome. This is due to direct toxic effect of vancomycin on mast cells, degrading them and releasing massive amount of histamine. Delayed hypersensitivity reactions include urticaria and anaphylaxis (Bruniera et al., 2015).

16.4.9 Toxicity of streptogramins

Streptogramins (quinupristin and dalfopristin) in ratio of 30:70 were introduced into clinical practice for the treatment of vancomycin-resistant *S. aureus* infections. The most common side effects reported so far are infusion related, causing pain and inflammation at site of injection. This side effect can be minimized by controlling the rate of infusion or by infusing the drug through a central venous catheter. Myalgias and arthralgias have also been reported more commonly in patients with hepatic insufficiency (Yılmaz and Özcengiz, 2017) (Fig. 16.4).

16.4.10 Toxicity of linezolid

Linezolid is a synthetic antimicrobial agent belonging to the oxazolidinone group of drugs. It was approved by the U.S. Food and Drug Administration (FDA) for the treatment of several resistant strains of microbes. The minor side effects include GI side effects, rashes, and headaches. Long-term

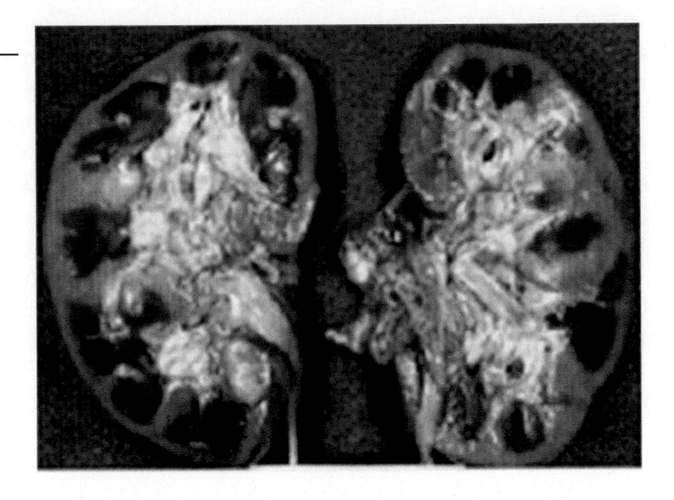

FIGURE 16.4

Acute tubular necrosis in aminoglycoside-induced nephrotoxicity (Blumenthal et al., 2019).

use of linezolid may lead to side effects like lactic acidosis, peripheral neuropathies, and optic neuritis. Hematological side effects occur in 2.4% of the treated patients, specifically people on long-term therapy of the drug. Myelosuppression, thrombocytopenia, leukopenia, and anemia have more commonly been observed (Fig. 16.5). Linezolid should be reserved as an alternative agent for treatment of infections caused by multiple-drug-resistant strains. It should not be used when other agents are likely to be effective (e.g., community-acquired pneumonia, even though it has the indication). Indiscriminant use and overuse will hasten selection of resistant strains and the eventual loss of this valuable newer agent (Goodman & Gilman, 2012).

16.4.11 Toxicity of folic acid synthesis inhibitors

Sulfonamides were one of the first antibiotics employed in humans systemically to treat various bacterial infections. Sulfonamides are derivatives of p-aminobenzene sulfonamide (sulfanilamide) (Fig. 16.6). These drugs are further classified into antibacterial sulfonamides and non-antibacterial sulfonamides (Table 16.2). In 1932, the first drug containing sulfonamide moiety was patented with the name Prontosil. Gerhard Domagk tested the compound in mice with streptococcal infection and observed that the compound provided dramatic cure. Later Prontosil was used in a 10-month-old child with staphylococcal septicemia and provided radical cure. Domagk was awarded the Nobel Prize in Physiology or Medicine in 1939 (Lesch, 2007). There are numerous side effects of sulfonamides.

16.4.12 Hypersensitivity reactions

The most common hypersensitivity reactions are type 1 reactions, including rash, urticaria, purpura, scarlatinal, morbilliform, erysipeloid , erythema multiforme of Stevens-Johnson syndrome, erythema

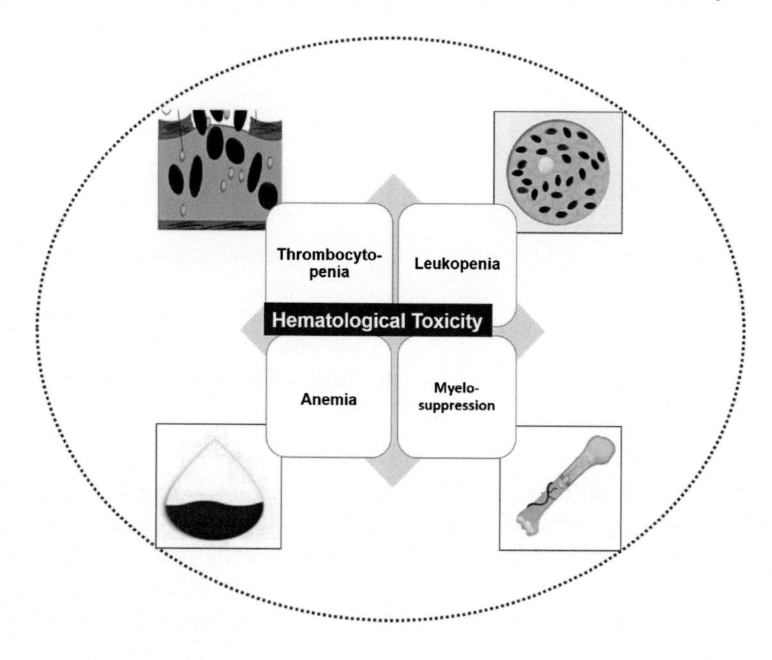

FIGURE 16.5

Hematological side effects of linezolid.

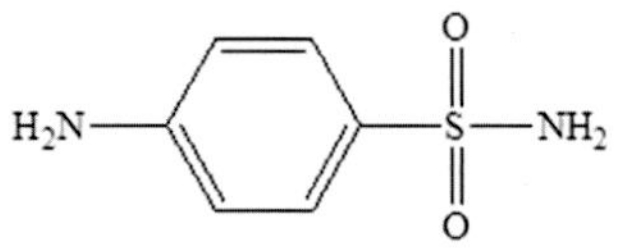

Sulfanilamide

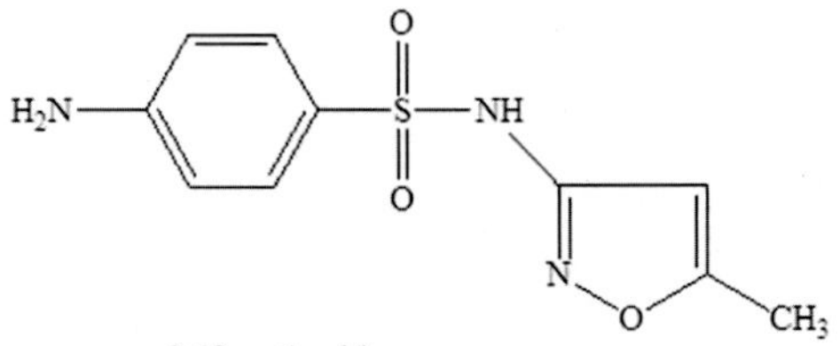

Sulfamethoxide

FIGURE 16.6

Chemical structure of sulfanilamide and sulfamethoxazole.

Table 16.2 Types of sulfonamides.

Anti-bacterial sulfonamides	Non-antibacterial sulfonamides
Absorbed and excreted rapidly includes 1. Sulfisoxazole ($t_{1/2}$ is 5–6) 2. Sulfamethoxazole ($t_{1/2}$ is 11 h) 3. Sulfadiazine ($t_{1/2}$ is 10 h)	Loop diuretics
Poorly absorbed from lumen 1. Sulfasalazine	Thiazide diuretics
Topical 1. Sulfacetamide 2. Silver sulfadiazine	Carbonic anhydrase inhibitors
Long—acting 1. Sulfadoxine ($t_{1/2}$ is 100–230 h)	Nonsteroidal antiinflammatory drugs (COX 2 selective)
	Sulfonyl ureas Antivirals 5-Hydroxytryptamine receptor antagonist

nodosum, Behcet's syndrome, photosensitivity, and exfoliative dermatitis. These reactions are mostly caused by the metabolites of sulfonamides. The parent sulfonamide compound or active metabolite may either bind with native protein and stimulate cellular and humoral responses, or may bind to cellular proteins and cause cytotoxicity, or may stimulate T cells thereby activating an immune response without antigen presentation. In the liver the N4 nitrogen of sulfonamides becomes acetylated and is believed to be responsible to act as happens and initiate an immune response. The hypersensitivity reactions occur mostly in the first week of therapy, especially in previously sensitized individuals. Drug fever is a common side effect accompanied by malaise and pruritus. The nonbacterial sulfonamides don't share the same molecular structure as antibacterial sulfonamides, hence have not been reported to cause hypersensitivity reactions (Wulf and Matuszewski, 2013).

16.4.13 Other side effects

The most common side effects with all sulfonamides are anorexia and vomiting, occurring in 1%–2% of patients, and the mechanism of these effects is central. Sulfonamides are contraindicated in newborn infants as they displace bilirubin from plasma albumin. As the liver is not fully developed in newborns, so displaced bilirubin becomes deposited in the subthalamic nuclei and basal ganglia, thereby causing encephalopathy called kernicterus. These drugs are also contraindicated in lactating mothers because of their excretion in milk. Incidence of development of crystals in urinary tract in elderly treated with less-soluble sulfonamides is high. Acquired immune deficiency syndrome patients are also at high risk for the development of crystalluria. Glucose 6 phosphate dehydrogenase deficient people are at high risk for the development of acute hemolytic anemia, therefore sulfonamides are administered with great caution in such patients. Of the patients on sulfonamide therapy, 0.1% develop agranulocytosis, especially those receiving sulfadiazine.

16.4.14 Toxicity of DNA synthesis inhibitors

The most common adverse effects reported by administration of quinolones and fluoroquinolones in 17% of the patients are nausea, vomiting, and abdominal discomfort. Ciprofloxacin (500 mg–1 g for 5–10 days) is the first-line treatment for dysentery according to current World Health Organization guidelines. However, it has also been reported to cause *Clostridium difficile* colitis.

Central nervous system side effects are the next most common reported, occurring in about 11% of the patients on fluoroquinolones. These mainly include dizziness and headache. Other psychiatric disorders, including cognitive disabilities, disorientation, memory impairment delirium, seizures, and hallucinations, have been reported (Fig. 16.7). Patients receiving nonsteroidal antiinflammatory drugs (NSAIDS) or theophylline along with fluoroquinolones have higher incidence of development of psychiatric disorders. It has been reported that ciprofloxacin and pefloxacin increase the serum

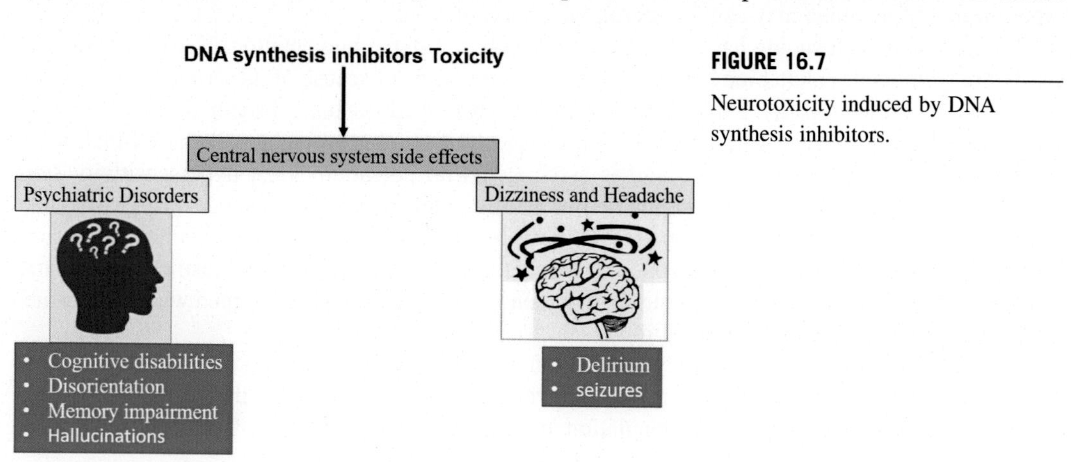

FIGURE 16.7

Neurotoxicity induced by DNA synthesis inhibitors.

concentrations of these drugs by inhibition of the metabolism of these drugs. Quinolones administered along with NSAIDs have been reported to increase the displacement of γ-aminobutyric acid from its receptors. The incidence of serious complications of ruptures or tears in the arteries because of fluoroquinolones is rare, but has been reported. QT interval prolongation, specifically by sparfloxacin and less commonly by morifloxacin and gatifloxacin, has been reported. Thereby oral or intravenous use of fluoroquinolones has been indicated to be high risk in patients with history of high blood pressure, blockage, or aneurysm of blood vessels or in elderly patients. The FDA has issued warnings about the musculoskeletal adverse effects of fluoroquinolones, including tendinopathy and tendon rupture. In 2018, the FDA also issued a serious warning that fluoroquinolones cause blood glucose disturbances. Gatifloxacin has been found to be associated with both hyper- and hypoglycemia. The use of fluoroquinolones for uncomplicated infections like acute sinusitis, uncomplicated urinary tract infections, and acute bronchitis has been restricted by the FDA. Some rarely reported reactions of various fluoroquinolones include elevated serum transaminase levels, eosinophilia, and leukopenia. Fluoroquinolones have been contraindicated in children. However, where risk outweighs benefit as in case of children with complicated urinary tract infections, typhoid fever, cystic fibrosis, meningoencephalitis, sepsis in neutropenic children, or multidrug-resistant tuberculosis, fluoroquinolones are recommended and safety and efficacy data are also available (Bacci et al., 2015).

16.4.15 Toxicity of antiviral agents

There are two major classes of antiviral drugs, nonretroviral and retroviral agents. The nonretroviral drugs mostly are employed for the treatment of infections caused by herpes simplex (type 1 and 2), varicella zoster, cytomegalovirus, influenza, and hepatitis B and C viruses (Table 16.3).

16.4.16 Toxicity of antiretroviral agents

The antiretroviral therapy suppresses the infections caused by human immunodeficiency virus (HIV). In 1964, the first antiretroviral agent, zidovudine, was synthesized by Horwitz, and in 1972 its efficacy to inhibit replication of retrovirus was identified by Osterag. Later, many nucleoside reverse transcriptase inhibitors were synthesized and subjected to clinical trials and marketed. These include lamivudine, emtricitabine, tenofovir, zalcitabine, didanosine, abacavir, and stavudine. Among these, lamivudine, emtricitabine, and tenofovir are the least toxic with very few side effects. Prolonged exposure to emtricitabine may cause hyperpigmentation of skin.

Chronic use of zidovudine has been associated with nail hyperpigmentation. Hepatic toxicity is rare but may be fatal. Zalcitabine has been removed from market because of its serious toxic effects, mainly peripheral neuropathy. Didanosine is also less prescribed because of the availability of less-toxic agents. It is associated with exacerbation of pancreatitis and neuropathies, both of which are a result of mitochondrial toxicity. Abacavir is associated with potentially fatal hypersensitivity syndrome, occurring in 2%–9% of the population treated with this drug. Stavudine has been associated with development of peripheral neuropathies in 71% of the patients treated with this drug as monotherapy. Dose reductions are recommended to combat the side effects. The drug also causes hepatic steatosis and lactic acidosis, more commonly in patients when stavudine is combined with didanosine. It is also strongly associated with lipoatrophy (fat wasting).

The nonnucleoside reverse transcriptase inhibitors include nevirapine, efavirenz, and etravirine. Mild macular or popular rash has been associated with nevirapine. Incidence of life-threating Stevens-Johnson syndrome is only in 0.3% population treated. Fifty-three percent of patients have been

Table 16.3 Adverse effects of nonretroviral agents.

Antiviral drug	Side effects
Acyclovir	
1. Topical	Mucosal irritation, burning when applied to genital lesions
2. Oral	GI effects: diarrhea, nausea, rash, headache, renal insufficiency, neurotoxicity
3. Intravenous	Phlebitis, rash, hypotension, nephrotoxicity, neurotoxicity (1%–4%) includes tremor, myoclonus, interstitial nephritis
Famciclovir	
2. Oral	Headache, diarrhea, nausea, urticaria, rash, hallucinations or confusion especially in elderly, chronic administration is tumorigenic
Penciclovir	
1. Intravenous	
2. Topical	Reactions at site of application
Ganciclovir	
1. Oral	Myelosuppression, neutropenia (15%–40%), thrombocytopenia (5%–20%), CNS disorders
Valganciclovir	
1. Oral 2. Intravenous	Headache, gastrointestinal disturbance
Cidofovir	
1. Topical	Pain at site of application, burning, pruritus
2. Intravenous	Nephrotoxicity including glycosuria, proteinuria, metabolic acidosis, azotemia
Foscarnet	
1. Intravenous	Fever, rash, nausea, vomiting, hepatotoxicity, pain, and phlebitis at site of injection. Increased creatinine levels, nephrotoxicity, hypocalcemia, nephrogenic diabetes insipidus, interstitial nephritis, CNS side effects include headache (25% patients), irritability, tremors, and hallucinations
Fomivirsen	
1. Intravitreal	Cataracts, iritis, vitritis, increased intraocular pressure in 15%–20% patients
Idoxuridine	
1. Topical	Edema of eyelids, pruritus, pain, and inflammation
Amantadine and rimantadine	
1. Oral	CNS: nervousness, difficulty concentrating, insomnia, light-headedness. High doses cause delirium, hallucinations, seizures,

Continued

Table 16.3 Adverse effects of nonretroviral agents.—cont'd

Antiviral drug	Side effects
	cardiac arrhythmias, and coma. GIT: nausea and loss of appetite. Teratogenic
Oseltamivir	
1. Oral	Abdominal discomfort, nausea, emesis, and local irritation
Zanamivir	
1. Inhalation	Wheezing and bronchospasm
Interferons	
1. Intramuscular/ subcutaneous	Acute influenza-like syndrome: fever, chills, arthralgia, myalgia, nausea, vomiting. Depression, myelosuppression, neurotoxicity manifested by somnolence, confusion, behavioral disturbances, and seizures. Autoimmune disorders include thyroiditis and hypothyroidism. CVS effects include hypotension and tachycardia, proteinuria and azotemia, interstitial nephritis, autoantibody formation, and hepatotoxicity
Ribavirin	
1. Aerosol 2. Oral	Rash, wheezing, conjunctival irritation Bone marrow suppression, hyperuricemia, increased bilirubin levels
Adefovir	
1. Oral	Nephrotoxicity: azotemia, hypophosphatemia, acidosis, glycosuria, and proteinuria Hepatitis
Lamivudine	
1. Oral	Occasional increase in serum transaminases
Entecavir	
1. Oral	Headache, fatigue, dizziness, and nausea Exacerbation of hepatitis B if therapy discontinued
Telbivudine	
1. Oral	Increased creatinine kinase, nausea, diarrhea, fatigue, myopathy, and myalgia
Clevudine	
1. Oral	Myopathy

reported to develop central nervous system side effects with efavirenz, and therapy may be discontinued in 5% patients. Rash may occur in 27% patients treated with efavirenz and is also most commonly associated with etravirine treatment as well.

HIV protease therapy includes drugs like saquinavir, indinavir, ritonavir, nelfinavir, fosamprenavir, lopinavir, atazanavir, tipranavir, and darunavir. All these drugs are associated with gastrointestinal side effects. These effects are dose dependent and include nausea, vomiting, loose stools, abdominal pain, and alteration in taste. Ritonavir and lopinavir combination is associated with long-term risk of atherosclerosis. Fosamprenavir may cause severe rash in treated patients. Indinavir therapy causes nephrolithiasis and crystalluria.

16.4.17 Toxicity of antiprotozoal chemotherapy

Artemisinin-based combination therapies (ACTs) have been recently very commonly employed for the treatment of malarial infections all around the world. They are generally well-tolerated drugs in adults but use in children less than 5 kg and pregnant females is contraindicated. In fetuses and children, they have been associated with brain, liver, and bone marrow toxicities. Conventional antimalarial therapy includes chloroquine, quinine, mefloquine primaquine, piperaquine, atovaquone, and pyrimethamine.

16.4.17.1 Chloroquine

Chloroquine in high doses (>1 g/kg) has been associated with retinopathy and ototoxicity. Prolonged treatment in therapeutic doses has been associated with widening of QRS interval and T wave abnormalities.

16.4.17.2 Quinine

Quinine has been associated with cinchonism, hypotension, and hypoglycemia. Cinchonism symptoms include tinnitus, deafness, visual disturbances, dysphoria, headache, nausea, and vomiting. Quinine causes hemolytic anemia in patients who are glucose 6 phosphate dehydrogenase deficient. Mefloquine administration has been associated with neuropsychiatric disorders occurring in 10% of patients receiving the therapy. Methemoglobinemia is the most common side effect of primaquine. This drug may also cause granulocytopenia, hypertension, arrhythmias, and CNS side effects. Atovaquone is generally well tolerated. Cessation of therapy may be required with side effects like abdominal pain, vomiting, diarrhea, rash, and headache.

16.4.17.3 Pyrimethamine

Antimalarial doses of pyrimethamine cause skin rashes and hematopoiesis. High doses are teratogenic and cause megaloblastic anemia. Proguanil is safe in pregnancy, and therapeutic doses are associated with occasional nausea and diarrhea (Alkadi, 2007).

16.4.17.4 Metronidazole

Metronidazole is commonly employed for treatment of trichomoniasis, amebiasis, giardiasis, and *Clostridium difficile* associated colitis. Apart from the common side effects associated with GIT, metronidazole can cause serious neurotoxic side effects like encephalopathy, convulsions, incoordination, paresthesia, and numbness (Fig. 16.8). In such cases therapy is discontinued. Therapy is also discontinued if hypersensitivity reactions develop. Increased pelvic pressure, dysuria, and cystitis

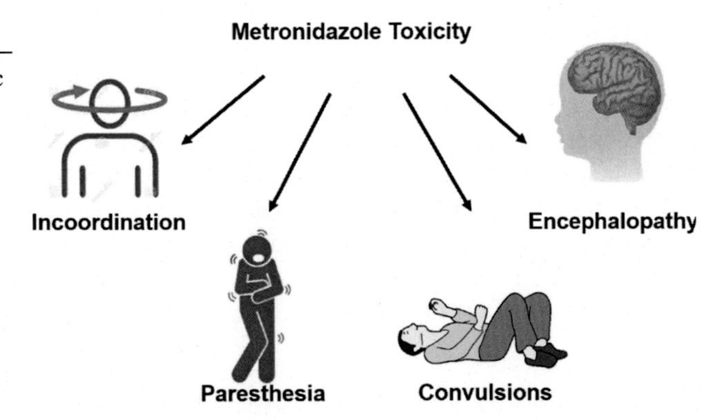

FIGURE 16.8

Metronidazole-associated serious neurotoxic adverse events.

have also been reported. In high doses, the drug has been found to be carcinogenic and mutagenic in rodents.

16.5 Toxicity of antifungal agents

Fungal infections are difficult to treat and usually require long-term treatment. Pathogenic fungi can cause infections systemically and topically, and, on this basis, antifungal therapy has been divided into two major categories: systemic and topical antifungals.

16.5.1 Amphotericin B

Amphotericin B is one of the most widely clinically used systemic antifungal agents. Most of the amphotericin side effects are formulation related. The conventional amphotericin (C-AMB, formulated by complexing with bile salt deoxycholate) results in infusion-related reactions. These reactions are worst with amphotericin B colloidal dispersion (ABCD) and least with liposomal amphotericin B (L-AMB). C-AMB causes azotemia in 80% of treated patients whereas L-AMB is least nephrotoxic. The most common side effects are nausea, headache, malaise and phlebitis at site of infusion.

16.5.2 Flucytosine

Flucytosine, an oral antifungal drug, majorly depresses bone marrow thereby causing hematological disturbances.

16.5.3 Imidazoles and triazoles

Imidazoles and triazoles are available in intravenous, topical, and oral formulations. Imidazoles include ketoconazole, clotrimazole, miconazole, econazole, oxiconazole, sertaconazole, and butoconzole. Triazoles include fluconazole, itraconazole, posaconazole, voriconzole, isavuconazole, and terconazole. Major side effects of this class arise due to interaction with other drugs. All drugs variably inhibit hepatic cytochromes thus increasing the incidence of interactions with other drug classes.

16.5.4 Itraconacole

Itraconacole has been associated with severe hepatic toxicity. Intravenous drug may lead to congestive heart failure in patients with compromised cardiac function. Fluconazole and voriconazole are teratogenic, leading to skeletal and cardiac deformities in newborns. Intravenous administration of voriconazole may lead to visual and auditory hallucinations, but symptoms decrease with time. Posaconazole is well tolerated but causes fetal bone malformation, thereby is contraindicated in pregnancy. Topical econazole and miconzole cause erythema, burning stinging, and itching. Miconazole may also cause skin rash in some treated patients.

16.5.5 Echinocandin

Echinocandins include caspofungin, micafungin, and anidulafungin. Adverse effects of these drugs are rare and rarely lead to discontinuation of therapy. All three drugs are teratogenic. Griseofulvin causes headache and incidence in 15% of patients. Hematological disturbances, hepatotoxicity, gastrointestinal and CNS side effects (neuritis, vertigo, lethargy, fatigue, and mental confusion) occur with griseofulvin administration, however, incidence varies in population studies. Griseofulvin also inhibits the hepatic cytochrome P450 enzymes. Renal side effects include albuminuria. Oral terbinafine rarely causes fatal hepatotoxicity and hypersensitivity reactions.

16.6 Conclusion

This chapter focuses on the major and severe toxicities caused by different antibiotics. Antibiotic resistance and toxicities are major limiting factors in the use of the antibiotics. The challenges in combating diseases caused by pathogenic microorganisms include overcoming the toxicities of the drugs employed in therapy. The information in the current chapter will be useful in limiting the toxic outcomes of various antibiotic therapies.

References

Alkadi, H.O., 2007. Antimalarial drug toxicity: a review. Chemotherapy 53 (6), 385–391. https://doi.org/10.1159/000109767.

Aminov, R.I., 2010. A brief history of the antibiotic era: lessons learned and challenges for the future. Frontiers in Microbiology 1 (134), 1–7. https://doi.org/10.3389/fmicb.2010.00134.

Bacci, C., Galli, L., de Martino, M., Chiappini, E., 2015. Fluoroquinolones in children: update of the literature. Journal of Chemotherapy 27 (5), 257–265. https://doi.org/10.1179/1973947815Y.0000000054.

Blumenthal, K.G., Peter, J.G., Trubiano, J.A., Phillips, E.J., 2019. Review antibiotic allergy. The Lancet 393 (10167), 183–198. https://doi.org/10.1016/S0140-6736(18)32218-9.

Brinker, A.D., Wassel, R.T., Lyndly, J., Serrano, J., Avigan, M., Lee, W.M., Seeff, L.B., 2009. Telithromycin-associated hepatotoxicity: clinical spectrum and causality assessment of 42 cases. Hepatology 49 (1), 250–257. https://doi.org/10.1002/hep.22620.

Bruniera, F.R., Ferreira, F.M., Saviolli, L.R.M., Bacci, M.R., Feder, D., Pedreira, M.D.L.G., Fonseca, F.L.A., 2015. The use of vancomycin with its therapeutic and adverse effects: a review. European Review for Medical and Pharmacological Sciences 19 (4), 694–700.

Demoly, P., Romano, A., 2005. Update on beta-lactam allergy diagnosis. Current Allergy and Asthma Reports 5 (1), 9–14. Retrieved from. http://www.ncbi.nlm.nih.gov/pubmed/15659257.

Dhiman, R.K., Chawla, Y.K., 2005. Herbal medicines for liver diseases. Digestive Diseases and Sciences 50 (10), 1807–1812. https://doi.org/10.1007/s10620-005-2942-9.

El-Gamal, M.I., Brahim, I., Hisham, N., Aladdin, R., Mohammed, H., Bahaaeldin, A., 2017. Recent updates of carbapenem antibiotics. European Journal of Medicinal Chemistry 131, 185–195. https://doi.org/10.1016/j.ejmech.2017.03.022.

Goodman, L.S., Gilman, A., 2011. Goodman and Gilman's pharmacological basis of therapeutics.

Guo, D., Cai, Y., Chai, D., Liang, B., Bai, N., Wang, R., 2010. The cardiotoxicity of macrolides: a systematic review. Die Pharmazie 65 (9), 631–640. https://doi.org/10.1691/ph.2010.0644.

Imani, S., Buscher, H., Marriott, D., Gentili, S., Sandaradura, I., 2017. Too much of a good thing: a retrospective study of β-lactam concentration-toxicity relationships. Journal of Antimicrobial Chemotherapy 72 (10), 2891–2897. https://doi.org/10.1093/jac/dkx209.

Krewski, D., Acosta, D., Andersen, M., Anderson, H., Bailar, J.C., Boekelheide, K., Zeise, L., 2010. Toxicity testing in the 21st century: a vision and a strategy. Journal of Toxicology and Environmental Health Part B: Critical Reviews 13 (2–4), 51–138. https://doi.org/10.1080/10937404.2010.483176.

Kuruvilla, M.E., Khan, D.A., 2015. Antibiotic allergy. In: Mandell, Douglas, and Bennett's Principles and Practice of Infectious Diseases, pp. 298–303. https://doi.org/10.1016/B978-1-4557-4801-3.00023-0.

Lesch, J.E., 2007. The First Miracle Drugs : How the Sulfa Drugs Transformed Medicine. Oxford University Press. Retrieved from. https://global.oup.com/academic/product/the-first-miracle-drugs-9780195187755#.XL9cX9ylGXE.mendeley.

Lockwood, A.M., Cole, S., Rabinovich, M., 2010. Azithromycin-induced liver injury. American Journal of Health-System Pharmacy 67 (10), 810–814. https://doi.org/10.2146/ajhp080687.

Martínez-Salgado, C., López-Hernández, F.J., López-Novoa, J.M., 2007. Glomerular nephrotoxicity of aminoglycosides. Toxicology and Applied Pharmacology 223 (1), 86–98. https://doi.org/10.1016/j.taap.2007.05.004.

Rolain, J.M., Baquero, F., 2016. The refusal of the Society to accept antibiotic toxicity : missing opportunities for therapy of severe infections. Clinical Microbiology and Infections 22 (5), 423–427. https://doi.org/10.1016/j.cmi.2016.03.026.

Schoenenberger, R.A., Haefeli, W.E., Weiss, P., Ritz, R.F., 2009. Association of intravenous erythromycin and potentially fatal ventricular tachycardia with Q-T prolongation (torsades de pointes). BMJ 300 (6736), 1375–1376. https://doi.org/10.1136/bmj.300.6736.1375.

Shepherd, G.M., 2003. Hypersensitivity reactions to drugs: evaluation and management. New York Mount Sinai Journal of Medicine 70 (2), 113–125. Retrieved from. http://europepmc.org/abstract/MED/12634903.

Sogebi, O.A., Adefuye, B.O., Adebola, S.O., Oladeji, S.M., Adedeji, T.O., 2017. Clinical predictors of aminoglycoside-induced ototoxicity in drug-resistant tuberculosis patients on intensive therapy. Auris Nasus Larynx 44 (4), 404–410. https://doi.org/10.1016/j.anl.2016.10.005.

Stanat, S.J.C., Carlton, C.G., Crumb, W.J.C., Agrawal, K.C., Clarkson, C.W., 2003. Characterization of the inhibitory effects of erythromycin and clarithromycin on the HERG potassium channel. Molecular and Cellular Biochemistry 1–7.

Wareham, D.W., Wilson, P., 2002. Chloramphenicol in the 21st century. Hospital Medicine 63 (3), 157–161. Retrieved from. http://europepmc.org/abstract/MED/11933819.

Wulf, N.R., Matuszewski, K.A., 2013. Sulfonamide cross-reactivity: is there evidence to support broad cross-allergenicity? American Journal of Health-System Pharmacy 70 (17), 1483–1494. https://doi.org/10.2146/ajhp120291.

Yılmaz, Özcengiz, G., 2017. Antibiotics: pharmacokinetics, toxicity, resistance and multidrug efflux pumps. Biochemical Pharmacology 133, 43–62. https://doi.org/10.1016/j.bcp.2016.10.005.

Zaunbrecher, V., Beryt, E., Parodi, D., Telesca, D., Doherty, J., Malloy, T., Allard, P., 2017. Has toxicity testing moved into the 21st century ? A survey and analysis of perceptions in the field of toxicology. Eniviron Health Perspect 125 (8), 1–10.

CHAPTER

Carbapenems and *Pseudomonas aeruginosa*: mechanisms and epidemiology

17

Adriana Silva[1,2,3], Vanessa Silva[1,2,3], Gilberto Igrejas[2,3,4,5], Patrícia Poeta[1,2,3]

[1]*Veterinary Sciences Department, University of Trás-os-Montes and Alto Douro, Quinta de Prados Vila Real, Portugal;* [2]*MicroART- Microbiology and Antibiotic Resistance Team, University of Trás-os-Montes and Alto Douro, Quinta de Prados Vila Real, Portugal;* [3]*Associated Laboratory for Green Chemistry (LAQV-REQUIMTE), University NOVA of Lisboa, Lisboa, Caparica, Portugal;* [4]*Department of Genetics and Biotechnology, Functional Genomics and Proteomics' Unit, University of Trás-os-Montes and Alto Douro, Vila Real, Portugal;* [5]*Functional Genomics and Proteomics Unit, University of Tras-os-Montes and Alto Douro (UTAD), Vila Real, Portugal*

17.1 Introduction

Antibiotics have revolutionized medicine and countless lives have been saved, however, the use of these drugs has been accompanied by the fast development of resistance strains. Antibiotic resistance occurs as a consequence of natural selection and a rapid evolution of bacterial resistance compromises public health all over the globe (Liu et al., 2017; Hancock and Speert, 2000). *Pseudomonas aeruginosa* is a universal environmental bacterium and one of the significant bacterial pathogens that causes nosocomial infections worldwide. This bacterium can be the cause of many infections in immunocompromised and hospitalized patients. Consequently, the developing prevalence of this infection severely compromises the selection of proper treatment. Its successful treatment is becoming more difficult and compromised by the possible development of resistance to antimicrobials, such as carbapenem and colistin (Driscoll et al., 2007; Goli et al., 2016; Wang et al., 2010). The emergence of gram-negative opportunist *P. aeruginosa* causes development of resistance to most available antibiotics, including β-lactams, carbapenems, fluoroquinolones, aminoglycosides, and polymyxins. However, carbapenems resistance and colistin resistance have raised a great deal of interest in science community. These antibiotics had an excellent clinical utility in treatment of infections and are considered the last-line of defense against *P. aeruginosa* (Kafil, 2015; Goli et al. 2016; Labarca et al., 2016). This review will focus on antimicrobial resistance, virulence factors, and biofilm formation of *P. aeruginosa*.

17.2 Pseudomonas

In 1894, the genus *Pseudomonas* was described by Migula and belonging to the Gamma-Proteobacteria subclass, Pseudomonadales order, and Pseudomonadaceae family. This genus was

Antibiotics and Antimicrobial Resistance Genes in the Environment. https://doi.org/10.1016/B978-0-12-818882-8.00017-6

formed by many distinguishing species and the classification of them was based on phenotypic characteristics and phylogenetic classification based on their 16S rRNA gene sequences (Igbinosa and Igbinosa, 2015; Peix et al., 2009). They are one of the most ubiquitous bacterial genera in the natural world, showing a metabolic and physiologic versatility. A great diversity of species has been isolated from diverse ecological niches such as soil, waters, various plants and vegetables, extreme environments, clinic specimens, among others. This colonizing capacity in diverse niches is based in their capacity to synthesize many different virulence factors (Igbinosa and Igbinosa, 2015; Silby et al. 2011). Different species that belong to this genus went through several reclassifications, however; *P. aeruginosa* is the most common pseudomonal species and one that generates a greater clinical interest. *P. aeruginosa* barely causes disease in healthy patients but may turn out to be one of the main opportunistic pathogens in nosocomial infections (Peix et al., 2009).

17.2.1 Characteristics of *P. aeruginosa*

P. aeruginosa is an aerobic gram-negative bacillus, rod-shaped, asporogenous, and has a singular polar flagellum that aids in its locomotion. It has minimal growth requirements, uses varied carbon sources for growth, is able to survive in concentrated salt solutions, and tolerates a temperature ranging from 20 to 42°C (R. and R.B. 1998). *P. aeruginosa* is a ubiquitous microorganism that is the result of several factors including its capacity to colonize and survive in multiple environmental niches such as soil, surface waters, plants and vegetables, and can utilize numerous environmental compounds as energy sources. More important, their ability to survive in these conditions has made them colonizers in hospital environments (R. and R.B. 1998; Igbinosa and Igbinosa, 2015). Most of these strains produce at least two pigments, but the characteristic that most distinguishes *P. aeruginosa* from the other pseudomonads is a fluorescent yellow pigment and a blue pigment named pyocyanin; these pigments together give the characteristic green color observed when grown on agar. These characteristics have generally been used as an easy way for identification of bacterial isolates and show an important role in *P. aeruginosa* nutrition (Radó et al. 2017; Morrison and Wenzel, 2015; Peix et al. 2009; Vasil,1986). The genome of *P. aeruginosa* is uniquely large in the prokaryotic world, providing an unusually high proportion of proteins involved in regulation, transport, and virulence functions, which explain the high nutritional versatility and adaptive capacity of this bacteria to host environment (Mesaros et al., 2007).

17.3 Pathogenesis of *P. aeruginosa*

P. aeruginosa has gradually become a main cause of nosocomial infections, a source of diseases in patients with impaired host defenses and one of the prevalent pathogens in intensive care units (ICUs) (Rossolini and Mantengoli, 2005). Nosocomial infection, or “hospital acquired infection,” is a term used for any infection acquired by a patient into medical care who was accepted for a reason other than that infection, and the infection was absent or incubating at the period of admission. Bacteria are the cause of 90% of nosocomial infections. This type of infection affects a huge number of patients worldwide and affects people in both developed and underdeveloped countries, resulting in high mortality and morbidity rates of hospitalized patients (Silva et al. 2014; Danasekaran et al. 2014).

The most common nosocomial infections are respiratory tract infections, surgical wound infections, gastrointestinal tract infections, urinary tract infections, bloodstream infections, and sepsis. Several studies have shown that infection rates are higher in patients with increased susceptibility such as being immunocompromised, of certain age (newborn and elderly), chronic hemodialysis patients, and chemotherapy patients (Rossolini et al. 2005; Silva et al., 2014). According to reports from the WHO, "The burden of health care-associated infection worldwide (2016)" about 15% of all patients hospitalized in ICUs acquire these types of infections. The prevalence of nosocomial infection in developed countries is between 3.5% and 12% (a usual prevalence of 7.1% in Europe) and in underdeveloped countries it varies between 5.7% and 19.1%. The incidence of long-term infections in low-income regions is higher than in high-income regions (Khan et al., 2017). With increasing infections, there is an intensification in extended hospital stays, long-term incapacity, antimicrobial resistance, socioeconomic disturbance, and mortality. Risk factors that may be determinant in these increasing infections are inappropriate use of antimicrobial drugs, nonadherence to precautionary measures, patient conditions and poorly developed surveillance systems, and nonexistent control approaches (Danasekaran et al., 2014; Khan et al., 2017). The dissemination of new infections can occur in three ways: from environment (air, water, and food can be contaminated), from health care professionals, and from hospital waste management. Attention to clear preventive strategies may reduce transmission rates, like using infection control practices and appropriate methods for antimicrobial use (Khan et al., 2017; Khan et al., 2015; Saloojee and Steenhoff, 2001). Efficient and appropriate surveillance methods can help health care institutions minimize the chance of nosocomial infection (Khan et al., 2017; Khan et al. 2015; Mesaros et al. 2007).

17.3.1 Virulence factors

The pathogenesis of *P. aeruginosa* is not correlated to a single virulence factor, but associated with a diversity of different virulence factors that contribute to the bacterial invasion, adhesion, resistance, and toxicity. The versatility of *P. aeruginosa* is a consequence of this capability to produce a high diversity of both cell-associated and extracellular virulence factors in reaction to environmental conditions (Faraji et al. 2016; Choi et al., 2002). Virulence factors can be classified and divided into different groups and depend on their mode of action, infection model, and production of various secreted and cell-associated virulence genes inclusive of exotoxins and enzymes.

Interactions between *P. aeruginosa* virulence factors and the host immune response determine the severity and type of infection. This pathogen is a quiescent colonizer, causing chronic infections, or an extremely virulent invader through acute infections. In chronic infections, *P. aeruginosa* is capable to provoke infections in compromised patients with diseases like cystic fibrosis and chronic obstructive pulmonary disease. In acute infections, *P. aeruginosa* can trigger infections in patients with predisposing factors, such as HIV-positive, cancer, bone marrow transplant, and burn victims or patients with surgical wounds (Veesenmeyer et al., 2010; Delden, 2004; Huber et al., 2016). *P. aeruginosa* owns an arsenal of cell-associated (flagella, pili, lectins, alginates/biofilms, lipopolysaccharides) and extracellular (proteases, hemolysins, cytotoxins, pyocyanins, siderophores, exotoxin A, exoenzyme S, exoenzyme U, among others) virulence factors that are important both in acute infections and chronic infections (Faraji et al., 2016; Delden, 2004; K. et al., 2015; Strateva and Mitov, 2017).These virulence factors are coordinated by a cell density regulatory mechanism named quorum sensing. It is an intercellular, small compound that is composed of an acyl-homoserine lactone, called an autoinducer, which is released in the environment by gram-negative bacteria, like *P. aeruginosa*. The quorum sensing plays a fundamental role in the regulation of expression of a wide diversity of virulence factors, biofilm formation, survival, and colonization by phenotypic alterations at premature stages of infection. Currently, three quorum sensing systems (autoinducers) have been identified in *P. aeruginosa*, the *Pseudomonas* quinolone signal (PQS), which is composed of two systems, *las* and *rhl*. These PQS systems act in a hierarchical method, with *las* system regulating both *rhl* and the production of quinolones. *P. aeruginosa* uses this mechanism not only to modulate virulence factor production but to regulate a variety of biological processes important to adaption to the metabolic stresses and survival in a community (Al-wrafy et al., 2017; Veesenmeyer et al., 2010; Gellatly and Hancock, 2013). *P. aeruginosa* exhibits several virulence factors (Table 17.1), as well as multiple antibacterial resistance mechanisms that have contributed to growing rates of antibacterial resistance (Al-wrafy et al., 2017).

17.4 Biofilms

The term *biofilm* refers to an aggregate of motile bacteria, like *P. aeruginosa*, that possesses flagella that enable free swimming (planktonic) and adhere to each other on a living or nonliving surface. Attached cells produce extracellular polymeric substances, including proteins, exopolysaccharides, extracellular DNA (eDNA), and metabolites. The maturation of biofilms commonly involves cell-to-cell communication over quorum sensing and dispersion of planktonic cells in the biofilms that colonize in new surfaces. These are typical features of all bacterial biofilms, over the process of infection (Fig. 17.1). To define infections caused by biofilms, we may follow the criteria: (1) the bacteria require to be adherent to a surface (plastic, metal, glass, and environmental) in response to a diversity of environmental signals; (2) the examination of infection tissue shows bacteria living in cell clusters, covered in an extracellular matrix; (3) the infection is confined to a specified location; and (4) the infection is hard or impossible to eradicate with antibiotics (Mah, 2012; Driscoll et al., 2007; Parsek and Singh, 2003; Taylor et al., 2014). There are several differentiating characteristics between biofilm and planktonic cells, but the most significant difference is the fact that biofilm cells are much less susceptible to antimicrobial agents, increasingly decline responses to antibiotic treatment. This antibiotic treatment eliminates the planktonic cells, but the sessile forms are resistant and continue to proliferate within the biofilm (Newman et al., 2017; Mulcahy et al. 2008; Chadha, 2014). Biofilms are related with an emergence of antibiotic resistance bacteria, and with cells sitting close to each other may be ideal for horizontal gene transfer. This promotes development and genetic range of natural microbial association. The transfer of DNA occurs within the biofilm community by three major mechanisms: transformation, transduction, and conjugation (Olsen, 2015; Chadha, 2014). Resistance to antibiotics in biofilms is shown by an arrangement of distinct factors: poor antibiotic penetration conferred by the extracellular polysaccharide matrix, incidence of cells showing a resistant phenotype, and presence of nongrowing cells or cells that generated stress responses under hostile chemical conditions in the biofilm matrix. The rising resistance of bacterial biofilm to certain antibiotics allows biofilm-based infections to become chronic infections (Newman et al., 2017; Balcázar et al. 2015). Bacterial biofilms have been implicated in the pathogenesis (80% of bacterial infections in immunocompromised humans involve biofilm-associated microorganisms) and are recognized to colonization in biotic and abiotic superficies such as indwelling medical devices and are involved in many

Table 17.1 Some *Pseudomonas aeruginosa* virulence factors important in acute and chronic infections.

Virulence factor	Type of infections	Biological action
Pili	Acute	Responsible for adherence to cell membranes and other surfaces, formation of biofilms, initiates an inflammatory response
Flagella	Acute	Role in motility, initiates an inflammatory response and mediates initial surface interactions
Exotoxin A	Acute	Cytotoxic by inhibiting protein synthesis (macrophages and T cell mitogen)
Phospholipase C	Acute	Hemolysis, tissue damage and destroys surfactant
Exoenzyme S	Acute	Adherence to epithelium and cytotoxic in host immune cells (apoptosis and necrosis)
Pigments (Pyoverdin and pyochelin)	Chronic	Inhibit ciliary beat (dysfunction in the respiratory tract), toxic to other bacterial species and human cells, enhances oxidative and exerts proinflammatory effects that damage host cells
Alginate	Chronic	Adherence to epithelium; protective factor to biofilms and antibiotics; inhibits antibody and complement binding

Adapted from R. and R.B. 1998

human infections, infectious kidney stones, bacterial endocarditis, cystic fibrosis, biliary tract infections, periodontitis, ophthalmic infections, and other chronic diseases (Driscoll et al., 2007; Parsek and Singh, 2003; Römling and Balsalobre, 2012). The hardiness of biofilms and their resistance to antibiotics has led to a development problem in health care settings, and so, *P. aeruginosa* has turned into one of the most relevant model organisms for the study of biofilm-associated infections (Parsek and Singh, 2003; Taylor et al. 2014).

17.5 Antibiotic resistance

Antibiotics are chemical substances, produced by microorganisms that have ability to inhibit the evolution or destroy bacteria, fungus, and parasite infections. Soon after the discovery of antibiotics (1940) it became evident that bacteria could become resistant to them. Resistance is the skill of an organism to ward off the effects of an antagonistic chemical substance and develop defenses against bacteria, and the successful use of any therapeutic agent is implicated by the progress of tolerance or resistance (Brown, 1982; Davies and Davies, 2010; Martinez, 2014). The disproportionate use and misuse of antibiotics have resulted in a rise in antibiotic resistance among several human pathogens, causing an impact in human and animal health. This uncontrolled use has a direct effect on the possibility of treating infections and compromising treatments that require immunosuppression such as transplantation, intubation, implants of prostheses, catheterization advanced surgical procedures, or anticancer chemotherapy. *P. aeruginosa* has been noticed as a pathogen related with nosocomial

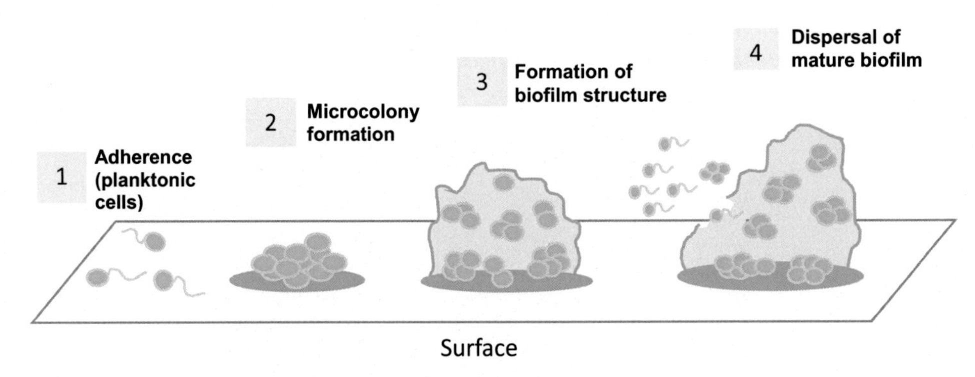

FIGURE 17.1

Biofilm formation progresses. (1) adherence of motile planktonic cells attached to a surface involves the action of pili and flagella (planktonic cells); (2) aggregate cells grow into a microcolony and adherence becomes stronger due to the initiation of matrix production; (3) formation of a mature biofilm structure; (4) some cells from the outer surface of the biofilm colony becoming motile and disperse.

infections and that demonstrates resistance to a variety of antibiotics. In 2017, the WHO printed a list of antibiotic-resistance "priority pathogens" and divided them into three categories according to the urgency of necessity for new antibiotics: Priority 1 (critical), 2 (high), and 3 (medium). *P. aeruginosa* have become resistant to many antibiotics, including third-generation cephalosporins and carbapenems, so these bacteria were included in the most critical group (WHO 2017). Also in 2017, ECDC published a surveillance report on antibiotic resistance in Europe, showing that 30.8% of the *P. aeruginosa* isolates were resistant to at least one antimicrobial group, and a high percentage of *P. aeruginosa* resistant to fluoroquinolones (20.3%), piperacillin ± tazobactam (18.3%), carbapenems (17.4%), ceftazidime (14.7%), and aminoglycosides (13.2%) was also detected (Fig. 17.2). Resistance to two or more antimicrobials was common in 18.3% of all isolates. Antibiotic resistance will continue to be a problem in dealing with *P. aeruginosa* infections, so research dedicated to the development of new antibiotics and therapeutic approaches (Mesaros et al., 2007; Hancock and Speert, 2000) is urgently needed.

17.6 Antibiotic resistance mechanisms in *P. aeruginosa*

P. aeruginosa infections are becoming harder to treat because this pathogen displays a naturally resistance to multiple classes of antibiotics and the number of multidrug-resistance strains is increasing worldwide. This resistance to antibiotics in *P. aeruginosa* can be via intrinsic, adaptive, and acquired mechanisms (Moradali et al., 2017; Bonomo and Szabo, 2006; Pang et al. 2018). The intrinsic antibiotic resistance of a bacterial species to an antibiotic refers to its ability to resist the efficiency of a particular antibiotic over intrinsic structural or functional characteristics and is attributed to a restricted outer membrane permeability, efflux pumps, and production of antibiotic-inactivating enzymes such as β-lactamases. In the acquired resistance, bacteria are susceptible to an antibiotic; the two mechanisms that explain this development of antibiotic resistance are through mutational changes or acquirement

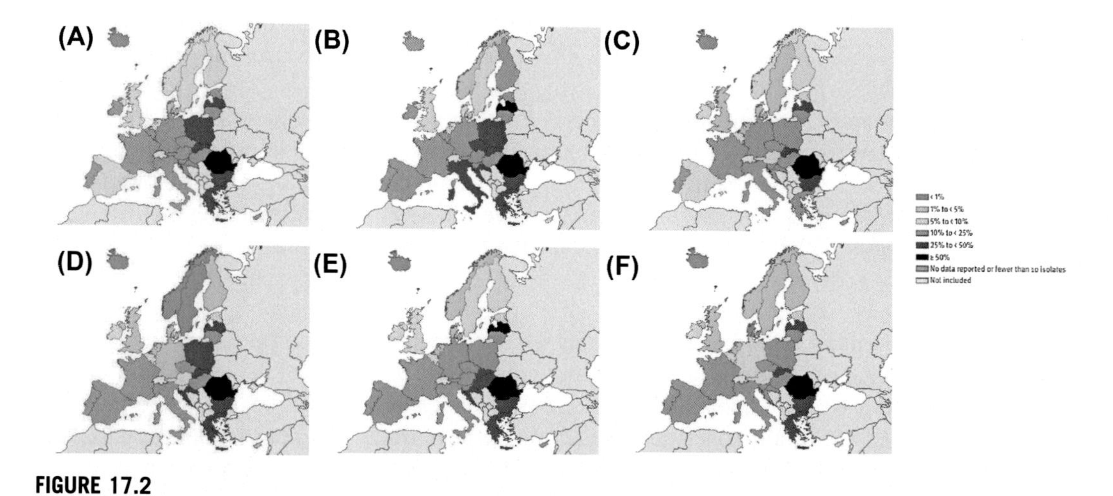

FIGURE 17.2

Percentage (%) of *P. aeruginosa* isolates with resistance in Europe (2017). (A) piperacillin ± tazobactam; (B) fluoroquinolones; (C) ceftazidime; (D) aminoglycosides; (E) carbapenems; (F) combined resistance (three or more antimicrobial groups among piperacillin ± tazobactam, fluoroquinolones, ceftazidime, aminoglycosides, and carbapenems).

Adapted by ECDC 2017(Surveillance report of antimicrobial resistance in Europe 2017 by European Centre for Disease Prevention and Control (ECDC))

of resistance genes via horizontal gene transfer. Adaptive resistance is characterized by as an unstable and transient form of resistance, which increases the capability of a bacterium to survive antibiotic attack due to transient alterations in gene and/or protein expression in reply to an environmental stress. This occurred via adaptive multiple mutational mechanisms and genetic reversion (Breidenstein et al. 2011; Pang et al. 2018; Martinez 2009, 2014; El Zowalaty et al., 2015; Blair et al., 2015; Moradali et al. 2017). The understanding of this increase and the mechanisms of antimicrobial resistance are important to optimize treatments against pseudomonal infections and to employ strict infection control measures. Consequently, the detection of carbapenem-resistant and colistin-resistance strains becomes critical for limiting the spread of the underlying resistance mechanisms, since carbapenems and colistin are last-line antibiotics when pathogens are resistant to all other antibiotics (El Zowalaty et al., 2015; Meletis et al., 2012; Ghasemian et al., 2018).

17.7 Resistance to carbapenems

Carbapenems are bactericidal β-lactam antimicrobial agents with proven efficacy in severe infections with an exceptionally broad spectrum of activity against numerous pathogens. These antibiotics are the most effective antimicrobial agents against gram-positive and gram-negative bacteria and have been used clinically to treat a variety of infections. Since the 1980s, carbapenem antibiotics, including imipenem, meropenem, and doripenem, have been considered a last defense against many drug-resistant bacterial infections and are currently the choice of antibiotic for multidrug-resistant *P. aeruginosa* infections (Codjoe and Donkor, 2018; Zhanel et al., 2007;

Meletis et al., 2012; Deanna J. Buehrle, et al. 2017). *P. aeruginosa* strains that have carbapenem resistance mechanisms severely compromise the proper treatments, since this resistance is commonly associated with intrinsic resistance to other antibiotic classes. The mechanisms of resistance to carbapenems in gram-negative bacteria *P. aeruginosa* are multifactorial and are a result of an interaction between factors, such as mutation in the outer membrane protein OprD, over-expression activity of the efflux pumps MexAB-OprM, acquired Amber class B metallo-beta-lactamase (MBLs) production, and chromosomal β-lactamase (AmpC) overproducer (Mirsalehian et al., 2017; Wang et al. 2010; Meletis et al., 2012; Pai et al., 2001). This resistance is frequently mediated by OprD loss or reduced protein expression that could confer a resistance to imipenem but also confer a reduced susceptibility to meropenem. Porin protein in outer membranes like OprD is a channel for entry of amino acids and carbapenems. Overexpression of efflux pumps or active efflux is a responsible mechanism for transport antibiotics from inside to outside of the cell. The MexAB-OprM system is expressed in almost all *P. aeruginosa* isolates and leads to resistance to multiple classes of antibiotics such quinolones, antipseudomonal penicillins, and antipseudomonal cephalosporins. In *P. aeruginosa* strains overexpressing MexAB-OprM systems, meropenem susceptibility may decrease, but imipenem susceptibility is usually not affected. MBLs are an important group of β-lactamases that can hydrolyze carbapenems and also extended spectrum cephalosporins. The main classes that have been identified in *P. aeruginosa* strains include IMP, VIM, SPM, NDM, SIM, KPC, and GIM. These classes lead to resistance to imipenem and meropenem plus the anti-pseudomonal cephalosporins, including cefepime and antipseudomonal penicillins. AmpC is chromosomally encoded in *P. aeruginosa* and can hydrolyze the carbapenems, however, mutations can occur in the components of AmpC leading to stable hyperproduction of AmpC β-lactamases in *P. aeruginosa* isolates and cause a high-level resistance to many classes of β-lactam antibiotics (ticarcillin, piperacillin, and third-generation cephalosporins) (Mirsalehian et al., 2017; Bonomo and Szabo, 2006; Zhanel et al., 2007; Wang et al., 2010). Several reports of carbapenem resistance showed that the prevalence and rates of resistance in *P. aeruginosa* strains is high. This resistance diverges greatly among different areas, which may reflect differences in infection control policies (Labarca et al., 2016). In this review we demonstrate the emergence of carbapenemases in *P. aeruginosa* in the world among hospital environment isolates and in different areas in the world, such as Europe (Germany, Spain, France, and Croatia), Asia (China, Turkey, and Iran), and Africa (Kenya and Tanzania). A database was created in which study area location, period in which samples were collected, sample collection, number of isolates tested for carbapenem resistance, and carbapenem resistance prevalence were included (Table 17.2). Studies performed in different areas in the world on *P. aeruginosa* carbapenem-resistance isolates that were published from 2006 to 2016, included 2538 patients carrying *P. aeruginosa* from hospitals distributed in Europe, Asia, and Africa. Carbapenem-resistance rates ranged from 13.7% to 100% among *P. aeruginosa*, showing the overall trend to be on the rise over the time, the different distribution among various geographic areas and varying in differences in infection control policies. In Europe, in studies performed by Katchanov et al. 2018, Estepa et al. 2017, Rojo-Bezares et al. 2016 and Bubonja-sonje and Matovina 2015, 656 clinical isolates of *P. aeruginosa* were acquired for blood and clinical samples during 2008–2016 from Germany, Spain, France, and Croatia. Carbapenem-resistance isolates of *P. aeruginosa* were detected in 419 patients (63.8%) treated in these four countries. Higher rates of resistance to carbapenem were reported in all the different hospitals in Europe, and a high number of patients colonized or infected with carbapenem-resistant bacteria highlights the clinical relevance of these pathogens in health care centers. In Asia, in studies performed by Wang et al. 2010, Mirsalehian et al. 2017, Neyestanaki et al. 2017 and Farajzadeh Sheikh et al. 2019, 1225 clinical isolates of *P. aeruginosa* were acquired from burn and clinical samples during 2006–2016. Among the 1225 *P. aeruginosa* isolates collected throughout Asia, carbapenem-resistance isolates were detected in 525 patients (46.6%) in China and Iran. Between the years 2006 and 2016, two types of samples were studied, one with a variety of clinical specimens and one containing only samples of patients who were burned. From 2006 to 2012, in both China and Iran, the prevalence of carbapenem-resistant strains was approximately 40%, with no significant increase in prevalence. The opposite occurred from 2013 to 2016, when there was a large increase in the prevalence of carbapenem-resistant strains (90.7%) in Iran. Thus, it is concluded that this resistance has been an exponential increase and as a consequence it has become a severe public health problem and also a great challenge. In Africa, in studies performed by Terzi et al. 2015, Pitout et al. 2008, and Mushi et al. 2014, 661 clinical isolates of *P. aeruginosa* were acquired for clinical samples during 2007–2012. Among the *P. aeruginosa* isolates collected throughout Asia, 146 strains (22%) were confirmed positive for presence of carbapenems resistance. These values are not considered very significant since we are in the presence of a poorly developed country in terms of infection control policies. Studies conducted globally highlighting the emergence of carbapenem resistance among immuno-compromised patients and an intensive use of carbapenems has caused an increase in carbapenem resistance by acquisition of different mechanisms. Since carbapenems are important agents for the treatment of infections caused by *P. aeruginosa*, the development of this resistance reduces the effective therapeutic options.

17.8 Resistance to colistin

Colistin (polymyxin E) is a lipid antibiotic that was discovered in the 1940s and has broad activity against most clinically relevant gram-negative bacteria, such as *P. aeruginosa.* It was also classified as a last line of defense antimicrobial (Lim et al., 2015; Willmann et al., 2017). The mechanism of action of colistin is essentially based on the electrostatic interaction between positively charged colistin and the negativity charged lipid A subunits present in the structure of lipopolysaccharide (LPS). The initial binding to an anionic LPS component displaces divalent cations, calcium and magnesium, from the outer membrane of *P. aeruginosa.* Therefore, it causes cell permeability changes in the cell envelope and leak of cell contents, which is followed by the reinstatement of the cytoplasmic membrane. This antibiotic also has the capacity to bind and neutralize the lipopolysaccharide molecule of bacteria, giving it antiendotoxin activity (Fig. 17.3) (Goli et al., 2016; Ghasemian et al., 2018; Martis and Leroy, 2014; Lim et al., 2015; Rhouma et al., 2016). The mechanisms of resistance to colistin can be developed through adaptive or mutational mechanisms. Colistin resistance is commonly due to altering the bacterial outer membrane, which is colistin site of action. Reduction of calcium and magnesium contents is an adaptive resistance mechanism and its controlled by two-component regulatory systems (*pho*P-*pho*Q and *pmr*A-*pmr*B). Decreased binding to the outer membrane by lipopolysaccharide remodeling caused changes in these two-component regulatory systems, which results in a less anionic lipid A and stops or decreases this primary interaction with a different mechanism. Resistance caused by mutation, such as efflux pump overexpression (MexAB-OprM and MexXY-OprM), contributes to intrinsic multidrug resistance in *P. aeruginosa* (Lim et al., 2015; Goli et al., 2016). Colistin is, in some

Table 17.2 Reports of carbapenem-resistance infection caused by *Pseudomonas aeruginosa isolates* in the world.

Location	Samples	Period	Number of isolates	Carbapenem resistance isolates	Carbapenem resistance isolates %	References
Germany	Clinical	2015–16	119	66	55.5%	Katchanov et al. (2018)
Spain	Clinical	2008–11	61	55	90%	Estepa et al. (2017)
France	Blood	2011–13	434	260	60%	Rojo-Bezares et al. (2016)
Croatia	Clinical	2009–10	38	38	100%	Bubonja-sonje and Matovina (2015)
China	Clinical	2006–07	645	258	40%	Wang et al. (2010)
Iran	Burn	2011–12	130	53	40.8%	Mirsalehian et al. (2017)
Iran	Clinical	2013–16	236	214	90.7%	Farajzadeh Sheikh et al. (2019)
Turkey	Clinical	?	18	9	50%	Terzi et al. (2015)
Kenya	Clinical	2006–07	416	57	13.7%	Pitout et al. (2008)
Tanzania	Clinical	2007–12	227	80	35%	Mushi et al. (2014)

cases, the only therapeutic option to treat *P. aeruginosa* infections, so isolates that develop resistance to colistin are a thoughtful concern and may have devastating effects if no other treatment options are available to fight the infection (Lim et al., 2015).

Nowadays the prevalence of colistin resistance is low, however, despite most of the studies reporting resistance rates below 10%, there are some studies reporting high resistance rates. The increasing use of colistin to treat *P. aeruginosa* infections is the cause of the development of colistin resistance in some countries globally. This prevalence may vary between geographical regions with different treatment strategies and over time. According to Nation et al. 2014, in some countries, like Japan and South Africa, clinicians do not have access to colistin; on the other hand, in a few areas of the world, like Europe and Australia, clinicians have only the parenteral formulation of colistin. In the United States, Brazil, Malaysia, and Singapore, clinicians can use either colistin or the polymyxin B parenteral for treatment. It became crucial to obtain information related to colistin resistance in the world to create the appropriate guidelines that can help to control colistin resistance spread. In Iran, in a study performed by Goli et al. 2016, 100 nonrepetitive clinical isolates of *P. aeruginosa* were acquired for clinical samples during January to June 2014. These isolates were testing according to the guidelines (CLSI) for antimicrobial susceptibility testing. Two colistin-resistance strains were isolated

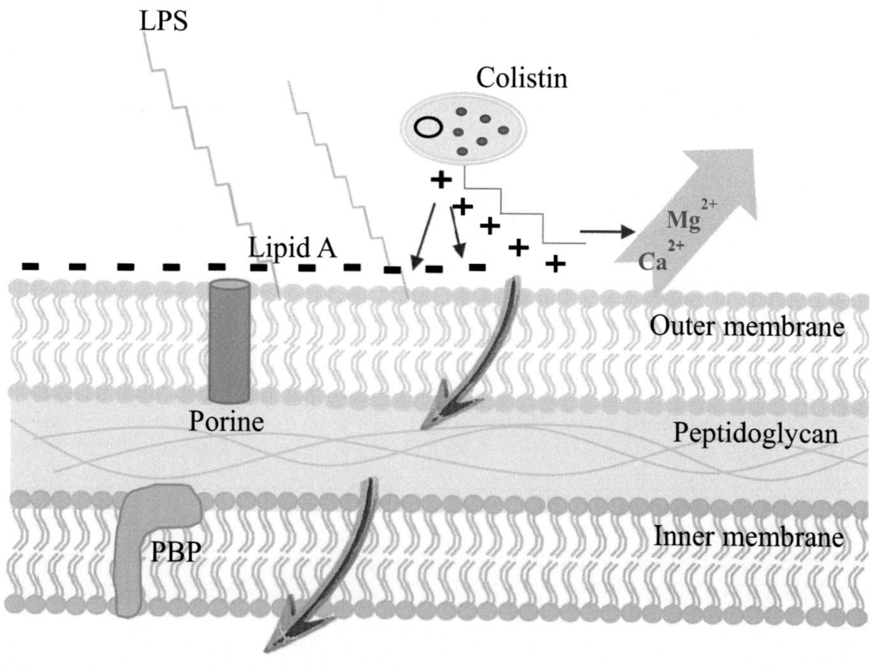

FIGURE 17.3

Mechanism of colistin on *P. aeruginosa* membrane. The cationic cycle decapeptide structure of colistin binds with the anionic LPS molecules by dislocating calcium (Ca^{2+}) and magnesium (Mg^{2+}) ions from the outer cell membrane, leading to permeability changes in the cell by binding to the hydrophobic lipid A element of LPS. PBP: penicillin-binding protein.

from two patients. However, a study in the northwest of Iran reported a higher rate of resistance to colistin (14.9%) and neighboring countries of northwest Iran showed resistance varying from 0% to 31.7%. Differences between these reports are due to variances between accessibility of colistin and policies associated with the use of this antibiotic in hospitals of these countries (Goli et *al.*, 2016). Although *P. aeruginosa* is a problem in medical environments, rising levels of antimicrobial agents employed by veterinary medicine makes animals probable reservoirs of resistant bacteria. A study was done by Liu et al. 2017 in Taiwan and carried out based on 58 *P. aeruginosa* isolates disease-free of oral cavity of snakes (captive, wild, and clinical cases). These isolates were testing according to the guidelines (CLSI) for antimicrobial susceptibility testing; 32 isolates (55%) were found to be unsusceptible to colistin. This study highlights the complex problem of resistance because *P. aeruginosa* are not uncommon in snakes' oral cavity and these results are probably from contact with unspecified antibacterial peptides (Liu et al., 2017).

Resistance to colistin has been found and a wide range of studies on colistin resistance should be performed to deal with the antibiotic resistance era and use colistin effectively (Ko et al., 2017).

17.9 Conclusions

P. aeruginosa is a uniquely problematic nosocomial pathogen and frequently resistant to a broad range of antibiotics. The prevalence of this strain appears to be increasing worldwide and has become a serious problem (Strateva and Yordanov, 2019).

A better consideration of the host bacteria *P. aeruginosa* as an integrated system is essential for the progress of advanced therapeutic strategies to fight against this pathogen that has a significant impact on public health. There is an urgent requirement call for implementation of global antimicrobial resistance surveillance and monitoring programs to help track antimicrobial resistance profiles in *P. aeruginosa* (Pang et al., 2018; El Zowalaty et al., 2015). Preventive measures are currently being put in place; however, the improvement is slow, and medicine has many trials ahead before it will resolve this problem. An advance in fighting and containing this opportunistic pathogen is hopeful for the future; awareness, greater understanding and research directed toward combating *P. aeruginosa* are urgently needed (Jenny and Kingsbury, 2018).

Acknowledgments

This work was supported by the Associate Laboratory for Green Chemistry-LAQV, which is financed by national funds from FCT/MCTES (UID/QUI/50006/2019). Vanessa Silva is supported by FCT under the PhD grant SFRH/BD/137947/2018.

References

Al-wrafy, F., et al., 2017. Pathogenic Factors of *Pseudomonas aeruginosa* — the Role of Biofilm in Pathogenicity and as a Target for Phage Therapy (Czynniki Patogenności *Pseudomonas aeruginosa* — Rola Biofilmu W Chorobotwórczości I Jako Cel Terapii Fagowej), pp. 78—91.

Balcázar, J.L, Subirats, J., Borrego, C.M, 2015. The role of biofilms as environmental reservoirs of antibiotic resistance. *Frontiers in Microbiology* 6, 1—9. https://doi.org/10.3389/fmicb.2015.01216.

Blair, J.M.A., et al., 2015. Molecular mechanisms of antibiotic resistance. Nature Reviews Microbiology 13 (1), 42—51. https://doi.org/10.1038/nrmicro3380. Nature Publishing Group.

Bonomo, R.A., Szabo, D., 2006. Mechanisms of multidrug resistance in acinetobacter species and *Pseudomonas aeruginosa*. Clinical Infectious Diseases 43 (Suppl. 2), S49—S56. https://doi.org/10.1086/504477.

Breidenstein, E.B.M., de la Fuente-Núñez, C., Hancock, R.E.W., 2011. *Pseudomonas aeruginosa*: all roads lead to resistance. Trends in Microbiology 19 (8), 419—426. https://doi.org/10.1016/j.tim.2011.04.005.

Brown, L., 1982. Antibiotics AU. Journal of American College Health Association. Taylor & Francis 30 (5), 227—229. https://doi.org/10.1080/07448481.1982.9938897.

Bubonja-sonje, M., Matovina, M., 2015. Mechanisms of carbapenem resistance in multidrug-resistant clinical isolates of *Pseudomonas aeruginosa* from a Croatian hospital. Microbial Drug Resistance 21 (3), 1—9. https://doi.org/10.1089/mdr.2014.0172.

Buehrle, D.J., Ryan, K., Shields, a, B., et al., 2017. Crossm aeruginosa Bacteremia : risk factors for mortality and microbiologic treatment failure. Antimicrobial Agents and Chemotherapy 61 (1), 1—7.

Chadha, T., 2014. Bacterial biofilms: survival mechanisms and antibiotic resistance. Journal of Bacteriology & Parasitology 05 (03), 5—8. https://doi.org/10.4172/2155-9597.1000190.

Choi, J.Y., et al., 2002. Identification of virulence genes in a pathogenic strain of *Pseudomonas aeruginosa* by representational difference analysis. Journal of Bacteriology 184 (4), 952–961. https://doi.org/10.1128/jb.184.4.952-961.2002. American Society for Microbiology.

Codjoe, F.S., Donkor, E.S., 2018. Medical sciences carbapenem Resistance : a review. Medical Sciences 1–28. https://doi.org/10.3390/medsci6010001.

Danasekaran, R., Mani, G., Annadurai, K., 2014. Prevention of healthcare-associated infections: protecting patients, saving lives. International Journal of Community Medicine and Public Health 1 (1), 67. https://doi.org/10.5455/2394-6040.ijcmph20141114.

Davies, J., Davies, D., 2010. Origins and evolution of antibiotic resistance. Microbiology and Molecular Biology Reviews 74 (3), 417–433. https://doi.org/10.1128/MMBR.00016-10.

Delden, C., 2004. Virulence factors in. Pseudomonas 2, 3–45. https://doi.org/10.1007/978-1-4419-9084-6_1 (Figure 1).

Driscoll, J.A., Brody, S.L., Kollef, M.H., 2007. The epidemiology, pathogenesis and treatment of *Pseudomonas aeruginosa* infections. Drugs 67 (3), 351–368. https://doi.org/10.2165/00003495-200767030-00003.

El Zowalaty, M.E., et al., 2015. *Pseudomonas aeruginosa*: arsenal of resistance mechanisms, decades of changing resistance profiles, and future antimicrobial therapies. Future Microbiology 10 (10), 1683–1706. https://doi.org/10.2217/fmb.15.48.

Estepa, V., et al., 2017. Characterisation of carbapenem-resistance mechanisms in clinical *Pseudomonas aeruginosa* isolates recovered in a Spanish hospital. Enfermedades Infecciosas y Microbiología Clínica 35 (3), 141–147.

Faraji, F., et al., 2016. Molecular detection of virulence genes in *Pseudomonas aeruginosa* isolated from children with Cystic Fibrosis and burn wounds in Iran. Microbial Pathogenesis 99, 1–4. https://doi.org/10.1016/j.micpath.2016.07.013. Elsevier Ltd.

Farajzadeh Sheikh, A., et al., 2019. Molecular epidemiology of colistin-resistant *Pseudomonas aeruginosa* producing NDM-1 from hospitalized patients in Iran. Iranian Journal of Basic Medical Sciences 22 (1), 38–42. https://doi.org/10.22038/ijbms.2018.29264.7096.

Gellatly, S.L., Hancock, R.E.W., 2013. *Pseudomonas aeruginosa*: new insights into pathogenesis and host defenses. Pathogens and Disease 159–173. https://doi.org/10.1111/2049-632X.12033.

Ghasemian, A., Shafiei, M., Hasanvand, F., 2018. Carbapenem and colistin resistance in Enterobacteriaceae : worldwide spread and future perspectives. Reviews in Medical Microbiology (4), 173–176. https://doi.org/10.1097/MRM.0000000000000142.

Goli, H.R., et al., 2016. Emergence of colistin resistant *Pseudomonas aeruginosa* at Tabriz hospitals , Iran. Iranian Journal of Microbiology 8 (1), 62–69.

Hancock, R.E.W., Speert, D.P., 2000. Antibiotic resistance in *Pseudomonas aeruginosa*: mechanisms and impact on treatment. Drug Resistance Updates 3 (4), 247–255. https://doi.org/10.1054/drup.2000.0152.

Huber, P., et al., 2016. *Pseudomonas aeruginosa* renews its virulence factors. Environmental Microbiology Reports 1–20. https://doi.org/10.1111/1758-2229.

Igbinosa, I., Igbinosa, E., 2015. The Pseudomonads as a versatile opportunistic pathogen in the environment, pp. 822–831. Available at: http://www.microbiology5.org/microbiology5/book/822-831.pdf.

Jenny, M., Kingsbury, J., 2018. iMedPub journals properties and prevention : a review of *Pseudomonas aeruginosa* abstract, pp. 1–8.

K, N., et al., 2015. Virulence genes and antibiotic resistance profile of pseudomonas aeruginosa Isolates in Northwest of Iran. Journal of Pure and Applied Microbiology 9, 383–389 (SpecialEdition1 PG-383-389). Available at: http://www.microbiologyjournal.org/NS -.

Kafil, H.S., 2015. Review Colistin , mechanisms and prevalence of resistance. Current Medical Research and Opinion (January). https://doi.org/10.1185/03007995.2015.1018989.

Katchanov, J., et al., 2018. Carbapenem-resistant Gram-negative pathogens in a German university medical center : prevalence, clinical implications and the role of novel β -lactam/β -lactamase inhibitor combinations, pp. 1–14.

Khan, H. A., Ahmad, A., Mehboob, R, 2015. Nosocomial infections and their control strategies.

Khan, H.A., Baig, F.K., Mehboob, R., 2017. Nosocomial infections: epidemiology, prevention, control and surveillance. Asian Pacific Journal of Tropical Biomedicine 7 (5), 478–482. https://doi.org/10.1016/j.apjtb.2017.01.019. Elsevier B.V.

Ko, K.S., Choi, Y., Lee, J., 2017. Old drug, new findings : colistin resistance and dependence of Acinetobacter baumannii. Precision and Future Medicine.

Labarca, J.A., et al., 2016. Carbapenem resistance in *Pseudomonas aeruginosa* and Acinetobacter baumannii in the nosocomial setting in Latin America Carbapenem resistance in *Pseudomonas aeruginosa* and Acinetobacter baumannii in the nosocomial setting i. Critical Reviews in Microbiology. https://doi.org/10.3109/1040841X.2014.940494, 7828(October 2017).

Lim, L.M., et al., 2015. Resurgence of colistin: a review of resistance, toxicity, pharmacodynamics, and dosing. Pharmacotherapy 30 (12), 1279–1291. https://doi.org/10.1592/phco.30.12.1279.Resurgence.

Liu, P., et al., 2017. Colistin resistance of *Pseudomonas aeruginosa* isolated from snakes in taiwan. Hindawi 2017, 1–5. https://doi.org/10.1155/2017/7058396.

Mah, T.-F., 2012. Biofilm-specific antibiotic resistance. Future Microbiology 07 (09), 1061–1072. https://doi.org/10.2217/fmb.12.76.

Martinez, J.L., 2009. The role of natural environments in the evolution of resistance traits in pathogenic bacteria. Proceedings of the Royal Society B: Biological Sciences 276 (1667), 2521–2530. https://doi.org/10.1098/rspb.2009.0320.

Martinez, J.L., 2014. General principles of antibiotic resistance in bacteria. Drug Discovery Today: Technologies 11 (1), 33–39. https://doi.org/10.1016/j.ddtec.2014.02.001. Elsevier Ltd.

Martis, N., Leroy, S., 2014. Colistin in multi-drug resistant *Pseudomonas aeruginosa* blood-stream infections A narrative review for the clinician. The Journal of infection. https://doi.org/10.1016/j.jinf.2014.03.001, 33.

Meletis, G., et al., 2012. Mechanisms responsible for the emergence of carbapenem resistance in *Pseudomonas aeruginosa*. Hippokratia 303–307.

Mesaros, N., et al., 2007. *Pseudomonas aeruginosa*: resistance and therapeutic options at the turn of the new millennium. Clinical Microbiology and Infections 13 (6), 560–578. https://doi.org/10.1111/j.1469-0691.2007.01681.x.

Mirsalehian, A., et al., 2017. Determination of carbapenem resistance mechanism in clinical isolates of *Pseudomonas aeruginosa* isolated from burn patients, in Tehran, Iran. Journal of Epidemiology and Global Health 7 (3), 155–159. https://doi.org/10.1016/j.jegh.2017.04.002. Ministry of Health, Saudi Arabia.

Moradali, M.F., Ghods, S., Rehm, B.H.A., 2017. *Pseudomonas aeruginosa* lifestyle: a paradigm for adaptation, survival, and persistence. Frontiers in Cellular and Infection Microbiology 7 (February). https://doi.org/10.3389/fcimb.2017.00039.

Morrison, a J., Wenzel, R.P., 2015. Epidemiology of infections due to *Pseudomonas aeruginosa*. Reviews of Infectious Diseases 6 (Suppl. 3), S627–S642. https://doi.org/10.1093/clinids/6.Supplement_3.S627 (October).

Mulcahy, H., Charron-Mazenod, L., Lewenza, S., 2008. Extracellular DNA chelates cations and induces antibiotic resistance in *Pseudomonas aeruginosa* biofilms. PLoS Pathogens 4 (11). https://doi.org/10.1371/journal.ppat.1000213.

Mushi, M.F., et al., 2014. Carbapenemase genes among multidrug resistant gram negative clinical isolates from a tertiary hospital in Mwanza , Tanzania. Biomed Research International 2014.

Nation, R.L, Velkov, T., Li, J., 2014. Colistin and polymyxin B : peas in a pod, or chalk and Cheese?,, 59, 88–94. https://doi.org/10.1093/cid/ciu213.

Newman, J.W., Floyd, R.V., Fothergill, J.L., 2017. The contribution of *Pseudomonas aeruginosa* virulence factors and host factors in the establishment of urinary tract infections. FEMS Microbiology Letters (June), 1–11. https://doi.org/10.1093/femsle/fnx124.

Olsen, I., 2015. Biofilm-specific antibiotic tolerance and resistance. European Journal of Clinical Microbiology & Infectious Diseases 34 (5), 877–886. https://doi.org/10.1007/s10096-015-2323-z.

Pai, H., et al., 2001. Carbapenem resistance mechanisms in *Pseudomonas aeruginosa* clinical isolates. Antimicrobial Agents and Chemotherapy 45 (2), 480–484. https://doi.org/10.1128/AAC.45.2.480-484.2001.

Pang, Z., et al., 2018a. Antibiotic resistance in *Pseudomonas aeruginosa*: mechanisms and alternative therapeutic strategies. Biotechnology Advances. Elsevier 37 (1), 177–192. https://doi.org/10.1016/j.biotechadv.2018.11.013.

Pang, Z., et al., 2018b. Antibiotic resistance in *Pseudomonas aeruginosa*: mechanisms and alternative therapeutic strategies. Biotechnology Advances. Elsevier 37, 177–192. https://doi.org/10.1016/j.biotechadv.2018.11.013 (November 2018).

Parsek, M.R., Singh, P.K., 2003. Bacterial biofilms: an emerging link to disease pathogenesis. Annual Review of Microbiology 57 (1), 677–701. https://doi.org/10.1146/annurev.micro.57.030502.090720.

Peix, A., Ramírez-Bahena, M.H., Velázquez, E., 2009. Historical evolution and current status of the taxonomy of genus Pseudomonas. Infection, Genetics and Evolution 9 (6), 1132–1147. https://doi.org/10.1016/j.meegid.2009.08.001.

Pitout, J.D.D., et al., 2008. Metallo- b -lactamase-producing *Pseudomonas aeruginosa* isolated from a large tertiary centre in Kenya. Clinical Microbiology and Infections 14 (8), 755–759. https://doi.org/10.1111/j.1469-0691.2008.02030.x. European Society of Clinical Infectious Diseases.

R, W., R.B, D., 1998. *Pseudomonas aeruginosa* and other related species. Thorax 53 (3), 213–219. Available at: http://www.embase.com/search/results?subaction=viewrecord&from=export&id=L28172375%5Cnhttp://sfx.library.uu.nl/utrecht?sid=EMBASE&issn=00406376&id=doi:&atitle=Pseudomonas+aeruginosa+and+other+related+species&stitle=Thorax&title=Thorax&volume=53&issue=3&sp.

Radó, J., et al., 2017. Characterization of environmental pseudomonas aeruginosa using multilocus sequence typing scheme. Journal of Medical Microbiology 66 (10), 1457–1466. https://doi.org/10.1099/jmm.0.000589.

Rhouma, M., Beaudry, F., Letellier, A., 2016. International Journal of Antimicrobial Agents Resistance to colistin : what is the fate for this antibiotic in pig production ? International Journal of Antimicrobial Agents 48 (2), 119–126. https://doi.org/10.1016/j.ijantimicag.2016.04.008. Elsevier B.V.

Rojo-Bezares, B., et al., 2016. Characterization of carbapenem resistance mechanisms and integrons in pseudomonas aeruginosa strains from blood samples in a French hospital. Journal of Medical Microbiology 65 (4), 311–319. https://doi.org/10.1099/jmm.0.000225.

Römling, U., Balsalobre, C., 2012. Biofilm infections, their resilience to therapy and innovative treatment strategies. Journal of Internal Medicine 272 (6), 541–561. https://doi.org/10.1111/joim.12004.

Rossolini, G.M., Mantengoli, E., 2005. 'Treatment and control of severe infections caused by multiresistant *Pseudomonas aeruginosa*. Clinical Microbiology and Infections 11, 17–32. https://doi.org/10.1111/j.1469-0691.2005.01161.x. European Society of Clinical Infectious Diseases.

Saloojee, H., Steenhoff, A., 2001. The health professional's role in preventing nosocomial infections. Postgraduate Medical Journal 77 (903), 16–19. https://doi.org/10.1136/pmj.77.903.16.

Silby, M.W., et al., 2011. Pseudomonas genomes: diverse and adaptable. FEMS Microbiology Reviews 35 (4), 652–680. https://doi.org/10.1111/j.1574-6976.2011.00269.x.

Silva, N., Igrejas, G., Poeta, P., 2014. High-throughput genomic technology in research of virulence and antimicrobial resistance in microorganisms causing nosocomial infections. Journal of Integrated Omics 4 (2), 44–56. https://doi.org/10.5584/jiomics.v4i2.167.

Strateva, T., Mitov, I., 2017. Contribution of an arsenal of virulence factors to pathogenesis of *Pseudomonas aeruginosa* infections. Annals of Microbiology. https://doi.org/10.1007/s13213-011-0273-y (8328).

Strateva, T., Yordanov, D., 2019. *Pseudomonas aeruginosa* – a phenomenon of bacterial resistance. Journal of Medical Microbiology 1133–1148. https://doi.org/10.1099/jmm.0.009142-0 (2009).

Taylor, P.K., Yeung, A.T.Y., Hancock, R.E.W., 2014. Antibiotic resistance in *Pseudomonas aeruginosa* biofilms: towards the development of novel anti-biofilm therapies. Journal of Biotechnology 191, 121–130. https://doi.org/10.1016/j.jbiotec.2014.09.003. Elsevier B.V.

Terzi, H.A., et al., 2015. Investigation of oprd porin protein levels in carbapenem-resistant *Pseudomonas aeruginosa* isolates. Jundishapur Journal of Microbiology 8 (12), 10–14. https://doi.org/10.5812/jjm.25952.

Vasil, M.L., 1986. Pseudomonas aeruginosa: biology, mechanisms of virulence, epidemiology. The Journal of Pediatrics.

Veesenmeyer, J.L., Hauser, A.R., Lisboa, T., R, J., 2010. *Pseudomonas aeruginosa* virulence and therapy: evolving translational strategies. Critical Care Medicine 37 (5), 1777–1786. https://doi.org/10.1097/CCM.0b013e31819ff137. Pseudomonas).

Wang, J., et al., 2010. Molecular epidemiology and mechanisms of carbapenem resistance in *Pseudomonas aeruginosa* isolates from Chinese hospitals. International Journal of Antimicrobial Agents 35 (5), 486–491. https://doi.org/10.1016/j.ijantimicag.2009.12.014. Elsevier B.V.

Willmann, M., et al., 2017. Crossm rapid and consistent evolution of colistin resistance in extensively drug-resistant pseudomonas aeruginosa during morbidostat culture. Antimicrobial Agents and Chemotherapy 61 (9), 1–16.

Zhanel, G.G., et al., 2007. Comparative review of the carbapenems. Drugs 67 (7), 1027–1052.

CHAPTER

Environmental and public health effects of antibiotics and AMR/ARGs

18

Saadia Andleeb, Mahnoor Majid, Sumbal Sardar
Atta-ur-Rahman School of Applied Biosciences (ASAB), National University of Sciences and Technology (NUST), Islamabad, Pakistan

18.1 Introduction to antimicrobial resistance in the environment

Mankind has greatly progressed in terms of therapeutics and medicine. Pharmaceuticals have become crucial for the survival and maintenance of public health, which ultimately determines the quality and standard of life. There are numerous natural and synthetic compounds that serve to prevent and cure various kinds of mild to serious life-threatening conditions. There is no doubt regarding the benefits of these active compounds, but during the last 20 years, their concentrations in the soil and water ecosystems have been gradually surging due to the advancements in the field of environmental sciences (Carvalho and Santos, 2016). The high percentage of antibiotics in the environmental samples out of all the other pharmaceuticals is alarming because of the emergence of multidrug resistance microbes and the misuse of antibiotics, which all pose serious threats to the health of human and animals (Hanna et al., 2018). Antimicrobials are specified as those compounds that have the ability to inhibit or cease the growth of microorganisms such as bacteria, fungi, algae, protozoa, and archaea. Antibiotics are those antimicrobials that particularly act on the bacteria and fungi found in the hosts of human and animals (Nankervis et al., 2016). They can be naturally produced by various plants and fungi or can be manufactured as well by adding various beneficial additives. Mostly used as a therapeutic aid, they are also sold as pesticides in the agricultural sector. They are further classified into various categories depending upon their chemical structures and mode of action, where some of them inhibit the cell wall synthesis of the target cell, cause disruption of cell membrane or inhibition of protein synthesis and nucleic acid synthesis (Orthobullets.com, 2019). The impact of these antibiotics on environment is perilous as these drugs are not fully metabolized in the human body and they get excreted in large concentration via urine and feces. These unaltered active compounds thus, after passing through various routes, get to wastewater treatment plants (WWTPs) (Manaia et al., 2018). Numerous research studies have indicated the emergence of resistant microbes in the wastewater treatment plants, and as these treatment plants are not conventionally built to eliminate the incoming antibiotics, they get released directly into the receiving niche causing serious concerns for the surrounding habitats (Halling-Sørensen et al., 1998). The presence of antibiotic resistant bacteria in the wastewater is a prime factor for the further spread of the resistant genes in the surroundings. Various procedures have

Antibiotics and Antimicrobial Resistance Genes in the Environment. https://doi.org/10.1016/B978-0-12-818882-8.00018-8

been suggested for the effective elimination of the antibiotics in WWTPs such as coagulation, oxidation, microfiltration, etc., but, due to their high costs and maintenance, they are not typically employed (Homem and Santos, 2011). As well, the unmonitored disposal of drugs, the application of reclaimed water and animal manure in agricultural sector also contribute to the spread of antibiotic resistant genes and bacteria in the environment. The presence of antibiotic resistant bacteria in plants can prove to be a major factor in transferring and strengthening the resistant strains in human hosts (Bengtsson-Palme and Larsson, 2015).

18.2 Antibiotic resistance in the environment

The ability of a microorganism to multiply and reproduce in the presence of antibiotics is commonly termed as the antibiotic resistance. Since the discovery of antibiotics, the phenomenon of resistance has been a constant threat to human and animal health. The intrinsic genes present in the genetic makeup of the microorganisms encode the resistance toward the antibiotics. The process is natural as the microbes also produce antibiotics themselves. Overuse and inappropriate prescriptions have played a vital role in the selection of drug-resistant microorganisms (Neu, 1992). The impact of the antibiotic-resistant strains on natural resources and environment is getting precarious and immediate action plans need to be devised to manage the concerning situation (Grenni et al., 2018). There are a number of ways that contribute to the contamination of environment, which include the untreated urban waste water, wastewater coming from hospitals, untreated water containing effluents from the livestock farms, and aquaculture consisting of one or a few resistant strains of enteric bacteria. Besides these, the sludge obtained from sewage that is used in recycled form for agricultural lands also contains traces of certain antibiotics and antibiotic-resistant bacteria. Similarly, animal manure also contains a significant amount of veterinary antibiotics. Various research studies indicate the adverse effects of antibiotics on the algal and bacterial communities. These communities further have the capability to create an ecological cascade impact on higher tropic levels that rely on them as a potential food source (Bengtsson-Palme et al., 2017). The presence of antimicrobial-resistant bacteria (AMR) and antibiotic-resistant genes (ARGs) indicates the contamination of the environment for which effective characterizations of the AMRs and ARGs are pertinent.

18.3 Global antimicrobial-resistance action plan

Resistance against the antimicrobial agents is a very serious concern that has shaken the entire foundation of therapeutics. The antibiotics are essential for the prevention and management of various health conditions. The prophylactic, empiric, and definitive therapies are all based on the optimal utilization of antibiotics. They are the prerequisites for surgical treatments and procedures so as to minimize the risks of infections. However, due to the misuse of these wonder drugs, the situation has become complicated within recent years and it is feared that if effective global action plans are not implemented, the world will soon enter into a postantibiotic era where a simple infection could kill us again. In order to find a solution for this health crisis, the World Health Assembly conceived a global action plan on antimicrobial resistance that entails five major objectives (World Health Organization, 2015).

1. Employment of effective programs and trainings to increase awareness about the resistance against antimicrobials.
2. Implementation of research and surveillance to enhance the knowledge base.
3. Efforts to minimize the incidence of infections by introducing effective preventive measures.
4. Optimization of antimicrobials in human and animals.
5. Development of a sustainable economic plan to meet the needs of all the countries.
6. Increment in investments for new therapies, tools, medicines, and diagnostics (World Health Organization, 2015).

18.4 Food and Agriculture Organization antimicrobial resistance development framework

It has been estimated that the global economy might lose about six trillion USD by the year 2050 if the crisis of antimicrobial resistance is not taken care of. Antimicrobial resistance not only risks the lives of human and animals but it also threatens the economy, development, and security of communities due to the surge in untreatable infections. The situation is more severe in underdeveloped countries where the people who are already deprived of balanced nutrition have to face the additional burden of drug resistance. This puts the sustainable development goals of such countries in peril. The primary reason for the huge effect of antimicrobials is their spread across the globe as the resistant genes can cross borders and affect everyone equally. Therefore, the need to devise a "One Health" solution has become crucial to managing this pressing situation. As the AMR is not restricted to a particular region, holistic approaches must be implemented to address this crisis. The World Health Organization (WHO) with the support and assistance of the Food and Agriculture Organization (FAO) and World Organization for Animal Health (OIE), have taken the initiative of a global action plan to minimize the spread of antimicrobial resistance. FAO works as a main player in the food and agriculture sector to mitigate the effects of AMR. Moreover, the organization plays a role in spreading public awareness and approaches to manage AMR in livestock production, crop farming, and aquaculture. The FAO framework specifically works on the governance and policy-making body to address the hurdles and challenges and provides guidance to the local governments to systematically improve the AMR polices on national levels (Food and Agriculture Organization, 2018).

18.5 Antimicrobial resistance, National Action Plan, Pakistan

Antimicrobial resistance has affected every country equally. In Pakistan, the AMR has risen within a few years due to the misuse and inappropriate prescriptions by the clinicians. On the country level, the Minister of State, Ministry of National Health Services Regulations & Coordination jointly devised a "National Strategic Framework for Containment of Antimicrobial Resistance" through effective consultations that took place in 2016. To further work on the case, a workshop was arranged in March 2017 to translate the AMR strategic framework into an AMR National Action Plan (NAP). Professionals from all the concerned sectors participated to endorse the "One Health" solution approach. The AMR NAP is a commitment of the government of Pakistan on WHA68.7 resolution on AMR. The entire plan is based on the objectives laid out by the WHO and fully supports the cause of an effective solution (AMR National Action Plan, Pakistan, 2017).

18.6 Key drivers of AMR/ARGs in the environment

The massive utilization of antibiotics in clinical and animal practices has led to the intrinsic danger of selecting the ARGs. These ARGs also promote resistance to other chemicals besides the resistance toward conventional and synthetic antibiotics. Recent studies have indicated the role of the ARGs in conferring resistance to many different kinds of harmful chemicals as well (Levy, 2002). A number of factors drive resistance toward chemicals such as antimicrobials, heavy metals, and biocides. Antimicrobials further consist of subclasses of antibiotics, antifungals, antivirals, and antiparasitics that cause resistance in the environment. Moreover, biocides such as disinfectants and surfactants along with plant derived and xenobiotics (octanol, hexane, and toluene) are known to select for resistance genes (Friedman, 2015).

18.6.1 Horizontal gene transfer

Horizontal gene transfer is one of the primary factors that play a role in the spread of AMR/ARGs in the environment. The phenomenon is defined as the acquisition of foreign genes or mobile genetic elements between different genera. It mostly arises owing to the inherent characteristics of microorganisms or due to the extrinsic environmental triggers, therefore, it is regarded as a nonrandom genome innovation process (Kristiansson et al., 2011; Jain, 2003). The process entails the uptake of foreign DNA followed by its genetic recombination and regulatory integration, finally establishing itself in the host population. Horizontal gene transfer (HGT) also holds significant importance as it establishes itself in the exchange community, which is defined as those organisms that have the ability to share genes by HGT without needing to be in close proximity to each other (Skippington and Ragan, 2011). HGT is considered as an integral component of bacterial evolution since the discovery of the ability of bacteria that it can acquire resistance to multiple antibiotic resistances that have been incorporated in their hereditary fingerprints (Heuer and Smalla, 2007). It has been suggested that the ARGs are portions of plasmids that are self-transferrable or transposable elements and can be genetically transferred between related or unrelated bacteria. It is fascinating that nature has enabled bacterial species to release antibiotics in the environment so as to compete with other bacteria (Djordjevic et al., 2013). This mechanism of survival mostly arises when the antibiotic resistant gene becomes trapped in a mobile genetic element, for instance, a phagemid, transposon, integron, or a chromosomal island. These mobile genetic elements then serve as the key factors to introduce the natural antibiotic-resistant genes into those bacterial species that are clinically relevant from various environmental sources. Approximately 20% of a bacterial genome can be acquired from other species, which is possible due to the horizontal gene transfer (Ochman et al., 2000). Microbial communities acquire ARGS due to a number of factors, including selection pressure. The presence of antibiotics in the environment creates a pressure, ultimately causing stress to the microbes, resulting in the horizontal gene transfer (Alanis, 2005). The antibiotic resistance genes in pathogenic bacteria occur in clinical settings, food chains, and local communities. The selective pressure exerted by the antibiotics that are persistent in the environment holds a key responsibility for the spread of the resistant genes via the HGT. Besides this, the intrinsic genes present in abundance in sewage, manure, and hospital water all promote the pollution of the environmental sources with the resistance genes (Heuer et al., 2011).

18.6.2 Soil-borne resistance

Soil is an essential habitat for the survival of various organisms and plants. It is responsible for the degradation of pollutants, thus acting as a bioreactor for the transformation of nutrients. Various studies have indicated that soil receives incoming antibiotics due to the consistent applications of manure and sewage sludge as fertilizers. Therefore, soil serves a prominent hotspot for the spread of antibiotic-resistant genes among the indigenous microorganisms (Thiele-Bruhn, 2003). Some current research also hypothesizes that antibiotic-resistant genes and various genetic determinants have been detected in immaculate environments as well as in populations of human and animals that have never been exposed to antibiotics. This proposes that the antibiotic resistant genes that have successfully integrated into the genetic elements gained the ability to sustain and spread in the environment without the need of antibiotics ((Martinez, 2009; Chee-Sanford et al., 2009). Another factor that contributes toward the development of antimicrobial resistance in the presence of antibiotics is the high density of microbes in the soil environment, which facilitates the rapid genetic exchange ultimately enhancing the resistance (Ding and He, 2010). In various agricultural countries, it is a common practice to incorporate animal wastes into the soil to increase its nutrient capacity. This procedure is the main entry point for the antibiotic-resistant microbes and genetic-resistant elements into the environment. Moreover, the microbes such as enteric bacteria, depending upon their species and temperature, can survive in the environment for months, whereas certain genetic determinants also have the ability to stay active in the soil niches (Marti et al., 2013).

18.6.3 Animal husbandry

Apart from the HGT and soil-borne resistance, another major determinant of antibiotic resistance in the environment is the animal husbandry. The guts of animals are not able to fully adsorb the antibiotics, therefore, the residues are excreted in the urine and feces of the animals. The wastes produced by the animals do not undergo the secondary treatment like human waste does, due to which a high concentration of the antibiotic residues from the animal wastes enter into the environment (Sarmah et al., 2006). A typical example of the spread of resistance in the environment by the animal management sector is known as vegetation, aquaculture, and cage (VAC). The VAC is a recycling farm that is present throughout Asia. It consists of a vegetable field, an aquaculture pond, and caged animals. The waste from the animals such as pigs, cattle, chickens, and ducks is transported directly to the aquaculture pond and the vegetable fields. The waste and sewage from the livestock is not treated and is directly employed for the fertilization of fish ponds and vegetables. Moreover, the manure from the animals is a good source for the eutrophication of the fish ponds, therefore, the VAC system is considered to be an economical method of farming (Hop, 2003). However, the major drawback of this type of recycling farm is the spread of antibiotic resistance in the environment as antibiotics are used for the health maintenance of the animals. But due to the poor adsorption capacity of the animals, the antibiotics are released into their excreta. Thus, the farm waste circulating in VAC contains high concentrations of antibiotic resistant genes, which ultimately spread into the surrounding environment (Heuer and Smalla, 2007). In another study, it was indicated that the livestock in the VAC system, particularly pigs, were found to be highly resistant to antibiotics nalidixic, tetracycline, and enrofloxacin in the enteric bacteria (Dang et al., 2011). This proposed that when the livestock became exposed to high concentrations of antibiotics, the risk of colonization with the antibiotic resistant microbes and genes increases in the guts of the animals. Furthermore, the resistant genes can be easily transferred to other environments through water and food. The gastrointestinal bacteria are the main source of antibiotic-resistance selection and they happen to be present in the animal manure. This manure is basically a reservoir for the antibiotics and the resistant bacteria, which then spread in the environment (Suzuki and Hoa, 2012). Efforts are needed to minimize the spread of the antibiotic resistance from the livestock into the surrounding environment. WHO has developed a special program, the Advisory Group on Integrated Surveillance of Antimicrobial Resistance, whose main objective is to mitigate the public health impact of AMR in the food-producing animals. WHO has also joined hands with FAO to employ surveillance of food-borne pathogens in the livestock industry, which will somehow fill the missing information gaps leading toward the development of efficient national policies for better animal husbandry systems with appropriate hygiene measures (Manges et al., 2007).

18.6.4 Wastewater and sludge

Wastewater plays an important role in the dissemination of antibiotic-resistant genetic elements and microorganisms in the environment. Moreover, drinking water coming from surface water sources provides a route for the resistance genes to be introduced into the natural bacterial ecosystems. As various contaminants such as detergents, antimicrobials, disinfectants, heavy metals, and industrial wastes are received in the water, they gradually accumulate, leading to the facilitated spread of resistant organisms in the environment. Selection pressure then acts as a triggering factor for the rampant spread of antibiotic resistance (Baquero et al., 2008). The presence of antibiotics has been confirmed by various studies in the wastewater, sludge, pond water, and animal effluents in the industrial countries such as China. Sulfamethazine (75%), oxytetracycline (64%), tetracycline (60%), sulfadiazine (55%), and sulfamethoxazole (51%) were found to be frequent (Suzuki and Hoa, 2012). The WWTPs serve to treat the polluted water coming from the industries and agricultural sector. However, these treatment plants have become major reservoirs of antibiotic-resistant microorganisms. The high concentration of industrial wastes contributes toward the selection of resistant strains, which overall become a reason for the spread of antibiotic resistance (Li et al., 2010).

18.7 Environmental pathways for antibiotic resistance

Environment plays a very important role in the spread of antibiotic resistance. It provides various regulators for controlling resistance-driving agents, including antibiotics and biocides (Singer et al., 2016). There are various pathways that help in the dissemination of antibiotic resistance, which are described below.

18.7.1 Spread of antimicrobial resistance via municipal and industrial water

Along with the natural resistance of microorganisms, misuse and senseless use of antibiotics have also caused an increase in the resistance. This resistance mostly spreads by the wastewater. Other than for treatment of human and animal diseases, antibiotics are also used for the growth of animals (Dincer and Yigittekin, 2017) and they enter into water bodies via different routes, including WWTPs, effluents, and infiltration of water for agriculture purposes (Tahrani et al., 2016). Municipal waste water is a reservoir of antibiotics as human and other animals excrete various chemicals including antibiotics in the urine. These chemicals, including antibiotics, are excreted in their active state, which might be biodegradable. Sometimes they get absorbed into the sewage sludge or may enter into the effluents without changing their active state (Singer et al., 2016). Antibiotics are not efficiently removed by the conventional method of water treatments and these wastewater plants are considered as hot spots for the dissemination of antibiotic resistance (Lood et al., 2017). The large number of microorganisms in these water bodies results in the spreading of antibiotic resistance through horizontal and vertical gene transfer (Zarei-Bayg et al., 2019). Along with conventional methods, some advanced technologies were also applied to compare the results of different methodologies. One of the studies reported in 2007 by Watkinson et al. indicated that dominant antibiotics present in waste water were ciprofloxacin, cephalexin, cefaclor, sulfamethoxazole, and trimethoprim. Conventional and advanced, both of these systems remove antibiotics with an average rate of 92%, but, still, antibiotics were found in the effluents of activated sludge after treatment (Watkinson et al., 2007). Another study, reported from Tunisia in 2016, also indicated the presence of several antibiotics in different water samples such as wastewater treatment, industrial wastewater, and seawater. Aminoglycosides and phenicol antibiotics were detected in almost all reported samples. Highest concentrations of neomycin (16.4 ng/mL) and kanamycin B (7.5 ng/mL) were observed in the wastewater influents. It was also concluded from the reported study that wastewater plants did not completely remove antibiotics (Tahrani et al., 2016). Similarly, concentrations of antibiotics were observed in four sewage treatment plants in China and Japan. Significant amount of antibiotics were detected in the influents and effluents. Among 12 antibiotics, clarithromycin (1.1−1.6 μg/L) and levofloxacin (3.6−6.8 μg/L) were present in highest concentrations in both of the plants (Ghosh et al., 2016). It has been reported that trace amounts of antibiotic-resistance DNA were present in the wastewater treatment plant's product, and as a result, this resistant DNA reentered into the environment and caused resistance development. Studies carried out by researchers at the University of California found that even the presence of low amounts of antibiotics results in resistance development against several microbes (Zarei-Baygi et al., 2019). Another quantitative analysis of antibiotic-resistance genes was carried out by researchers in Poland who analyzed the treated and untreated water samples collected from wastewater treatment plants and molecularly characterized the genes' resistance to β-lactamase and tetracycline via RT PCR technique. They found that *blaOXA* and *blaTEM* and *tetA*, respectively, were dominant among all the tested genes resistant to tetracycline and β-lactamase (Osiäska et al., 2019). Finally, it may be concluded from the above discussions that wastewater, including industrial and domestic water, plays a huge role in the dissemination of antibiotic resistance and the proliferation of ARGs. For now, even the wastewater treatment plants cannot completely resolve this issue, and they also provide an opportunity for spreading of resistance and provide a gateway for the dissemination of resistant microbes.

18.7.2 Transmission of antimicrobial resistance via livestock

Antimicrobials play a very significant role in the maintenance of good health of animals, but high exposure of animals to antimicrobial agents can cause the dissemination of ARGs (Magouras et al., 2017). It was estimated that about 63,151 tons of antibiotics were used in livestock in 2010, and this value will increase to 67% by 2030 (Van Boeckel et al., 2015). Use of antimicrobial agents in veterinary practices is considered as one of the basic routes in the transmission of AMR and antibiotic resistance. For spreading of resistance from animals to humans, some pathogens follow direct contact while others follow different pathways such as food-borne routes like *Salmonella enterica*, *Campylobacter coli/jejuni*, and *Yersinia enterocolitica*. Environment and fauna become a reservoir of antibiotic resistance and also serve as the source of proliferation of antibiotic-resistant bacteria and their spread among humans and animals. It is because of the fact that antibiotic residues and bacteria get released from food-animal production with manure and reenter into the environment where they promote the development of resistance (Wegener, 2012). Use of livestock manure as a fertilizer and the overuse of antibiotics in aquaculture are two important ways of spreading antibiotic resistance (Magouras et al., 2017). Antibiotics used in food-producing animals are similar to those used in humans and can select for resistance by animals. Cross transmission of resistant bacteria and resistance genetic elements can also occur easily (Tang et al., 2017). One of the experimental studies in United States confirmed the presence of gentamycin-resistance genes in *Enterococci* isolated from animals, and the same genes were also present in the food products of the same animals. It was observed that similar resistance patterns were also shown by *Enterococci* isolated from human and retailed food of different regions (Donabedian et al., 2003). A study from Nigeria confirms the presence of resistant *E.coli* isolated from poultry forms. Various resistance genes were found in the isolates, including *bla-TEM*, *sul2*, *sul3*, *aadA*, *tetA*, *tetB*, etc. These results provide evidence that livestock production farms are important reservoirs of ARGs (Adelowo et al., 2014). Therefore, livestock veterinarians and farmers should play a responsible role for controlling antimicrobial resistance and antimicrobial usage.

18.7.3 Spreading of antibiotic resistance via gray water and recycled water

Gray water can be defined as domesticated water coming from hand basins, washing machines, and kitchens (Jefferson et al., 2000). Gray water is very important in the countries facing the problem of water scarcity. Although it is not contaminated with feces, various pathogenic microbes are present in it such as *Staphylococcus aureus* and *Pseudomonas aeruginosa* (Al-Gheethi et al., 2016). Treated gray water was analyzed for the presence of antibiotic-resistant bacteria at the molecular level as well as phenotypically in the central Negev Desert, Israel. Tetracycline-resistant bacteria was found to be the most prevalent among the water samples. Twenty-four tetracycline-resistance strains were detected among them; 14 carried the *tet39* gene, which were phylogenetically similar in different environmental and clinical samples (Troiano et al., 2018). In this regard, it is very important to test the quality of gray water in order to stop the dissemination of antibiotic-resistant pathogens in the environment.

Reclaimed or recycled water is very important and used for various purposes. But it also acts as reservoir for the spread of AMR and ARGs. The microbial community selects the antibiotic residues present in the recycled water, for development of antibiotic-resistance genes. One of the studies carried out in the United States provides experimental evidence that reclaimed water contains antibiotic-resistant genes. Three different recycling water systems were analyzed for the presence of antibiotic resistance, including those resistant to sulfonamides, macrolides, tetracycline glycopeptide, and methicillin. In addition to this, water samples were tested for the presence of pathogens such as *Legionella pneumophila*, *Escherichia coli*, and *P. aeruginosa* (Fahrenfeld et al., 2013). Similarly, ARGs were reported in 6 out of 11 reclaimed water samples in Australia. These six samples contained ARGs showing resistance to methicillin and sulfonamide. The percentage of resistance was 9% and 45%, respectively. In comparison to river water, antibiotic resistance is more prevalent in recycled water (Barker-Reid et al., 2010). Therefore, it is very important to monitor reclaimed water for the presence of antibiotics and ARGs in order to reduce antimicrobial resistance.

18.8 Ways to reduce antimicrobial resistance in the environment

Antimicrobial resistance is commonly associated with the overuse, misuse, and abuse of antimicrobial agents. In order to ensure the persisting efficiency of antibiotics, every member of society should play a concerted role. This is the only way to reduce antibiotic resistance. Effective strategies are required to address the problem of antimicrobial resistance, some of which are discussed below.

18.8.1 Minimization of antibiotic use in humans, animals, and plants

Antimicrobial stewardship programs (ASPs) should be followed while consuming antibiotics in order to reduce antibiotic resistance. The purpose of ASPs is to reduce the inappropriate use of antimicrobials and antibiotics. Second, is to promote the principle of optimal antibiotics use and to minimize the collateral damage of antibiotics (CUNHA, 2018). Judicious use of antibiotics is required because it is commonly observed that antibiotic resistance is associated with the use of antibiotics because every existing pathogen develops resistance against the antibiotics commonly used in the clinics. Among them, extended spectrum β-lactamases are commonly encountered in hospitals. After the discovery of new drugs, resistant pathogens have been developed in the clinical settings where the use of antimicrobials is more prevalent (Cantas et al., 2013). Up to 40% of antibiotic prescriptions for acute respiratory tract infection are unnecessary. This includes conditions such as pharyngitis, sinusitis, tonsillitis, the common cold, and pneumonia (Barlam et al., 2016). In the community we can reduce the process of antibiotic resistance by minimizing the use of antibiotics, as shown by the study from Finland that demonstrated that during a 4-year period erythromycin resistance was reduced from 16.5 to 8.6 among the group of *Streptococcus pneumonia* (Seppala et al., 1997). According to the U.S. Centers for Disease Control and Prevention (CDC), one in three antibiotics prescriptions is unnecessary. Therefore, the U.S. White House released a National Action Plan for Combating Antibiotic-Resistant Bacteria in 2015 in order to reduce the inappropriate use of antibiotics by 2020. According to the CDC, health care professionals and patients must follow the rules in order to reduce the use of antibiotics. Outpatient health care providers can evaluate their prescribing habits and implement antibiotic stewardship activities, such as watchful waiting or delayed prescribing, when appropriate, into their practices. Patients can talk to their health care providers about when antibiotics are needed and when they are not. These conversations should include information on patients' risk for infections by antibiotic-resistant bacteria (National Vaccine Advisory Committee, 2016). Similarly, use of antibiotics and antimicrobial agents should be minimized in livestock and in plants to combat antibiotic resistance. In the United States, about 80% of the antibiotics are used in the farming, agriculture, and aquaculture (Hollis and Ahmed, 2013). The total sale of antibiotics for therapeutic purpose in the field of veterinary medicine in 10 European countries during 2007 varied from 18 to 180 mg/kg biomass (Grave et al., 2010). Antibiotics are given to the animals to promote growth, due to which bacteria are exposed to low quantity of antibiotics for a long period of time and as a result, antibiotic resistant strains develop in the animals. For this, WHO recommends the farmers and food industry to not use antibiotics as a growth promoter. Similarly, antibiotics should not be used to prevent diseases in healthy animals (World Health Organization, 2019). Inappropriate use of antibiotics in plants and agriculture promotes the emergence of antibiotic resistance as well. Use of antimicrobial agents in plants are usually distributed in the soil and water system and have a great impact on the environment, animal and human resistomes

(Thanner et al., 2016). One of the studies from China detected antibiotic-resistance genes *tetX*, *blaCTX-M*, and *sul1* and *sul2* in the endophytic bacteria isolated from pak choi (*Brassica chinensis* L.) when the plant was exposed to tetracycline, cephalexin, and sulfamethoxazole in order to promote its growth (Zhang et al., 2017). This indicates the potential risk of overuse of antibiotics. In conclusion, the prescription of inappropriate antibiotics should be reduced in humans, animals and in plants. Antibiotics should only be taken when it is required and taken exactly as prescribed. Be antibiotic aware.

18.8.2 Efficient treatment of wastes to eliminate the antimicrobial residues (urban, industrial, agricultural, hospitals, solid and liquid waste)

The presence of antibiotics in wastewater is very important to manage and to eliminate as they can provide a reservoir for spreading of antibiotic resistance even in small concentrations. The presence of antimicrobial agents in the wastewater poses a great threat to water bodies, including streams, oceans, and rivers. Antibiotic resistance bacteria and antibiotic resistance genes in the environment, particularly in wastewater, are of particular interest as they are dangerous for human health (Su et al., 2015). Therefore, efficient strategies are required to treat wastewater for the removal of antimicrobial residues. Ozone and UV irradiation are used to kill resistant bacteria in wastewater by damaging its DNA and proteins (Jager et al., 2018). Chlorination is also an effective method of reduction of ARGs but this needs to be explored. A study from China demonstrates that different levels of chlorine have different effects on the removal of ARGs such as erythromycin- and tetracycline-resistance genes (Yuan et al., 2015). Aerobic and anaerobic-aerobic sequence bioreactors effectively reduce ARGs when compared to aerobic reactors (Christgen et al., 2015). Anaerobic membrane bioreactors are also used in order to eliminate ARGs because they are considered as a sustainable, low-energy-consuming process for treatment of wastewater. Another study from Singapore detected a significant reduction in the level of ARGs and bacteria in wastewater after treatment in the anaerobic membrane bioreactor (Le et al., 2018). Efficient biosolid treatment is required because it also acts as an effective reservoir for spreading of ARGs. Pyrolysis is one of the effective methods for treating biosolid waste. It effectively reduces the bacterial biomass and ARGs as well as converts waste material into energy-rich compounds such as py-oil and py-gas (Kimbell et al., 2018). Similarly, phytoremediation also works well for removal of antibiotics and thus can be helpful in the reduction of antibiotic resistance. By translocation, antibiotics get absorbed into the plant roots and then get transferred to the stems. *Acrostichum aureum* and *Rhizophora apiculate* were checked for their phytoremediation effects and these plants significantly reduced antibiotics levels in the contaminated sediments in shrimp farms (Hoang et al., 2013). Overall, it may be proposed that antibiotic resistance is a serious problem but if all the responsible members of society play their roles, then this problem can be brought under control.

18.8.3 Efforts to minimize the use of antibiotics by improving hygiene practices

There is a famous saying that prevention is better than cure. Therefore, prevention of infections by maintaining proper hygienic conditions in turn will reduce the usage of antibiotics, and as a result combating antibiotic resistance will be possible. Poor control measure of infections such as hand hygiene and precaution is one of the important reasons of cross transmission of resistance bacteria

(Pittet, 2004). Since 2009, WHO committed to a resolution to combat transmission of AMR by starting the campaign Save LIVES. The aim of this program is to improve hand hygiene worldwide so as to stop the transmission of infections and infection-causing agents. One way to reduce AMR transfer is by maintaining proper hygienic conditions so this will stop the transfer of resistant bacteria via hands of health care workers (particularly) and thus, this will reduce the entry of antibiotic-resistance bacteria into vulnerable patients (Kilpatrick and O'Conner, 2014). There is experimental evidence that suggests the reduction of hospital acquired infections by maintaining proper hand hygiene practices by health care workers (Rosenthal et al., 2005; Pessoa-Silva et al., 2007). Along with hand hygiene, cleaning of gloves, gowns, and uniforms should also be considered as the uniforms of health care providers are colonized with Multi-drug resistant microorganisms (MDR) (Lee et al., 2013). One of the studies from the United States reports the vancomycin-resistant *Enterococci* contamination of gloves, gowns, and stethoscopes (Zachary et al., 2001). Similarly, the health care environment also harbors the multidrug-resistant pathogens and can survive there for long period of time (Chemaly et al., 2014). Cross-sectional study from Brazil demonstrates that environmental surfaces from health care units carried methicillin-resistant *S. aureus* (Ferreira et al., 2011). Therefore, resources should be directed for the effective cleaning and disinfection of health care environments and surfaces.

18.8.4 Spreading awareness and educating the masses

Combating antibiotic resistance will become easier if awareness is spread among the members of society. As antibiotics are commonly prescribed and given to anyone having infections, therefore, it is necessary to disseminate information regarding the use of antibiotics and antibiotic resistance in order to change the attitudes and behavior of our young generations toward use of antibiotics and other antimicrobial agents. The E-Bug project has been introduced in 18 European countries, aiming to educate young people about the appropriate use of antibiotics, transmission of infections, hygiene, and vaccines, and thus to make them realize that it is very important to control antibiotic resistance (McNulty et al., 2011). Similarly, World Antibiotics Week takes place every November (12–18) to increase the global awareness of antibiotic resistance and to encourage the general public, health workers, and prescribers to ensure the best practices for reducing the emergence of antibiotic resistance (WHO). Different surveys conducted on awareness of antibiotics show that the general public is not aware of antibiotic resistance. A cross-sectional questionnaire-based study reported from Eastern Romania demonstrated that there is reduced awareness regarding antibiotic resistance and self-medication in the general public (Topor et al., 2017). So, it is very important to spread knowledge about antibiotics. Antibiotics save lives, but they are not always the answer as they do not work on viruses. When antibiotics are not needed, they won't help but instead may cause minor to severe side effects, as well as antibiotic resistance.

18.9 Ethics regarding the use of antibiotics in the environment

Antibiotics endanger the survival of a healthy and sustainable environment. Various studies have reported the presence of several antibiotics such as fluoroquinolones and sulfonamides in the environmental reserves where they cause fluctuation in the processing of nitrogen, leading toward the buildup of nitrogen in the environment, which is extremely dangerous for various species. Moreover, the excessive use of antibiotics also poses a serious threat to the maintenance of biodiversity as it can be life threatening to commensal microbes and other flora and fauna (Roose-Amsaleg and Laverman, 2015). The ethical dilemmas that originate as a result of inadequate prescriptions by the clinicians play an important role in the spread of antibiotic resistance in the environment in addition to the adverse side effects to the patients. In 1970, Van Rensselaer Potter conceptualized the term *bioethics*. The focal point of his theory was the dependency of the human survival upon adequate resource allocation. He further emphasized ethics as a sole priority for the support system in ecological fields (Kuhse and Singer, 2013). But, this concept by Potter was not largely adopted and later it was referred to as *biomedical ethics*, which is limited to clinicians and patients individually, while environmental ethics deals with the various aspects of environment and its safe preservation. Still, there is a crucial need for combining the two concepts as the drugs prescribed to patients are not just limited to the individual but rather they go into the environment and become a serious concern of antibiotic resistance spread (Gruen and Ruddick, 2009).

18.10 Ethical facets of antibiotics resistance

There are various problems that arise due to the growing resistance against the commonly used antibiotics. Depending upon the complexity, composition, and level of utilization, the ethical dilemmas differ from each other. Fundamentally, they can be further categorized into four major areas.

18.10.1 Ethics regarding aantibiotic use for controlling infectious diseases

Incorrect prescriptions and lack of awareness about the use of antibiotics among the general masses exacerbate the dilemma of resistance. Moreover, the timely and credible reporting about the patients' conditions with contagious bacterial infections still remains controversial as it leads to the social separation of patients from the community. Thus, the ethics exist but their proper implementation is often not properly followed. The rise in the number of dangerous bacterial infections as well as the inability of current drugs to treat them have made it imperative to address the ethical conflicts for the safety of the community. Due to the increased costs and limitations of resources in the health care sector, the situation can become more serious owing to the conflicts for the demand of more resource allocations (Hollis and Maybarduk, 2015).

18.10.2 Ethics regarding the fair distribution of global resources

Antibiotic resistance equally affects everyone, and it does not spare borders. However, those low- and middle-income countries that are already facing various other crises have to deal with this dilemma with more burden and stress. The health care management in underdeveloped countries is deplorable and it is critical to minimize the overuse of antibiotics. As well, it is an evident reality that the people in poor countries do not normally have access to high-quality drugs, therefore, the mortality rate is much higher due to the lack of availability of drugs than those who die from the resistance in developed and high-income countries. On a global scale, an immediate ethical response is needed to facilitate the low-capacity countries in dealing with the resources and antibiotic resistance (Basnyat, 2014).

18.10.3 Ethics regarding the use of antibiotics in veterinary practices

According to an estimate, almost one half of the total antibiotics are being utilized by the veterinary sector. There is no doubt that this sector saves a lot of money due to the use of antibiotics, but the ethical concerns related to the welfare of the animals and their health are rarely taken into consideration. If all the classes of antibiotics will be used for humans, then it would become impossible to treat certain infections in animals. Therefore, it is important to devise serious plans that should focus not only just on the rapid farming of the animals but also consider their welfare and health (Bengtsson and Greko, 2014).

18.10.4 Ethics regarding the use of antibiotics in environment

WHO and the U.S. Food and Drug Administration (FDA) are jointly working to devise consistent and sustainable solutions for antibiotic resistance and its rapid spread in the environment. The ethics of using correct and appropriate usage of these drugs is under debate and it is very important to make the population aware of the adverse side effects of the drugs. Therefore, not only just clinicians but the patients should understand their responsibility in managing the antibiotics effectively (Parsonage et al., 2017).

18.11 Alternative therapies to eradicate the AMR/ARGs

Antibiotics are consequential agents for effective control and management of harmful and infectious diseases. Within recent years, the abusive implementation of these drugs has created havoc, ultimately resulting in the emergence of multidrug-resistant microorganisms. Humans and animals absorb only a limited portion of these drugs, while a large portion of them get released into the aquatic and terrestrial environments, thus spreading resistance via various transmission mechanisms (Goulas et al., 2018). In order to effectively manage this situation, there is a pertinent need to look for alternative therapies, therefore, bringing order from the chaos. Some of the strategies that can be adopted are mentioned in this section of the chapter.

18.11.1 Environmental nanotechnology

The frequent misuse of antibiotics has caused a prominent decrease in the efficacy of the antimicrobials and serious concerns have been recognized by global health policy on this emerging issue. Microorganisms have developed various methods through which they acquire resistance against the antibiotics. These include the inactivation of the enzymes, rapid decrease in the permeability of the cells, protection of the targets, modified target sites, and increased efflux, along with some other mechanisms. Recently, it has been identified that the biofilm formation by some microbes can be induced by the antibiotics as well, though such phenomena are not caused or affected by the exposure to the antibiotics (Baptista et al., 2019). The implementation of nanomaterials in the fight against the antibiotic-resistant bacteria has become an important tool for the eradication of resistance from the environment. Nanomaterials are defined as those products that have at least one side in the nanoscale range (1–100 nm), which is different from the properties of the bulk material. Nanoparticles belonging to this category of materials have piqued the interest of researchers due to their capability to be used as effective vectors for those drugs used commonly for combating the resistant pathogens (Wang et al., 2017a,b). As a result of their vast surface-to-volume ratio and ultra-small sizes, they are considered as important carriers, and when these nanoparticles are functionalized with various biomolecules, their activities pronouncedly get enhanced. Nanoparticles such as Au, Al, Ag, Cu, Cr, Ce, Cd, Mg, Ni, Ti, and Zn produce great effects against the multidrug resistant bacteria, thus providing a suitable alternative to the available antibiotics (Hemeg, 2017). While the poor membrane transportation limits the potency of antibiotics, these nanoparticles loaded with the drugs can easily enter the cells through the process of endocytosis, thus easily entering into the intracellular components (Wang et al., 2017a,b). With the increasing therapeutic potential of these nanoparticles in the environmental sector, it has become crucial to understand the various mechanisms by which these agents impact the cell viability of bacteria. These agents have the capacity of not only just macro-targeting but also possess the micro-targeting mechanisms; once these mechanisms are fully understood, they can open new ways for the widescale use of these particles as potential vectors (Singh and Lillard, 2009).

18.11.2 Antimicrobial peptides

Nature has provided humankind with various alternatives and all possess similar effectiveness. Just like antibiotics have the ability to inhibit and terminate the growth of pathogens, there are various antimicrobials that also produce the same results. There are many organisms that produce antimicrobial peptides, including certain strains of microorganisms as well. These proteins possess innate immunity toward the bacterial infections as well. Approximately more than 800 antimicrobial peptides have been isolated from various sources and these discovered peptides have a wide spectrum of activities, including antiviral, antifungal, antibacterial, as well as antiprotozoal effects. These peptides serve as an excellent alternative to the resistant microbes, and thus antibiotics can be substituted with these natural compounds (Baltzer and Brown, 2011).

18.11.3 Bacteriophage therapy

The use of bacterial viruses to treat bacterial infections is an old practice, which recently became popular again due to the increase in antibiotic resistance. On a conventional basis, phage therapy comprises use of phages that can be found in various natural sources and then applying them at the site of infection to lyse the bacterial cell. But, with the advancements in biotechnology, scientists have developed new strategies to enhance the efficacy of the phages, which include the bioengineered phages and cocktails as well as phage proteins. The current investigations in the field of bacteriophages and their lytic proteins suggest that they can be used in combinations or as an alternative to antibiotics. More research still needs to be done to create an effective solution to antibiotic resistance with the employment of the purified phages in the pharmaceutical industries. Moreover, the use of phage therapy is environmentally friendly and causes no stress to the circulating strains in the environment, thus, not aggravating the resistance spread (Lin et al., 2017).

18.11.4 CRISPR approaches to environmental AMR

Amidst the antibiotic resistance, there is a new strategy that biologists have devised while looking out for alternatives for antibiotic resistance. A new gene-editing technology known as CRISPR-Cas has recently been introduced in genetic engineering. It is a bacterial immune system called clustered regularly interspaced short palindromic repeats-CRISPR—associated. Those genes that promote the antibiotic resistance are basically found on accessory genetic elements known as plasmids. These plasmids can transfer the resistant genes between various bacterial species, thus, spreading the resistance on a wide scale. Scientists used CRISPR system comprising a genetically engineered plasmid, which specifically targets the resistance genes for gentamicin. When this designed plasmid finds the genes for gentamicin, it cuts out the DNA, ultimately removing the resistance and making the bacteria susceptible to antibiotics again. This technology, if used properly, can help to remove the wide spread of antimicrobial resistance in the environment. The scope of this technology would create effective impact when connections between environmental hot spots and medical settings are more completely understood (Pursey et al., 2018).

18.12 Public health and AMR/ARGs

18.12.1 AMR/ARGs, a serious public health problem

Basically, the resistance of microbes such as bacteria and viruses toward antimicrobial agents is termed as antimicrobial resistance. Resistance to antibiotics is the most concerning type of AMR. Worldwide, AMR is regarded as chronic and poses great threats to global health (MacIntyre and Bui, 2017). The European Center for Disease Control (ECDC) estimates that about two million people become infected with antibiotic resistant bacteria every year in the United States and about 25,000 deaths occur per year (ECDC, 2019). Delayed treatment, persistent infections, and transmission of resistance to other organisms are some of the attributions of AMR (Jindal et al., 2015). Various clinical challenges have been arising due to the antimicrobial resistance, and multidrug-resistant health care acquired infections are one of them. The CDC estimated about 20,000 cases of vancomycin-resistant *Enterococcus* and 7000 cases of MDR *Pseudomonas aeruginosa* in patients every year in U.S. hospitals (Toner et al., 2015). Various bacterial species have been developing several resistance clades such as *E. coli*, *Salmonella typhi*, *S. aureus*, and *Clostridium difficile* (Bloom et al., 2018).

18.12.2 Primary and secondary AMR

Antimicrobial resistance may be primary or may be secondary. Primary AMR is basically the antimicrobial resistance that develops in the microbes naturally. Secondary AMR develops during the treatment when patients don't complete a full course of antibiotics (MacIntyre and Bui, 2017). Sometimes the environmental changes, such as in pH and light, as well as radiation, cause development of resistance in bacterial species (Ali et al., 2018). In order to stop the secondary antibiotic resistance, it is important to take medication exactly as prescribed because the antibiotics are effective only when taken regularly. If a course of antibiotics is stopped in the middle of the treatment, then it does not kill the bacteria properly and remaining bacteria will become resistant to the antibiotics (Fda.gov, 2019).

18.12.3 Diseases with potential of antimicrobial resistance

18.12.3.1 Influenza

Diseases that have potential AMR are considered as an important factor to impact health, for instance, influenza, plague, and smallpox. Seasonal influenza oseltamivir-resistant H1N1 emerged in Europe as a primary resistance in 2007 and its strains were spread globally within a year even in those areas where use of oseltamivir was not common (Hurt et al., 2012). However, until now oseltamivir- and zanamivir-resistant influenza strains have been rare and mostly limited to immune-deficient individuals. In case of influenza, deaths and infection are attributed to virus itself, but in some cases, secondary bacterial infection occurs, and thus bacterial resistance also contributes to the overall impact (MacIntyre and Bui, 2017). In intensive care units, around 30%–50% of deaths are attributed to bacterial coinfections during influenza (Papanicolaou, 2013). During 2009, in the United States, from 838 influenza samples, 33% showed evidence of bacterial coinfections, among which 39% were *S. aureus* and 58% were methicillin resistant (Randolph et al., 2011). Other AMR bacterial infections contributing to morbidity and mortality during influenza are infections that are caused by multidrug-resistant bacterial strains including vancomycin-resistant *Enterococcus* (VRE), *Pseudomonas*, *Acinetobacter*, and drug-resistant *Pneumococci* (MacIntyre and Bui, 2017; Morris et al., 2017).

18.12.3.2 Plague

Plague is caused by a bacterium known as *Yersinia pestis*, which affects humans and other mammals. Usually, humans get infected when bitten by rodent fleas carrying the bacterium (Cdc.gov, 2019). Antimicrobial resistance is rare in plague-causing *Y. pestis*. In 1995, the first resistant strains were isolated, which carried out a self-transmissible plasmid that showed resistance to several antimicrobial agents (Welch et al., 2007). In 2018, a 24-year-old individual was diagnosed with *Y. pestis* and he was treated with streptomycin, but he did not achieve complete recovery. Later on, presence of MDR *Stenotrophomonas maltophilia* was detected in his lungs. This opportunistic infection actually occurred because of the immunosuppression caused by *Y. pestis* infection (Andrianaivoarimanana et al., 2018). Plague can be treated with antibiotics, but AMR could substantially increase the fatality rate of the disease.

18.12.3.3 Smallpox

Smallpox is one of the earliest viral infections that was eradicated around the globe by the efforts of WHO in 1980. Caused by the variola virus, smallpox used to be highly contagious and its mode of transmission constituted the infected droplets from the infected individual and his contact with healthy people. Since its eradication, there are no longer large outbreaks of the disease reported in the world, but two of its viral forms still exist in two laboratories in Russia and in the United States. Therefore, the danger of its spread from these scientific centers cannot be eliminated. Moreover, due to its fully sequenced genome, it can be synthetically manufactured at a science facility and due to the imminent danger of its outbreak in the future, many countries keep on preparing antiviral stocks that all lead toward the potential development of AMR if a smallpox pandemic should happen. Antibiotics, such as penicillin and cephalosporin, can be used for the treatment of secondary skin infections that arise due to smallpox and, thus, these would be potentially affected by the AMR as well (MacIntyre and Bui, 2017).

18.12.4 Preventive measures for reduction of AMR

There is no doubt that humankind has been brought great relief since the discovery of antibiotics as they not only minimize the burden of disease but also help in improving the quality of life. With the employment of these drugs, food safety was enhanced and life expectancy increased. Unfortunately, due to the emergence of highly resistant bacterial strains, all these benefits are at risk as the antibiotics no longer provide the effective management of the emerging diseases. The antimicrobial resistance is a major challenging issue for the world health authorities and efforts are continuously being introduced to tackle this deteriorating phenomenon. Some of the key preventive measures, proposed by WHO, which should be immediately implemented for the management of AMR, are as follows:

1. The foremost factor that needs to be dealt with regarding the management of antimicrobial resistance is educating the community and making them aware about the associated risks and dangers. For this, communicative programs should be planned as well as social media campaigns for effective understanding.
2. With the rising AMR, it has become necessary to devise national surveillance systems and connect them to regional and global systems so that policy-making bodies can manage the related health concerns properly.
3. For effective analysis of the AMR tackling strategy, it is important to collect as much data as possible about the antimicrobial agents in humans and animals. This will further facilitate the monitoring of various trends and their impacts on the respective populations.
4. For reducing the risk of infection, the hygienic conditions must be strengthened so that the risk of infection can be mitigated.
5. Introduce holistic-based approaches for the fostering of health care procedures. Moreover, the food-based industries should also be strengthened for promoting safe and healthy food products.
6. For the optimization of antimicrobial agents, following the guidelines of WHO, regulations should be implemented for effective licensures of antibiotics, their distribution, and dispensing.

18.13 Conclusion

Antimicrobial resistance and antimicrobial-resistance genes have become a serious health concern during the last few years. Moreover, due to the consistent utilization of antibiotics in the health, food, and livestock sectors, the situation has become graver because of their direct connection with each other. The environment is no more different then these fields as AMR sporadically spreads in the environment and the resulting contamination can cause serious health concerns. There is a need to research more and collect advanced scientific data so as to fill the gaps and better understand the potential harms presented by the resistant bacterial strains in the environment. AMR varies from country to country, however, as a mechanism, it is common across the globe. Therefore, holistic-based approaches that have relevant local interventions need to be implemented.

References

Adelowo, O.O., Fagade, O.E., Agers, Y., 2014. Antibiotic resistance and resistance genes in *Escherichia coli* from poultry farms, Southwest Nigeria. The Journal of Infection in Developing Countries 8, 1103–1112.

Al-Gheethi, A.A., Radin Mohamed, R.M.S., Efaq, A.N., Amir Hashim, M.K., 2016. Reduction of microbial risk associated with greywater by disinfection processes for irrigation. Journal of Water and Health 14, 379–398.

Alanis, A., 2005. Resistance to antibiotics: are we in the post-antibiotic era. Archives of Medical Research 36 (6), 697–705.

Ali, J., Rafiq, Q.A., Ratcliffe, E., 2018. Antimicrobial resistance mechanisms and potential synthetic treatments. Future Science OA 4 (4), FSO290.

Andrianaivoarimanana, V., Bertherat, E., Rajaonarison, R., Rakotondramaro, T., Rogier, C., Rajerison, M., 2018. Mixed pneumonic plague and nosocomial MDR-bacterial infection of lung: a rare case report. BMC Pulmonary Medicine 18 (1), 92.

Antimicrobial Resistance, National Action Plan, Pakistan, 2017. Ministry of National Health Services Regulations & Coordination Government of Pakistan, pp. 1–64.

Baltzer, S., Brown, M., 2011. Antimicrobial peptides – promising alternatives to conventional antibiotics. Journal of Molecular Microbiology and Biotechnology 20 (4), 228–235.

Baptista, P., McCusker, M., Carvalho, A., Ferreira, D., Mohan, N., Martins, M., Fernandes, A., 2019. Nano-strategies to fight multidrug resistant bacteria— "A battle of the titans". Frontiers in Mocrobiology 9, 1441.

Baquero, F., Martínez, J., Cantón, R., 2008. Antibiotics and antibiotic resistance in water environments. Current Opinion in Biotechnology 19 (3), 260–265.

Barker-Reid, F., Fox, E.M., Faggian, R., 2010. Occurrence of antibiotic resistance genes in reclaimed water and river water in the Werribee Basin, Australia. Journal of Water and Health 8, 521–531.

Barlam, T.F., Soria-Saucedo, R., Cabral, H.J., Kazis, L.E., 2016. Unnecessary Antibiotics for Acute Respiratory Tract Infections: Association with Care Setting and Patient Demographics. Oxford University Press.

Basnyat, B., 2014. Antibiotic resistance needs global solutions. The Lancet Infectious Diseases 14 (7), 549–550.

Bengtsson, B., Greko, C., 2014. Antibiotic resistance—consequences for animal health, welfare, and food production. Upsala Journal of Medical Sciences 119 (2), 96–102.

Bengtsson-Palme, J., Larsson, D., 2015. Antibiotic resistance genes in the environment: prioritizing risks. Nature Reviews Microbiology 13 (6), 396-396.

Bengtsson-Palme, J., Kristiansson, E., Larsson, D., 2017. Environmental factors influencing the development and spread of antibiotic resistance. FEMS Microbiology Reviews 42 (1).

Bloom, D.E., Black, S., Salisbury, D., Rappuoli, R., 2018. Antimicrobial resistance and the role of vaccines. Proceedings of the National Academy of Sciences 115 (51), 12868–12871.

Cantas, L., Shah, S.Q.A., Cavaco, L.M., Manaia, C.M., Walsh, F., Popowska, M., et al., 2013. A brief multidisciplinary review on antimicrobial resistance in medicine and its linkage to the global environmental microbiota. Frontiers in Microbiology 4, 96.

Carvalho, I., Santos, L., 2016. Antibiotics in the aquatic environments: a review of the European scenario. Environment International 94, 736–757.

Chee-Sanford, J., Mackie, R., Koike, S., Krapac, I., Lin, Y., Yannarell, A., Maxwell, S., Aminov, R., 2009. Fate and transport of antibiotic residues and antibiotic resistance genes following land application of manure waste. Journal of Environmental Quality 38 (3), 1086.

Chemaly, R.F., Simmons, S., Dale Jr., C., Ghantoji, S.S., Rodriguez, M., Gubb, J., et al., 2014. The role of the healthcare environment in the spread of multidrug-resistant organisms: update on current best practices for containment. Therapeutic Advances in Infectious Disease 2, 79–90.

Christgen, B., Yang, Y., Ahammad, S.Z., Li, B., Rodriquez, D.C., Zhang, T., et al., 2015. Metagenomics shows that low-energy anaerobic-aerobic treatment reactors reduce antibiotic resistance gene levels from domestic wastewater. Environmental Science & Technology 49, 2577–2584.

CUNHA, C.B., 2018. Antimicrobial stewardship programs (ASP): perspective on problems and potential. Rhode Island Medical Journal 101, 18–21 (2013).

Dang, S., Petersen, A., Van Truong, D., Chu, H., Dalsgaard, A., 2011. Impact of medicated feed on the development of antimicrobial resistance in bacteria at integrated pig-fish farms in Vietnam. Applied and Environmental Microbiology 77 (13), 4494–4498.

Dincer, S., Yigittekin, E.S., 2017. Spreading of antibiotic resistance with wastewater. In: Biological Wastewater Treatment and Resource Recovery IntechOpen.

Ding, C., He, J., 2010. Effect of antibiotics in the environment on microbial populations. Applied Microbiology and Biotechnology 87 (3), 925–941.

Djordjevic, S., Stokes, H., Chowdhury, P., 2013. Mobile elements, zoonotic pathogens and commensal bacteria: conduits for the delivery of resistance genes into humans, production animals and soil microbiota. Frontiers in Microbiology 4.

Donabedian, S.M., Thal, L.A., Hershberger, E., Perri, M.B., Chow, J.W., Bartlett, P., et al., 2003. Molecular characterization of gentamicin-resistant enterococci in the United States: evidence of spread from animals to humans through food. Journal of Clinical Microbiology 41, 1109–1113.

European Centre for Disease Prevention and Control, 2019 [Online]. Available: https://www.ecdc.europa.eu/en/home.

Fahrenfeld, N.L., Ma, Y., O'Brien, M., Pruden, A., 2013. Reclaimed water as a reservoir of antibiotic resistance genes: distribution system and irrigation implications. Frontiers in Microbiology 4, 130.

Fdagov, 2019. Combating Antibiotic Resistance [online] Available at: https://www.fda.gov/ForConsumers/ConsumerUpdates/ucm092810.html.

Ferreira, A.M., Andrade, D.d., Rigotti, M.A., Almeida, M.T. G.d., 2011. Methicillin-resistant *Staphylococcus aureus* on surfaces of an intensive care unit. Acta Paulista de Enfermagem 24, 453–458.

Food and Agriculture Organization, 2018. Antimicrobial Resistance Policy Review and Development Framework – A Regional Guide for Governments in Asia and the Pacific to Review, Update and Develop Policy to Address Antimicrobial Resistance and Antimicrobial Use in Animal Production. Bangkok, p. 64.

Friedman, M., 2015. Antibiotic-resistant bacteria: prevalence in food and inactivation by food-compatible compounds and plant extracts. Journal of Agricultural and Food Chemistry 63 (15), 3805–3822.

Ghosh, G., Hanamoto, S., Yamashita, N., Huang, X., Tanaka, H., 2016. Antibiotics removal in biological sewage treatment plants. Pollution 2, 131–139.

Goulas, A., Livoreil, B., Grall, N., Benoit, P., Couderc-Obert, C., Dagot, C., Patureau, D., Petit, F., Laouénan, C., Andremont, A., 2018. What are the effective solutions to control the dissemination of antibiotic resistance in the environment? A systematic review protocol. Environmental Evidence 7 (1).

Grave, K., Torren-Edo, J., Mackay, D., 2010. Comparison of the sales of veterinary antibacterial agents between 10 European countries. Journal of Antimicrobial Chemotherapy 65, 2037–2040.

Grenni, P., Ancona, V., Barra Caracciolo, A., 2018. Ecological effects of antibiotics on natural ecosystems: a review. Microchemical Journal 136, 25–39.

Gruen, L., Ruddick, W., 2009. Biomedical and environmental ethics alliance: common causes and grounds. Journal of Bioethical Inquiry 6 (4), 457–466.

Halling-Sørensen, B., Nors Nielsen, S., Lanzky, P., Ingerslev, F., Holten Lützhøft, H., Jørgensen, S., 1998. Occurrence, fate and effects of pharmaceutical substances in the environment - A review. Chemosphere 36 (2), 357–393.

Hanna, N., Sun, P., Sun, Q., Li, X., Yang, X., Ji, X., Zou, H., Ottoson, J., Nilsson, L., Berglund, B., Dyar, O., Tamhankar, A., Stålsby Lundborg, C., 2018. Presence of antibiotic residues in various environmental compartments of Shandong province in eastern China: its potential for resistance development and ecological and human risk. Environment International 114, 131–142.

Hemeg, H., 2017. Nanomaterials for alternative antibacterial therapy. International Journal of Nanomedicine 12, 8211–8225.

Heuer, H., Smalla, K., 2007a. Horizontal gene transfer between bacteria. Environmental Biosafety Research 6 (1–2), 3–13.

Heuer, H., Smalla, K., 2007b. Manure and sulfadiazine synergistically increased bacterial antibiotic resistance in soil over at least two months. Environmental Microbiology 9 (3), 657–666.

Heuer, H., Schmitt, H., Smalla, K., 2011. Antibiotic resistance gene spread due to manure application on agricultural fields. Current Opinion in Microbiology 14 (3), 236–243.

Hoang, T.T.T., Tu, L.T.C., Le, N.P., Dao, Q.P., 2013. A preliminary study on the phytoremediation of antibiotic contaminated sediment. International Journal of Phytoremediation 15, 65–76.

Hollis, A., Ahmed, Z., 2013. Preserving antibiotics, rationally. New England Journal of Medicine 369, 2474–2476.

Hollis, A., Maybarduk, P., 2015. Antibiotic resistance is a tragedy of the commons that necessitates global cooperation. Journal of Law Medicine & Ethics 43 (3_Suppl. 1), 33–37.

Homem, V., Santos, L., 2011. Degradation and removal methods of antibiotics from aqueous matrices – a review. Journal of Environmental Management 92 (10), 2304–2347.

Hop, L., 2003. Programs to improve production and consumption of animal source foods and malnutrition in vietnam. Journal of Nutrition 133 (11), 4006S–4009S.

Hurt, A.C., Hardie, K., Wilson, N.J., Deng, Y.M., Osbourn, M., Leang, S.K., Lee, R.T.C., Iannello, P., Gehrig, N., Shaw, R., 2012. Characteristics of a widespread community cluster of H275Y oseltamivir-resistant A (H1N1) pdm09 influenza in Australia. The Journal of Infectious Diseases 206 (2), 148–157.

Jager, T., Hembach, N., Elpers, C., Wieland, A., Alexander, J., Hiller, C., et al., 2018. Reduction of antibiotic resistant bacteria during conventional and advanced wastewater treatment, and the disseminated loads released to the environment. Frontiers in Microbiology 9.

Jain, R., 2003. Horizontal gene transfer accelerates genome innovation and evolution. Molecular Biology and Evolution 20 (10), 1598–1602.

Jefferson, B., Laine, A., Parsons, S., Stephenson, T., Judd, S., 2000. Technologies for domestic wastewater recycling. Urban Water 1, 285–292.

Jindal, A.K., Pandya, K., Khan, I.D., 2015. Antimicrobial resistance: a public health challenge. Medical Journal Armed Forces India 71 (2), 178–181.

Kimbell, L.K., Kappell, A.D., McNamara, P.J., 2018. Effect of pyrolysis on the removal of antibiotic resistance genes and class I integrons from municipal wastewater biosolids. Environmental Sciences: Water Research & Technology 4, 1807–1818.

Kilpatrick, C., O'Conner, H., 2014. Hand hygiene is central to tackling antibiotic resistance. Nursing Times 110 (18), 11-11.

Kristiansson, E., Fick, J., Janzon, A., Grabic, R., Rutgersson, C., Weijdegård, B., Söderström, H., Larsson, D., 2011. Pyrosequencing of antibiotic-contaminated river sediments reveals high levels of resistance and gene transfer elements. PLoS One 6 (2), e17038.

Kuhse, H., Singer, P. (Eds.), 2013. A Companion to Bioethics. John Wiley and Sons, New York, NY.

Le, T.H., Ng, C., Tran, N.H., Chen, H., Gin, K.Y.-H., 2018. Removal of antibiotic residues, antibiotic resistant bacteria and antibiotic resistance genes in municipal wastewater by membrane bioreactor systems. Water Research 145, 498–508.

Lee, C.R., Cho, I., Jeong, B., Lee, S., 2013. Strategies to minimize antibiotic resistance. International Journal of Environmental Research and Public Health 10, 4274–4305.

Levy, S., 2002. Active efflux, a common mechanism for biocide and antibiotic resistance. Journal of Applied Microbiology 92 (s1), 65S–71S.

Li, D., Yu, T., Zhang, Y., Yang, M., Li, Z., Liu, M., Qi, R., 2010. Antibiotic resistance characteristics of environmental bacteria from an oxytetracycline production wastewater treatment plant and the receiving river. Applied and Environmental Microbiology 76 (11), 3444–3451.

Lin, D., Koskella, B., Lin, H., 2017. Phage therapy: an alternative to antibiotics in the age of multi-drug resistance. World Journal of Gastrointestinal Pharmacology and Therapeutics 8 (3), 162.

Lood, R., Ertrk, G., Mattiasson, B., 2017. Revisiting antibiotic resistance spreading in wastewater treatment plants - bacteriophages as a much-neglected potential transmission vehicle. Frontiers in Microbiology 8, 2298.

MacIntyre, C.R., Bui, C.M., 2017. Pandemics, public health emergencies and antimicrobial resistance-putting the threat in an epidemiologic and risk analysis context. Archives of Public Health 75 (1), 54.

Magouras, I., Carmo, L.s.P., St+ñrk, K.D., Schpbach-Regula, G., 2017. Antimicrobial usage and-resistance in livestock: where should we focus? Frontiers in Veterinary Science 4, 148.

Manaia, C., Rocha, J., Scaccia, N., Marano, R., Radu, E., Biancullo, F., Cerqueira, F., Fortunato, G., Iakovides, I., Zammit, I., Kampouris, I., Vaz-Moreira, I., Nunes, O., 2018. Antibiotic resistance in wastewater treatment plants: tackling the black box. Environment International 115, 312–324.

Manges, A., Smith, S., Lau, B., Nuval, C., Eisenberg, J., Dietrich, P., Riley, L., 2007. Retail meat consumption and the acquisition of antimicrobial resistant *Escherichia coli* causing urinary tract infections: a case–control study. Foodborne Pathogens and Disease 4 (4), 419–431.

Marti, R., Scott, A., Tien, Y., Murray, R., Sabourin, L., Zhang, Y., Topp, E., 2013. Impact of manure fertilization on the abundance of antibiotic-resistant bacteria and frequency of detection of antibiotic resistance genes in soil and on vegetables at harvest. Applied and Environmental Microbiology 79 (18), 5701–5709.

Martinez, J., 2009. Environmental pollution by antibiotics and by antibiotic resistance determinants. Environmental Pollution 157 (11), 2893–2902.

McNulty, C.A., Lecky, D.M., Farrell, D., Kostkova, P., Adriaenssens, N., Koprivov+í Herotov+í, T., et al., 2011. Overview of e-Bug: an antibiotic and hygiene educational resource for schools. Journal of Antimicrobial Chemotherapy 66, v3–v12.

Morris, D.E., Cleary, D.W., Clarke, S.C., 2017. Secondary bacterial infections associated with influenza pandemics. Frontiers in Microbiology 8, 1041.

Nankervis, H., S Thomas, K., M Delamere, F., Barbarot, S., K Rogers, N., C Williams, H., 2016. Antimicrobials including antibiotics, antiseptics and antifungal agents. In: Scoping Systematic Review of Treatments for Eczema, seventh ed. Queen's Printer, Southampton, UK.

National Vaccine Advisory Committee, 2016. A call for greater consideration for the role of vaccines in national strategies to combat antibiotic-resistant bacteria: recommendations from the national vaccine advisory committee: approved by the National Vaccine Advisory Committee on June 10, 2015. Public Health Reports 131, 11–16.

Neu, H., 1992. The crisis in antibiotic resistance. Science 257 (5073), 1064–1073.

Ochman, H., Lawrence, J., Groisman, E., 2000. Lateral gene transfer and the nature of bacterial innovation. Nature 405 (6784), 299–304.

Orthobulletscom, 2019. Antibiotic Classification & Mechanism – Basic Science – Orthobullets [online] Available at: https://www.orthobullets.com/basic-science/9059/antibiotic-classification-and-mechanism.

Osiäska, A., Korzeniewska, E., Harnisz, M., Niest-Öpski, S., 2019. Quantitative occurrence of antibiotic resistance genes among bacterial populations from wastewater treatment plants using activated sludge. Applied Sciences 9, 387.

Papanicolaou, G.A., 2013. Severe influenza and *S. aureus* pneumonia: for whom the bell tolls? Virulence 4 (8), 666–668.

Parsonage, B., Hagglund, P., Keogh, L., Wheelhouse, N., Brown, R., Dancer, S., 2017. Control of antimicrobial resistance requires an ethical approach. Frontiers in Microbiology 8.

Pessoa-Silva, C.L., Hugonnet, S.p., Pfister, R., Touveneau, S., Dharan, S., Posfay-Barbe, K., et al., 2007. Reduction of health care associated infection risk in neonates by successful hand hygiene promotion. Pediatrics 120, e382–e390.

Pittet, D., 2004. The role of hospital hygiene in the reduction of antibiotic resistance. Bulletin de l'Academie nationale de medecine 188, 1269–1280.

Pursey, E., Sünderhauf, D., Gaze, W., Westra, E., van Houte, S., 2018. CRISPR-Cas antimicrobials: challenges and future prospects. PLoS Pathogens 14 (6), e1006990.

Randolph, A.G., Vaughn, F., Sullivan, R., Rubinson, L., Thompson, B.T., Yoon, G., Smoot, E., Rice, T.W., Loftis, L.L., Helfaer, M., 2011. Critically ill children during the 2009–2010 influenza pandemic in the United States. Pediatrics 128 (6), e1450.

Roose-Amsaleg, C., Laverman, A., 2015. Do antibiotics have environmental side-effects? Impact of synthetic antibiotics on biogeochemical processes. Environmental Science and Pollution Research 23 (5), 4000–4012.

Rosenthal, V.D., Guzman, S., Safdar, N., 2005. Reduction in nosocomial infection with improved hand hygiene in intensive care units of a tertiary care hospital in Argentina. American Journal of Infection Control 33, 392–397.

Sarmah, A., Meyer, M., Boxall, A., 2006. A global perspective on the use, sales, exposure pathways, occurrence, fate and effects of veterinary antibiotics (VAs) in the environment. Chemosphere 65 (5), 725–759.

Singer, A.C., Shaw, H., Rhodes, V., Hart, A., 2016. Review of antimicrobial resistance in the environment and its relevance to environmental regulators. Frontiers in Microbiology 7, 1728.

Singh, R., Lillard, J., 2009. Nanoparticle-based targeted drug delivery. Experimental and Molecular Pathology 86 (3), 215–223.

Skippington, E., Ragan, M., 2011. Lateral genetic transfer and the construction of genetic exchange communities. FEMS Microbiology Reviews 35 (5), 707–735.

Su, J.-Q., Wei, B., Ou-Yang, W.-Y., Huang, F.-Y., Zhao, Y., Xu, H.-J., Zhu, Y.-G., 2015. Antibiotic resistome and its association with bacterial communities during sewage sludge composting. Environmental Science & Technology 49 (12), 7356–7363.

Suzuki, S., Hoa, P., 2012. Distribution of quinolones, sulfonamides, tetracyclines in aquatic environment and antibiotic resistance in Indochina. Frontiers in Microbiology 3.

Tahrani, L., Van Loco, J., Ben Mansour, H., Reyns, T., 2016. Occurrence of antibiotics in pharmaceutical industrial wastewater, wastewater treatment plant and sea waters in Tunisia. Journal of Water and Health 14, 208–213.

Tang, K.L., Caffrey, N.P., Nobrega, D.B., Cork, S.C., Ronksley, P.E., Barkema, H.W., et al., 2017. Restricting the use of antibiotics in food-producing animals and its associations with antibiotic resistance in food-producing animals and human beings: a systematic review and meta-analysis. The Lancet Planetary Health 1, e316–e327.

Thiele-Bruhn, S., 2003. Pharmaceutical antibiotic compounds in soils – a review. Journal of Plant Nutrition and Soil Science 166 (2), 145–167.

Toner, E., Adalja, A., Gronvall, G.K., Cicero, A., Inglesby, T.V., 2015. Antimicrobial resistance is a global health emergency. Health security 13 (3), 153–155.

Topor, G., Grosu, I.-A., Ghiciuc, C.M., Strat, A.L., Lupuşoru, C.E., 2017. Awareness about antibiotic resistance in a self-medication user group from Eastern Romania: a pilot study. PeerJ 5, e3803.

Troiano, E., Beneduce, L., Gross, A., Ronen, Z., 2018. Antibiotic-resistant bacteria in greywater and greywater-irrigated soils. Frontiers in Microbiology 9.

Van Boeckel, T.P., Brower, C., Gilbert, M., Grenfell, B.T., Levin, S.A., Robinson, T.P., et al., 2015. Global trends in antimicrobial use in food animals. Proceedings of the National Academy of Sciences 112, 5649–5654.

Wang, L., Hu, C., Shao, L., 2017a. The antimicrobial activity of nanoparticles: present situation and prospects for the future. International Journal of Nanomedicine 12, 1227–1249.

Wang, Z., Dong, K., Liu, Z., Zhang, Y., Chen, Z., Sun, H., Ren, J., Qu, X., 2017b. Activation of biologically relevant levels of reactive oxygen species by Au/g-C_3N_4 hybrid nanozyme for bacteria killing and wound disinfection. Biomaterials 113, 145–157.

Watkinson, A.J., Murby, E.J., Costanzo, S.D., 2007. Removal of antibiotics in conventional and advanced wastewater treatment: implications for environmental discharge and wastewater recycling. Water Research 41, 4164–4176.

Wegener, H.C., 2012. Antibiotic resistance—linking human and animal health. In: Improving Food Safety through a One Health Approach: Workshop Summary. National Academies Press, p. 331.

Welch, T.J., Fricke, W.F., McDermott, P.F., White, D.G., Rosso, M.L., Rasko, D.A., Mammel, M.K., Eppinger, M., Rosovitz, M.J., Wagner, D., 2007. Multiple antimicrobial resistance in plague: an emerging public health risk. PLoS One 2 (3), e309.

World Health Organization, 2019. Smallpox [online] Available at: https://www.who.int/csr/disease/smallpox/en/.

World Health Organization, WHO, 2015. Global Action Plan on Antimicrobial Resistance, first ed. WHO Document Production Services, Geneva, Switzerland, pp. 1–28.

Yuan, Q Bin, Guo, M.T., Yang, J., 2015. Fate of antibiotic resistant bacteria and genes during wastewater chlorination: Implication for antibiotic resistance control. PLoS One 10 (3).

Zachary, K.C., Bayne, P.S., Morrison, V.J., Ford, D.S., Silver, L.C., Hooper, D.C., 2001. Contamination of gowns, gloves, and stethoscopes with vancomycin-resistant enterococci. Infection Control and Hospital Epidemiology 22 (9), 560–564.

Zarei-Baygi, A., Harb, M., Wang, P., Stadler, L.B., Smith, A.L., 2019. Evaluating antibiotic resistance gene correlations with antibiotic exposure conditions in anaerobic membrane bioreactors. Environmental Science & Technology 53 (7), 3599–3609.

Further reading

CDC, 2010. About Antimicrobial Resistance: A Brief Overview. U.S. Department of Health and Human Services, Centres for Disease Control and Prevention, Washington, DC. Available from: http://cdc.gov/drugresistance/about.html.

CHAPTER

Antibiotics resistance mechanism 19

Muhammad Naveed[1], Zoma Chaudhry[2], Syeda Aniqa Bukhari[2], Bisma Meer[3], Hajra Ashraf[3]

[1]*Department of Biotechnology, University of Central Punjab, Lahore, Punjab, Pakistan;* [2]*Department of Biochemistry and Biotechnology, University of Gujrat, Gujrat, Punjab, Pakistan;* [3]*Department of Biotechnology, Quaid-e-Azam University, Islamabad, Punjab, Pakistan*

19.1 What are antibiotics?

Antibiotics are vast range of medicines that mimic the growth of bacteria on onset of infection in living systems. Originally, they were termed as "classical antibiotics," but with passage of time, bacterial resistance prolonged and as a result new and advance forms of antibiotics have been discovered (Hancock, 1997). For the last 30 years, peptide antibiotics have been used due to less chance of resistance and rapid mechanism of action due to their action of direct cell membrane disruption (Fajardo and Martínez, 2008). Pathogens, especially bacteria, are greatly influenced by these antibiotics. With excessive use of antibiotics in adults, pathogens are showing resistance against antibiotics, which is an alarming situation and can weaken the immune system of an individual (Boman, 1995). Here, different mechanisms are discussed that are involved in producing resistance in bacterial species.

19.1.1 Human pathogens

The human body is complex and has a thriving ecosystem. The human body constitutes 10^{13} human cells and 10^{14} fungal, protozoan, and bacterial cells that are evidence of thousands of species of microbes residing in the human body. These microorganisms, referred as normal flora, reside in certain parts, i.e., mouth, vagina, skin, and large intestine. Additionally, human body gets infected by viruses. Some microbes are pathogenic, which means they cause diseases to humans. Pathogens are differentiated from normal flora in such a way that normal flora cause infection only in immunocompromised persons or if microbes are capable to have access to sterile body parts, e.g., peritonitis is caused by entry of gut flora in peritoneal cavity during bowel perforation. Pathogens don't need special conditions to attack when a person is injured or their immune system is weakened. Pathogens breach biochemical and cellular barriers and elicit responses from the host in which they reside by development of special mechanisms, so these microbes are able to survive and multiply (Alberts et al., 2002). Survival and multiplication of microbes is possible if a pathogen has the following properties:

1. Pathogens must be capable to colonize the host.
2. They must find niches that are nutritionally compatible.

Antibiotics and Antimicrobial Resistance Genes in the Environment. https://doi.org/10.1016/B978-0-12-818882-8.00019-X

3. Microbes are capable of subverting innate as well as adaptive immune responses of host.
4. Microbes' replication process uses resources of the host (Raff et al., 2002)

There are many types of pathogens that are capable of causing diseases in humans. Most common pathogens are bacteria and viruses. Viruses capable of causing diseases range from AIDS, to smallpox, to the common cold. Viruses basically constitute nucleic acid (DNA or RNA) enclosed in a protein coat. They use host machinery for replication, transcription, and translation. Bacteria are complex and larger than viruses. They are basically free-living cells that take nutrition from the host for performing metabolic functions. Most of the bacteria in our lives are beneficial while the minority of them are pathogens (Waterfield and Wren, 2004).

Some are eukaryotic organisms that cause infections. These constitute single-celled fungi, protozoa, as well as metazoans, i.e., parasitic worms that are complex and large. *Ascaris lumbricoides* is responsible for causing infestation in the gut and is a very common infectious disease. This nematode resembles its cousin, *Caenorhabditis elegans*. *C. elegans* is basically a model organism that is used for developmental biological and genetic research. Its size is 1 mm while *Ascaris* is 30 cm in length. Pathogens show diversity in size, nucleic acid content, and shape. Pathogens are capable of causing disease in different ways, making it complex to understand the mechanisms of infection (Waterfield and Wren, 2004).

19.1.2 How antibiotics work

How antibiotics work relies on their complex strategies to eliminate the bacterial cell. There is some similar content between bacteria and human cells, but there are many differentiations as well. Antibiotics work by targeting things that bacterial cells contain but absent in human cells. This can be done by either destroying genomic content or by targeting bacterial cell's organelles. A bactericidal-type antibiotic eliminates the whole bacteria, while the bacteriostatic-type antibiotic inhibits bacterial growth without killing the cells. After that, the human immunity is then required to clear the infection. The immune system comes in handy for eliminating all pathogenic parts (Reynolds, 1989).

Killing of pathogens ranges from nonspecific to specific targeting. But the pattern of killing remains always same. Antibiotics, being the "cell inspector," identify the differences between normal healthy human cells and bacterial cells. Sometimes this is done by the antigenic sites present on the surface of the bacterial walls. These sites were used by immune system to kill pathogen naturally, and are now utilized by antibiotics for target killing of bacteria inside the body. As well, the region of the human body that gets infected by antibiotics emits some chemical signals to the immune system. These signals are identified by antibiotics and are proven to be beneficial for the killing of pathogens from the body. The overall pattern of characteristic killing always remains same (Waterfield and Wren, 2004). For example, human body cells do not acquire cell walls, while some classes of bacteria do. The antibiotic penicillin works by keeping a bacterium from building a cell wall. Moreover, bacteria and human cells also show difference in their cell membrane structure and the mechanism they utilize to build proteins and copy DNA. Some antibiotics specifically dissolve the membranes of bacteria cells only. Others influence protein building sites and DNA replication ingredients that are specific to bacteria (Vannuffel and Cocito, 1996).

19.1.3 Broad-spectrum antibiotics

Antibiotics are divided into two subcategories. Narrow-spectrum antibiotics (NSAs) are those that perform active working in opposition to a selected group of bacterial and microbe types. NSAs are used for the exact infection when the contributory organism is acknowledged and will not affect a lot of microorganisms in the body as broad-spectrum antibiotics (BSAs) do. So, NSAs have less capability to root out a super infection. Examples of BSAs include Azithromycin as well as Clarithromycin, etc. (Ferretti et al., 2016). The phrase "broad-spectrum antibiotics or BSA" was initially used to delegate antibiotics that were efficient against both gram-positive as well as gram-negative bacteria, in contrast to penicillin, for example, which is helpful mainly against gram-positive bacteria, or streptomycin, which is vigorous chiefly against gram-negative organisms. The BSAs have an antimicrobial range that comprises some gram-positive and gram-negative organisms as well as larger viruses, certain rickettsiae, pleuropneumonia-like organisms, and protozoa (Almulhim and Alotaibi, 2018). The list of antibiotics, with use and abuse, is given in Table 19.1.

19.1.4 Vaccines block diseases

Vaccination against diverse form of viral as well as bacterial diseases is an integral component of the communicable disease control worldwide. The purpose of vaccination is to decrease socioeconomic burden along with incidence of a particular disease. High-quality immunization results in inhibition of transmission of several vaccine preventable diseases. Immunization plays an integral role in controlling diseases, proven, for example, by the complete elimination of smallpox globally. Polio also has nearly been eradicated, although there are still some countries suffering from this disease. It is known that when an infant is immunized at an early stage and the vaccination schedule is completely fulfilled, then the burden and incidence of vaccine preventable diseases is greatly reduced (Australia & ARCHIVE).

19.1.5 Example of malaria transmission blocking vaccines

In most poor countries of the world, vaccines effectively decrease the burden of disease at the economic level. These vaccines constitute numerous components that provide protection in various ways. Liver-stage vaccines are helpful in lowering the chance of individual being infected. Vaccines of asexual blood stage play a role in lowering the severity of disease and risk of death from infection. By targeting the sexual stage of parasites, transmission blocking vaccines are preventing malarial spread within communities. This has the potential to decrease the burden of malarial disease and death, especially where malaria is most prevalent, i.e., Africa. In Latin America as well as in Asia, transmission-blocking vaccines are eradicating malarial parasite disease. It also has a remarkable role to prevent spread of malarial parasites that are resistant to these vaccines (Carter, 2001).

19.2 Attack and nature of pathogens

The epithelial layer of different organs is covered with mucus, which aims to protect organs from foreign bodies including pathogens. But some variants of bacterial pathogens have developed though evolution that can cross the mucus and reach the epithelial layer. For colonizing host cells, adhesion is

Table 19.1 Description of different antibiotics with reference to specific drugs and role.

Antibiotics	Drugs	Functions	Side effects	References
Ampicillin	Carbenicillin Amoxil, Augmentin, Moxatag, dicloxacillin, and penicillin	Use to treat and prevent pneumonia, bladder infection, gonorrhea, gastrointestinal tract (GIT) infections, and meningitis	Severe stomach pain, flu symptoms, unusual weakness, body aches, diarrhea that is watery or bloody Not for people with kidney issues and asthma	Almulhim & Alotaibi (2018)
Amoxicillin	Amoxicot Amoxil, Apo-Amoxi, Trimox	Fights infections caused by bacteria. such as bronchitis, tonsillitis, gonorrhea, and pneumonia, infections of skin, throat, nose, and urinary tract	Abdominal or stomach cramps, bloating, pain in lower body, low blood pressure, irritated eyes, difficulty with swallowing and breathing, inflammation of the joints	Joshi & Milfred (1995)
Carbapenems	Merrem, Merrem Novaplus	Class of beta-lactam antibiotics works against many gram-positive and -negative bacteria. Treats skin tissue and stomach infections, good for intraabdominal infections	Clay-colored stools Blisters, wheezing Vomiting of blood	Joshi & Milfred (1995)
Trimethoprim	Primsol Trimpex *Proloprim*	To treat bladder, kidney, and ear infections caused by certain bacteria	Itching, shortness of breath, pale skin scaly skin fever with or without chills Severe loss of appetite Nausea	Tzialla et al. (2012)
Ticarcillin	Ticar	Used in the treatment of: febrile neutropenia, kidney infections, joint infection, bone infection, peritonitis, pelvic inflammatory disease, intraabdominal	Hypersensitivity reactions, development of hypokalemia, anemia, and neuromuscular excitability	Tzialla et al. (2012)

Continued

Table 19.1 Description of different antibiotics with reference to specific drugs and role.—cont'd

Antibiotics	Drugs	Functions	Side effects	References
		infection, and pneumonia		
Piperacillin	Pipracil	Treat or prevent bacterial infections	Swollen skin, fever, nausea, diarrhea, seizures, and extra muscle action	Tzialla et al. (2012)
Tetracyclines	Adoxa TT, Oraxyl, Vibramycin, Terramycin	Treat infections by susceptible microbes such as gram-positive and gram-negative bacteria, protozoans, mycoplasmata, chlamydiae, or rickettsiae	Bloody diarrhea, petechial rash fatigue, fever	Almulhim & Alotaibi (2018)
Quinolones	Cipro Cystitis Pack, Ciloxan, Otiprio, Cetraxal	Divided into many subclasses, used to treat typhoid fever, malaria, urinary tract infection, pneumonia, bronchitis, bladder infections, and anthrax	Nausea, fever, bloating, shortness of breath, diarrhea, and stomach ulcers	Tzialla et al. (2012)

key factor as it prevents the mechanical removal of pathogens (Kline et al., 2009). Pili are hair-like structures arising outward from bacteria. They are one of the structures involved in adhesion of pathogens to host cells. The base of a pilus is present in the outer membrane of bacteria while its tip helps in attachment to the host. UPEC are the pathogenic strains of *E. coli* that are responsible for kidney infection. These pathogens display P (pyelonephritis-associated) pili at their surface. Adhesion factor called PapG composes the tip of this pilus, which binds to the kidney epithelium's glycosphingolipids (Roberts et al., 1994).

Some other strains of UPEC express another type I pili at their surface that specifically binds to D-mannosylated receptors (Lillington et al., 2014). Other gram-negative bacteria, e.g., *Neisseria meningitidis*, display type IV pili at their surface, which are made of pilin protein. Two types of pili have been described in gram-positive species. The first type is sortase-assembled pili. This type of pilus is formed by the catalytic activity of sortase. The second type is type IV-like pili. These pili resemble gram-negative bacteria's type IV pili (Melville and Craig, 2013).

Besides pili, bacteria have a variety of factors present on their surface that help in their attachment to a variety of host molecules, which may be transmembrane proteins or the components of extracellular matrix (Mattick, 2002). In addition, bacteria causing diarrheal diseases in children and food-borne diseases use a special type of mechanism to contact their host. They inject an effector called Tir, which after insertion into the plasma membrane of the host serves as a receptor for the attachment of bacterial protein intimin. After attachment, bacteria use M cells or macrophages to get engulfed through phagocytosis where bacteria that are resistant to the host defense mechanism not only prevent acidification of phagosome but also prevent its fusion with lysosome. In spite of using this phagocytic process, some bacteria use two other mechanisms like zipper and trigger (Cossart and Roy, 2010), which are activated through signaling. Zipper mechanism involves adhesion of bacterial protein with host protein, i.e., integrin or adhesins, while trigger mechanism results in formation of membrane projections called ruffles that fuse with the cell surface and entrap nearby bacteria (Pizarro-Cerdá and Cossart, 2006, 2009).

Conclusively, bacteria, after internalization, expose different virulence genes that have roles in disease progression and spread (Ribet and Cossart, 2015).

19.3 Mechanism of action of antibiotics

Antibiotics are very specific for their mode of action for killing targeted microbes. This whole process is done by different ways of specific shifting inside the cell of the microorganism. Some of them include:

19.3.1 Impairment synthesis of cell wall

Microbial cells are enclosed in a cell wall of peptidoglycan, which is composed of chains of long sugar polymers. The peptidoglycan endures crisscross arrangement of glycan strands by transglycosidases, and the peptide chains expand from the sugar, making bonds from one peptide to another (Kahne et al., 2005). The D-alanyl-alanine segment of peptide chain is cross associated by a series of glycine residues accompanying penicillin-binding proteins (PBPs) (Reynolds, 1989). This cross-linking strengthens the cell wall. β-lactams as well as the glycol peptides of antibiotics restrain cell wall synthesis.

The PBPs are the most imperative targets of β-lactam agents. It has been conjectured that the β-lactam ring impersonates the D-alanyl D-alanine section of the peptide region that is usually bound by PBP. The PBP network with β-lactam ring then does not present for synthesis of new peptidoglycan. The interference of peptidoglycan sheet leads to lysis of bacteria (Džidić et al., 2008). The glycol peptides attach to D-alanyl D-alanine segment of precursor peptidoglycan subunit at peptide region. The impulsive antibiotics avert binding of this D-alanyl subunit to PBP, and therefore block cell wall synthesis (Grundmann, Aires-de-Sousa, Boyce and Tiemersma, 2006).

19.3.2 Impairment of protein biosynthesis

The macromolecule known as a ribosome synthesizes proteins by translation through mRNA. Protein biosynthesis is catalyzed by ribosomes with the aid of some cytoplasmic protein manufacturing factors. The bacterium70S ribosome is self-possessed with bonding of ribonucleoprotein subunits, 30 and 50S subunits (Yoneyama and Katsumata, 2006). Antibiotics hinder biosynthesis of protein by aiming at the 30S or 50S subunit of microbe ribosome (Vannuffel and Cocito, 1996).

The aminoglycosides (AGs) and tetracyclines are known as positively charged antibiotics, which attach on outer membrane of bacterium as negatively charged, resulting in formation of large pores in the membrane and, in consequence, permit antibiotic diffusion inside the bacterium cell

(Yoneyama and Katsumata, 2006). The major objective of action is to reach ribosomes through a cytoplasmic sea requiring energy requiring active bacterial transportation mechanism, which necessitates oxygen as well as an active proton. The antibiotics have synergism with other ones, which inhibit cell wall biosynthesis to permit larger penetration of antibiotics at low dosages within the cell (Nelson et al., 2019). It acts on 16S r-RNA of the 30S subunit avoiding binding of t-RNA to the A site and thus impulsive execution of mRNA. On the other hand, antibiotics like chloramphenicol and macrolides interact onto the conserved region of peptidyl transferase pocket of 23S r-RNA of the 50S subunit (Farhadi et al., 2019). Hence, they block the protein biosynthesis by binding of t-RNA to the A site of the bacterial ribosome, resulting in a premature disconnection of deficient peptide chains (Lambert, 2005).

19.3.3 Impairment of DNA replication

The mastermind of antibiotics possesses all the weak points of the bacterial territory. The main part is the genome of bacterium, targeted for killing by antibiotics by producing nicks in the genomic strand. For example, in gram-positive bacteria, topoisomerase IV nicks and splits daughter DNA strands right after DNA replication. The antibiotics such as fluoroquinolones (FQs) block the enzyme activity of bacterial DNA gyrase, which nicks dsDNA, initiating negative supercoiling and then ligating the nicked ends. This is essential to avoid extreme positive supercoils of DNA strands to permit replication or transcription. The DNA gyrase of bacterium consists of two subunits of A and B, respectively. A subunit responsible for introducing nicks in DNA, B subunit promotes negative supercoiling, and thus A subunit ligates the strands. The FQ antibiotics bind to A subunit with such high affinity that they hinder the subunits with strand nicking and ligation function. This results in the premature DNA replication, leading to destruction of bacterium (Yoneyama and Katsumata, 2006).

19.3.4 Alteration of cell membrane

Plasma membrane, as a phospholipid bilayer, works for the integrity of bacterium. The lipid portion of membrane is resistant to charged compounds. However, channels are likely to present in the bilayer for the passive transportation of many ions, sugars, and amino acids; mysterious channels are known as porins. These molecules are there in periplasm, the section from cytoplasm to outer cell membranes. The periplasm encloses the peptidoglycan layer for responses of extracellular chemical signals as well as for many proteins needed for substrate binding or hydrolysis. The periplasm is considered to subsist in a gel-like condition in spite of a liquid state because of concentrated environment of peptidoglycan and proteins present in it. Because it is located between the cytoplasm and cell membranes, signals are received by substrate binding capacity of the bacterial cell. As a result, transportation and signaling of proteins occur. This biochemical signaling pathway of bacterium is destroyed by antibiotics by changing the integrity of the bacterial plasma membrane. Alteration in the membrane by inducing serious structural damage helps in the killing of bacterial cells from the body (Kohanski et al., 2010).

Polymyxin is an example of an antibiotic that can cause the bacterial cell membrane to modify its functionality. On binding to bacterial lipopolysaccharide (LPS), it disrupts both the outer and inner membranes. By activating detergent mode of action, the hydrophobic tail of the drug causes membrane damage. Detachment of the hydrophobic tail of polymyxin acquiesces polymyxin nonapeptide, which is responsible for binding of drug to LPS without destroying the bacterial cell. The main purpose of this action is to promote permeability of the membrane to antibiotics causing major membrane disorganization (Uphoff and Drexler, 2011).

19.3.5 Antimetabolite activity

Antimetabolites are artificial defenders, meaning they are analogs of natural immunity providing compounds. Their major work is to interfere in the metabolic routes of bacteria for their proper destruction. Most of these antibiotics deal with the inhibition of nucleic acid synthesis. Some of them disturb the folate manufacturing process, resulting in the targeted killing of the bacteria cells (Avendaño and Menendez, 2015). Many antimetabolites are used for destruction of specific metabolic routes. Sulfanilamide are a good example of bacterial-specified metabolism killing. These antimetabolites work by disrupting metabolic pathways that are present in bacterial cells but not in human cells. Other examples that include target nucleic acid metabolite killing are antibiotics such as azathioprine, mercaptopurine, and thioguanine, which are antagonists of purines, and fluorouracil and floxuridine antagonists of pyrimidine. Cytarabine, being a precursor destroyer, obstructs with dihydrofolate reductase, which is essential for the production of tetrahydrofolate as well as folic acid, which is required for DNA formation (Fig. 19.1) (Bhattacharjee, 2016).

19.4 Era of dose to cure

From dosage to cure, antibiotics exhibit the class of specification for destruction of bacterium. This includes number of prescription and mg or mL of the drug required for special thermokinetic activity in the body (Almulhim and Alotaibi, 2018). Main classes of antibiotics with their dosage concentrations are provided in Table 19.2.

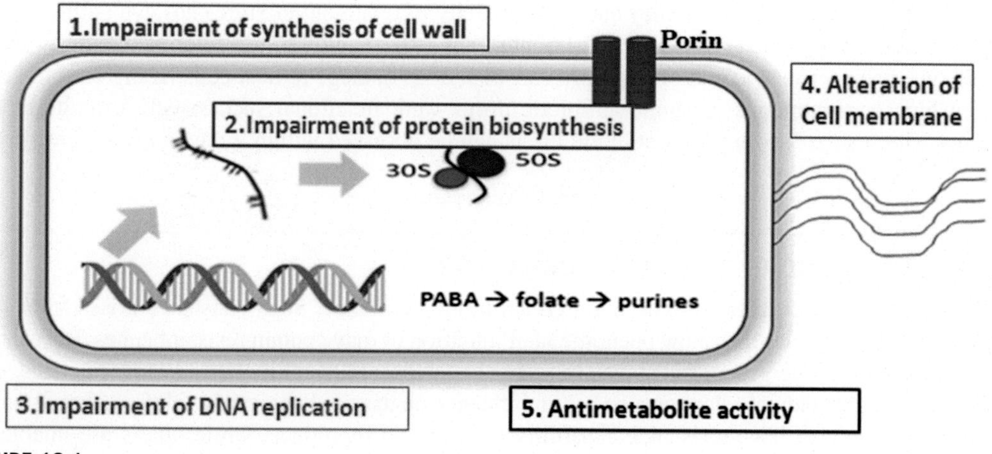

FIGURE 19.1

General mechanism of action for antibiotics.

Table 19.2 List of major antibiotics and their curing mechanisms with particular dosage level.

Antibiotics	Dosage	Curing mechanism
Quinolone	Oral suspension 250mg/5 mL, 500mg/ 5 mL tablet 100, 250, 500, 750 mg	Inhibits DNA synthesis
Rifampicin	10 mg/kg/dose once daily or maximum 5 times per week	Inhibits RNA synthesis
Penicillin	125–250 mg used every 6–8 h for 10 days	Inhibits cell wall synthesis
Ampicillin	Applied in order of 250 mg, 500 mg, 125 mg/5 mL, 250 mg/ 5 mL, 125 mg, 1 g, 2 g, 10 gas per recommended by doctor	Inhibits cell wall synthesis Alternate cell membrane Inhibits DNA synthesis
Tetracycline	Strength of 500 mg, 250 mg, 125 mg/5 mL, 100 mg	Inhibits protein synthesis

19.5 Toxicity of antibiotics

In spite of their discerning killing ability, antibiotics still cause critical adverse reactions in hosts, mostly due to imperfect drug regimens. The main factors are pharmacokinetics or pharmacodynamics models, resistome analyses, and toxicity of antibiotics as resistance mechanisms. Multidrug efflux pumps have additional physiological purpose in stress variation as well as virulence of bacterial cells (Sun et al., 2015; Yılmaz and Özcengiz, 2017). Toxic effects of current antibiotic drugs may be veiled by demonstration of the infection they cure. The toxicity of treatment mixtures has cytopathic effects on prime cultures of human cells. In arrangement of any two antibiotics, the combined toxicity status and the interface between extent of exposure causes serious damage to host cells (Kang et al., 1997). Diminishing concentrations of single antibiotic drugs were functional successively to cultures for 14 days. The scale of toxicity for antibiotic drugs can be observed within 7 days of exposure (Berry et al., 1995).

19.6 Emergence of antibiotic resistance

Antibiotic resistance is referred to as acquired antibiotic resistance. Such resistance arises from susceptible bacterial isolates that show phenomena of mutation or may contain resistant genes. Resistance against antibiotics, emergence, and spread rely on when exposure to antibiotics takes place. Thus, it is has been found that natural selection also has an influence on the mechanism of evolution of resistance. Those organisms that have resistance capability survive and reproduce, while others are unable to compete and reproduce so they become extinct. But in fact, emergence of antibiotic resistance is quite a difficult phenomenon that is predicted to occur when an abnormal resistant trait in a bacterial pathogen emerges. Emergence is basically revealed when frequency of the abnormal resistant trait is high so that resistance that occurs first might be enigmatic (Cantón and Morosini, 2011).

19.6.1 Pathway toward self-medication

Antibiotics are mostly purchased as drugs globally. In developing countries, antibiotics are used for treating infectious diseases that are responsible for mortality of individuals. Self-medication refers to the uptake of medicines to cure disorders that are self-diagnosed without consulting with a doctor. Almost 50% of antibiotics used are not prescribed by a doctor. When self-medicated antibiotics are used this has serious health consequences. This practice is common in developing counties where poverty, insufficient health care facilities are present and there is a lack of medical expertise. Self-medication of antibiotics is linked with risk of inadequate drug usage. It predisposes patients to interact with drugs, mask symptoms of particular disease and then microbial resistance development takes place. When disease symptoms appear to be improved then inadequate usage of drugs constitutes inappropriate dose, shorter treatment duration, medicine sharing and avoiding treatment. Various multidrug-resistant strains of bacteria have emerged that show resistance against a wide range of antibiotics, raising concerns about antibiotic resistant globally. Thus, resistance to antibiotics is often responsible for extended duration of stays in hospital, prolonged illnesses, the need for more expensive medical treatment, and sometimes even death. After decades of economic development and growth in developing countries, infectious diseases are still a high burden. The consequences of self-medication with antibiotics and magnitude of the problem has not been identified in many regions. There is need to plan as well as implement strategies as well as interventions to overcome spread of resistance against antibiotics (Nepal and Bhatta, 2018).

19.6.2 Genetic mutations among bacteria and microorganisms

One of the factors that results in antibiotic resistance is mutation among bacteria and microorganisms. Variations in environmental conditions lead to evolution of living organisms. This promotes genetic variant selection within species over passage of time. This process was called natural selection by Charles Darwin. In the process of natural selection an individual is not particularly the target of the response to any selection event for increasing possibility of its survival, rather variations occur randomly within the genome of organism. This alteration might or might not have any effect on increasing fitness on a microorganism. These variations are called as mutations. Mutation rate is a fixed range from 1 in 109 to 1 in 1010. This mutation is in replicated base pairs.

Mutations are of several types: there is chance of substitution of one nucleotide to another. This is pint mutation or base substitution. Due to this substitution, translation of protein that is terminated is known as non-sense mutations or results in an amino acid that is specified by a given codon known as missense mutations. Frameshift mutations also occur that constitute deletion or insertion of nucleotides. This variation results in shifting of reading the frame to left or right, used by ribosomes to perform translation (Fig. 19.2) (Watson, 1997).

Basically, mRNA is transcribed from the coding strand of DNA and then proceeds to translation. Mutation in the first and second position of triplet codon results in change of amino acids as compared to the codon at third position. This is known a wobble. If variation is such that the amino acid gets

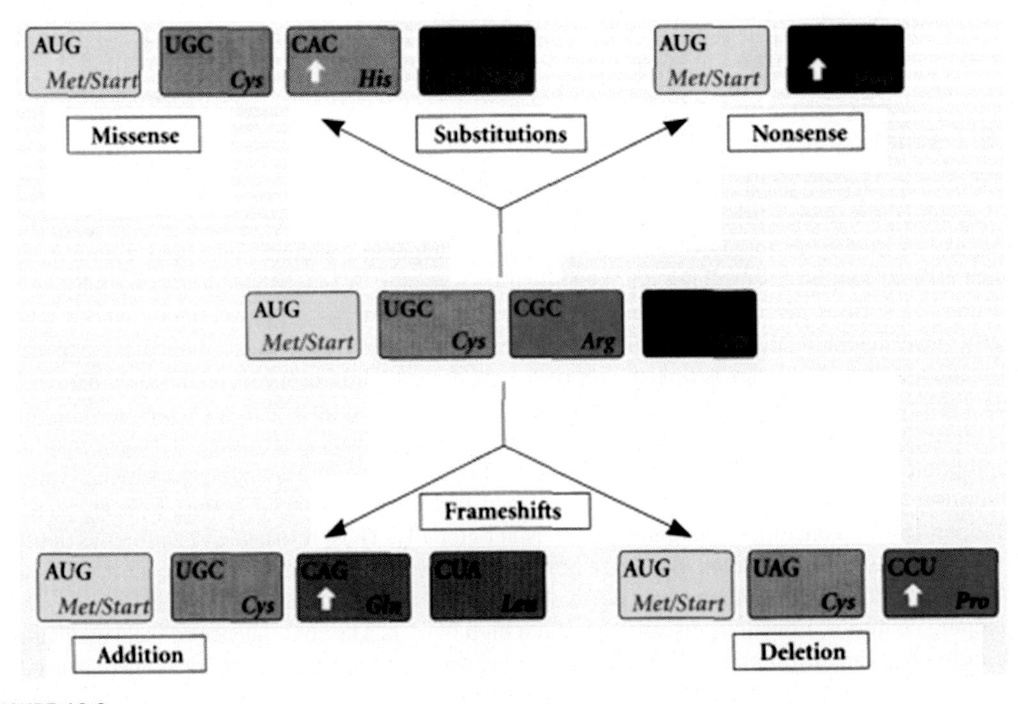

FIGURE 19.2

Types of genetic mutations.

altered, then question arises about its effect on individual's survival regarding changes in environment (Watson, 1997).

Antibiotic resistance in bacteria is rising. In 2013, a report released by the U.S. Centers for Disease Control and Prevention that showed that 23,000 people died just because of bacterial resistance to antibiotics in United States. By variations in genes, certain species of bacteria such as mycobacteria develop resistance against drugs. To predict molecular processes that are responsible for antibiotic resistance of *Mycobacterium smegmatis*, several cultures of this were prepared (Gomez et al., 2017). *M. smegmatis* is of family of that species that caused tuberculosis. Bacteria, when exposed to the low concentration of antibiotics, observed less effect of microbes killing relatively low. Thus, this is helpful to monitor the sensitive bacteria that are killed by exposure to antibiotics along with isolating the cells in which mutants are predicted. Outgrowth of the antibiotic-resistant mutants in several cultures revealed that a single mutation was carried by each individual in several compartments of ribosome. Ribosomes are molecular machines of complex nature that result in formation of proteins. Ribosomal mutations result in resistance to several antibiotic classes that not only have effects on ribosomes but also the mutants that are secretes. It is also responsible for increasing resistance to stresses that are nonantibiotics, particularly membrane stress and heat shock. Due to mutations, *M. smegmatis* fitness increased against several drugs. Mutations make bacterial strains less sensitive to antibiotics as well as nonantibiotic stresses. Understanding the phenomena of resistance to antibiotics is helpful in developing and optimizing new drugs to treat bacterial infections (Gomez et al., 2017).

19.6.3 Phenotypic resistance to antibiotics

Sometimes bacteria become resistant to antibiotics without any genetic mutations. Most common are persistence phenomena, drug indifference, and growth in biofilms. However, certain situations also occur where bacteria showed variations regarding to their susceptibility against antibiotics that rely on metabolic state. Phenotypic resistance is attributed to growth arrest that impedes the work of antibiotics that are bactericidal, i.e., beta-lactams. Association between phenotype of susceptibility to antibiotics and the metabolic state of bacterial populations are complex phenomena. Resistance also increases by increasing the special induction of mechanisms that are responsible for resistance. Susceptibility to antibiotics varies because of changes in metabolism of bacteria. Susceptibility to antibiotics is under metabolic control. The two-way road on which susceptibility to antibiotics is change and bacteria show more resistant to antibiotics transiently and conversely (Davies and Davies, 2010).

19.6.4 Development of resistance

Bacterial machinery is smarter than it looks. Bacteria develop the "thief door" to protect them from the antibiotic environment and become resistant. This is done by two types of resistant properties exhibited by microorganisms. For example, in intrinsic resistance, there is an antibiotic responsible for elimination of cell walls of the bacterium. If bacteria do not possess a cell wall, the antibiotic drug will have no effect on it, hence it becomes resistant. On the other hand, a bacterium that was earlier inclined to an antibiotic resistance is known to have acquired resistance.

19.6.5 Enzymatic degradation of antibacterial drug

Many antibiotics possess esters and amides like chemical bonds that are hydrolytically susceptible, whose reliability is vital to biological activities. Not astonishingly then, there are numerous enzymes produced by bacteria and other microbes that have progressed to targets and split these exposed bonds between them, consequently, fulfilling the purpose of ambushing antibiotic activities. Most important of them are amidases that destroy the bonds of h-lactam ring of the penicillin as well as cephalosporin modules of antibiotic drugs. The macrolide antibiotics, for example, erythromycin, inhibit the peptide exit passageway of the large ribosomal subunit. Macrolides are a cyclized structure due to presence of an ester bond. This structure helps them to inhibit the thioesterase segment of the polyketide synthetase in order to avoid formation of a final ring of enzymes to generate 6-deoxyerythronolide B and, hence, restrict protein biosynthesis (Wright, 2005).

19.6.6 Alteration of bacterial proteins that are antimicrobial target

Alteration causes delusion to avoid the action of antibiotics is the most favorite method used by a bacterium. Data from experimental strains demonstrated that resistance can be established in every group of antibiotics, nevertheless this is the action strategy used by antibiotics. Target site deviations result in the mutated changes in bacterium gene or change in the antibiotic pathway. For example, genetic mutations caused by pathway alteration is the important ingredient during RNA polymerase synthesis as well as DNA gyrase development, giving rise to resistance rifamycins plus quinolones, correspondingly. On As well, achievement of resistance may encompass transmission of resistance genes from other gene pool by conjugation and transformation. Acquisition of *mecA* genes produces methicillin resistance in *Staphylococcus aureus* as well as in enterococci showing resistance to glycopeptides (Lambert, 2005).

19.6.7 Change in the membrane permeability

Gram-negative bacteria develop resistance to antibiotics by establishing a mechanism of reduced permeability in their outer membrane. Bacterium's outermost hydrophobic membrane permits hydrophilic particles to cross only by its aqueous apertures. The transmembrane pores have a trimer geometry formed by each monomer that works as an aqueous channel. Diameter of these channels is 1–1.2 nm. Changes in the entire number of pores by bacterium as well as alteration in their function reduce the chances of diffusion of antibiotics' entrance. This mechanism is applied as a cross resistance and is favorable to many subclasses of antibiotic drugs. One of the prime features of this type of resistance is that such cases are not always detected in clinical applications. The species with such resistance are enterobacteria, including, *Enterobacter*, *Klebsiella*, *Salmonella*, and *Serratia*. In case of *P. aeruginosa*, resistance results from a discrepancy in protein D2 in 12%–15% of the strains. Reduced permeability is predominantly operative when accompanying another application of resistance permitting the bacteria to prompt a higher level of resistance (Van Nguyen and Gutmann, 1994).

19.6.8 Genetic mutation

Antibiotics' main work is simple, like targeting cellular processes, hindering growth, and triggering cell death. However, if bacteria are exposed to antibiotics below the dosage essential to destroy all bacterial cells in a populace, they can metamorphose by genetic mutations and develop resistance against antibiotic treatment. This whole process of mutating bacteria possesses a natural assortment for resistance-conversing mutations. These genetic mutations also are enriched by implementation of a plasmid responsible for encoding mutant genes in a chromosome itself, responsible for resistance against antibiotics (Richardson, 2017).

Antibiotic resistance that develops from chromosomal mutations is specific to a certain antibiotic. Mutations in genes encoding ribosomal machinery promote resistance to numerous structurally as well as mechanistically disparate classes of antibiotics. This enhances survival from heat shock protein and other membrane stress. These mutant genes affect ribosome assembly causing large-scale changes at transcriptomic and proteomic levels. This includes downregulation of the catalase KatG, which is required for isoniazid sensitivity, and upregulation of WhiB7, which promotes antibiotic resistance. These ribosomal mutations promote progression of specific target-based resistance. Furthermore, suppressor mutations are developed by wild-type growth factors (Gomez et al., 2017).

19.6.9 Modify or bypass the antibiotic's target

Changes in the geometry and chemical formula of the target in bacterial cells can also stop the antibiotic interaction with the target. Otherwise, the bacteria can add a chemical group's side chain in the structure of the drug, in this way defending it from the antibiotic. Occasionally bacteria can harvest a diverse variant of geometry it needs. For example, Vancomycin-resistance by bacteria makes a cell wall associated to predisposed bacteria. The antibiotic is not capable to break through this type of cell wall (Walsh, 2000).

19.7 Transfer of resistant genes among pathogens

Currently, the major concern globally is increase antimicrobial resistance (AMR) that threatens human life as mortality rates and costs are increasing but treatment of bacterial infection is limited because of AMR. Natural antibiotics have been used for many of years (Barlow and Hall, 2002; Bhullar et al., 2012). Such strains are produced that have natural ability to compete with the harmful bacteria either by inhibiting or completely removing them (Aminov, 2009).Various studies have illustrated the antibiotic resistance genes (ARGs) responsible for AMR (Perron et al., 2015). Major factors involving resistance to antibiotics are either mutation or transfer of ARGs through horizontal gene transfer (HGT). HGT is considered to be a most effective factor in developing AMR. HGT is old mechanism, but continuous used of antibiotics by humans increases resistant strains and AMR mechanisms (von Wintersdorff et al., 2016).

19.7.1 Mechanisms of horizontal gene transfer

Bacteria recruit a number of mechanisms like conjugation, transformation, and transduction to move genes horizontally. Naturally transformation is a complicated process; it involves several multicomponent cell structures like type IV pili, type II secretion systems (T2SS) for the uptake of foreign DNA (Krüger and Stingl, 2011). DNA uptake involves two steps: DNA movement from the surface into the cytoplasmic membrane, and then crossing the membrane by membrane channel. This mechanism of transfer was confirmed by visualization of transformation in *Helicobacter pylori* (Stingl et al., 2010). However, there is risk of uptake damaging DNA like *Haemophilus*, which take DNA that is capable of recombination with host genetic material (Ambur et al., 2007). Transduction is a process by which DNA is transferred through bacteriophages in a horizontal way; besides transferring horizontal genes, bacteriophages also transfer mobile genetic elements (Novick et al., 2010). Conjugation is a process in which DNA is transferred from one bacterial cell to another either through direct contact or by sex pili. It utilizes T4SSs for the transfer of DNA in a horizontal way (Alvarez-Martinez and Christie, 2009). Currently, nanotube-mediated transfer has been performed with greatest frequency of transfer to neighboring cells (Ficht, 2011).

19.8 Antibiotics resistant bacterial infections

19.8.1 Methicillin-resistant *Staphylococcus aureus*

Methicillin-resistant *Staphylococcus aureus* (MRSA) infection is caused by the bacterium *Staphylococcus* (staph). This bacterium acquired resistance against a variety of antibiotics. *Staphylococcus* bacteria reside in the nose and on skin. Normally it is not harmful, but if it multiplies uncontrollably then this leads to MRSA infection. Infection starts when a cut is induced in skin. It is very contagious and transmits directly through infection in an individual. MRSA infection also may be transmitted from objects with which an infected individual was in contact (Carissa Stephens, 2017).

Methicillin resistance was shown in a clinical study to basically appear from methicillin-hydrolyzing β-lactamase expression and by expression of variant form of PBP2 that acquired less penicillin-binding affinity and increased rates for releasing bound drug in contrast to normal PBP2. But the main process of methicillin resistance in *S. aureus* occurs by expressing foreign PBP, PBP2a that showed resistance against methicillin action but had capability to carry transpeptidation reactions that are also referred to as cross-linking reactions of the host PBPs. Formation of PBP2a properly regulated at low level, but formation elevated when mutation was induced in regulatory genes (Stapleton and Taylor, 2002).

19.8.2 Vancomycin-resistant enterococci

Enterococcus is basically a gram-positive, round-shaped bacterium that resides in the gut but has the ability to cause infection in any body part. It has shown resistance to several antibiotics but earlier physicians prescribed vancomycin for treating enterococcal infections. Recently it has been found that some enterococci also showed resistance against vancomycin. Basically, two species that are problematic are vancomycin-resistant Enterococcus faecium as well as vancomycin-resistant Enterococcus faecalis. E. faecium is referred to as common species of vancomycin-resistant Enterococci (VRE). These bacteria belong to a different genus. Resistance to vancomycin occurs when sensitive Enterococcus contains a special piece of DNA called a plasmid that allows bacteria to show resistance against vancomycin. New strains are named as VRE. VRE strains also have capability for transmission of vancomycin resistance to unrelated bacteria such as MRSA). VRE is resistant to more than one antibiotic. Due to VRE, 30% enterococcus infections occur (Cetinkaya et al., 2000).

19.8.3 Drug-resistant *streptococcus pneumoniae*

Streptococcus pneumoniae is the main causative agent of pneumonia, sepsis, bacteremia, meningitis, and otitis media. *S. pneumoniae* has acquired increased resistance to various classes of antibiotics. Almost 22% of isolates of pneumoniae showed resistance against clindamycin (Cherazard et al., 2017). The process through which *S. pneumoniae* maintains resistance against b-lactam antibiotics includes variation among one or more penicillin-binding proteins that show adverse effect among PBP's affinity against penicillin as well as b-lactam antibiotics. This process of resistance occurred by natural transformation mechanism. In this process a particular genome that encodes variations is selected from other bacteria named *pneumococci* and added into their own DNA (Jacobs, 1999).

19.8.4 Drug-resistant *mycobacterium tuberculosis*

Tuberculosis (TB) is recognized as an infectious disease and is a global public health concern. According to the latest report published by the World Health Organization (WHO), in 2012 almost 8.6 million incident cases of TB were reported and 1.3 million morbidity cases occurred due to disease. In 2012, almost 450,000 multidrug resistant (MDR)-TB cases were reported along with 170,000 individuals' death. MDR-TB was caused by *Mycobacterium tuberculosis*, which showed resistance against isoniazid and rifampicin. These 2 y drugs are key factors for treating disease. Since 2006, more strains resistant against *M. tuberculosis* that are referred to as extensively drug resistant (XDR)-TB have been recognized. Such strains also show resistance to fluoroquinolone and among second-line drugs that are injected—amikacin, kanamycin, and capreomycin—along with MDR resistance. A good understanding of anti-TB drugs' process, as well as development of resistance against results in identification of novel drug targets and more appropriate processes to predict resistance against drugs (Palomino and Martin, 2014) are needed.

19.8.5 Carbapenem-resistant enterobacteriaceae

Enterobacteriaceae refers to a bacterial family that resides in bowels of individual without causing illness. Carbapenems act as powerful antibiotics that are used for treating serious infections. Some Enterobacteriaceae showed resistance against these antibiotics, predicting that they will be ineffective to cope with infections that may form. These are called carbapenem-resistant Enterobacteriaceae. Sometimes it is possible that bacteria spread out of the bowel and result in causing infection, e.g., urinary tract infections (Control and Prevention, 2013).

19.8.6 Multidrug-resistant *Pseudomonas aeruginosa*

Pseudomonas aeruginosa is major cause of nosocomial infections. Infections that occur by *P. aeruginosa* are more severe as well as life-threatening. They are tough to treat due to susceptibility is limited against antimicrobial agents and antibiotic resistance evolution is at high frequency, which results in adverse outcomes. Antibiotic resistance in *P. aeruginosa* species increases due to de novo arrival of resistance in particular individuals after antimicrobial exposure along with spreading of infection from one patient to another. Resistance accumulates with increased exposure to antibiotics; as well, exposure to cross-resistance between agents contributes to MDR *P. aeruginosa*. Such condition exists in patients with cystic fibrosis; along with *P. aeruginosa*, persistent infection arises with sequential evolution of resistance against several types of antibiotic agents. Such MDR *P. aeruginosa* strains transfer from one patient to another and sometimes result in outbreaks along with cystic fibrosis (Aloush et al., 2006).

19.8.7 Multi-drug-resistant *Acinetobacter baumannii* infection

The genus *Acinetobacter* is linked with hospital-acquired infections as well as health care–associated infections (HAI), most commonly in developing countries. *Acinetobacter* genomic species became a major problem in 1990, causing pneumonia, meningitis as well as infections in catheters in patients in private and public hospitals. *Acinetobacter* emergence is attributed not only to characteristics of bacteria like long period of survival on medical equipment as well as on surfaces but also is linked with transmission from one person to another. Resistance in *Acinetobacter* genomic species has caused a major problem that resulted in making treatment of *Acinetobacter* spp. difficult. Carbapenems are used as the first option to treat HAI induced by *Acinetobacter* but the rate of effectiveness reduces with the passage of time. Increments predicted in resistance to carbapenems are linked with drugs abusive prescription. *Acinetobacter* susceptibility improving for carbapenems is attributed to a stewardship program but evidence of effectiveness is not very strong (Abbo et al., 2005).

19.8.8 ESBL producing *enterobacteriaceae* resistance

In this modern era of medicine, generation of resistance against antibiotics is inevitable by Enterobacteriaceae and area major reason in treating infections of bacteria in hospital as well as in community. Beta-lactamase-producing bacteria (BLPB) are a key factor contributing to polymicrobial infections. They have ability of direct impact of pathogens to cause infection along with an indirect impact by production of beta-lactamase enzyme. These enzymes have capability to produce Enterobacteriaceae families and are responsible for higher mortality as well as morbidity rates. Extended spectrum β-lactamases (ESBLs) are basically enzymes that are formed by gram-negative bacterial varieties conferring elevated level of resistance against antibiotics. ESBLs are considered as plasmid-mediated enzymes conferring resistance against a broad range of β-lactams. At first, resistance appears to cephalosporins that are in third generation among gram-negative rods, mainly occurring by dissemination of several types such as TEM- and SHV-type ESBLs, that are point mutants of classic TEM, CTX, SHV as well as enzymes that have extended specificity for substrate. ESBLs producing strains of Enterobacteriaceae resistance are very problematic in patients that are in hospitals or reside in near community (Thenmozhi et al., 2014).

19.8.9 Antibiotic-resistant *Neisseria gonorrhoeae*

Antibiotic-resistant *Neisseria gonorrhoeae* have emerged as well as spread quickly. Resistance is shown against several classes of antibiotics, i.e., penicillin, fluoroquinolones, and tetracycline. In recent years, resistance has also been predicted against cefixime, which is basically oral third-generation cephalosporin. Since 2010, cefixime has not been prescribed as a first-line treatment, following guidelines from the WHO that recommend that such antibiotics not be used when more than 5% *N. gonorrhoeae* strains showed resistance. Ceftriaxone that is injectable, along with oral azithromycin, is regarded as the last antibiotic remaining that is prescribed as first-line treatment, although various other antibiotics have also been evaluated for safety as well as to recognize efficacy for *N. gonorrhoeae* treatment. By evaluation it is predicted that new varieties of antibiotics are not available and antibiotics management is urgently required for preservation of efficacy.

Recent strategy for management aims to reduce the overall burden of *N. gonorrhoeae* infection by expansion of screening processes along with treatment of hosts, but its outcome showed uncertain resistance (Fingerhuth et al., 2016).

19.9 Future perspectives

19.9.1 Gene target medications to avoid antibiotic resistance

With new innovations in the medical field, gene therapy provides a new route of escape from disease without any severe effect on human body. The resistant properties of pathogen growth and failure in a number of latent novel antibiotics to recuperate natural merchandise. Most antibiotics are formed as a result of secondary metabolites formed by microorganisms or plants. To avoid perversity, an antibiotic manufacturer anchorages resistance gene habitually present within the identical biosynthetic gene cluster (BGC) accountable for engineering the antibiotic. Prevailing withdrawal tools are exceptional at spotting BGCs or unaffected genes in all-purpose, but offer little benefit in arranging and recognizing gene constellations for compounds active in contradiction of specific as well as novel targets. The Antibiotic Resistant Target Seeker stimulate target-focused genome removal methods, antibiotic gene knot prophecies, and "essential gene screening" to deliver an collaborating understanding of identified and reputed targets in BGCs (Alanjary et al., 2017).

19.9.2 Continuous change in nature of antibiotics

While it is understandable that unwarranted and impulsive use of antibiotics pointedly underscores the development of resistant straining, antibiotic resistance is similarly pragmatic in bacteria of inaccessible residences unlikely to be obstructed by human interposition. Together antibiotic biosynthetic genes as well as resistance conferring genes have been acknowledged to develop billions of eons since, long already irrefutable use of antibiotics. Hereafter it looks like antibiotics and antibiotics resistance factors have selected each other for a starring role in nature, which often escapes our attention. So novel types of antibiotics should be introduced in the market in order to avoid any continuous resistance from pathogens (Sengupta et al., 2013).

19.9.3 RNA-mediated gene silencing of pathogens

In spite of using drugs with various functional groups, why do we not look for drugs that cause gene silencing? Gene silencing by RNAi mediation is a new method of treating medication problems. Conventional antibacterial drug expansion has been futile in trying to replace the armamentarium struggling with the resistance problem, and novel resolutions are straightaway required. In bacteria, there are mesmerizing transformations of trans- and cis-antisense sequences analysis that inhibit translation. The monitoring regions inside a single mRNA can encompass adjacent cis-antisense sequences that produce an intramolecular antisense sense binding fold. The folded structure guides the ribosome-binding site (RBS), and starts CRISPR mechanism against resistance. RNA-mediated destruction of phages are copious, extensive, and intricate in basic cellular processes and adaptive responses (Stach and Good, 2011).

19.9.4 Nanomedicines

Despite a collection of dynamic antibiotics, bacterial contaminations, remarkably those fashioned by nosocomial bacteria, still persist as a prominent factor of indisposition and transience everywhere in the globe. So regardless of all existing therapies, the new era of technology demands new ways to remove hurdles of resistance. This can be done by using nanocarriers as a source of drug implication. Nanomedicine plays its role in the target killing of bacteria in a much more specified way. We hope for many improvements in the form of nanomedicines as an inventive tool for battling high degrees of antibiotic resistance. Antimicrobial activity of metal as well as metal oxide nanoparticles (NPs) has been comprehensively used. The microbes are eradicated either by microbicidal properties of the NPs, like the discharge of free metal ions terminating in cell membrane impairment, DNA interfaces, or free radical groups by microbiostatic effects together with assassination of the host's immune system (Hemeg, 2017).

References

Abbo, A., Navon-Venezia, S., Hammer-Muntz, O., Krichali, T., Siegman-Igra, Y., Carmeli, Y., 2005. Multidrug-resistant acinetobacter baumannii. Emerging Infectious Diseases 11 (1), 22.

Alanjary, M., Kronmiller, B., Adamek, M., Blin, K., Weber, T., Huson, D., Ziemert, N., 2017. The Antibiotic Resistant Target Seeker (ARTS), an exploration engine for antibiotic cluster prioritization and novel drug target discovery. Nucleic Acids Research 45 (W1), W42–W48.

Alberts, B., Johnson, A., Lewis, J., Raff, M., Roberts, K., Walter, P., 2002. The Molecular Genetic Mechanisms that Create Specialized Cell Types *Molecular Biology of the Cell*, fourth ed. Garland Science.

Almulhim, A.S., Alotaibi, F.M., 2018. Comparison of broad-spectrum antibiotics and narrow-spectrum antibiotics in the treatment of lower extremity cellulitis. International Journal of Health Sciences 12 (6), 3.

Aloush, V., Navon-Venezia, S., Seigman-Igra, Y., Cabili, S., Carmeli, Y., 2006. Multidrug-resistant *Pseudomonas aeruginosa*: risk factors and clinical impact. Antimicrobial Agents and Chemotherapy 50 (1), 43–48.

Alvarez-Martinez, C.E., Christie, P.J., 2009. Biological diversity of prokaryotic type IV secretion systems. Microbiology and Molecular Biology Reviews 73 (4), 775–808.

Ambur, O.H., Frye, S.A., Tønjum, T., 2007. New functional identity for the DNA uptake sequence in transformation and its presence in transcriptional terminators. Journal of Bacteriology 189 (5), 2077–2085.

Aminov, R.I., 2009. The role of antibiotics and antibiotic resistance in nature. Environmental Microbiology 11 (12), 2970–2988.

Australia, N. L. o., & Archive, A. G. W. Vaccine Preventable Diseases.

Avendaño, C., Menendez, J.C., 2015. Medicinal Chemistry of Anticancer Drugs. Elsevier.

Barlow, M., Hall, B.G., 2002. Phylogenetic analysis shows that the OXA b-lactamase genes have been on plasmids for millions of years. Journal of Molecular Evolution 55 (3), 314–321.

Berry, M., Gurung, A., Easty, D., 1995. Toxicity of antibiotics and antifungals on cultured human corneal cells: effect of mixing, exposure and concentration. Eye 9 (1), 110.

Bhattacharjee, M.K., 2016. Antimetabolites: Antibiotics that Inhibit Nucleotide Synthesis *Chemistry of Antibiotics and Related Drugs*. Springer, pp. 95–108.

Bhullar, K., Waglechner, N., Pawlowski, A., Koteva, K., Banks, E.D., Johnston, M.D., Wright, G.D., 2012. Antibiotic resistance is prevalent in an isolated cave microbiome. PLoS One 7 (4), e34953.

Boman, H.G., 1995. Peptide antibiotics and their role in innate immunity. Annual Review of Immunology 13 (1), 61–92.

Cantón, R., Morosini, M.-I., 2011. Emergence and spread of antibiotic resistance following exposure to antibiotics. FEMS Microbiology Reviews 35 (5), 977–991.

Carissa Stephens, R., CCRN, C.P.N., 2017. MRSA (Staph) Infection.

Carter, R., 2001. Transmission blocking malaria vaccines. Vaccine 19 (17–19), 2309–2314.

Cetinkaya, Y., Falk, P., Mayhall, C.G., 2000. Vancomycin-resistant enterococci. Clinical Microbiology Reviews 13 (4), 686–707.

Cherazard, R., Epstein, M., Doan, T.-L., Salim, T., Bharti, S., Smith, M.A., 2017. Antimicrobial resistant Streptococcus pneumoniae: prevalence, mechanisms, and clinical implications. American Journal of Therapeutics 24 (3), e361–e369.

Control, C.f.D., Prevention, 2013. Carbapenem-resistant Enterobacteriaceae (CRE).

Cossart, P., Roy, C.R., 2010. Manipulation of host membrane machinery by bacterial pathogens. Current Opinion in Cell Biology 22 (4), 547–554.

Davies, J., Davies, D., 2010. Origins and evolution of antibiotic resistance. Microbiology and Molecular Biology Reviews 74 (3), 417–433.

Džidić, S., Šušković, J., Kos, B., 2008. Antibiotic resistance mechanisms in bacteria: biochemical and genetic aspects. Food Technology and Biotechnology 46 (1).

Fajardo, A., Martínez, J.L., 2008. Antibiotics as signals that trigger specific bacterial responses. Current Opinion in Microbiology 11 (2), 161–167.

Farhadi, F., Khameneh, B., Iranshahi, M., Iranshahy, M., 2019. Antibacterial activity of flavonoids and their structure–activity relationship: an update review. Phytotherapy Research 33 (1), 13–40.

Ferretti, J., Stevens, D., Fischetti, V., 2016. *Streptococcus Pyogenes*: Basic Biology to Clinical Manifestations. University of Oklahoma Health Sciences Center, Oklahoma City (OK) (Chapter: Mechanisms of Antibiotic Resistance).

Ficht, T.A., 2011. Bacterial exchange via nanotubes: lessons learned from the history of molecular biology. Frontiers in Microbiology 2, 179.

Fingerhuth, S.M., Bonhoeffer, S., Low, N., Althaus, C.L., 2016. Antibiotic-resistant Neisseria gonorrhoeae spread faster with more treatment, not more sexual partners. PLoS Pathogens 12 (5), e1005611.

Gomez, J.E., Kaufmann-Malaga, B.B., Wivagg, C.N., Kim, P.B., Silvis, M.R., Renedo, N., Fishbein, S., 2017. Ribosomal mutations promote the evolution of antibiotic resistance in a multidrug environment. Elife 6, e20420.

Grundmann, H., Aires-de-Sousa, M., Boyce, J., Tiemersma, E., 2006. Emergence and resurgence of meticillin-resistant *Staphylococcus aureus* as a public-health threat. The Lancet 368 (9538), 874–885.

Hancock, R.E., 1997. Peptide antibiotics. The Lancet 349 (9049), 418–422.

Hemeg, H.A., 2017. Nanomaterials for alternative antibacterial therapy. International Journal of Nanomedicine 12, 8211.

Jacobs, M.R., 1999. Drug-resistant Streptococcus pneumoniae: rational antibiotic choices. The American Journal of Medicine 106 (5), 19–25.

Joshi, N., Milfred, D., 1995. The use and misuse of new antibiotics: a perspective. Archives of Internal Medicine 155 (6), 569–577.

Kahne, D., Leimkuhler, C., Lu, W., Walsh, C., 2005. Glycopeptide and lipoglycopeptide antibiotics. Chemical Reviews 105 (2), 425–448.

Kang, F., Zhu, S., Zhang, S., 1997. The toxicity of gentamicin on corneal cells in culture. Chinese Journal of Ophthalmology 33 (5), 366–369.

Kline, K.A., Fälker, S., Dahlberg, S., Normark, S., Henriques-Normark, B., 2009. Bacterial adhesins in host-microbe interactions. Cell Host & Microbe 5 (6), 580–592.

Kohanski, M.A., Dwyer, D.J., Collins, J.J., 2010. How antibiotics kill bacteria: from targets to networks. Nature Reviews Microbiology 8 (6), 423.

Krüger, N.J., Stingl, K., 2011. Two steps away from novelty—principles of bacterial DNA uptake. Molecular Microbiology 80 (4), 860–867.

Lambert, P.A., 2005. Bacterial resistance to antibiotics: modified target sites. Advanced Drug Delivery Reviews 57 (10), 1471–1485.

Lillington, J., Geibel, S., Waksman, G., 2014. Biogenesis and adhesion of type 1 and P pili. Biochimica et Biophysica Acta (BBA) – General Subjects 1840 (9), 2783–2793.

Mattick, J.S., 2002. Type IV pili and twitching motility. Annual Reviews in Microbiology 56 (1), 289–314.

Melville, S., Craig, L., 2013. Type IV pili in Gram-positive bacteria. Microbiology and Molecular Biology Reviews 77 (3), 323–341.

Nelson, D.W., Moore, J.E., Rao, J.R., 2019. Antimicrobial resistance (AMR): significance to food quality and safety. Food Quality and Safety 3 (1), 15–22.

Nepal, G., Bhatta, S., 2018. Self-medication with antibiotics in WHO Southeast asian region: a systematic review. Cureus 10 (4).

Novick, R.P., Christie, G.E., Penadés, J.R., 2010. The phage-related chromosomal islands of Gram-positive bacteria. Nature Reviews Microbiology 8 (8), 541.

Palomino, J., Martin, A., 2014. Drug resistance mechanisms in *Mycobacterium tuberculosis*. Antibiotics 3 (3), 317–340.

Perron, G.G., Whyte, L., Turnbaugh, P.J., Goordial, J., Hanage, W.P., Dantas, G., Desai, M.M., 2015. Functional characterization of bacteria isolated from ancient arctic soil exposes diverse resistance mechanisms to modern antibiotics. PLoS One 10 (3), e0069533.

Pizarro-Cerdá, J., Cossart, P., 2006. Bacterial adhesion and entry into host cells. Cell 124 (4), 715–727.

Pizarro-Cerdá, J., Cossart, P., 2009. Listeria monocytogenes membrane trafficking and lifestyle: the exception or the rule? Annual Review of Cell and Developmental 25, 649–670.

Raff, M., Alberts, B., Lewis, J., Johnson, A., Roberts, K., 2002. Molecular Biology of the Cell, fourth ed. National Center for Biotechnology InformationÕs Bookshelf.
Reynolds, P.E., 1989. Structure, biochemistry and mechanism of action of glycopeptide antibiotics. European Journal of Clinical Microbiology & Infectious Diseases 8 (11), 943–950.
Ribet, D., Cossart, P., 2015. How bacterial pathogens colonize their hosts and invade deeper tissues. Microbes and Infection 17 (3), 173–183.
Richardson, L.A., 2017. Understanding and overcoming antibiotic resistance. PLoS Biology 15 (8), e2003775.
Roberts, J.A., Marklund, B.-I., Ilver, D., Haslam, D., Kaack, M.B., Baskin, G., Normark, S., 1994. The Gal (alpha 1-4) Gal-specific tip adhesin of *Escherichia coli* P-fimbriae is needed for pyelonephritis to occur in the normal urinary tract. Proceedings of the National Academy of Sciences 91 (25), 11889–11893.
Sengupta, S., Chattopadhyay, M.K., Grossart, H.-P., 2013. The multifaceted roles of antibiotics and antibiotic resistance in nature. Frontiers in Microbiology 4, 47.
Stach, J.E., Good, L., 2011. Synthetic RNA silencing in bacteria–antimicrobial discovery and resistance breaking. Frontiers in Microbiology 2, 185.
Stapleton, P.D., Taylor, P.W., 2002. Methicillin resistance in *Staphylococcus aureus*: mechanisms and modulation. Science Progress 85 (1), 57–72.
Stingl, K., Müller, S., Scheidgen-Kleyboldt, G., Clausen, M., Maier, B., 2010. Composite system mediates two-step DNA uptake into *Helicobacter pylori*. Proceedings of the National Academy of Sciences 107 (3), 1184–1189.
Sun, J., Deng, Z., Yan, A., 2015. Bacterial multidrug efflux pumps: mechanisms, physiology and pharmacological exploitations (vol. 453, pg 254, 2014). Biochemical and Biophysical Research Communications 465 (1), 165-165.
Thenmozhi, S., Moorthy, K., Sureshkumar, B., Suresh, M., 2014. Antibiotic resistance mechanism of ESBL producing Enterobacteriaceae in clinical field: a review. International Journal of Pure and Applied Bioscience 2 (3), 207–226.
Tzialla, C., Borghesi, A., Perotti, G., Garofoli, F., Manzoni, P., Stronati, M., 2012. Use and misuse of antibiotics in the neonatal intensive care unit. Journal of Maternal-Fetal and Neonatal Medicine 25 (Suppl. 4), 27–29.
Uphoff, C.C., Drexler, H.G., 2011. Elimination of Mycoplasmas from Infected Cell Lines Using Antibiotics *Cancer cell Culture*. Springer, pp. 105–114.
Van Nguyen, J., Gutmann, L., 1994. Resistance to antibiotics caused by decrease of the permeability in gram-negative bacteria. Presse Medicale (Paris, France: 1983) 23 (11), 527–531, 522.
Vannuffel, P., Cocito, C., 1996. Mechanism of action of streptogramins and macrolides. Drugs 51 (1), 20–30.
von Wintersdorff, C.J., Penders, J., van Niekerk, J.M., Mills, N.D., Majumder, S., van Alphen, L.B., Wolffs, P.F., 2016. Dissemination of antimicrobial resistance in microbial ecosystems through horizontal gene transfer. Frontiers in Microbiology 7, 173.
Walsh, C., 2000. Molecular mechanisms that confer antibacterial drug resistance. Nature 406 (6797), 775.
Waterfield, N.R., Wren, B.W., 2004. Invertebrates as a source of emerging human pathogens. Nature Reviews Microbiology 2 (10), 833.
Watson, D.A., 1997. What is behind antibiotic resistance? Laboratory Medicine 28 (5), 324–327.
Wright, G.D., 2005. Bacterial resistance to antibiotics: enzymatic degradation and modification. Advanced Drug Delivery Reviews 57 (10), 1451–1470.
Yılmaz, Ç., Özcengiz, G., 2017. Antibiotics: pharmacokinetics, toxicity, resistance and multidrug efflux pumps. Biochemical Pharmacology 133, 43–62.
Yoneyama, H., Katsumata, R., 2006. Antibiotic resistance in bacteria and its future for novel antibiotic development. Bioscience Biotechnology & Biochemistry 70 (5), 1060–1075.

CHAPTER

Microbial risk assessment and antimicrobial resistance

20

Sarfraz Ahmed[1], Muhammad Ibrahim[2], Fiaz Ahmad[3], Hafsa Anwar Rana[2], Tazeen Rao[2], Wajiha Anwar[2], Muhammad Younus[4], Waqas Ahmad[5], Shahid Hussain Farooqi[5], Asma Aftab[6], Munawar Hussain[7], Muhammad Khalid[7], Ghulam Mustafa Kamal[7]

[1]*Department of Basic Siences, University of Veterinary and Animal Sciences, Narowal, Punjab, Pakistan;* [2]*Department of Biochemistry, Bahauddin Zakariya University, Multan, Pakistan;* [3]*Central Cotton Research Institute, Multan, Pakistan;* [4]*Department of Basic Sciences, University of Veterinary and Animal Sciences, Narowal, Punjab, Pakistan;* [5]*Department of Clinical Sciences, University of Veterinary and Animal Sciences Lahore, Narowal Campus, Narowal, Pakistan;* [6]*Research Centre for CO_2 Capture (RCCO2.C), Department of Chemical Engineering Universiti Teknologi PETRONAS Tronoh, Perak, Malaysia;* [7]*Department of Chemistry, Khwaja Fareed University of Engineering & Information Technology, Rahim Yar Khan, Pakistan*

20.1 Introduction

In the field of pathogenic microorganisms threats that are related to health of humans, risk evaluation is described as a part of risk investigation process; the other parts are risk management and risk communication (Commission, 1999). Risk management query is necessary to be described for the evolution of risk evaluation. This query specifies the pathogenic microorganisms' risk that is to be examined and with the interconnected result, for instance, the danger of human contagion. Description of risk query is the necessary part of the entire procedure. This query urges the identification of several factors, for example, the concerned country, time span for which the risk is examined, and abnormality of interest. When the risk query is identified, the manuscript framework is used to figure out the whole process that is necessary for identification and evaluation of the results of pathogenic microorganisms' hazard. The Codex Aliment Arius Commission (CAC), an organization of international standards for the trade of food stuffs at the international level, and EU Scientific Committee for Food stuffs used the framework of risk evaluation based on four steps as listed below (Carattoli et al., 2002): (1) risk identification; (2) exposure assessment; (3) hazard characterization; and (4) risk characterization.

(1) Risk identification is the initial step of the entire framework. In connection with resistance against antimicrobials, the microbiological risk can be a pathogenic microbe with developed resistance against a specific group or class of antimicrobial, for instance, vancomycin, which has developed resistance against *Enterococcus fascism*; quinolone, which has developed resistance against *Campylobacter*; or *Salmonella* Typhimurium, which has developed resistance against a number of drugs. Other feasible microbe-associated risks include existence of a resistant gene,

Antibiotics and Antimicrobial Resistance Genes in the Environment. https://doi.org/10.1016/B978-0-12-818882-8.00020-6

such as, CMY-2 gene, which is responsible for resistance to *Salmonella* (Rankin et al., 2002; Food and Organization, 2002).

(2) Disclosure assessment, which directs the estimation of rate of hazard occurring and the quantity of the hazard to which an animal or human is exposed. This obeys multiple steps, including development of pathways of exposure, assemblage of facts and figures, and progress of the disclosure model (Lammerding and Fazil, 2000). The widespread presence of hazard at each stage is modeled as the amount of risk (Buchanan et al., 2000).

(3) The results of animal or human hazard are examined in the third component of assessment of risk. which is hazard characterization. The results of hazard also take into account both extremity (such as, infection, ailment, demise) and time duration of harmful effects; although there are a number of aspects that affect the consequences, these aspects can be related to pathogenic microorganisms, food stuffs, and hosts. Dose−response modeling is a quantitative study−based approach that is used for the prediction of consequences after the ingestion of a certain quantity of hazard of interest and is typically generated from data from experimental feeding trials, outbreak investigations, or trials involving surrogate hosts or pathogens (Haas et al., 1999; Cassin et al., 1998).

(4) In the last step, characterization of risk, the integration of hazard characterization and exposure assessment is done, which are the components of risk assessment, to give an overall estimation of risk. Overall estimation of risk is sometimes quantified as probability of infection per human per year or also called annual number of cases per year. Risk assessment may be of two types, quantitative or qualitative. The process is the same for each type, i.e., first, the pathway of risk is identified, then the data is collected, and lastly the risk is assessed. In qualitative assessments risk is estimated in words, such as high, medium, or low. All the related data and information, including numerical data, are used in drawing a final result. Normally, qualitative assessment is done before the quantitative assessment and it requires some mathematical resources. The quantitative assessment is required or not as indicated by final results. Quantitative risk assessment biological processes are described by using different techniques of mathematics. So, the risk estimation is numerical. All the relevant data is used to build up the model. Mathematical data of microbial risk assessment is a direct result of data availability. Data-associated uncertainty and variability are the cause of complexity of model. Variability gives the description of natural variation of process (for instance, number of resistant pathogenic microbes per gram of the cattle waste material). Uncertainty is lack of knowledge, for instance, the consequence of small size of sample. Incorporated variability and uncertainty give variable and uncertain final risk estimation. So, this provides more information, such as, confidence intervals (Vose, 2008).Therefore, significant uncertainties may be associated with a model or data. Microbial risk assessment is used to collect the data to minimize the uncertainties. The degree to which the uncertainty is associated with the risk assessment parameters of the model can be quantified by uncertainty or sensitivity analysis. Thus, when inputs of most sensitive model have a high uncertainty, then more sources for data collection will be needed to generate a risk estimate with low uncertainty. Using the data collected from new sources will generate the model with low uncertainty covering the final outcome of risk estimate, thus giving more accurate and precise information for making the final decision. The required information sources for risk assessment are depicted graphically in Fig. 20.1, where each of the boxes represents a database or a set of tools that could be used for risk assessment.

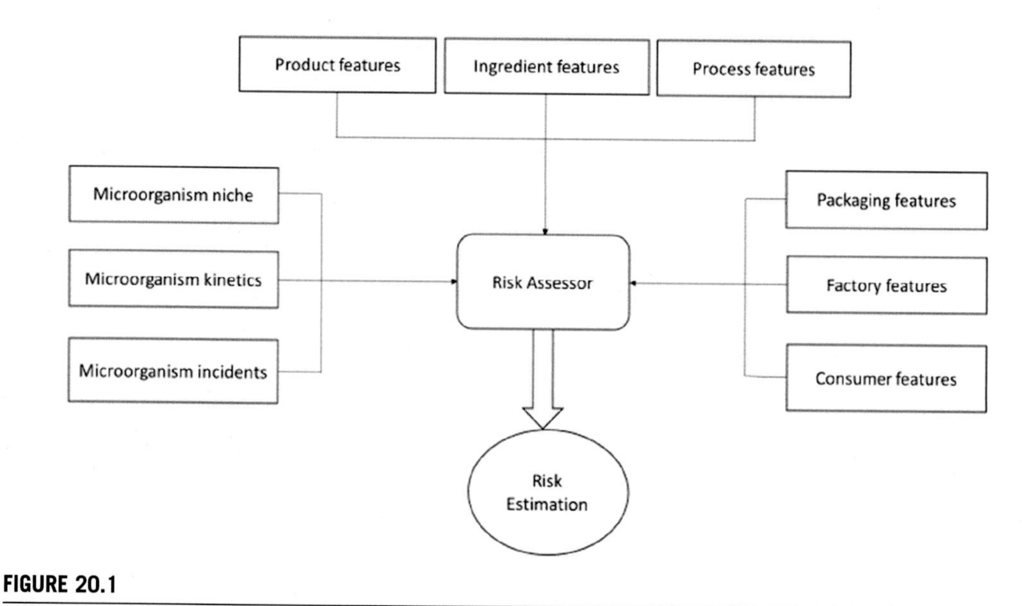

FIGURE 20.1

Sources of information for risk assessor.

Adapted from Nauta, M. J. 2002. Modelling bacterial growth in quantitative microbiological risk assessment: is it possible? International Journal of Food Microbiology, 73, 297–304.

20.2 Risk assessment of antimicrobial resistance in food safety

Food safety inspection–related microbial risk assessment (MRA) started in the 1800s (Brown and Stringer, 2002). MRA is an advanced, defined, and well-characterized process (Forsythe, 2008; Haas et al., 1999; Council, 1983).

This is the infancy period of antimicrobial resistance risk assessment (ARRA). The models that are proposed for the ARRA and MRA originated for health risk estimation of any hazard such as radiation or chemicals. These methods are developed for the pathways of environmental risks such as contaminated food–related illness. Exposure estimation of risk process is managing a framework that is provided for current health risk estimations (Council, 1983).The elements of risk evaluation/assessment described are identification of hazard, assessment of dose response, assessment of exposure, and characterization of risk (Council, 1994).The scientific rules that are the base of the paradigm remain recognizable across most of the health risk evaluations. Why does the ARRA model for food safety explain the generalized MRA for risk analysis of food safety? First, the risk of health due to use of antimicrobials in animals is inherently indirect (Salisbury et al., 2002). Second, it is difficult to characterize the nature of hazard as a single agent (Salisbury et al., 2002). Three correlated hazards evaluated are:

(i) Antibiotics,
(ii) Antibiotic-resistant bacteria,
(iii) Genetic determinants for antibiotic resistance (antibiotic resistance genes).

After exposure to such hazards, the harmful results are:

(i) The exposure of antibiotic-resistant pathogenic bacteria.
(ii) Spreading of antibiotic-resistant pathogenic bacteria, human exposure OR infection results.
(iii) Transmission of antibiotic-resistant genes (ARGs) to the other bacteria.

The list of feasible harmful effects must include the accumulative impact by the loss of therapeutic selection of antimicrobials, i.e., harmful effects expressed as social value. The concern is about therapeutic choice and is suggested in debate in food. Moreover, it increases risk assessment and management problems. It is clear that MRA and ARRA are the two overlapping processes. ARRA is not as simple as MRA, in which only a specific portion of interest (in bacteria) is resistant to specific antimicrobial. Also it includes the accumulative outcome from the loss of choice of antimicrobial drugs. This consequence extends the issues of both risk assessment and risk management that recommend that hazard is concept of using same antimicrobial drugs in humans and animals. Intelligibly, both MRA and ARRA are overlapping procedures, however, ARRA is not merely an MRA in which the portion of the bacteria of our interest is resistant to a specific antimicrobial drug (NRC, 1983).

20.2.1 Assessment of exposure

The process through which duration, intensity, and frequency of the human connection with harmful agents are determined is called assessment of exposure (Council, 1994; Barza and Travers, 2002). The assessment of exposure for any type of chemical, biological, or physical agent is the wide-ranging cognitive agreement in the stages of risk assessment. For example, the exposure assessment starts with the judgment of nanoscopic factors that may accelerate the liberation of the resistant pathogenic bacteria, followed by the conduction and fate of the resistant pathogenic bacteria in the macroscopical environmental routes "from farmstead to the fork," and ends with the nanoscopic factors that may increase the chances for the bacteria to colonize the humans. Both the MRA and the ARRA have the characteristic of transmissibility: the humans not only participate as unprotected population but moreover the part of fate and transport system in the exposure, for instance, contact of person-to-person could factor importantly in risk exposure of the medical drug-resistant commensal bacterial populations (Bezoen et al., 1999; Organization, 2014).

1. A concept-based model for the direct contaminated food consumption pathway of antimicrobial-repellent infections caused by pathogenic bacteria. The persons who eat the food contaminated with pathogenic microbes and are infected by different bacteria leading to the ailments of gastrointestinal tract (GIT). This mechanism depends on pathogens or microbes causing the infection. The last point of the antimicrobial repellent infection is sometimes not detected until the failure of treatment with a respective antimicrobial drug. This mechanism looks like MRA, in which pathogenic bacteria are of great importance. In ARRA the main focus is on drug repellent pathogens. This figure does not show competing human-oriented pathways of contamination and infection (Fig. 20.2).
2. A concept-based theoretical model for the indirect poisonous food-related antimicrobial-repellent hospital-acquired infection by the commensal bacterium is represented in Fig. 20.3. In this conceptual model, ill individuals might be hospitalized before colonization by a commensal

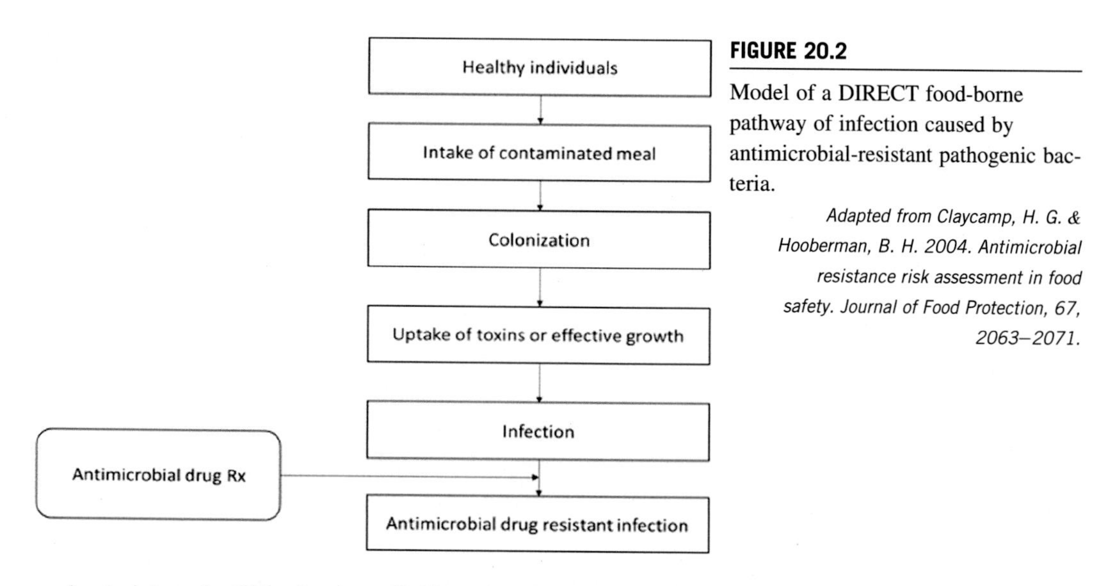

FIGURE 20.2

Model of a DIRECT food-borne pathway of infection caused by antimicrobial-resistant pathogenic bacteria.

Adapted from Claycamp, H. G. & Hooberman, B. H. 2004. Antimicrobial resistance risk assessment in food safety. Journal of Food Protection, 67, 2063–2071.

bacterial strain. Colonization of affected individuals may occur (1) by using contaminated food material (intake) or (2) through systemic self-contamination by contact with the respiratory tube or urinary tube. In another pathway, a healthy individual may be admitted to hospital just after

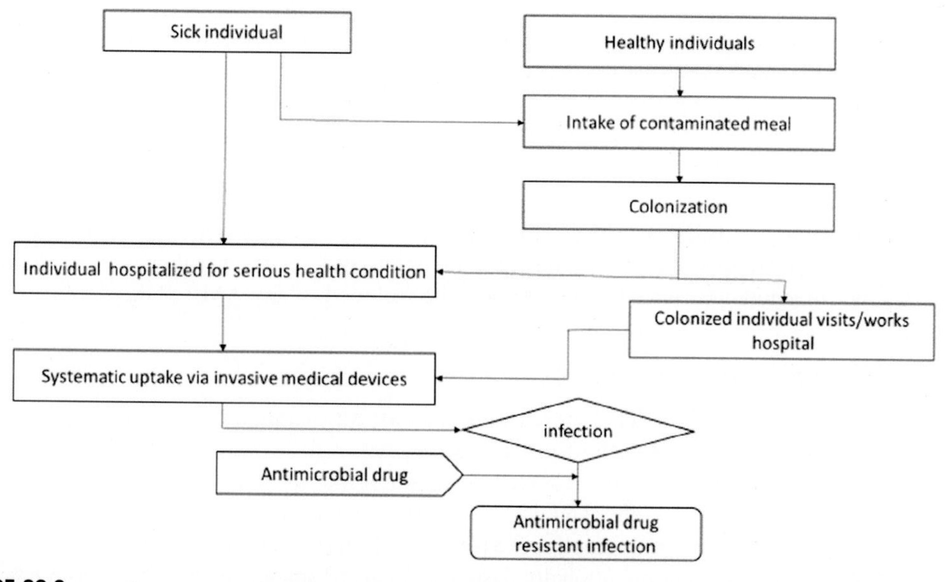

FIGURE 20.3

Model for an indirect contaminated food-related antimicrobial-repellent hospital acquired infection.

Adapted from Claycamp, H. G. & Hooberman, B. H. 2004. Antimicrobial resistance risk assessment in food safety. Journal of Food Protection, 67, 2063–2071.

colonization by coincidence. Staff of hospital and the visitors can also increase contamination to the medical appliances or can increase the possibility of infection.

20.3 Antimicrobial resistance risk assessment in water and sanitation

Antimicrobial resistance (AMR) has become the world's top health issue in the 21st century, but there is an indication of returning toward the preantibiotics due to the lack of serious and immediate awareness (Fehr et al., 2017). The incapability to cure infectious illness negatively affects the public health and ultimately increases the world's economic load (Su et al., 2017). After the identification of such disaster, the World Health Organization (WHO) designed a global action plan on AMR (GAP on AMR), which was later endorsed by the United Nations and followed by national AMR strategies design (Wi et al., 2017). The GAP on AMR is needed for the optimization of antimicrobial drug usage for the health of humans and animals, urging the development of rules and awareness for risk evaluation from the residues of antimicrobials and wastewater. It is notable that the GAP on AMR details the hazards and surging for the understanding of role of liberation of AMR microbes that are present in food materials, water, and environment and farm grounds. Thus for now, no specific policy has been advised regarding sewer water and reuse. The European Commission (EC) published "A European One Health Action Plan Against Antimicrobial Resistance (AMR)" just 2 years after WHO's publication of GAP on AMR.

In this innovative strategy, the EC is invited to talk about the lack of knowledge regarding AMR in environment and, specifically, its transmission. The EC calls the states to participate in research in Section 3.6:

1) The proliferation and spread of resistant microbes and antimicrobials in the environment
2) Risk evaluation strategies for the protection of animal health and human health
3) The designing of such techniques that can remove antimicrobials from used water and surrounding environment. The objective of strategy was the reduction in the proliferation of AMR. Expansion of the EC plan has the objective of this perspective as the EC plan is the one and only such major plan that is involved in recognition of many issues that need to be solved. The definition of sanitation is the treatment of human waste and it is important in the reduction of global load of infectious diseases that are caused by antibiotic resistant pathogenic microbes. Strong water or sanitation-related strategies must be included in any AMR plan.

20.3.1 Wastewater treatment plants

Wastewater treatment plants (WWTPs) are an important aspect of modern urban life, consisting of different processes for the degradation of organic waste, removal of phosphorus and nitrogen, and reduction of pathogen burden before the release of used water into the surrounding environment. The used water management systems gather and then treat different waste materials (Pruden, 2013). In the last century, WWTP's fundamental structure was improved and enhanced the water and sanitation management system that was not intentionally designed for AMR management. The interpretation of the destiny of antibiotic resistant bacteria (ABR) and ARGs via the water sanitation and cleanliness practices now has become severe for the rehabilitation of used water for irrigation and drinking purposes (Pruden et al., 2012).

20.3.2 Water treatment techniques and management strategies as barriers for AMR spreading

AMR risk associated with water and sanitation can be managed through the treatment of sewer water. Currently, extensive research is being done on the destiny of ABR and ARGs through the treatment of sewer water. The main challenge is to collect data and information that are accessible and informative to those who deal with water and its sanitation. Many risk-based and other AMR monitoring strategies have been designed. However, different studies involve different methods to evaluate the performance of the process. The effectiveness of wastewater treatment for decreasing the amplification of down-regulated AMR markers can be measured by effects of the water sanitation and sludge on surrounding environments, including clean water and soil (Amos et al., 2014; Burch et al., 2014; Czekalski et al., 2014; Ashbolt et al., 2018). The pathogens can cause different transformations like oxidation, adsorption, decomposition, and photo transformation in antimicrobial compounds (Huijbers et al., 2015).

20.4 Risk assessment of antibiotic resistance transmission through environment to humans

20.4.1 Transmission and entry of antibiotic-resistant bacteria to humans

The ABR can be transmitted from surroundings to human body via different ways, but particularly through exposure to contaminated water, air, food items, or via touch. However, the exact pathway of entry of ABR is not known. There are two logical ways for the consideration of their modes of entry. The first one is exogenous infections and the second is endogenous infections. In exogenous mode of action, the infection-causing agent directly enters the organ or tissue of the body. Whereas in the endogenous mode of action the infection-causing agent invades the organ or tissue and this is followed by the colonization of mucosa or skin. In such case the vectors can remain a part of the commensal population for a long time until the condition of the host allows them to proliferate or cause invasion. Fig. 20.4 represents some examples of ABR and their corresponding entry ways that they use to enter and cause colonization of the human body, though they don't cause illness. These examples are selected on the basis of their habitats and the place of infection with which they are associated (Huijbers et al., 2015).

20.4.2 Important steps for risk assessment

When somebody comes in contact with hazardous material (that can cause harm) then harm results. Risk is a combined measurement of both probability and the acuteness of harm. The risk assessment requires an approach that includes identification of hazard, characterization of hazard, assessment of exposure, and characterization of risk (Table 20.1). Risk assessment that is applied for issues of public health are sometimes very challenging. When it includes microbes and the genes of microbes this can seriously damage the accuracy of risk estimation process. Attempts are being made to improve food safety by making advancements in this field. The process of setting guidelines for the assessment of risks resulting from vectors' transmission from the surrounding to humans is in the developmental phase. The process of risk assessment depends upon a number of variables, many of which cannot be

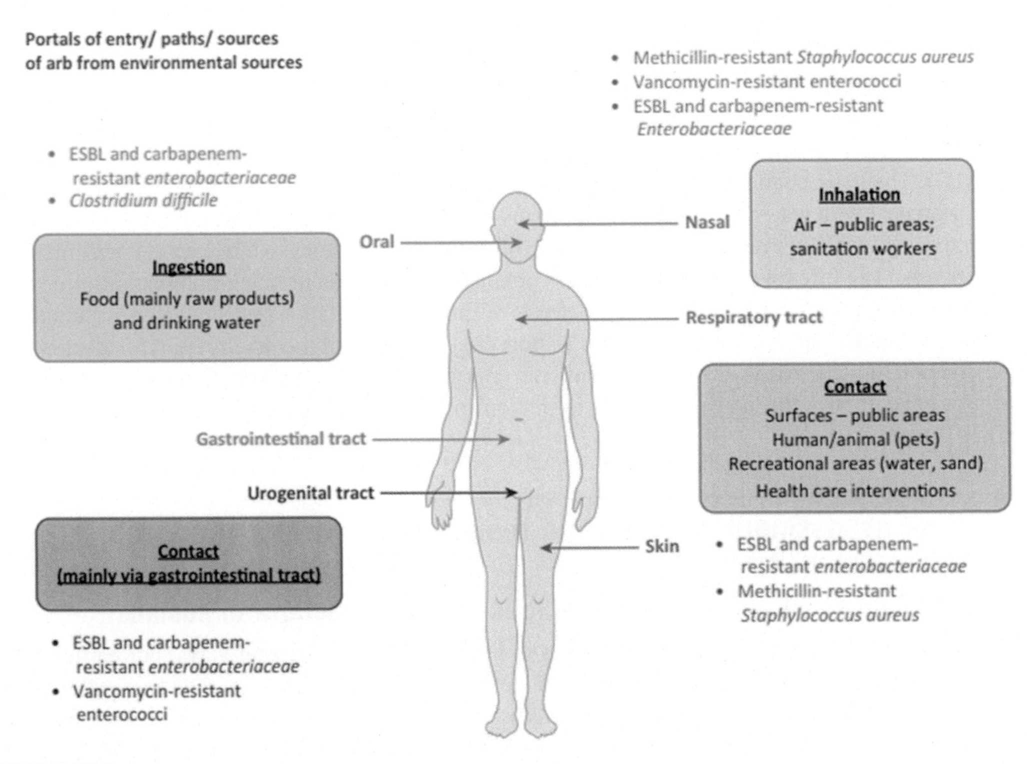

FIGURE 20.4

Entry and transmission of ABR to humans from surroundings.

Adapted from Manaia, C. M. 2017. Assessing the risk of antibiotic resistance transmission from the environment to humans: non-direct proportionality between abundance and risk. Trends in Microbiology, 25, 173–181.

quantified. Table 20.1 outlines the elements that are critical for the assessment of risk factors that are critical to assess the level of risk that is associated with the environment as a source of ABR transmission to the humans. In the subsections some parameters are discussed that must be identified to assess the risk (Huijbers et al., 2015).

20.4.2.1 Characterization of environmental compartment

In a specific part of environment, the identification and quantification of ABR and ARGs is a potential source of transmission of antibiotic resistance. The screening is reinforced by deep-rooted methods for the detection and characterization of ABR and ARGs, for example, quantitative PCR (qPCR) and metagenomic analysis. However, theoretically, if a specific vector bacterium is present in a sufficient quantity that is able to colonize a human, then the specific part of environment will have a chance of transfer to humans. Its presence may be risky even in small numbers depending on the vector type transmitting it. Moreover, the level to which a specific part of environment is exposed is used to measure the probability that the specific vector bacteria will colonize humans (product of exposure and infective dose) (Huijbers et al., 2015).

Table 20.1 Critical factors that determine the level of risk associated with antibiotic-resistant bacteria (ABR) and antibiotic-resistance genes (ARGs) in the environment (Manaia, 2017).

Risk assessment steps	Description about steps	Critical factors
Identification of hazard (A) Is a potential source for ABR and ARGs	Identification of ABR and ARGs related to contaminant resistome, i.e., related to human activities	Receptors of human or animal environment having the bacteria for proliferation and horizontal gene transfer Direct or indirect contact with the environment (e.g., wastewater, soil irrigation with wastewater, entertainment areas, urban wildlife etc.
Hazard characterization (B) The potential source is identified in A, which nurtures ABR to colonize humans, i.e., vector?	Qualitative or quantitative evaluation of ABR occurrence is able to colonize human and the assessment of associated adverse effects	ABR are able to colonize, proliferate, and invade (in tissues, organs of the human body) ABR can combat last defense antibiotics or multidrug resistant (MDR)
Exposure assessment (C) How probable is the colonization of vectors identified in B?	The qualitative or quantitative evaluation of probability that vectors (from environmental sources) may affect somebody	Whether the degree of exposure to source is identified in A is very high Whether the bacteria identified in B are highly communicable (by water, food, air, fomites, person to person, pests) A high capability to colonize human at a very low infectious dose
Risk characterization (D) How dangerous is the risk? $D = A \times B \times C$	Estimation of adverse outcomes that are likely to occur, based on the combination of hazard characterization and exposure assessment	

20.4.2.2 Characterization of carrier bacteria

Carrier bacteria are another important component for risk assessment; level of strain and species identification may be required for good characterization, to reinforce prediction of related features. For example, related features that are required for risk assessment are capability of a species member that can colonize and affect the human (respective organ or tissue), typical transmission pathways to humans, and the method of entrance to the host body (Fig. 20.3). Features that are related to the genetic framework can also vary from strain to strain; include capability to mobilize and come into the ARG (site of phage modification, transposons, or carrier) and ARG pool that already exist. If the quantity of carrier bacterium is unknown that can colonize or pass infection to humans, then all the information has no value. For pathogenic microbes, this quantity is known as infection -causing dose and this corresponds to the number of bacteria that are required to cause infection in the host body (Leggett et al., 2012). All these parameters are important for a good model of risk assessment. This is why the infection-causing dose and the means of transmission of many ABR (e.g., *Pseudomonas aeruginosa*, *Klebsiella pneumoniae*, *Acinetobacter baumannii*, *Enterococcus faecium*, or *Enterococcus faecalis*) are not still known. This is known for few bacteria groups, such as for *Staphylococcus aureus* (>105 cells), for *Mycobacterium tuberculosis* (>10 cells), or for enterohemorrhagic strains of *Escherichia coli* (>10 cells) (Schmid-Hempel and Frank, 2007; Fewtrell and Kay, 2015). The available values of infection-causing dose suggest that often a very minute amount of vectors could be able to harm humans. The dose of vector bacteria that could start colonization in a host can be extremely low. However, infection-causing dose depends on the mechanism of pathogenicity invading the immune system (Schmid-Hempel and Frank, 2007); in case of ABR-related vectors, the factors, for example, the capability of colonization and the potential for the horizontal gene transfer can act as determinants. Sometimes the individuals with maximum ABR infection risk have compromised the primary barriers of their immune system. It is concluded from these arguments that for some of the vectors the quantity that could endanger the health of a host immediately or after a long time could be very minute.

20.4.2.3 Quantification of ABR and ARGs: a limitation for accuracy of risk assessment

Imagine that the infection-causing dose for all clinically related vectors is known, then the next question will be how. In the first step it is essential to collect quantity-based data related to ABR of interest, and this is the basis for the estimation of exposure assessment as this would be the basis for estimating the exposure assessment value (Table 20.1). One of the limitations of this feature is that the sensitivity and specific nature of presently used methods like qPCR and metagenomics are not sufficient for the detection of a specific carrier. The usage of culture-based or selective probing (e.g., fluorescence in situ hybridization techniques) might be required as a complementary approach. For such type of assessments it is not sufficient to have values of relative abundance or prevalence, in most cases it is shown in metagenomics-based or quantitative PCR-based studies (Christgen et al., 2015; Narciso-da-Rocha et al., 2014). The absolute value is required for risk estimation, not the value of relative abundance. It may be possible that any list of microbial cells is shown in the form of average value. Sometimes bacteria found in the form of aggregates can reach their clinically relevant dose, although their average abundance is low, i.e., below the infection-causing dose. The arguments show that risk produced by the environment could be greater than the expected level.

20.5 Antimicrobial resistance risk assessment in environment

20.5.1 Environmental transmission pathways

Antibiotic-resistant bacteria, like carbapenemase-producing Enterobacteriaceae, extended-spectrum β-lactamase-producing Enterobacteriaceae, methicillin-resistant *S. aureus*, and vancomycin-resistant *Enterococcus* spp., have been detected in different parts of environment, including soil, water, and air (Walsh et al., 2011; Ahammad et al., 2014; Kümmerer, 2009). Remnants of antibiotics, such as aminoglycosides, fluoroquinolones, macrolides, and tetracyclines, have been broadly identified (Woappi et al., 2016). In the environment antibiotic resistance may arise from animals, humans, or the surrounding environment (Bondarczuk et al., 2016). Fecal waste, sewer water, livestock farming, and wild animals are the main sources of contamination of ABR and ARGs in the environment. The combined liberation of both antibiotics and resistant bacteria makes the assessment of the role that the unconsumed antibiotics play in the generation of resistance in the surrounding environment difficult. Shifting of pathways of transmission of humans and animals toward aquatic and terrestrial pollution, respectively, is a complete distortion. The widespread use of sewage sludge as a fertilizer of soil plays a major role in spread of resistant genes, bacteria, and antibiotics (Larsson, 2014). Direct discharge from antibiotic manufacturing industries contaminates the surrounding environment with a high level of antibiotics and antibiotic resistance in microbes (Bengtsson-Palme and Larsson, 2015).

20.5.2 Risk analysis process

Risk analysis is defined in the OIE Code as "The process composed of hazard identification, risk assessment, risk management and risk communication." This term is often used to outline the whole procedure of adequately solving a risk matter. It includes both assessing and managing the risk jointly with proper communication between risk evaluators and collaborators (Vose et al., 2001). A classic risk analysis process proceeds as:

a) Risk managers establish the framework of policy and outline the types of risk to be solved, understanding and ranking the risk among other issues. In meeting with experts and risk assessors, an outline for assessing the risk is formed. The framework of policy also explains the different risk management options that might be taken into the account by the administrative and guiding framework. The policy must also explain the process of risk evaluation and quantification and the risk acceptance level.

b) Risk manager identifies the risk issue and takes possible actions to resolve it.

c) In a meeting with risk evaluators, experts, and other stakeholders, a plan is made for the preassessment of the risk giving a solution of how the risk is to be evaluated.

d) Risk evaluators carry out a prequalitative assessment (scoping study) and advise the management to make quantitative risk assessment feasible.

e) From the scoping study of risk, the managers make a determination about the severity of risk. When the risk is feasible and sufficient then the risk managers direct the risk assessors for complete risk assessment (both qualitative and quantitative).

f) The risk assessment report is reviewed at many stages until the refined report of risk assessment has been made, then can be accessed by the public. Particularly, this feature of the process is very important for ensuring the transparency and efficiency of the whole process.

g) Then the final results of risk assessment are used by the risk manager for the determination of a well-defined strategy. Moreover, the suitable actions are taken for risk management.
h) Then the risk management decision is presented to the public with maximum feasible clarity by the regulatory authority.
i) The next step must be the implementation and organization of these decisions by the risk manager. Moreover, other steps (follow-up) are taken for the evaluation of effects of these decision as compared to assumed results.
j) Data that is achieved by follow-up has to be assessed for further feasible modification in risk analysis strategy (Vose et al., 2001).

20.6 Methods and management of risk assessment in relation to antimicrobial resistance

20.6.1 Policy of risk regulation

The risk regulation policy is a current expression explained as "the framework of regulatory policy for monitoring, assessing, managing and measuring the risks that are involved in the antimicrobials usage." An unfavorable ancestor to the risk investigation process is the evolution and public details of such a strategy structure. The strategic structure suggests the policy skeleton which explains the vision at the back of control and controlling risks concern among the usage of antimicrobials into food-producing animals. This also explains the involvement of risk evaluation among the vindication of modern drug use, the various restrictions of over usages and unfavorable impact on usage of the drugs (Vose et al., 2001).

The policy frame can also additionally tackle the extra value regarding definitive antimicrobial drugs for which there are no positive choice therapies available to treat infectious diseases. Furthermore, to give an explanation for the spread over risk reduction movements, management may choose legislative yet regulatory restrictions. The frame may seek to give an explanation for the impact over doubt concerning the risk management decision. (Vose et al., 2001).

The policy skeleton addresses measure of a long-term impact, or may additionally do some cut-off duration and discount aspects to acknowledge the decreased value of a drug remedy. However, the policy skeleton does not always avoid risk management out of thinking about potent risk management preferences that might also be outside the modern area of the regulatory authority (Vose et al., 2001). The prerequisites of policy permit the pharmaceutical industries and or the health care expert bodies to take an advance look at the prediction of the modern or future antimicrobial products with clearly gained objectives. Policy skeleton ensures transparency at some stage in the risk management section in evaluation of a risk analysis. People may function according to the policy skeleton to reduce the risk at major level (Vose et al., 2001).

20.6.2 Components of risk management

Risk administration is performed by risk managers who hold a comprehensive knowledge about the policy and technical history in assessing risk. The definition according to OIE is that the risk management consists of the following steps:

1. Risk evaluation

   The system estimating the risk including the stage of safety concerns.

2. Option evaluation

   The system identifies evaluation probability, measures within the discipline in conformity with limits of the risk-related importation and degree of safety. The working efficiency is the quantity according to which an option reduces the possibility and/or magnitude regarding unfavorable biological, yet financial, consequences. Evaluation of working efficiency about the choices elected is an iterative method that includes their incorporation among the risk evaluation observed. The assessment for probability usually focuses on technical, operational, and financial elements affecting the implementation about the risk administration options (Vose et al., 2001).

3. Execution

   The standard procedure is followed to address the risk administration and selection and make sure that management measures have been put into place (Vose et al., 2001)

4. Examining and analysis

   The ceaseless procedure by means of which the risk administration measures are constantly audited in accordance with insuring that they are attaining the outcomes intended (Vose et al., 2001).

20.7 Risk evaluation and regulation

20.7.1 Risk evaluation

Risk evaluation is described within the OIE Code as follows: "The assessment on the probability or the biological and financial penalties about entry, establishment, and thoroughness concerning a pathogenic viceregent inside the tract regarding an importing country." There is a range of techniques in imitation of assessing the magnitude of a hazard and then the cost of potent risk reduction options. These may be largely classified in three types: qualitative, semi-quantitative, and quantitative hazard assessments. Whatever strategy is taken, the risk evaluation ought to be designed in accordance with addressing the particular query posed through the risk managers. The risk evaluation technique is generally subdivided into four components: risk release evaluation, exposure evaluation, final result evaluation, and risk approximation (Vose et al., 2001) (Fig. 20.5).

20.7.1.1 Release evaluation

It is defined into the OIE Code as "Description of the organic pathways essential for the usage of an antimicrobial among animals in accordance with its release and resistance determinants of a specific environment, or estimation of the likelihood on which the whole procedure is happening both qualitatively and quantitatively" (Vose et al., 2001).

20.7.1.2 Exposure evaluation

It is defined in the Office International des Epizooties Code as "explaining the organic routes essential because of exposure on animals and human beings in conformity with the risks launched out of a partial source, or estimating the chance on the revelation occurring, both qualitatively and quantitatively" (Vose et al., 2001).

20.7.1.3 Final result evaluation

It is defined into the OIE Code as like "Description on the alliance among particular exposures in conformity with an organic agent and the consequences regarding these exposures." A causal procedure needs to explore the risks exposures damaging health and environmental entities, which might also result in socio-economic consequences. The end result evaluation describes the potent consequences on a partial exposure or estimates the chance of them occurring. This calculation can also be both qualitative and quantitative (Vose et al., 2001).

20.7.1.4 Risk approximation

It is defined into the OIE Code as follows "Integration over the consequences from the release evaluation, exposure evaluation, and final result evaluation to produce standard measures about risks related together with the risks recognized at the outset. Thus, risk determination takes into score the complete risk course from risk recognized followed by undesirable outcome.". The coverage framework will supply suggestions in accordance with the hazard assessors as to how to investigate the whole impact on someone's risk problem and hazard reducing strategies. For example, disposing of an antimicrobial from veterinary usage might also mean using another antimicrobial in its place and

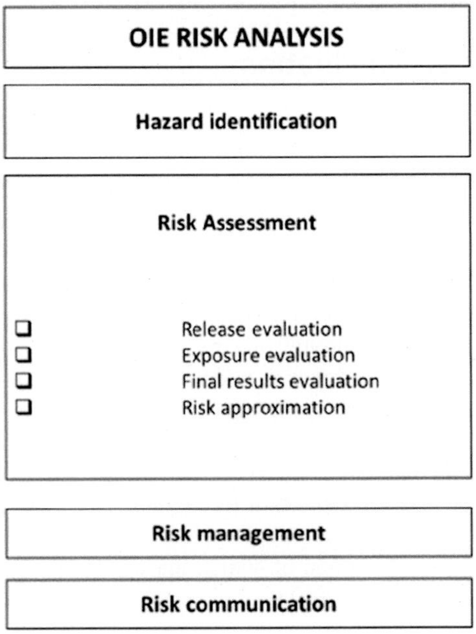

FIGURE 20.5

OIE model of risk analysis.

Adapted from Claycamp, H. G. & Hooberman, B. H. 2004. Antimicrobial resistance risk assessment in food safety. Journal of Food Protection, 67, 2063–2071.

perhaps with worse consequences. Despite minor impacts, whether positive and negative, being addressed, the risk administration approach may be sub-optimal (Vose et al., 2001). The preliminary levels of planning in risk evaluation can be described as below:

1) The risk problem into query is formally expressed following to check that either the members agree upon as the trouble to be addressed. The potent mechanisms or pathways by which the risk may result in an adverse impact are additionally described. The risk evaluation team understands the problem which can further be defined using some extra flow diagrams. At that point, the layout is purely conceptual and therefore there is no need for data. The motive diagrams are placed with centeral concept to make facts standing useful as feasible risk administration choices exist in general system. It can be beneficial with a wide participation of stakeholders and applicable experts (Vose et al., 2001).
2) A preliminary information search is carried out in imitation of assessing elements concerning the system that may be appropriately quantified. Components would possibly include, for example: the occurrence of resistant microorganisms within feces, water, and carcasses; the parceling via animal species, time, or geographical region of usage concerning an antimicrobial, the frequency regarding the usage of the antimicrobial among human medicinal drug, and the fitness status of those taking the antimicrobial. At that stage, that is enough to be aware of about the presence of data. Requests for data up to expectation may assist in quantifying the elements of the system, done in accordance with stakeholders or applicable experts.

The useful data that is not available immediately should be given strong consideration, but with the help of some research efforts that could become accessible within a suitable period. The meaning that constitutes a realistic duration will mirror the immanency and clarity about the risk problem in question. It might also be suitable to think about finishing a risk evaluation swiftly to assist selection makers perceive the instant movements, recognizing a reassessment about the risk problem when more data is accessible to change the preliminary moves that had been taken (Vose et al., 2001).

3) A comment of the system such as perceived via the risk evaluation team, collectively from the facts accessible to quantify the elements of that provision that can supply essential guidance. It may explain that risk administration options can be precise assessed because of their effectiveness. It may additionally guide the hazard assessor related to the manufacturing as quantitative risk assessment, so that amount would remain primarily based on information of whether a certain a model ought to be validated in some way or not. It is the mixture of possible risk administration options which include information that ought to be accessible and to be examined for those options for the description of risk evaluation. If the system is no longer sufficiently well understood and or inadequate information is available to meaningfully quantify the model, this may lead to produce a characteristic risk analysis. However, quantification on definitive factors of the system can also additionally be possible, which ought to allow the assessment of a confirmed quantity on risk administration options (Vose et al., 2001).

The risk evaluation model may remain as easy as feasible to assist the extent regarding risk administration choices being considered. The model shape can also no longer consist of a whole access evaluation about the risk situation. Flexibility among the strategy according to modeling will decrease the endeavor required to outturn the evaluation and to control the wide variety or type of assumptions that can also be done using the model. However, the model may additionally no longer be beneficial in addressing mean questions that occur upon the identical risk problem and can also no longer assist other stakeholders to make a contribution to successfully manage the risk. It can also be tough to show propriety among models where distinct model constructions have been ancient collectively with pretty exclusive assumptions (Vose et al., 2001).

20.8 Future perspectives

MRA always requires the data that can remain as of much specific types. Although many researchers seek advice from hazard assessors before a project starts, nevertheless there exists ambiguity related to the sort of information that is required for MRA. Unfortunately, for antimicrobial resistance data there is no specific list on the basis of genera because the needed data is based on the risk question. However, to increase the demands, the means of data published should be improved (Pleydell and Migura, 2002).

From the discussion of many researchers, it is obvious that many of them would consider their data collection to be perfect for usage in MRA. However, besides suitable data, the developments regarding MRAs may be hindered by several factors. Some data boundaries recognized are no longer solely by means of risk assessors, but also additionally via the research community of the antimicrobial resistance. Implications of certain information barriers should also be addressed between disciplines to motivate consideration. However in partial situations, microbial risk assessments are being introduced along with laboratory or research work that is field-based (Snary et al., 2003).

20.9 Conclusion

In the 21st century, the pool of microorganisms that are antimicrobial resistant in humans, environment, and the food-producing species remains a problem. However, scientific investigations of this issue as in other areas of human health and animals may be eased by usage of microbial risk assessment. Although various information about antimicrobial hindrance may be difficult to obtain, challenging to collate, unsure, or not documented thoroughly. Industry and policy-makers should recognize the deficiencies in quality information connected with MRAs and its hazards of vast uncertainties. Ideally, risk evaluation has to be an iterative process; importantly is that it is capable to discover information desired in order to remain routinely up to date about the latest information. A descriptive excellent strategy may excel the statistics for the model to get generated in a best way to attain information, however, it should importantly focus on research.

References

Ahammad, Z.S., Sreekrishnan, T., Hands, C., Knapp, C., Graham, D., 2014. Increased waterborne bla NDM-1 resistance gene abundances associated with seasonal human pilgrimages to the Upper Ganges River. Environmental Science & Technology 48, 3014–3020.

Amos, G., Hawkey, P., Gaze, W., Wellington, E., 2014. Waste water effluent contributes to the dissemination of CTX-M-15 in the natural environment. Journal of Antimicrobial Chemotherapy 69, 1785–1791.

Ashbolt, N., Pruden, A., Miller, J., Riquelme, M., Maile-Moskowitz, A., 2018. Antimicrobial resistance: fecal sanitation strategies for combatting a global public health threat. In: Pruden, A., Ashbolt, N., Miller, J. (Eds.), Global Water Pathogens Project, Part 3 Bacteria. Unesco, Michigan State University, E. Lansing, MI. http://www.waterpathogens.org.

Barza, M., Travers, K., 2002. Excess infections due to antimicrobial resistance: the "Attributable Fraction". Clinical Infectious Diseases 34, S126–S130.

Bengtsson-Palme, J., Larsson, D.J., 2015. Antibiotic resistance genes in the environment: prioritizing risks. Nature Reviews Microbiology 13, 396.

Bezoen, A., VAN Haren, W., Hanekamp, J., 1999. Emergence of a Debate: AGPs and Public Health. *Human Health and Growth Promoters (AGPs) Reassessing the Risk*. Heidelberg Appeal Foundation, Amsterdam, The Netherlands. *ISBN*, 90-76548.

Bondarczuk, K., Markowicz, A., Piotrowska-Seget, Z., 2016. The urgent need for risk assessment on the antibiotic resistance spread via sewage sludge land application. Environment International 87, 49–55.

Brown, M., Stringer, M., 2002. Microbiological Risk Assessment in Food Processing. Woodhead Publishing.

Buchanan, R.L., Smith, J.L., Long, W., 2000. Microbial risk assessment: dose-response relations and risk characterization. International Journal of Food Microbiology 58, 159–172.

Burch, T.R., Sadowsky, M.J., Lapara, T.M., 2014. Fate of antibiotic resistance genes and class 1 integrons in soil microcosms following the application of treated residual municipal wastewater solids. Environmental Science & Technology 48, 5620–5627.

Carattoli, A., Tosini, F., Giles, W., Rupp, M.E., Hinrichs, S.H., Angulo, F., Barrett, T., Fey, P.D., 2002. Characterization of plasmids carrying CMY-2 from expanded-spectrum cephalosporin-resistant Salmonella strains isolated in the United States between 1996 and 1998. Antimicrobial Agents and Chemotherapy 46, 1269–1272.

Cassin, M.H., Lammerding, A.M., Todd, E.C., Ross, W., Mccoll, R.S., 1998. Quantitative risk assessment for *Escherichia coli* O157: H7 in ground beef hamburgers. International Journal of Food Microbiology 41, 21–44.

Christgen, B., Yang, Y., Ahammad, S., Li, B., Rodriquez, D.C., Zhang, T., Graham, D.W., 2015. Metagenomics shows that low-energy anaerobic– aerobic treatment reactors reduce antibiotic resistance gene levels from domestic wastewater. Environmental Science & Technology 49, 2577–2584.

Claycamp, H.G., Hooberman, B.H., 2004. Antimicrobial resistance risk assessment in food safety. Journal of Food Protection 67, 2063–2071.

Commission, C.A., 1999. Principles and Guidelines for the Conduct of Microbiological Risk Assessment. CAC/GL-30.

Council, N.R., 1983. Risk Assessment in the Federal Government: Managing the Process. National Academies Press.

Council, N.R., 1994. Science and Judgment in Risk Assessment. National Academies Press.

Czekalski, N., Díez, E.G., Bürgmann, H., 2014. Wastewater as a point source of antibiotic-resistance genes in the sediment of a freshwater lake. The ISME Journal 8, 1381.

Fehr, A., Lange, C., Fuchs, J., Neuhauser, H., Schmitz, R., 2017. Health Monitoring and Health Indicators in Europe.

Fewtrell, L., Kay, D., 2015. Recreational water and infection: a review of recent findings. Current environmental health reports 2, 85–94.

Food, Organization, A., 2002. Risk Assessments for Salmonella in Eggs and Broiler Chickens: Microbiological Risk Assessment Series. World Health Organization.

Forsythe, S.J., 2008. The Microbiological Risk Assessment of Food. John Wiley & Sons.

Haas, C.N., Rose, J.B., Gerba, C.P., 1999. Quantitative Microbial Risk Assessment. John Wiley & Sons.

Huijbers, P.M., Blaak, H., DE Jong, M.C., Graat, E.A., Vandenbroucke-Grauls, C.M., De Roda Husman, A.M., 2015. Role of the environment in the transmission of antimicrobial resistance to humans: a review. Environmental Science & Technology 49, 11993–12004.

Kümmerer, K., 2009. The presence of pharmaceuticals in the environment due to human use–present knowledge and future challenges. Journal of Environmental Management 90, 2354–2366.

Lammerding, A.M., Fazil, A., 2000. Hazard identification and exposure assessment for microbial food safety risk assessment. International Journal of Food Microbiology 58, 147–157.

Larsson, D.J., 2014. Pollution from drug manufacturing: review and perspectives. Philosophical Transactions of the Royal Society B: Biological Sciences 369, 20130571.

Leggett, H.C., Cornwallis, C.K., West, S.A., 2012. Mechanisms of pathogenesis, infective dose and virulence in human parasites. PLoS Pathogens 8, e1002512.

Manaia, C.M., 2017. Assessing the risk of antibiotic resistance transmission from the environment to humans: non-direct proportionality between abundance and risk. Trends in Microbiology 25, 173–181.

Narciso-Da-Rocha, C., Varela, A.R., Schwartz, T., Nunes, O.C., Manaia, C.M., 2014. bla_{TEM} and vanA as indicator genes of antibiotic resistance contamination in a hospital–urban wastewater treatment plant system. Journal of global antimicrobial resistance 2, 309–315.

Nauta, M.J., 2002. Modelling bacterial growth in quantitative microbiological risk assessment: is it possible? International Journal of Food Microbiology 73, 297–304.

Nrc, U., 1983. Risk Assessment in the Federal Government: Managing the Process, vol. 11. National Research Council, Washington DC, p. 3.

Organization, W.H., 2014. Antimicrobial Resistance: Global Report on Surveillance. World Health Organization.

Pleydell, E., Migura, L.G., 2002. Investigation of Persistence of Antimicrobial Resistant Organisms in Livestock Production. The International Conference on Antimicrobial Agents in Veterinary Medicine (AAVM).

Pruden, A., 2013. Balancing Water Sustainability and Public Health Goals in the Face of Growing Concerns about Antibiotic Resistance. ACS Publications.

Pruden, A., Arabi, M., Storteboom, H.N., 2012. Correlation between upstream human activities and riverine antibiotic resistance genes. Environmental Science & Technology 46, 11541–11549.

Rankin, S.C., Aceto, H., Cassidy, J., Holt, J., Young, S., Love, B., Tewari, D., Munro, D.S., Benson, C.E., 2002. Molecular characterization of cephalosporin-resistant *Salmonella enterica* serotype newport isolates from animals in Pennsylvania. Journal of Clinical Microbiology 40, 4679–4684.

Salisbury, J.G., Nicholls, T.J., Lammerding, A.M., Turnidge, J., Nunn, M.J., 2002. A risk analysis framework for the long-term management of antibiotic resistance in food-producing animals. International Journal of Antimicrobial Agents 20, 153–164.

Schmid-Hempel, P., Frank, S.A., 2007. Pathogenesis, virulence, and infective dose. PLoS Pathogens 3, e147.

Snary, E., Kelly, L.A., Clifton-Hadley, F., Liebana, E., Wooldridge, M., Reid, S., Threlfall, J., Lindsay, E., Hutchison, M., Davies, R., 2003. Assessing the risk of the transfer of antimicrobial resistance genes between bacteria in stored and spread farm wastes. Research in Veterinary Science 74, 5.

Su, J.-Q., An, X.-L., Li, B., Chen, Q.-L., Gillings, M.R., Chen, H., Zhang, T., Zhu, Y.-G., 2017. Metagenomics of urban sewage identifies an extensively shared antibiotic resistome in China. Microbiome 5, 84.

Vose, D., 2008. Risk Analysis: A Quantitative Guide. John Wiley & Sons.

Vose, D., Acar, J., Anthony, F., Franklin, A., Gupta, R., Nicholls, T., Tamura, Y., Thompson, S., Threlfall, E., Van Vuuren, M., 2001. Antimicrobial resistance: risk analysis methodology for the potential impact on public health of antimicrobial resistant bacteria of animal origin. Revue Scientifique et Technique-Office International des Epizooties 20, 811–819.

Walsh, T.R., Weeks, J., Livermore, D.M., Toleman, M.A., 2011. Dissemination of NDM-1 positive bacteria in the New Delhi environment and its implications for human health: an environmental point prevalence study. The Lancet Infectious Diseases 11, 355–362.

Wi, T., Lahra, M.M., Ndowa, F., Bala, M., Dillon, J.-A.R., Ramon-Pardo, P., Eremin, S.R., Bolan, G., Unemo, M., 2017. Antimicrobial resistance in Neisseria gonorrhoeae: global surveillance and a call for international collaborative action. PLoS Medicine 14, e1002344.

Woappi, Y., Gabani, P., Singh, A., Singh, O.V., 2016. Antibiotrophs: the complexity of antibiotic-subsisting and antibiotic-resistant microorganisms. Critical Reviews in Microbiology 42, 17–30.

CHAPTER

Environmental risk assessment of antibiotics and AMR/ARGs

21

Muhammad Ashfaq[1], Muhammad Zaffar Hashmi[2], Arooj Mumtaz[1], Deeba Javed[1], Noor Ul Ain[1], Sana Shifaqat[1], Muhammad Saif Ur Rehman[3]

[1]*Department of Chemistry, University of Gujrat, Gujrat, Pakistan;* [2]*Department of Chemistry, COMSATS University Islamabad, Pakistan;* [3]*Department of Chemical Engineering, Khawaja Fareed University of Engineering & Information Technology, Rahim Yar Khan, Pakistan*

21.1 Introduction

Frequent use of pharmaceuticals, their ubiquitous detection in different environmental matrices, and the long-term side effects to the ecosystem have urged environmentalists worldwide to call them "emerging contaminants" (ECs) of the environment (Christen et al., 2010; Lombardo-Agüí, Cruces-Blanco et al., 2014). The first report that highlighted the occurrence of these ECs in the environment was in 1977 when metabolites of clofibrate and aspirin were detected in effluent of a wastewater treatment plant in the United States (Hignite and Azarnoff, 1977; Cardoso et al., 2014). Since then it has been well recognized by the scientists that these ECs are responsible for the contamination of water (Heberer, 2002; Cardoso et al., 2014). These ECs are also of serious concern to human population as long-term exposure to very small concentrations can have lethal effects on health, in addition to the toxicity of their transformation products that may also be produced due to possible interaction of these chemicals with each other under different environmental conditions (Jones et al., 2005; Oliveira et al., 2015). Pharmaceuticals are made stable for their therapeutic action, so they show different degrees of degradation patterns inside biological systems (break down into different parts), due to which many of them partially are excrete unchanged or in conjugated form and these then can exert different types of harmful effects on the environment (Klavarioti et al., 2009; Sui et al., 2010).

The class of compounds within the pharmaceuticals that are used widely for the treatment of both human and animal infectious diseases is called "antibiotics." The word *antibiotic* is derived from "antibiosis" meaning "against life." Antibiotics are usually derived from microorganism and are also lethal for them (Etebu and Arikekpar, 2016). These may be natural or synthetic organic chemotherapeutic agents having severe effect on life even at very low concentration (may inhibit the microbial growth or kill them). On the basis of this quality, they are either bactericidal (kill bacterial) or bacteriostatic (those inhibit microbial growth).

In this modern world of development and technology, more and more antibiotics are being produced with increasing population to meet demand and are also causing environmental problems (Awad et al., 2014). In both developed and developing countries, the production as well as consumption of

Antibiotics and Antimicrobial Resistance Genes in the Environment. https://doi.org/10.1016/B978-0-12-818882-8.00021-8

antibiotics is increasing rapidly because antibiotics are the essential medicines used in chemotherapy, transplantation, and surgical intervention. According to a report, the annual production of antibiotics for veterinary and human use is approximated to be 100,000 to 200,000 ton worldwide. Much of this portion goes into the environment through many routes, resulting in accumulation of these bioactive compounds in the environment (Shi et al., 2014; Xu et al., 2014). Antibiotics not only pollute the environment through their effects on the surrounding environment, but they can also carry antibiotic resistant genes (ARGs) along with them. These antibiotic resistant genes are transported to the other environmental matrices, including soil where they continue to persist and aggravate their effects. These accumulated antibiotic resistant genes (ARGs) can move to the deeper soils with the passage of time. They are not only pollutants of the soil or environment but are considered as potential human health risks (Tang et al., 2015). Antibiotics are also used as a growth promoter in breeding stock and livestock, so antibiotics have both therapeutic and nontherapeutic usage for the livestock (Europea, 1996).

This chapter includes the sources of the pharmaceutical chemicals in the environment. The focus is on antibiotics risk assessment and antimicrobial resistant genes, considering: their release in the environment, types of antibiotics, their discharge and effects on the biotic or abiotic factors, what are the sources for the contamination, what are the sites that are most affected by the antibiotics, what types of risks are rising, what parameters should be adopted to avoid these risks, and how to control the antibiotic contamination in the environment. Dissemination of antibiotics and antimicrobial resistant genes can be cleared from the diagram.

21.2 Classification of antibiotics

Antibiotics used against bacteria and other microbial communities belong to the groups of tetracyclines, macrolides, penicillins, aminoglycosides, sulfonamides, and fluoroquinolones (Béahdy, 1974). The classifications of antibiotics based on their chemical affinity with action site are described as under.

Class of antibiotic	Antibiotic label	Site of action
Inhibitors of cell wall synthesis		
β-Lactams	Penicillins (piperacillin, penicillin, cloxacillin, ampicillin), cephalosporins (ceftazidime, cefaclor)	Binding proteins with penicillin
Glycopeptides	Vancomycin	Terminal units of peptidoglycan (D-Ala-D-Ala dipeptide)
Lipopeptides	Polymyxin B	Outer membranes of lipopolysaccharide

—cont'd

Class of antibiotic	Antibiotic label	Site of action
Others	Alafosfalin	Terminal units of peptidoglycan (D-Ala-D-Ala dipeptide)
	D-cycloserine	Alanine racemase and D-alanine
	Fosfomycin	Uridine diphosphate (UDP)-N-acetylglucosamine-3-enolpyruvyltransferase
	Bacitracin	C_{55}-isopropyl pyrophosphate
Inhibitors for DNA synthesis		
Fluoroquinolones	Nalidixic acid, Ciprofloxacin, levofloxacin, Sparfloxacin, Norfloxacin	Topoisomerase IV, DNA gyrase (topoisomerase II)
Sulfonamides	Sulfamethazine, sulfapyridine, Sulfamethoxazole, sulfadiazine, Sulfamerazine	DHPS (used in folate synthesis) competitive inhibitor
Others	Novobiocin	DNA gyrase
Inhibitors for RNA synthesis		
Rifamycins	Rifampicin, rifabutin, rifaximin	RNA polymerase (DNA-dependent)
Resistomycins	Resistomycin, resistoflavin	RNA polymerase
Inhibitors for protein synthesis		
Tetracyclines	Oxytetracycline, doxycycline, tetracycline, Demecycline, Minocycline	30s ribosome (inhibition of binding aminoacyl tRNA with ribosome)
Aminoglycosides	Tobramycin, gentamicin, amikacin, streptomycin, Spectinomycin	30s ribosome (mistranslation and mismatching)
Macrolides	Erythromycin, clarithromycin, midecamycin,	50s ribosome (peptidyl-tRNA dissociation)

Continued

—cont'd

Class of antibiotic	Antibiotic label	Site of action
	roxithromycin, spiramycin, Azithromycin	
Amphenicols	Chloramphenicol, thiamphenicol, florfenicol	50s ribosome (avoid elongation)
Lincosamides	Clindamycin, Lincomycin	50s ribosome (stimulate peptidyl-tRNA dissociation)
Pleuromutilins	Tiamulin	50s ribosome (forbids proper positioning at the end of tRNA)
Others	Thiostrepton	50s ribosome (inhibition of mRNA)
Intercalators (DNA replication)		
Anthracyclines	Doxorubicin, epirubicin, idarubicin	Intercalation of RNA/DNA and topoisomerase II
Others	Actinomycin D	Intercalation of G-C base pair
	Mithramycin	Intercalation of G-C rich DNA strand
	Tetracenomycin	Intercalation of DNA
Anaerobic DNA inhibitors		
Nitrofurans	Furazolidone, nitrofurantoin	Reduced form is highly reactive
Nitroimidazole	Ornidazole	Damages bacterial DNA
Others		
	Antimycin A	Reductase at cytochrome C
	Bafilomycin	Inhibition of proton transfer across membrane
	Monensin	Ionophore of membrane
	Netropsin	Inhibit DNA replication
	Nonactin	Ionophore of membrane
	Salinomycin	Ionophore of membrane
	Staurosporine	Inhibits binding of ATP with kinase
	Streptogramin	Inhibits DNA/RNA synthesis

—cont'd

Class of antibiotic	Antibiotic label	Site of action
	Tunicamycin	Inhibit synthesis of glycoprotein
	Valinomycin	Ionophore of membrane

Adapted from Wong, W. R., A. G. Oliver, et al. (2012). Development of antibiotic activity profile screening for the classification and discovery of natural product antibiotics. Chemistry & biology 19(11): 1483–1495.

21.3 Inhibitors of cell wall synthesis

21.3.1 β-Lactams

They contain a β-Lactam ring in their structure and include penicillins, cephalosporins, carbapenems, and cephamycins. They are among the category of most-prescribed medicines in the world (Calderwood, 2014). β-Lactams behave like bacteriostatics as well as bactericidals by the inhibition of transpeptidase and carboxypeptidase (essential enzymes) and act by the inhibition of cell division and growth, respectively (Williamson et al., 1986).

21.3.2 Glycopeptides

These are effective against gram-positive bacteria and are considered as a last line of defense. They contain aglycone moiety, which is responsible for its pharmacological activity along with two sugar molecules that define its hydrophilicity (Kang and Park, 2015).

21.3.3 Lipopeptides

These are effective against gram-positive bacteria and include calcium-dependent antibiotics like friulimicins, laspartomycin, and daptomycin (Strieker and Marahiel, 2009). Daptomycin is an example of an acidic lipopeptide antibiotic whose inhibitory action is due to the presence of calcium ions, so it is also called a calcium-dependent antibiotic (Ball et al., 2004).

21.4 Inhibitors for DNA synthesis

21.4.1 Sulfonamides

Sulfonamides are the important category of drugs having various pharmacological effects including antibacterial, anticarbonic anhydrase, hypoglycemic, diuretic and antithyroid activities. Their derivatives also have in vivo and in vitro antitumor effects (Scozzafava et al., 2003).

21.4.2 Fluoroquinolones

Fluoroquinolones have a peculiar mode of action. Their primary types are effective against only gram-negative bacteria, whereas the present types are broad spectrum (Walker, 1999). Their postantibiotic effects and pharmacokinetics are adequate against facultative and aerobic gram-negative bacilli. They resemble the β-lactam antibiotics (Lode et al., 1998).

21.5 Inhibitors for RNA synthesis

21.5.1 Rifamycins

These include the well-known broad-spectrum antibiotic rifampicin. It is highly potent against pathogens and is used as an antituberculosis agent (Campbell et al., 2001). Derivatives of rifamycins are rifampin and rifaximin (Huang et al., 2013).

21.5.2 Resistomycins

Resistomycin is a derivative of tetrahydroxyanthrone. It was isolated from *Streptomyces sulphureus* culture (Shiono et al., 2002). It is a pentacyclic polyketide metabolite and has various pharmacological effects (Jakobi and Hertweck, 2004), acting through inhibition of RNA synthesis.

21.6 Inhibitors for protein synthesis

21.6.1 Tetracyclines

These antibiotics include tetracycline, oxytetracycline, and chlortetracycline (Ji et al., 2016). Tetracyclines are effective against atypical life like rickettsiae, mycoplasmas, chlamydiae, and protozoans' parasites. These are also used to prevent malaria (Chopra and Roberts, 2001). These are broad-spectrum antibiotics effective against gram-negative as well as gram positive bacteria (Grossman, 2016).

21.6.2 Aminoglycosides

These antibiotics are effectively used for human as well as veterinary therapy (Li et al., 2016). These are broad-spectrum antibiotics used against gram-negative and gram-positive bacteria. They also give positive responses against multidrug-resistant tuberculosis (Garneau-Tsodikova and Labby, 2016). Their structure resembles the carbohydrates, with an aminocyclitol ring that contains amine and hydroxyl substituents. Amikacin, dibekacin, isepamicin, arbekacin, and netilmicin are semisynthetic aminoglycosides (Xu et al., 2017).

21.6.3 Macrolides

Macrolides represent a class of drugs having 12 macrocyclic rings of lactone of different sizes along with amino sugar (deoxy) (Kanoh and Rubin, 2010; Dinos, 2017). These drugs show various activities like antifungal activity and work as immunosuppressants and prokinetics (Kanoh and Rubin, 2010). Macrolides are effective against some gram-positive and a restricted number of gram-negative bacteria, so they have a limited spectrum of action. Some semisynthetic macrolides are clarithromycin, roxithromycin, dithromycin, and azithromycin (Sharma and Kasthuri, 1996).

21.6.4 Amphenicols

Chloramphenicol, thiamphenicol, and florfenicol are amphenicol derivatives (Huang et al., 2017). They bind with ribosomal peptidyl transferases and inhibit the formation of protein (Tereshchenkov et al., 2018). Amphenicols are broad-spectrum antibiotics and cure many gram-positive and gram-negative bacterial infections (Li et al., 2019a,b).

21.6.5 Lincosamides

These arise from natural product lincomycin. Some derivatives are semisynthetic, including clindamycin and pirlimycin (Morar et al., 2009). They bind with the 50S ribosomal unit for the inhibition of protein synthesis. Lincosamides are active especially against gram-positive bacteria, are bacteriostatic, and sometimes may be bactericidal (Spížek and Řezanka, 2017).

21.6.6 Pleuromutilins

These were first isolated from *Pleurotus mutilus* and *Pleurotus passeckerianus.* Its derivatives contain C-14 chain having heterocyclic moiety. It works against gram-positive bacteria (Li et al., 2019a,b). Infections of skin and respiratory tract are cured by this antibiotic (Novak and Shlaes, 2010). Pleuromutilins include derivatives like tiamulin and valnemulin (Brown and Dawson, 2015).

21.7 Intercalators (DNA replication)

21.7.1 Anthracyclines

Doxorubicin, daunorubicin, and epirubicin are active cytotoxic agents. These are used in the treatment of hematological malignancies and solid tumors (Cortes-Funes and Coronado, 2007). These antibiotics have effects on both DNA and RNA synthesis. Class I anthracyclines include adriamycin, carminomycin, and puromycin, which are DNA inhibitors; Class II anthracyclines are RNA synthesis inhibitors, including aclacinomycin, marcellomycin, and musettamycin (Crooke et al., 1978).

21.8 Anaerobic DNA inhibitors

21.8.1 Nitrofurans

These antibiotics include furaltadone, furazolidone, and nitrofurantoin (Edhlund et al., 2006). Antibacterial activity is due to a nitro group on the 2-substituted furan at 5 position. At 2 position a carbonyl group is present. If thiophene or pyrrole is present other than furan (heterocyclic ring system), then no therapeutic effectiveness is present (Dodd et al., 1950).

21.8.2 Nitroimidazoles

Infections caused by anaerobic bacteria are treated with these antibiotics. Imidazole is used to cure tetanus, gynecology sepsis, and abscesses (Wilcox, 2017). Metronidazole and 5-nitroimidazole are its two metabolites (Bähr and Ullmann, 1983).

21.9 Risk assessment of antibiotics

The environmental risks associated with failure of proper antibiotic treatment include occurrence of most of the partially decomposed antibiotics in the environment and its clinical outcome not being properly reported (Ashbolt et al., 2013). Hazard quotient (HQ) and risk quotient (RQ) values are the basic methods accepted and adopted internationally for risk assessment studies (Hernando et al., 2006). The selective toxicity of antibiotics can have a significant impact on human health, the environment, and aquatic community (Robinson et al., 2005; Kim and Aga, 2007). Antibiotics are increasingly causing major ecological risks; their concentrations are measured in terms of risk quotient (Ashfaq et al., 2017). Risk quotient is basically used for evaluating the risk caused by the contaminant in the environment (Wu et al., 2014). Risk quotient can be calculated by taking the ratio of the measured environmental concentrations (MEC) to the predicted no-effect concentration (PNEC) for the specific pollutants (Al Aukidy, Verlicchi et al., 2014; Yang et al., 2016).

$$RQ = MEC/PNEC$$

Where, predicted no-effect concentration can be calculated by taking the ratio of no observed effect concentration (NOEC) to 100—100 is for the acute or chronic toxicity that is used to derive predicted no-effect concentration (Yan et al., 2013).

$$PNEC = NOEC/100$$

Accurate and cost-effective risk assessment can be done with HQ method (Raybould et al., 2011). HQ value is the ratio between the possible exposure of a substance and its concentration level, commonly used for the environmental risk assessment to characterize the risk level (Chapman et al., 1997). The HQ value greater than one indicates a high risk level (Sofuoglu and Kavcar, 2008) and environmental impact can be expected for that specific antibiotic (Li et al., 2013). The HQ value less than 0.01 suggests no risks (Hu et al., 2018), between 0.01 and 0.1 indicates a low risk level, while HQ value from 0.1 to 1 represents a medium risk level (Chen and Zhou, 2014). Recently the exceeding HQ value of ciprofloxacin suggested a greater environmental risk for bacteria, algae, and rural population in Hyderabad, India. (HQ value developed for treated sewage water for bacteria = 29.0, algae = 25.4) (Mutiyar and Mittal, 2014). The investigation and discussion reveal that while the HQ method is significantly used to measure the concern level, it is not suitable to measure the risk (EPA, 1989).

21.9.1 Toxic unit

A toxic unit (TU) is the ratio between potential exposure of a component and level at which concentration effect is expected (Gómez-Gutiérrez et al., 2007). The toxicity level of a mixture can be determined with the addition of individual TUs for each component (Di Toro and McGrath, 2000). The TU approach reveals that both amoxicillin and spiramycin play a significant role to increase algal growth and toxicity of microcystis, which enhances threats to aquatic life (Liu et al., 2012).

21.9.2 Terrestrial or soil risk

Antibiotics are consumed by the humans and animals for betterment of health, but due to improper digestion, 30%–40% of antibiotics are released along with their waste (Sun et al., 2017). Antibiotics mainly accumulate into soil by different means and cause serious ecological risks (Yang et al., 2016).

These antibiotics are taken up by plants and their presence can be detected in vegetables, which results in the serious ecological risks (Li et al., 2014) as well as effects on plant growth by antibiotic uptake (Rehman et al., 2015). According to research in China, ciprofloxacin was more abundant than all of the fluoroquinolones and posed higher risks to the environment (Li et al., 2014). Among the sources of antibiotics to the soil are the organic fertilizers and irrigation with the wastewater that contain residues of antibiotics, mainly tetracycline and quinolones, either in the waste of animals or in human waste (Zhang et al., 2015; Sun et al., 2017). Antibiotics applied to the plants are also contributing antibiotic contamination; high concentrations of antibiotics can be detected in greenhouse soils (Li et al., 2015). RQ values of antibiotics found in the soil samples from different regions are found in Table 21.1.

21.9.3 Aquatic risks

Antibiotics for humans as well as veterinary antibiotics are contaminating the surface water, ground water, sediments, effluents, and biota (Cardoso et al., 2014). A major source is the discharge from the pharmaceutical industrial effluents to the main streams (Rehman et al., 2015). Use of antibiotics is affecting the organisms in aquatic life, not only the fishes but also microorganisms like cyanobacteria, algae, etc., especially in the aquaculture (Andrieu et al., 2015). Near the disposal point of the antibiotics into wastewater, the RQ values (>1) show high risk to the environment (Zhang et al., 2017). Not only the aquaculture and wastewater are contaminated but the surface water is also contaminated by antibiotics. Recent studies on risk assessment in China show high risk to algae from norfloxacin, tetracycline, sulfonamide, ciprofloxacin, and ofloxacin (Li et al., 2017). Studies on tap water revealed that it is also not safe from the antibiotics; about 17 types of antibiotics were present in tap water according to research in China. Out of 17, only four showed high levels (thiamphenicol, dimetridazole, sulfamethazine, and clarithromycin (Leung et al., 2013). Low-temperature and low-flow conditions in aquatic environment can increase the persistence of antibiotic residues (Yan et al., 2013). This leads to development of ARGs. When we talk about the offshore waters, they are also not safe from the antibiotic residues; their sources are not yet confirmed, but these residues are also posing hazardous risks to the environment (Zhang et al., 2013). The risk quotient values for antibiotic residues against different species are presented in Table 21.2 as reported in previous studies (Yan, Yang et al. 2013).

Table 21.1 Data showing RQ values of antibiotics found in soil samples from different regions.

	Min	Median	Max	References
Norfloxacin (NOR)	<0.01	>0.01	>0.1	Wu et al. (2014)
	<0.5	<0.5	1	Sun et al. (2017)
	0.1	1	>1	Li et al. (2015)
Ciprofloxacin (CIP)	<0.001	**<0.1**	**>1**	Wu et al. (2014)
	<0.5	>1	2.5	Sun et al. (2017)
	<1	1	>1	Li et al. (2015)
Lomefloxacin (LOM)	<0.001	<0.01	>0.1	Wu et al. (2014)
	>0.01	<0.1	0.1	Li et al. (2015)
Enrofloxacin (ENR)	0.001	0.1	1	Wu et al. (2014)
	<0.5	<0.5	<0.5	Sun et al. (2017)
	<0.1	1	>1	Li et al. (2015)

Table 21.2 Risk quotient values for the antibiotic residues found in the waters of different regions.

Antibiotics	Risk level by risk quotient (RQ)		
	Fish	**Daphnia**	**Green algae**
Sulfadiazine (SD)	Low<0.01	Low <0.1	Low <0.001
Sulfapyridine (SP)	<0.001	Medium >0.1	<0.0001
Sulfamethoxazole (SMX)	Low <0.1	Medium >0.1	<0.001
Sulfathiazole (ST)	<0.001	Low <0.01	<0.0001
Sulfamethazine (SMT)	Low = 0.01	Medium = 0.1	<0.001
Sulfaquinoxaline (SQ)	Low = 0.01	Low < 0.1	<0.001

Adapted from Yan, C., Yang, Y., et al. (2013). Antibiotics in the surface water of the Yangtze Estuary: occurrence, distribution and risk assessment. Environmental Pollution 175: 22–29.

During our previous studies (Ashfaq, Khan et al. 2016, 2017), the hospital wastewater and pharmaceutical industries' impact on wastewater was found to be highly risky and very high RQ values were observed against different aquatic species (Tables 21.3 and 21.4).

21.10 Antimicrobial resistance genes

When microorganisms encounter an antimicrobial drug, antimicrobial resistance appears in them (Organization, 2017). Microbial communities can easily evolve to deal with the changes in the environment, due their small size, small genome (limited number of accessory genes), and fast generation time (Michael et al., 2014). The major cause of AMR is the indiscriminate and significant use of antimicrobial drugs (i.e., antivirals, antimalarials, antifungals, antibiotics, and anthelmintics) in agriculture, animal or human health care against microbes (Holmes et al., 2016). *Klebsiella pneumoniae* is showing antibiotic resistance due to gene mutation and it cannot be treated with old drugs (Holt et al., 2015).

21.10.1 Classification of antibiotic resistance genes and their subtypes

The ARGs may be classified as: aminoglycoside RG; subtype aacC2, *str*S, *str*B, macrolide-lincosamide-streptogramin (MLSB) RG; subtype *erm*A, *erm*B, *mef*A, chloramphenicol RG; subtype *fex*A, *fex*B, *fol*R, *cfr, cml*A, *cat*A1, sulfonamide RG; subtype *sul*1, *sul*2, *sul*3, multidrug RG; vancomycin RG, tetracycline RG; subtype, *tet*A, *tet*B, *tet*C, *tet*M, *tet*O, *tet*T, *tet*W, and beta-lactams RG; subtypes $bla_{TEM\text{-}1}$, bla_{AmpC}, bla_{CTX}, bla_{VIM}, bla_{SHV} (Guo, Li et al., 2017; Zhu et al., 2017; Zhao et al., 2019).

21.10.2 Human exposure to ARGs at coastal site

The presence of antibiotic resistant genes in environment is the most emerging concern for environmental pollution (Czekalski et al., 2015). The information about the mechanism of development of resistance varies from natural to anthropogenic. The natural sources mainly attribute to the natural sources of antibiotics and antimicrobial agents. Humans may get exposed to ARGs by swimming at coastal sites (Leonard et al., 2015). The ingestion of antibiotics directly or uptake from air or soil as well as human immigration are major sources of ARG spread (MacPherson et al., 2009). The overdose of antimicrobials and antibiotics is controlled in humans through regular dosage amounts, but for

Table 21.3 Risk assessment of different pharmaceuticals in hospital wastewater of Lahore, Pakistan.

		M hospital	
Compound	**Species**	**MEC (μg/L)**	**RQ value**
Ofloxacin	*Vibrio fisheri*	66	3300
	Pseudomonas putida		66,000
	Fish		124
	Daphnia		46
	Green algae		3300
	Pseudokirchneriella subcapitata		6000
Ciprofloxacin	Fish	18	<0.1
	Daphnia		<0.1
	Green algae		1800
	Microcystis aeruginosa		3600
Moxifloxacin	Fish	224	<0.1
	Daphnia		0.2
	Green algae		0.7

MEC = Measured environmental concentration.
Adapted from Ashfaq, M., K. N. Khan, et al. (2016). Occurrence and ecological risk assessment of fluoroquinolone antibiotics in hospital waste of Lahore, Pakistan. Environmental Toxicology and Pharmacology 42: 16–22.

animals, crops, and fishery, it is not so strictly controlled, and these pharmaceuticals are being introduced in large amounts to get better efficiency in production (Bengtsson-Palme and Larsson, 2016).

The ARGs are the natural defense present in pathogenic bacteria against the natural antibiotics but well pronounced for those used in clinical treatment (Czekalski et al., 2015). Recently Jia et al. detected 194 different types of ARGs in a wastewater sample (Jia et al., 2017). The ARGs and mobile genetic element (MGE) studies have shown coresistance toward antibiotics and the metals (As, Cd, Co, Cr, Cu, Hg, Ni, Pb and Zn) that have antibiotic activity (Zhao et al., 2019).

21.10.3 ARGs from sludge, WWTPs and added to soil organic fertilizer

The ABR microorganisms and ARGs are present in the soil. The different types of soil have different bacterial colonies and have capacity to increase ARGs in the environment; this effect is enhanced when such soil is subjected to sludge as fertilizer (Chen et al., 2016). Irrespective of whether or not the sludge is treated, the existence of ARGs is equally likely (Zhang et al., 2018). The use of wastewater for irrigation also adds ABR microorganisms and ARGs to the soil; their concentration is higher in soil and they are present even if the water is treated before use (Pan and Chu, 2018). For example, the use of chicken manure, sludge, and chemical fertilizers is equally likely to enhance ARGs. The use of sludge increases ARGs content in soil 100 times more than the chemical fertilizers (Graham et al., 2016). The only antibiotic above the limit of detection (most abundant) (Qiu et al., 2019) was sulfonamides in

Table 21.4 Risk assessment of different pharmaceuticals in wastewater near pharmaceutical industrial units.

Compound	Specie	MEC (μg/L)	Risk quotient
Ofloxacin	*Vibrio fisheri*	81	4050
	Pseudomonas putida		81,000
	Fish		153
	Daphnia		56
	Green algae		4050
	Pseudokirchneriella subcapitata		7364
Ciprofloxacin	Fish	2.2	<0.1
	Daphnia		<0.1
	Green algae		220
	Microcystis aeruginosa		440
Moxifloxacin	Fish	17	<0.1
	Daphnia		<0.1
	Green algae		<0.1

Adapted from Ashfaq, M., K. N. Khan, et al. (2017). Ecological risk assessment of pharmaceuticals in the receiving environment of pharmaceutical wastewater in Pakistan. Ecotoxicology and Environmental Safety 136: 31–39.

Swiss lakes, which are highly controlled for ARGs (Czekalski et al., 2015). The ARGs in water reservoirs and sediments are comparable and the organic foods present in markets have been found to have 134 ARGs and on average 1.76×10^8 to 1.59×10^9 copies per sample. Their abundance is eight times higher than inorganic fertilizer food (Zhu et al., 2017). The soil sample has 5×10^7 copies/g, of which *intI*1 7.5×10^5 copies/g, bla_{TEM} 4×10^5 copies/g, and others are below limit of detection. The bacteria decrease 10^{4-6} in leaves and roots and 10^{2-4} in fruit, for bacteria, but for ARGs only 100 copies/g is present (Cerqueira et al., 2019). ARGs estimated here show complex a community relationship in wastewater treatment plants (WWTPs) (Guo, Li et al., 2017). The concentration of ARBs in sediments is more than in river water depending upon the amount of WWTP discharge; *erm*B is 4.01 log GC/mL and 8.640 log GC/mL, bla_{TEM} 2.2 log GC/mL and 6.5 log GC/Ml, *tetM* 2.7 log GC/

Table 21.5 The variation of ARGs in soil layers.

ARG		Surface layer	Deep layer
β-Lactam resistance genes	bla_{AmpC}	7.76×10^{-4}	1.78×10^{-2}
	Bla_{TEM-1}	1.78×10^{-2}	6.76×10^{-1}
Aminoglycoside resistance genes	*str*S	4.37×10^{-2}	2.19×10^{-1}
	*str*B	2.51×10^{-5}	1.10×10^{-3}

Adapted from Tan, L., F. Wang, et al. (2019). Antibiotic resistance genes attenuated with salt accumulation in saline soil. Journal of Hazardous Materials 374: 35-42; Tan, L., F. Wang, et al. (2019). Antibiotic resistance genes attenuated with salt accumulation in saline soil. Journal of Hazardous Materials 374: 35–42.

mL and 6.5 GC/mL and *qrn*S 0–2 log GC/mL and 4–5 log GC/Ml in water and sediments, respectively (Brown et al., 2019).

21.10.4 Mode of ARGs spread within microbial communities in soil

The proliferation of ARGs is associated with plasmid, horizontal gene transfer (HGT) (Alexander et al., 2016) between the same niche, MGEs (Zhang et al., 2019), and pathogenic integrons (Cerqueira et al., 2019). Four types of mechanisms of ABR were studied; antibiotic deactivation (40.9%), efflux pump (42.7%), cellular protection (15%), and transposase (Huang et al., 2019; Zhang et al., 2019). The soil properties have correlation with ARGs, core genome, MGE: positive correlation with soil salinity (Tan et al., 2019); negative correlation with metal and fertility of soil (−0.373 and −0.342, respectively) (Zhang et al., 2019). Transposase, MGEs, and integrin show similar negative trends with ARG abundance (Tan et al., 2019) (Table 21.5).

21.10.5 Spread of ARGs within bacterial communities

The different types of soil have different mineral compositions and metal contents influence microbial community growth in the soil. There is a direct correlation between antibiotic concentration in source

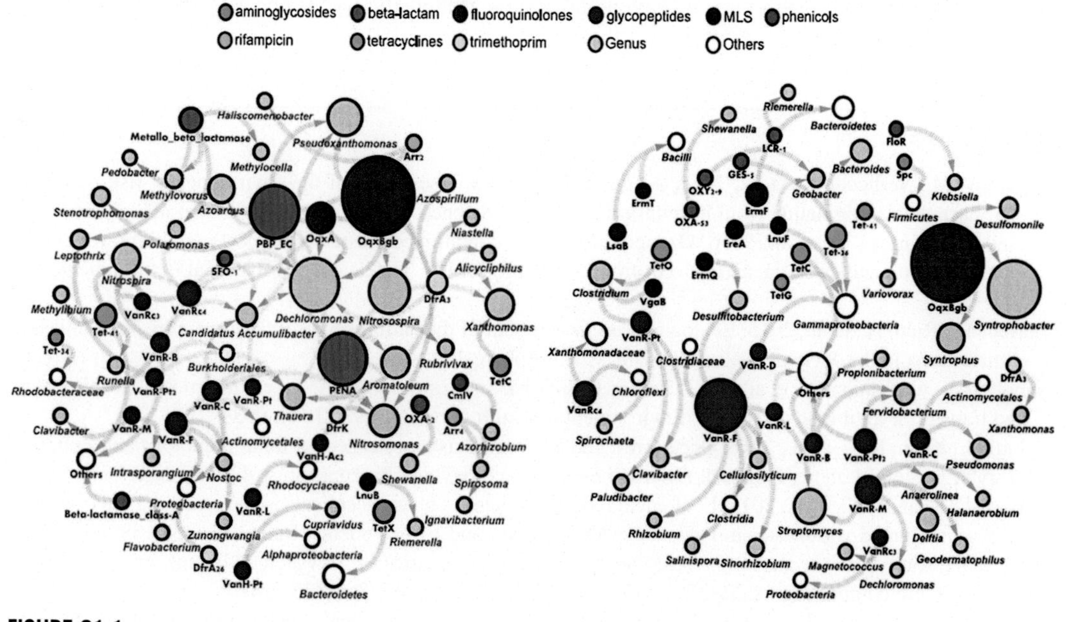

FIGURE 21.1

Interlinks between the ARGs and the bacteria species in the form of a network pipeline is shown in this figure. It identifying the source and spread of ARGs in bacterial communities in the form of metagenomics resistance.

Adapted from Guo, J., J. Li, et al. 2017. Metagenomic analysis reveals wastewater treatment plants as hotspots of antibiotic resistance genes and mobile genetic elements. Water Research, 123:468–478.

water, ponds, sediments, and abundance of ARGs (Yuan et al., 2019). The growth of microbial community is influenced by the presence of antimicrobial elements and ARGs in the colony; there are various reported modes of ARG transfer within the community and MDRG in microbes, which leads to a complex interlinked network of ARGs spread in microbial communities (Guo, Li et al., 2017). There is a correlation between the MGE and the ARG acquired by bacterial communities. For example, aminoglycosides, MLSB, beta lactam, and tetracycline RG are correlated with the bacterial communities Bacteroidetes, Firmicutes, and Fusobacteria (Zhou et al., 2018) (Fig. 21.1).

21.10.6 Reported life loss and recent protective measures for AMR

Recently, the emergence of parasites has led to increase in treatment failures (Talisuna et al., 2004), such as *Plasmodium falciparum* showing resistance to antimalarial drugs, so advancements in antimalarial drugs and protective measures to prevent the drugs from developing resistance are needed (Blasco et al., 2017). In Europe, the resistant bacteria cause annual financial and human loss of more than 1.6 billion euro and 25,000 deaths, respectively (ECDC, 2009). If no proper preventive measures are implemented soon, the death rate is predicted to increase up to 10 million per year after 30 years (de Kraker et al., 2016). To control the impact and spread of AMR in *Neisseria gonorrhoeae*, a global action plan has been organized by WHO to increase the response to prevent gonorrhoeae infection (Organization, 2012). The economic and environmental crises arising due to AMR are well known around the globe. The health care management bodies around the world are working on the development of advanced schemes to deal with them (Organization, 2014).

References

Al Aukidy, M., Verlicchi, P., et al., 2014. A framework for the assessment of the environmental risk posed by pharmaceuticals originating from hospital effluents. The Science of the Total Environment 493, 54–64.

Alexander, J., Knopp, G., et al., 2016. Ozone treatment of conditioned wastewater selects antibiotic resistance genes, opportunistic bacteria, and induce strong population shifts. The Science of the Total Environment 559, 103–112.

Andrieu, M., Rico, A., et al., 2015. Ecological risk assessment of the antibiotic enrofloxacin applied to Pangasius catfish farms in the Mekong Delta, Vietnam. Chemosphere 119, 407–414.

Ashbolt, N.J., Amézquita, A., et al., 2013. Human health risk assessment (HHRA) for environmental development and transfer of antibiotic resistance. Environmental Health Perspectives 121 (9), 993–1001.

Ashfaq, M., Khan, K.N., et al., 2016. Occurrence and ecological risk assessment of fluoroquinolone antibiotics in hospital waste of Lahore, Pakistan. Environmental Toxicology and Pharmacology 42, 16–22.

Ashfaq, M., Khan, K.N., et al., 2017. Ecological risk assessment of pharmaceuticals in the receiving environment of pharmaceutical wastewater in Pakistan. Ecotoxicology and Environmental Safety 136, 31–39.

Awad, Y.M., Kim, S.-C., et al., 2014. Veterinary antibiotics contamination in water, sediment, and soil near a swine manure composting facility. Environmental Earth Sciences 71 (3), 1433–1440.

Bähr, V., Ullmann, U., 1983. The influence of metronidazole and its two main metabolites on murine in vitro lymphocyte transformation. European Journal of Clinical Microbiology 2 (6), 568–570.

Ball, L.-J., Goult, C.M., et al., 2004. NMR structure determination and calcium binding effects of lipopeptide antibiotic daptomycin. Organic and Biomolecular Chemistry 2 (13), 1872–1878.

Béahdy, J., 1974. Recent Developments of Antibiotic Research and Classification of Antibiotics According to Chemical Structure. Advances in Applied Microbiology, vol. 18. Elsevier, pp. 309–406.

Bengtsson-Palme, J., Larsson, D.J., 2016. Concentrations of antibiotics predicted to select for resistant bacteria: proposed limits for environmental regulation. Environment International 86, 140–149.

Blasco, B., Leroy, D., et al., 2017. Antimalarial drug resistance: linking Plasmodium falciparum parasite biology to the clinic. Nature Medicine 23 (8), 917.

Brown, P., Dawson, M.J., 2015. A Perspective on the Next Generation of Antibacterial Agents Derived by Manipulation of Natural Products. Progress in Medicinal Chemistry, vol. 54. Elsevier, pp. 135–184.

Brown, P.C., Borowska, E., et al., 2019. Impact of the particulate matter from wastewater discharge on the abundance of antibiotic resistance genes and facultative pathogenic bacteria in downstream river sediments. The Science of the Total Environment 649, 1171–1178.

Calderwood, S., 2014. Beta-lactam Antibiotics: Mechanisms of Action and Resistance and Adverse Effects. UpToDate [database on the internet]. UpToDate, Waltham (MA).

Campbell, E.A., Korzheva, N., et al., 2001. Structural mechanism for rifampicin inhibition of bacterial RNA polymerase. Cell 104 (6), 901–912.

Cardoso, O., Porcher, J.-M., et al., 2014. Factory-discharged pharmaceuticals could be a relevant source of aquatic environment contamination: review of evidence and need for knowledge. Chemosphere 115, 20–30.

Cerqueira, F., Matamoros, V., et al., 2019. Distribution of antibiotic resistance genes in soils and crops. A field study in legume plants (*Vicia faba* L.) grown under different watering regimes. Environmental Research 170, 16–25.

Chapman, P., Cano, M., et al., 1997. Workgroup Summary Report on Contaminated Site Cleanup decisions." Ecological Risk Assessment of Contaminated Sediments. SETAC Press, Pensacola, Florida, pp. 83–114.

Chen, K., Zhou, J., 2014. Occurrence and behavior of antibiotics in water and sediments from the Huangpu River, Shanghai, China. Chemosphere 95, 604–612.

Chen, Q., An, X., et al., 2016. Long-term field application of sewage sludge increases the abundance of antibiotic resistance genes in soil. Environment International 92, 1–10.

Chopra, I., Roberts, M., 2001. Tetracycline antibiotics: mode of action, applications, molecular biology, and epidemiology of bacterial resistance. Microbiology and Molecular Biology Reviews 65 (2), 232–260.

Christen, V., Hickmann, S., et al., 2010. Highly active human pharmaceuticals in aquatic systems: a concept for their identification based on their mode of action. Aquatic Toxicology 96 (3), 167–181.

Cortes-Funes, H., Coronado, C., 2007. Role of anthracyclines in the era of targeted therapy. Cardiovascular Toxicology 7 (2), 56–60.

Crooke, S., Duvernay, V., et al., 1978. Structure-activity relationships of anthracyclines relative to effects on macromolecular syntheses. Molecular Pharmacology 14 (2), 290–298.

Czekalski, N., Sigdel, R., et al., 2015. Does human activity impact the natural antibiotic resistance background? Abundance of antibiotic resistance genes in 21 Swiss lakes. Environment International 81, 45–55.

de Kraker, M.E., Stewardson, A.J., et al., 2016. Will 10 million people die a year due to antimicrobial resistance by 2050? PLoS Medicine 13 (11), e1002184.

Di Toro, D.M., McGrath, J.A., 2000. Technical basis for narcotic chemicals and polycyclic aromatic hydrocarbon criteria. II. Mixtures and sediments. Environmental Toxicology & Chemistry: International Journal 19 (8), 1971–1982.

Dinos, G.P., 2017. The macrolide antibiotic renaissance. British Journal of Pharmacology 174 (18), 2967–2983.

Dodd, M., Cramer, D., et al., 1950. The relationship of structure and antibacterial activity in the nitrofurans. Journal of the American Pharmaceutical Association 39 (6), 313–318.

ECDC, E., 2009. The Bacterial Challenge–Time to React a Call to Narrow the Gap between Multidrug-Resistant Bacteria in the EU and Development of New Antibacterial Agents." Solna. ECDC & EMEA Joint Press Release.

Edhlund, B.L., Arnold, W.A., et al., 2006. Aquatic photochemistry of nitrofuran antibiotics. Environmental Science & Technology 40 (17), 5422–5427.

EPA, U., 1989. Risk Assessment Guidance for Superfund." Human Health Evaluation Manual Part A.

Etebu, E., Arikekpar, I., 2016. Antibiotics: classification and mechanisms of action with emphasis on molecular perspectives. International Journal of Applied Microbiology and Biotechnology Research 4, 90–101.

europea, U.e.C., 1996. Technical Guidance Document in Support of Commission Directive 93/67/EEC on Risk Assessment for New Notified Substances and Commission Regulation (EC) N. 1488/94 on Risk Assessment for Existing Substances. Office for Official Publications of the European Communities.

Garneau-Tsodikova, S., Labby, K.J., 2016. Mechanisms of resistance to aminoglycoside antibiotics: overview and perspectives. MedChemComm 7 (1), 11–27.

Gómez-Gutiérrez, A., Garnacho, E., et al., 2007. Screening ecological risk assessment of persistent organic pollutants in Mediterranean sea sediments. Environment International 33 (7), 867–876.

Graham, D.W., Knapp, C.W., et al., 2016. Appearance of β-lactam resistance genes in agricultural soils and clinical isolates over the 20 th century. Scientific Reports 6, 21550.

Grossman, T.H., 2016. Tetracycline antibiotics and resistance. Cold Spring Harbor Perspectives in Medicine 6 (4), a025387.

Guo, J., Li, J., et al., 2017. Metagenomic analysis reveals wastewater treatment plants as hotspots of antibiotic resistance genes and mobile genetic elements. Water Research 123, 468–478.

Heberer, T., 2002. Tracking persistent pharmaceutical residues from municipal sewage to drinking water. Journal of Hydrology 266 (3–4), 175–189.

Hernando, M.D., Mezcua, M., et al., 2006. Environmental risk assessment of pharmaceutical residues in wastewater effluents, surface waters and sediments. Talanta 69 (2), 334–342.

Hignite, C., Azarnoff, D.L., 1977. Drugs and drug metabolites as environmental contaminants: chlorophenoxyisobutyrate and salicylic acid in sewage water effluent. Life Sciences 20 (2), 337–341.

Holmes, A.H., Moore, L.S., et al., 2016. Understanding the mechanisms and drivers of antimicrobial resistance. The Lancet 387 (10014), 176–187.

Holt, K.E., Wertheim, H., et al., 2015. Genomic analysis of diversity, population structure, virulence, and antimicrobial resistance in Klebsiella pneumoniae, an urgent threat to public health. Proceedings of the National Academy of Sciences 112 (27), E3574–E3581.

Hu, Y., Yan, X., et al., 2018. Antibiotics in surface water and sediments from Hanjiang River, Central China: occurrence, behavior and risk assessment. Ecotoxicology and Environmental Safety 157, 150–158.

Huang, J.S., Jiang, Z.-D., et al., 2013. Use of rifamycin drugs and development of infection by rifamycin-resistant strains of *Clostridium difficile*. Antimicrobial Agents and Chemotherapy 57 (6), 2690–2693.

Huang, S., Gan, N., et al., 2017. Simultaneous and specific enrichment of several amphenicol antibiotics residues in food based on novel aptamer functionalized magnetic adsorbents using HPLC-DAD. Journal of Chromatography B 1060, 247–254.

Huang, H., Zeng, S., et al., 2019. Diverse and abundant antibiotics and antibiotic resistance genes in an urban water system. Journal of Environmental Management 231, 494–503.

Jakobi, K., Hertweck, C., 2004. A gene cluster encoding resistomycin biosynthesis in Streptomyces resistomycificus; exploring polyketide cyclization beyond linear and angucyclic patterns. Journal of the American Chemical Society 126 (8), 2298–2299.

Ji, Y., Shi, Y., et al., 2016. Thermo-activated persulfate oxidation system for tetracycline antibiotics degradation in aqueous solution. Chemical Engineering Journal 298, 225–233.

Jia, S., Zhang, X.-X., et al., 2017. Fate of antibiotic resistance genes and their associations with bacterial community in livestock breeding wastewater and its receiving river water. Water Research 124, 259–268.

Jones, O.A., Lester, J.N., et al., 2005. Pharmaceuticals: a threat to drinking water? TRENDS in Biotechnology 23 (4), 163–167.

Kang, H.-K., Park, Y., 2015. Glycopeptide antibiotics: structure and mechanisms of action. Journal of Bacteriology and Virology 45 (2), 67–78.

Kanoh, S., Rubin, B.K., 2010. Mechanisms of action and clinical application of macrolides as immunomodulatory medications. Clinical Microbiology Reviews 23 (3), 590–615.

Kim, S., Aga, D.S., 2007. Potential ecological and human health impacts of antibiotics and antibiotic-resistant bacteria from wastewater treatment plants. Journal of Toxicology and Environmental Health, Part B 10 (8), 559–573.

Klavarioti, M., Mantzavinos, D., et al., 2009. Removal of residual pharmaceuticals from aqueous systems by advanced oxidation processes. Environment International 35 (2), 402–417.

Leonard, A.F., Zhang, L., et al., 2015. Human recreational exposure to antibiotic resistant bacteria in coastal bathing waters. Environment International 82, 92–100.

Leung, H.W., Jin, L., et al., 2013. Pharmaceuticals in tap water: human health risk assessment and proposed monitoring framework in China. Environmental Health Perspectives 121 (7), 839–846.

Li, W., Shi, Y., et al., 2013. Occurrence and removal of antibiotics in a municipal wastewater reclamation plant in Beijing, China. Chemosphere 92 (4), 435–444.

Li, X.-W., Xie, Y.-F., et al., 2014. Investigation of residual fluoroquinolones in a soil–vegetable system in an intensive vegetable cultivation area in Northern China. The Science of the Total Environment 468, 258–264.

Li, C., Chen, J., et al., 2015. Occurrence of antibiotics in soils and manures from greenhouse vegetable production bases of Beijing, China and an associated risk assessment. The Science of the Total Environment 521, 101–107.

Li, R., Zhao, C., et al., 2016. Photochemical transformation of aminoglycoside antibiotics in simulated natural waters. Environmental Science & Technology 50 (6), 2921–2930.

Li, Q., Gao, J., et al., 2017. Distribution and risk assessment of antibiotics in a typical river in North China Plain. Bulletin of Environmental Contamination and Toxicology 98 (4), 478–483.

Li, Y.G., Wang, J.X., et al., 2019a. Antibacterial activity and Structure– activity relationship of a series of newly synthesized pleuromutilin derivatives. Chemistry and Biodiversity 16 (2), e1800560.

Li, W., Chen, N., et al., 2019b. Switchable hydrophilicity dispersive solvent-based liquid-liquid microextraction coupling to high-performance liquid chromatography for the determination of amphenicols in food products. Food Analytical Methods 12 (2), 517–525.

Liu, Y., Gao, B., et al., 2012. Influences of two antibiotic contaminants on the production, release and toxicity of microcystins. Ecotoxicology and Environmental Safety 77, 79–87.

Lode, H., Borner, K., et al., 1998. Pharmacodynamics of fluoroquinolones. Clinical Infectious Diseases 27 (1), 33–39.

Lombardo-Agüí, M., Cruces-Blanco, C., et al., 2014. Multiresidue analysis of quinolones in water by ultra-high performance liquid chromatography with tandem mass spectrometry using a simple and effective sample treatment. Journal of Separation Science 37 (16), 2145–2152.

MacPherson, D.W., Gushulak, B.D., et al., 2009. Population mobility, globalization, and antimicrobial drug resistance. Emerging Infectious Diseases 15 (11), 1727.

Michael, C.A., Dominey-Howes, D., et al., 2014. The antimicrobial resistance crisis: causes, consequences, and management. Frontiers in Public Health 2, 145.

Morar, M., Bhullar, K., et al., 2009. Structure and mechanism of the lincosamide antibiotic adenylyltransferase LinB. Structure 17 (12), 1649–1659.

Mutiyar, P.K., Mittal, A.K., 2014. Risk assessment of antibiotic residues in different water matrices in India: key issues and challenges. Environmental Science and Pollution Research 21 (12), 7723–7736.

Novak, R., Shlaes, D.M., 2010. The pleuromutilin antibiotics: a new class for human use. Current Opinion in Investigational Drugs (London, England: 2000) 11 (2), 182–191.

Oliveira, T.S., Murphy, M., et al., 2015. Characterization of pharmaceuticals and personal care products in hospital effluent and waste water influent/effluent by direct-injection LC-MS-MS. The Science of the Total Environment 518, 459–478.

Organization, W.H., 2012. Global Action Plan to Control the Spread and Impact of Antimicrobial Resistance in Neisseria Gonorrhoeae.

Organization, W.H., 2014. Antimicrobial Resistance: Global Report on Surveillance. World Health Organization.
Organization, W.H., 2017. Critically Important Antimicrobials for Human Medicine: Ranking of Antimicrobial Agents for Risk Management of Antimicrobial Resistance Due to Non-human Use.
Pan, M., Chu, L., 2018. Occurrence of antibiotics and antibiotic resistance genes in soils from wastewater irrigation areas in the Pearl River Delta region, Southern China. The Science of the Total Environment 624, 145–152.
Qiu, W., Sun, J., et al., 2019. Occurrence of antibiotics in the main rivers of Shenzhen, China: association with antibiotic resistance genes and microbial community. The Science of the Total Environment 653, 334–341.
Raybould, A., Caron-Lormier, G., et al., 2011. Derivation and interpretation of hazard quotients to assess ecological risks from the cultivation of insect-resistant transgenic crops. Journal of Agricultural and Food Chemistry 59 (11), 5877–5885.
Rehman, M.S., Rashid, N., et al., 2015. Global risk of pharmaceutical contamination from highly populated developing countries. Chemosphere 138, 1045–1055.
Robinson, A.A., Belden, J.B., et al., 2005. Toxicity of fluoroquinolone antibiotics to aquatic organisms. Environmental Toxicology & Chemistry: International Journal 24 (2), 423–430.
Scozzafava, A., Owa, T., et al., 2003. Anticancer and antiviral sulfonamides. Current Medicinal Chemistry 10 (11), 925–953.
Sharma, S., Kasthuri, A., 1996. Newer macrolides. Medical Journal, Armed Forces India 52 (4), 269.
Shi, H., Yang, Y., et al., 2014. Occurrence and distribution of antibiotics in the surface sediments of the Yangtze Estuary and nearby coastal areas. Marine Pollution Bulletin 83 (1), 317–323.
Shiono, Y., Shiono, N., et al., 2002. Effects of polyphenolic anthrone derivatives, resistomycin and hypericin, on apoptosis in human megakaryoblastic leukemia CMK-7 cell line. Zeitschrift für Naturforschung C 57 (9–10), 923–929.
Sofuoglu, S.C., Kavcar, P., 2008. An exposure and risk assessment for fluoride and trace metals in black tea. Journal of Hazardous Materials 158 (2–3), 392–400.
Spížek, J., Řezanka, T., 2017. Lincosamides: chemical structure, biosynthesis, mechanism of action, resistance, and applications. Biochemical Pharmacology 133, 20–28.
Strieker, M., Marahiel, M.A., 2009. The structural diversity of acidic lipopeptide antibiotics. ChemBioChem 10 (4), 607–616.
Sui, Q., Huang, J., et al., 2010. Occurrence and removal of pharmaceuticals, caffeine and DEET in wastewater treatment plants of Beijing, China. Water Research 44 (2), 417–426.
Sun, J., Zeng, Q., et al., 2017. Antibiotics in the agricultural soils from the Yangtze river Delta, China. Chemosphere 189, 301–308.
Talisuna, A.O., Bloland, P., et al., 2004. History, dynamics, and public health importance of malaria parasite resistance. Clinical Microbiology Reviews 17 (1), 235–254.
Tan, L., Wang, F., et al., 2019. Antibiotic resistance genes attenuated with salt accumulation in saline soil. Journal of Hazardous Materials 374, 35–42.
Tang, X., Lou, C., et al., 2015. Effects of long-term manure applications on the occurrence of antibiotics and antibiotic resistance genes (ARGs) in paddy soils: evidence from four field experiments in south of China. Soil Biology and Biochemistry 90, 179–187.
Tereshchenkov, A.G., Dobosz-Bartoszek, M., et al., 2018. Binding and action of amino acid analogs of chloramphenicol upon the bacterial ribosome. Journal of Molecular Biology 430 (6), 842–852.
Walker, R.C., 1999. The Fluoroquinolones. Mayo Clinic Proceedings. Elsevier.
Wilcox, M.H., 2017. Nitroimidazoles, Metronidazole, Ornidazole and Tinidazole; and Fidaxomicin. Infectious Diseases. Elsevier, pp. 1261–1263 e1261.
Williamson, R., Collatz, E., et al., 1986. Mechanisms of action of beta-lactam antibiotics and mechanisms of non-enzymatic resistance. Presse Medicale (Paris, France: 1983) 15 (46), 2282–2289.

Wong, W.R., Oliver, A.G., et al., 2012. Development of antibiotic activity profile screening for the classification and discovery of natural product antibiotics. Chemistry & Biology 19 (11), 1483–1495.

Wu, X.-L., Xiang, L., et al., 2014. Distribution and risk assessment of quinolone antibiotics in the soils from organic vegetable farms of a subtropical city, Southern China. The Science of the Total Environment 487, 399–406.

Xu, J., Zhang, Y., et al., 2014. Distribution, sources and composition of antibiotics in sediment, overlying water and pore water from Taihu Lake, China. The Science of the Total Environment 497, 267–273.

Xu, Z., Xu, X., et al., 2017. Effect of aminoglycosides on the pathogenic characteristics of microbiology. Microbial Pathogenesis 113, 357–364.

Yan, C., Yang, Y., et al., 2013. Antibiotics in the surface water of the Yangtze Estuary: occurrence, distribution and risk assessment. Environmental Pollution 175, 22–29.

Yang, Y., Owino, A.A., et al., 2016. Occurrence, composition and risk assessment of antibiotics in soils from Kenya, Africa. Ecotoxicology 25 (6), 1194–1201.

Yuan, J., Ni, M., et al., 2019. Occurrence of antibiotics and antibiotic resistance genes in a typical estuary aquaculture region of Hangzhou Bay, China. Marine Pollution Bulletin 138, 376–384.

Zhang, R., Tang, J., et al., 2013. Antibiotics in the offshore waters of the Bohai Sea and the Yellow Sea in China: occurrence, distribution and ecological risks. Environmental Pollution 174, 71–77.

Zhang, H., Luo, Y., et al., 2015. Residues and potential ecological risks of veterinary antibiotics in manures and composts associated with protected vegetable farming. Environmental Science and Pollution Research 22 (8), 5908–5918.

Zhang, X., Zhao, H., et al., 2017. Occurrence, removal, and risk assessment of antibiotics in 12 wastewater treatment plants from Dalian, China. Environmental Science and Pollution Research 24 (19), 16478–16487.

Zhang, J., Sui, Q., et al., 2018. Soil types influence the fate of antibiotic-resistant bacteria and antibiotic resistance genes following the land application of sludge composts. Environment International 118, 34–43.

Zhang, Y.-J., Hu, H.-W., et al., 2019. Salinity as a predominant factor modulating the distribution patterns of antibiotic resistance genes in ocean and river beach soils. The Science of the Total Environment.

Zhao, Y., Cocerva, T., et al., 2019. Evidence for co-selection of antibiotic resistance genes and mobile genetic elements in metal polluted urban soils. The Science of the Total Environment 656, 512–520.

Zhou, Z.-C., Feng, W.-Q., et al., 2018. Prevalence and transmission of antibiotic resistance and microbiota between humans and water environments. Environment International 121, 1155–1161.

Zhu, B., Chen, Q., et al., 2017. Does organically produced lettuce harbor higher abundance of antibiotic resistance genes than conventionally produced? Environment International 98, 152–159.

CHAPTER 22

Nanobiotechnology-based drug delivery strategy as a potential weapon against multiple drug-resistant pathogens

Rizwan Ali[1], Tahira Batool[2], Bushra Manzoor[2], Hassan Waseem[3,5], Sajid Mehmood[4], Ayesha Kabeer[3], Zeshan Ali[3], Sahrish Habib[3], Umer Rashid[4,*], Muhammad Javed Iqbal[3,4,*]

[1]*Centers for Biomedical Engineering, University of Science and Technology of China, Hefei, China;* [2]*Institute of Biochemistry and Biotechnology, The University of Punjab, Pakistan;* [3]*Department of Biotechnology, University of Sialkot, Sialkot, Pakistan;* [4]*Department of Biochemistry and Biotechnology, University of Gujrat, Gujrat, Pakistan;* [5]*Environmental Microbiology Laboratory, Department of Microbiology, Quaid-i-Azam University, Islamabad, Pakistan*

22.1 Introduction

Antimicrobial agents are the molecules that have plant- or animal-based origin, derived from natural products with fine chemical modification or totally synthetic in origin. The first antimicrobial agent as medicine to come into use for the treatment of infection was penicillin in 1942. This discovery opened a new avenue to tackle life-threatening microbial diseases. Commonly, all the antimicrobial agents are classified into antibiotics, disinfectants, and antiseptics. Antiseptic agents are used initially to inhibit (static) or kill (cidal) microbes found on or in the living tissues. However, disinfectants are commonly used to destroy microbial population on the surface of objects.

Antimicrobial resistance was initially thought to be of clinical concern but the recent findings of its dissemination via environmental routes have presented it as a global threat (Waseem et al., 2019a,b). Abuse of antibiotics and their inappropriate disposal are a few of the most vital factors contributing in the ever-increasing threat of antimicrobial resistance (Waseem et al., 2019a,b). With the increase in the abundance of resistant organisms, there is a need to replace the conventional antimicrobial strategies with the latest technology-based inventions (Waseem et al., 2017). Materials science plays an innovative role in every discipline of life, including electronics, physics, biomedical engineering, 3D bioprinting, bio-inks, artificial organs, and skin printing (Rudramurthy et al., 2016). Nanomaterials and nanostructures, mostly metallic in origin, have been reported to overcome this global threat in several clinical trials (Wu et al., 2017). Antimicrobial effects of nanomaterials largely rely upon molecular interaction of nanoparticles or nanomaterials with microorganism. There is a need for the in-depth understanding of microbial populations and to design multidimensional nanostructures, especially nanoparticles for targeted drug delivery and to destroy the resistant microorganisms at the molecular or nanoscale level (Weist et al., 2016).

*Both authors contribute equally to this chapter.

Antibiotics and Antimicrobial Resistance Genes in the Environment. https://doi.org/10.1016/B978-0-12-818882-8.00022-X

It has been reported by the European Antimicrobial Resistance Surveillance Network that currently there is a decline in effectiveness of existing drugs against gram-negative bacteria, in combination with the third-generation antibiotics aminoglycosides, cephalosporins, and fluoroquinolones. To tackle this serious health crisis, various strategies have been investigated, and experiments are designed by combining the selective antibiotics and with nano-sized materials (Weist et al., 2016).

22.2 Nanostructures and nanomaterials

Various nanostructures and nanomaterials have been reported in the literature and preclinical trials to conquer the strengthening fort of multiple drug-resistant microbial communities. Antimicrobial compounds are encapsulating or conjugate in or on the surface of different metallic and nonmetallic nanoparticles, including silver, gold, magnetic, zinc oxide, iron, copper oxide, chitosan, and carbon base nanoparticles. This nanobiotechnology-based drug delivery strategy has been proved as a potential weapon against multiple drug-resistant (MDR) pathogens.

22.2.1 Silver (Ag) nanoparticles

The world is facing a consistently increasing challenge of MDR pathogenic microbes for humans. With the high dependency and utilization of antibiotics, microbial communities still acquire means to survive and multiply. This present scenario leads to the increased cost of prolonged treatment and high rates of mortality and morbidity. The experimental search for potential compounds against MDR bacteria is the priority area of research in the biomedical and pharmaceutical sectors. Silver nanoparticles could be a new hope and have been reported as a multifunction nano-weapon to resolve the imminent global threat (Rai et al., 2012).

Silver has been used in different formulations since ancient times. Silver nitrate is a crucial component for the synthesis of silver nanoparticles. Silver nitrate solution (0.5%) was used in the 19th century to treat burns, and it was believed that silver nitrate promotes epithelization on the surface of wounds (Castellano et al., 2007). It has been reported to also be used as eye drops and in dressings for skin grafts (Lansdown, 2002). Silver nanoparticles have bactericidal potential against *Escherichia coli, Staphylococcus aureus*, and *Pseudomonas aeruginosa*.

Silver sulfadiazine (silver in combination with sulfadiazine) is also used as a broad-spectrum antibacterial water-soluble cream for the treatment of burns and has proved to have effective bactericidal activity against *Klebsiella* sp., *E. coli*, *Pseudomonas* sp., and *S. aureus*. Silver sulfadiazine is believed to have a specific affinity for bacterial cellular components, including DNA, and also possesses transcription inhibition properties (Atiyeh et al., 2007; McDonnell et al., 2001).

The tailored nano-antibiotic approach was carried out to synthesize antibiotic functionalized nanoparticles that have the potential to beat bacterial resistance and also to counter cellular toxicity of mammalian cells by conjugating antibiotics on the silver-silica core-shell nanoparticle surface. Nanoparticles (coated with mesoporous silica) that were functionalized with ampicillin were observed to have impressive antimicrobial potential against resistant *E. coli*. In addition, these antibiotic-functionalized nanoparticles have the advantage not to interfere with cell death inducers and are the key molecule of the mitotic cell cycle, including prophase, anaphase, and metaphase (De Oliveira et al., 2017). Now it is clearly understood by the scientific community that superior functionalization and desired biological effects are successfully achieved by tailored synthetic approach (Fig. 22.1).

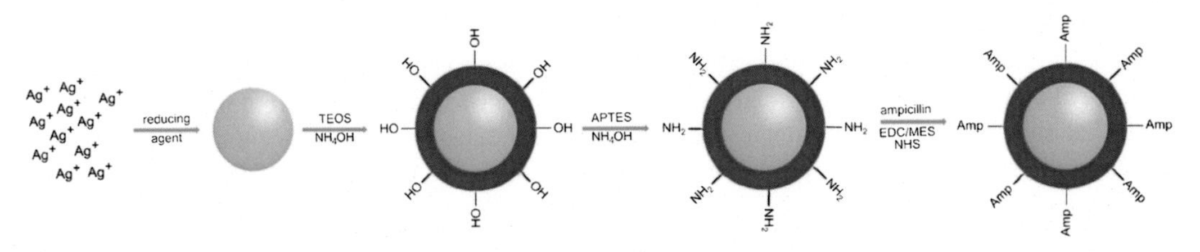

FIGURE 22.1

Synthesis steps of tailored silver nanoparticles functionalized with ampicillin.

Co-delivery of silver nanoparticles with antimicrobial compounds is proved to be more effective than its individual effect on gram-positive and gram-negative bacteria. The synergistic antibacterial activity has even proved to be very effective against most clinically resistant gram-negative microbial strains, including *Klebsiella pneumonia, E. coli, Enterobacter cloacae, Acinetobacter baumannii, and P. aeruginosa*. One of the effective compounds, ebselen (2-phenyl-1,2 benziso selenazol-3(2H)-one), can selectively target thioredoxin (Trx) against gram-positive pathogens. Trx can effectively transfer electrons through "thioredoxin reductase" (TrxR) from NAPDH to the substrate. This transfer of electrons is found to be crucial for the survival of bacteria and upregulates the production of reactive oxygen species (ROS) (Holmgren et al., 2002; Thangamani et al., 2015).

One of the recent studies in 2017 proved that the synergistic effect of silver with ebselen inhibits Trx-system (thioredoxin), and it also results in the rapid destruction of GSH (glutathione) pathway in gram-negative bacterial strains. It has been confirmed by several experiments that silver nanoparticles increase the bacterial sensitivity against various standard antibiotics (Holmgren et al., 2002). In one particular study, the synergistic effect of silver nanoparticles with four different antibiotics was observed, and it was noted that it leads toward the molecular blockage of Trx-system in bacteria (Zou et al., 2018).

Microorganisms use oxygen as an electron acceptor to activate energy metabolism. ROS are produced as a byproduct of aerobic metabolism that increases the potential threat of oxidative stress. Thiol-dependent antioxidant defense system plays a critical role in the survival of MDR microbes (Guevara-Flores et al., 2017). Thus this reported channel is one of the potential targets of antibiotics coupled with silver nanoparticles against MDR bacterial strains.

Antibiotics (ampicillin, kanamycin, norfloxacin, gentamicin) and all classes of bactericidal compounds (aminoglycoside, beta-lactam, and quinolone) can alter the bacterial redox physiology (Dwyer et al., 2014; Kohanski et al., 2007). However, the exact mechanism is still not clearly understood by the scientific community. Silver can enhance the antibacterial efficiency against gram-negative bacteria in tandem with ebselen and is correlated with ROS production (Zou et al., 2017).

A green approach for the synthesis of nanoparticles is attracting a great deal of attention from the scientific community because of its biological origin (Ali et al., 2017). Recently, nanoparticles stabilized by using plant gums were loaded with citrus fruit naringin (NRG) and flavonoid hesperidin (HDN). These nanoparticles were reported as a novel potential antimicrobial agent against MDR bacteria and brain-eating amebae. It was noted that HDN-loaded silver nanoparticles stabilized by gum acacia (GA-AgNPs-NRG) significantly suppresses 100% viability of ameba at 50 μg/mL concentration. It was also observed that these nanoparticles potentially inhibit the excystation and encystation by <85%. GA-AgNPs-NRG and GA-AgNPs-HDN have also exhibited significant antimicrobial potential against *E. coli* K1 and methicillin-resistant *Staphylococcus aureus* (MRSA). Importantly, when these nanoparticles were tested against human cells, they only showed 20% minimal cytotoxicity at the highest concentration of 100 μg/mL as compared to the standard 50 μg/mL concentration used for antimicrobial assay (Anwar et al., 2019). This novel formulation at nanoscale would be a contributing step forward for effective drug development and hold potential as a simple and urgent therapeutic agent for brain-eating ameba and MDR microbial strains.

22.2.2 Gold nanoparticles (AuNPs)

Gold nanoparticles (AuNPs) exhibit bactericidal effect (causing bacterial cell membrane distortion) by impeding ATPase activity, hindering ribosomal subunit attachment to transfer RNA (tRNA), and disrupting membrane potential (Baptista et al., 2018; Cui et al., 2012). Due to low toxicity to host organism, AuNPs act as carriers of antibacterial polymers and accelerate the bactericidal effect. AuNPs are functionalized with different molecules for the killing of MDR bacteria such as *E. coli, K. pneumoniae, and S. aureus*. The cationic-functionalized AuNPs are reported to be effective against gram-positive and gram-negative uropathogens. Moreover, AuNPs coated with N-acylated homoserine lactonase proteins (i.e., AiiA AuNPs) are said to exhibit enhanced activity against MDR bacteria compared to AiiA proteins alone (Baptista et al., 2018; Vinoj et al., 2015).

In addition to this, cationic-hydrophobic AuNPs provide a hopeful strategy to deal with MDR bacteria due to lack of toxicity, tunable in core size, versatility, and reduced resistance development (Kudgus, 2010; Li et al., 2014). The bacterial membrane integrity becomes compromised due to cationic-hydrophobic AuNPs (Li et al., 2014).

Moreover, the size of AuNPs affect its antimicrobial property substantially, i.e., an ultra-small size of gold nanoclusters (AuNCs) is associated with the better and enhanced interaction of AuNCs with bacteria, leading to metabolic imbalance, and killing of bacteria due to ROS generation (Ali et al., 2019; Zheng et al., 2017). AuNPs induce apoptosis-like death in *E. coli* by activating the caspase-like protein, i.e., RecA. RecA triggers ROS production and damages DNA, resulting in SOS response and RecA activation leading to cell death. Thus, RecA may act as a regulator of both apoptosis-like death and SOS repair system (Fig. 22.2) (Lee et al., 2018).

22.2.3 Copper oxide nanoparticles (CuO NPs)

Copper oxide nanoparticles (CuO NPs) serve as a potential antimicrobial weapon against MDR microbes due to the photocatalytic property of copper oxide (CuO). Cu^{2+} ions in CuO generate ROS in microbial cells, which consequently induce oxidative stress responsible for microbial DNA and membrane distortion, leading to microbial cell death (Baptista et al., 2018; Rudramurthy et al., 2016).

Besides this, CuO NPs are reported to be involved in the regulation of proteins taking part in nitrogen metabolism and electron transfer. Thus, CuO NPs affect the bacterial denitrification process. Moreover, in orthodontic treatment, CuO NPs that coated brackets hinder the adhesion and growth of *Streptococcus mutans* (Wang et al., 2017).

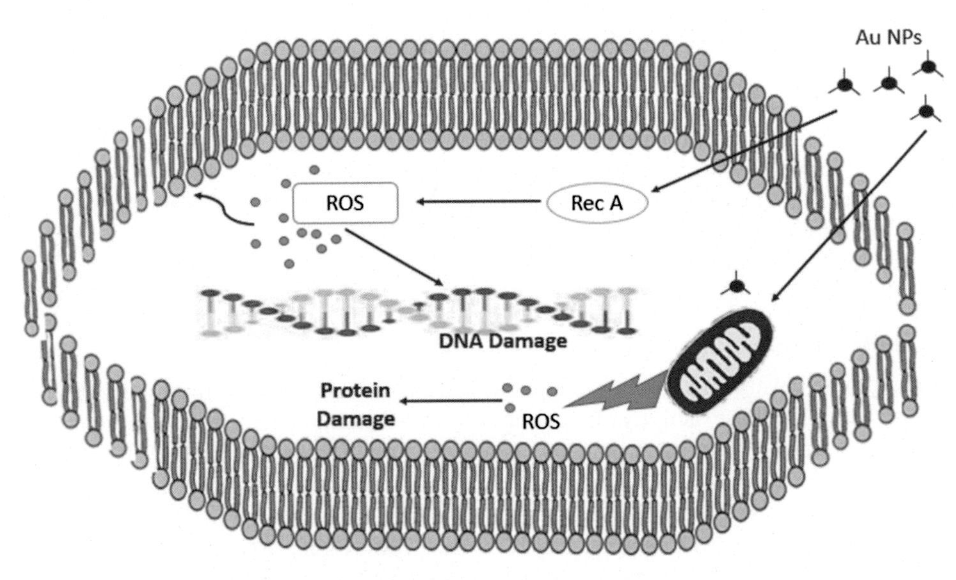

FIGURE 22.2

Mechanism of bacterial cell damage by gold nanoparticle.

Linoleic acid—capped Cu NPs are reported to inactivate bacterial enzymes, leading to hydrogen peroxide generation that is responsible for bacterial cell death. In addition to this, copper nanoparticles are reported to be involved in protein denaturation by interacting with thiol groups (-SH) of proteins in bacteria (Ingle et al., 2014).

22.2.4 Iron-containing nanoparticles (Fe_3O_4 NPs)

Iron-containing nanoparticles damage microbial cells by penetrating the microbial cell membrane and interfering with transmembrane electron transport. Iron is reported to be responsible for microbial cell decomposition, and NPs cause microbial cell aggregation, leading to the inactivation of microbes due to compression (Wang et al., 2017).

Like CuO NPs, iron-containing nanoparticles (Fe_3O_4 NPs) exhibit antimicrobial action through ROS. ROS includes radicals of superoxide (O^{2-}), hydrogen peroxide (H_2O_2), hydroxyl radicals (OH^-), and singlet oxygen ($^1O_2$), which generate oxidative stress and damage microbial proteins and DNA, causing microbial cell death (Behera et al., 2012; Rudramurthy et al., 2016).

Moreover, it is reported that electromagnetic attraction between microbes and metal nanoparticles is due to the presence of negative charges on microbes and positive charges on metal nanoparticles. Upon this attraction, the microbes become oxidized and die immediately (Prabhu et al., 2015).

22.2.5 Carbon-based nanoparticles

Carbon-based nanomaterials, which include carbon nanotubes, fullerenes, and graphene oxide, serve as a potential weapon against MDR microbes due to their antimicrobial property. The enhanced antimicrobial activity of carbon-based nanoparticles is associated with a decrease in nanoparticle size and increased surface area. In addition to this, the antimicrobial activity also relies upon the surface modifications and intrinsic properties of NPs (Kang et al., 2008; Rudramurthy et al., 2016).

Carbon-based NPs exhibit bactericidal effect through physical interaction with the cell membrane, oxidative stress, or respiratory chain impairment. Carbon-based NPs are active against *E. coli, Streptococcus* spp., *S. aureus,* and other gram-positive and gram-negative bacteria. Moreover, carbon-based NPs complexed with a silver (Ag) are reported to be active against MRSA and multidrug-resistant *A. baumannii* (Rudramurthy et al., 2016).

Functionalized carbon nanotubes (CNTs) have emerged as a novel therapeutic agent for antimicrobial drug delivery. The functional groups linked with CNTs prevent unnecessary absorption and desorption activities. As compared to traditional antibiotic therapies, CNTs are proved to be more efficient and cost-effective. The antimicrobial action of CNTs depends upon their length-dependent wrapping and diameter-dependent piercing, which are responsible for bacterial membrane potential failure, leading to the release of DNA and RNA outside the cell, resulting in complete eradication of bacteria (Mocan et al., 2017).

Furthermore, single-walled carbon nanotubes (SWCNTs) are reported to exhibit antimicrobial activity against *E. coli, Enterococcus faecium*, and *Salmonella enterica* (Dizaj et al., 2015). SWCNTs interact with the bacterial cell surface, causing cell membrane disturbances and electronic-structure-dependent bacterial oxidation that lead to the bacterial cell death (Rudramurthy et al., 2016; Vecitis, C.D. et al., 2010). Fullerenes C_{60} exhibit bactericidal action through a photochemical pathway. In this pathway, fullerenes, having closed-cage nanoparticle structure, absorb light, and shift into an excited state at the atomic level. Then, upon interacting with molecular oxygen after excitation, fullerenes generate ROS responsible for bacterial DNA damage that leads to bacterial cell death (Al-Jumaili et al., 2017).

Wrinkled surface graphene oxide (GO) films are reported to exhibit highly efficient antimicrobial property against *E. coli, S. aureus*, and *Mycobacterium smegmatis* due to their corrugated nature. GO films are suggested to form a surface trap around bacteria, resulting in significant bacterial cell membrane impairment (Zou et al., 2017). Graphene-based nanomaterials act as drug carriers due to their high specific surface area, and because of their inbuilt capability to cross cell membranes due to hydrophobic property. The high specific surface area provides numerous drug targeting attachment sites. Graphene-based nanomaterials are reported to load drugs via p–p stacking interactions, hydrophobic interactions, or hydrogen bonding. SN-38 drug is reported to be loaded to GO-PEG via p–p stacking (Zhao et al., 2017).

22.2.6 Magnetic nanoparticles

Magnetic nanoparticles (MNPs) are one of the best described nanoparticles for biomedical applications. These nanoparticles contain magnetic elements of iron, cobalt, nickel, chromium, etc., and their chemical compounds. Being superparamagnetic, they can be manipulated by an external magnetic field gradient. These nanoparticles exhibit some unique properties, like heat generation, when placed in an external alternating magnetic field. These properties make magnetic nanoparticles an excellent tool for analytical and therapeutic applications in biomedicine. Along with many other applications, MNPs can be selectively attached to a bioactive compound/drug for the targeted delivery (Tartaj et al., 2005). The most widely used magnetic nanoparticles are magnetite (Fe_3O_4) and maghemite (Fe_2O_3)

approved for medical use by the U.S. Food and Drug Administration (FDA). The most preferred MNPs are magnetite because of their higher saturation magnetizations (Huang et al., 2014).

MNPs' surface modifications are mostly required to make their surface water soluble and to allow surface conjugation of fundamental groups with biomolecules. Other modifications are required for targeted therapy like antibodies, and other functional groups such as carboxylic acid, thiol, or amine are conjugated to the surface of nanoparticles. These modifications help MNPs to bind and deliver the drug at the target site. MNPs can combine with a variety of substrates, biomolecules, and other nanoparticles to generate new antimicrobial agents that have antibacterial, antifungal, and antiviral properties. MNPs are superior to metallic nanoparticles in many aspects. Firstly, they can be controlled by an external magnetic field and can also be recycled using the external magnetic field. Secondly, these particles have superior antimicrobial activities compared to the simple metallic nanoparticles (Huang et al., 2014).

The antimicrobial activity of MNPs and other nanoparticles functions through three mechanisms. Firstly, by causing cell membrane damage, nanoparticles destroy the integrity of bacterial cell membrane leading to the death of microbes. Secondly, by releasing toxic metals, these metals bind and react with bacterial proteins, causing them to lose their functions. The third mechanism nanoparticles use is the generation of ROS. The working of these mechanisms is only possible when nanoparticles are in close proximity to the microbial cell membrane. Not all nanoparticles have such a high affinity to the bacterial membrane to cause death. Paramagnetic behavior of MNPs can solve this problem. The external magnetic field could be used to cause MNPs to adhere with membrane and cause the damage (Huang et al., 2014).

Antibacterial and antifungal properties of MNPs have gained considerable attention against multidrug resistance in microbes. Nanoparticles like iron oxide nanoparticles and chitosan nanoparticles have been shown to destroy bacterial biofilms that are associated with antibiotic resistance. Carboxymethyl chitosan coated on iron oxide MNPs is shown to kill 99% of *E.coli* and *S. aureus.* Iron oxide MNPs, when used with other metallic nanoparticles like silver nanoparticles, can significantly increase the antimicrobial activity of silver nanoparticles. Ag@Fe_3O_4 (ultrasmall Ag nanoparticles on Fe_3O_4 core) nanoparticles and Fe_2O_3@Ag (ultrasmall Fe_2O_3 nanoparticles on Ag core) nanoparticles are seen to have very high antibacterial and antifungal activity (Huang et al., 2014).

22.2.7 Chitosan nanoparticles

Chitosan is one of the crucial derivatives of chitin, and chitosan nanoparticles are excellent penetration enhancers and mucoadhesives. These nanoparticles pose antimicrobial properties and facilitate in both paracellular and transcellular transport of drugs. These properties make chitosan nanoparticles excellent carrier of antibiotics against MDR strains. Unlike many other nanoparticles, chitosan is biodegradable, biocompatible, and nontoxic (Chang et al., 2017). Chitosan is prepared by removing the acetate group from purified chitin. It is naturally occurring polysaccharide, highly basic and cationic, and has been approved from the FDA for drug delivery and other applications like tissue engineering (Baptista et al., 2018; Mohammed et al., 2017).

Chitosan can be modified to enhance stability, solubility, and mucoadhesion properties. For example, PVA-chitosan blend has been seen to improve the mechanical and barrier properties (tensile strength and water vapor permeability) of chitosan films. This modification results in cross-linked chitosan-PVA in the presence of glutaraldehyde, which is a semi-interpenetrating polymer and used for the controlled drug delivery of antibiotics like amoxicillin. Other common physical modifications include chitosan-poly (vinyl pyrrolidone) and chitosan-poly (ethyl oxide). The primary amine (-NH_2) groups can be subjected to several chemical modifications to achieve pharmaceutical applications. Chitosan-based silver nanoparticles have better and long-lasting antimicrobial activities against both gram-positive and gram-negative bacteria (Wei et al., 2009). An essential modification of chitosan is *N*-trimethyl chitosan chloride, which is a quaternized chitosan and has increased the solubility in intestinal media, and is used for delivery of hydrophilic macromolecular drugs (Mohammed et al., 2017). Quaternized chitosan nanoparticles possess cationic charges and interact with the bacterial anionic cell membrane and disrupt the membrane integrity. For example, trimethyl chitosan nitrate-capped silver nanoparticles have high antibacterial activity against clinically isolated multidrug-resistant *A. baumannii* strains (Chang et al., 2017). Cationic chitosan derivatives, *N*-(2-hydroxypropyl)-3-trimethylammonium chitosan chlorides, have been widely studied for antibacterial activities and are highly active against antibiotic-resistant bacteria and fungi (Hoque et al., 2016).

Chitosan nanoparticles can be modified with different ligands to enhance the recognition and uptake of nanocarriers into cells through receptor-mediated endocytosis. Common ligands include sugars (galactose, mannose, and trisaccharide), small compounds (folate), protein/peptides (OX26 antibodies, transferrin, RGD peptide), and other magnetic-sensitive pH-sensitive compounds. Sugars bind to the backbone of chitosan, for example, galactose with low molecular-mass chitosan is used to target Kupffer cells of the liver and has been evaluated both in vitro and in vivo. Likewise, mannose-bound chitosan is used to recognize mannose receptors on dendritic cells residing in tumors. So, these modifications can be done easily to target many types of cells in vivo or in vitro (Duceppe et al., 2010).

Chitosan nanoparticles can be loaded with antibiotics and can be administered in several ways for the targeted drug delivery and sustained release of the drug. These nanocarriers reduce side effects such as hepatic, hematologic, neurologic, or nephrologic problems. For example, amphotericin, which is used to treat fungal infections, has low solubility in the gastrointestinal tract that can lead to severe nephrotoxicity. Chitosan-based nanoparticles loaded with amphotericin reduce the toxicity (Duceppe et al., 2010). Freeze-dried powdered chitosan nanoparticles have been used for the targeted drug delivery of rifampicin, which is an antituberculosis antibiotic. The in vitro study has shown the sustained and targeted drug delivery of rifampicin from lungs. Studies also confirmed the minimal toxicity of chitosan nanoparticles in rat lungs when administered orally and inhaled by the animal (Rawal et al., 2017). Rifampicin-loaded mannose-conjugated chitosan nanoparticles can trigger macrophages in rats when injected in a vein, for effective management of visceral leishmaniasis (Chaubey et al., 2014). Cefazolin (a cephalosporins antibiotic)-loaded chitosan nanoparticles have very high antibacterial activity against MDR *P. aeroginosa, K. pneumoniae, and* extended-spectrum Beta-lactamase–positive *E. coli* (Jamil et al., 2016). Another vital drug delivery system where chitosan is used is ocular drug delivery system. The mucoadhesive chitosan-sodium Alg nanoparticles have been seen as promising in prolonged delivery of gatifloxacin as a topical ophthalmic antibiotic (Nagpal et al., 2010). Chitosan nanoparticles have enhanced the ability to penetrate biofilms of *S. aureus* and *P. aeruginosa.* These are excellent vehicles to combat bacterial resistance/antibiotic resistance (Darabpour et al., 2016). These nanoparticles also enhance the ability of essential oils to fight multiple drug resistance in bacteria.

22.3 Biological compatibility of nanoparticles

Practical applications of nanoparticles in clinical laboratories, technological developments, and drug designing are now numerous in research in recent years. The results are extremely small-sized particles of desired materials or elements having complete unique properties as compared to the same elements of actual size, which offer novel chemical and physical characteristics that are believed to play a key role in biomedical research including MDR. However, in parallel to various attractive attributes of nanoparticles, there are many concerns regarding the use of nanoparticles in clinical purposes and biological compatibility of nanoparticles is the prior one to tackle before implementation. As well, due to very complex nature, size, shape, and structural architecture, cellular toxicity is one of the leading challenges for researchers to tackle and to synthesize stable nanoparticles. It is crucial to shed light on the factors that influence the cellular toxicity and biocompatibility of nanoparticles for their safe and accurate delivery of desired compounds in the targeted organs.

22.3.1 Hemocompatibility and histocompatibility of nanoparticles

To examine the nanoparticles for MDR pathogens, it is important to understand their biocompatibility because nanoparticles easily enter the body and directly contact with cells and tissues. Blood-compatible response of nanoparticles is critical to address because nanoparticles are used as a vector for targeted delivery of drugs, gene delivery, gene therapy, and as biosensors, where nanoparticles directly contact with blood cells.

Blood cell coagulation experiments were performed to evaluate in vitro blood compatibility of nanoparticles by using cell aggregation and hemolysis assays. Hemolysis assay was performed on the surface of various synthetic nanoparticles, and it was noted that percentage of hemolysis is lower and nanoparticles possess a moderate level of compatibility (Chouhan et al., 2010; Li et al., 2012). In one of the study's analysis, chitosan-based nanoparticles were synthesized by using curcumin as a natural substrate for drug delivery. By using fresh human blood, hemolysis assay was performed, which indicated safe hemolytic ratio ($<5\%$) as per ISO-TR 7406 guidance and very little damage of erythrocytes was noted (Rejinold et al., 2011).

In one experiment, the in vivo histocompatibility of magnetite nanoparticles coated with human-like collagen protein (HLC-MNPs) was checked for their size. In this, 17 nm human-like collagen protein-coated Fe_3O_4 NPs were injected into mice followed by the injection of 0.5 mL hydrogels. After 1 week, the site of mice where Fe_3O_4 NPs coated with human-like collagen protein were injected showed reduced inflammatory response and exhibited improved histocompatibility (Chang et al., 2016).

22.4 In vivo and in vitro experimental analysis

In vivo and in vitro analyses for various nanomaterials were performed for their antimicrobial activities, drug delivery, and biocompatibility. There are few reported nanomaterials such as iron oxide magnetic nanoparticles approved by the FDA for in vivo drug deliveries and hyperthermia. Similarly, gold nanoparticles have been reported for being used in multiple in vivo studies of cancer treatments and also hold potential for MDR pathogens. Many nanoparticles cannot be used directly for in vivo drug delivery because of toxicity and lack of biocompatibility. But many studies have shown that the capping and modification of nanoparticles can make them compatible with in vivo drug delivery (Gnanadhas et al., 2013).

Multiple experiments have also been performed for in vitro analysis, for example, silver nanoparticles, iron oxide nanoparticles, gold nanoparticles, carboxylated dextran nanoparticles, and their superparamagnetic iron oxide nanoparticles (SPIONs) were studied for their antimicrobial activity against *S. aureus* and *Staphylococcus epidermidis* biofilm and have been reported as biocompatible for human tissues including liver. These results showed the biocompatibility and enhanced antimicrobial activities of silver ring-coated particles, consisting of SPION core (Mahmoudi et al., 2012). In vivo studies in mice have also proved that L-tyrosine polyphosphate nanoparticles, when loaded with silver–carbine complex, provide sustained drug release for several days against *P. aeruginosa* (Hindi et al., 2009).

22.5 Synthesis and characterization of nanoparticles

Physical, chemical, and biological methods have been used for the synthesis of desired metallic and nonmetallic nanoparticles (Fig. 22.3). Physical method for the preparation of nanoparticles is complicated, expensive, and time-consuming as compared to the other two approaches. In the physical approach, usually, nanostructures are synthesized by a top-down approach (Fedlheim et al., 2001). Other processes include high-energy ball milling, lithography, thermal evaporation deposition, electrospraying, melt mixing, and different techniques by the use of laser such as laser pyrolysis (Dhand et al., 2015). However, the chemical method and biological method almost resemble with each other but with some fine differences. Chemical approach is simple but it also demands expensive and highly purified chemicals. Few limitations have been reported by the use of chemical as they are not environmental

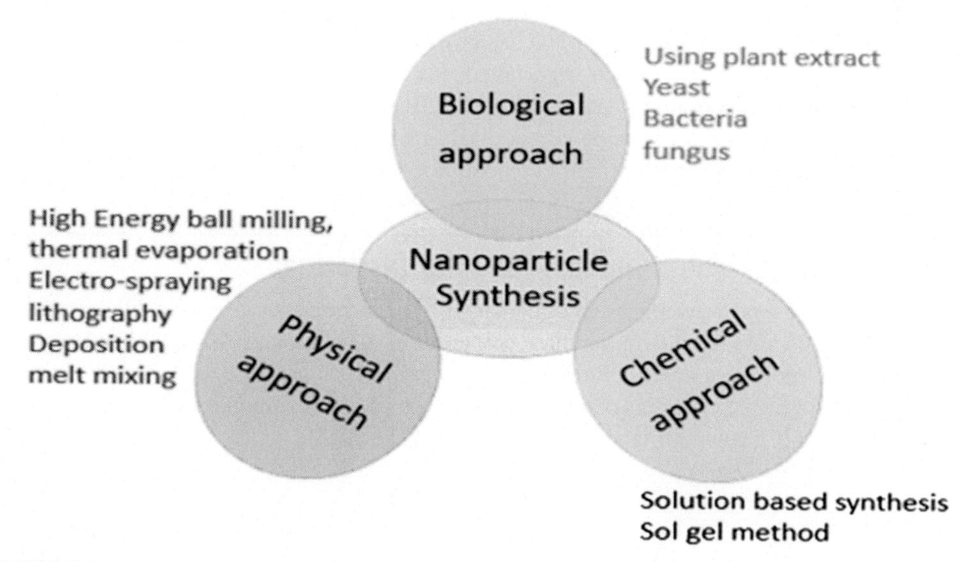

FIGURE 22.3

Different synthesis approaches of nanoparticles.

and biological friendly in nature. In comparison, biological approach is simple, time-saving, and environmental friendly in nature. Reducing and oxidizing agents are derived from the extracts of green plant sources; that is why it is also called a green approach for the synthesis of nanoparticles (Iqbal et al., 2018; Iravani et al., 2014).

When using the biological approach, plant material is selected to prepare its leaf or root extract and essential oils. Most of the plant extract is already reported in the literature to have significant antioxidant activity and also have prominent total phenolic and total flavonoid contents. Plant extracts could also be analyzed for an effective antimicrobial agent by the disk diffusion method (Shahwar et al., 2019). For the synthesis of silver nanoparticles, by using simple boiling method and by the addition of silver nitrate, silver nanoparticles are easily prepared after drying and purifying. However, unique properties of nanoparticles arise from their nanoscale dimensions that include size, shape, optical, electrical, and antimicrobial properties. Characterization of nanoparticles and nanomaterials is the fundamental step to determine the stability and ideal behavior of desired particles for targeted delivery. Characterization includes UV-visible spectroscopy, Fourier transform infrared analysis, X-ray diffraction analysis, energy dispersive spectroscopy, scanning electron microscopy, transmission electron microscopy, and zeta-sizer and zeta-potential approach to determine multiple features of nanoparticles that make them ideal candidates for MRD applications (Aras et al., 2014; Iqbal et al., 2018).

22.6 Drug delivery mechanisms

Photodynamic approach by using nanoparticles ensures the efficiency of multiple biological assays. To inactivate the pathogens efficiently, antimicrobial photodynamic therapy (aPDT) was used that is based on the use of light to activate nontoxic molecules, called a photosensitizer (PS). It results in the increased production of ROS that significantly kills bacterial population by inducing oxidative stress (Abrahamse et al., 2016). After the prolonged success of PDT in multiple biological experiments in the last decade, now it is translated into clinics and has proved very effective in both in vitro and in vivo assays (Cieplik et al., 2018). It has also been noted that fine modification of PS can increase the significance of antimicrobial efficiency (Kiesslich et al., 2013). However, it is necessary to thoroughly understand the structural makeup of targeted bacterial or biofilm composition so that targeted drug delivery by using nanostructure has optimum compatibility for maximum efficacy (Fig. 22.4).

22.7 Nanoparticles' mechanisms for drug targeting

Nanoparticles offer several advantages in drug targeting, sustained release, and biodegradability. Various factors play a vital role in defining these properties of nanoparticles. For the effective targeted drug delivery, these properties are size, surface charge, particle shape, dose, and surface properties. The specific effects of these properties depend upon the type of cells being investigated and the nature of the uptake process (Chavanpatil et al., 2006; Cooper et al., 2014).

22.7.1 Size

Size plays an important role in targeted drug delivery, efficacy, and biodistribution of the drug. It is considered that size parameters can play significant roles in the determination of cell interaction and

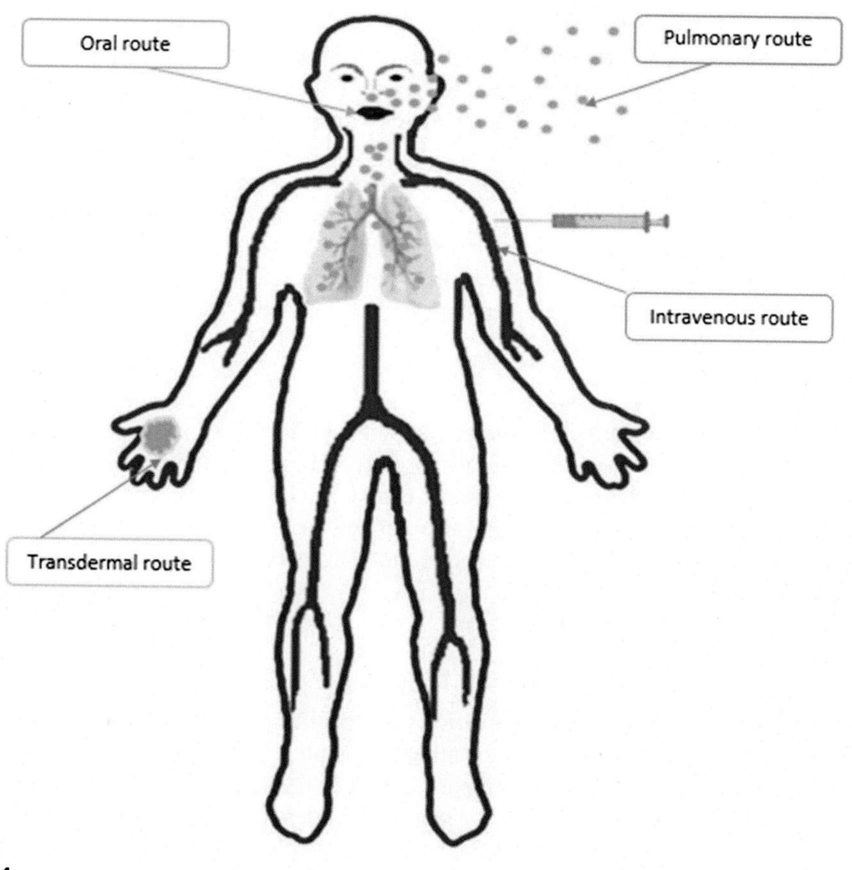

FIGURE 22.4

Nanoparticle-based therapeutics and delivery routes in human.

adhesion for various NPs. Different particle sizes define the designation of nanoparticles inside the body (Cooper et al., 2014). So, controlling the size of nanoparticles plays a defining role in the targeted drug delivery. For example, smaller nanoparticles are effectively taken up by Caco-2 cell line while larger-sized particles are effectively taken up by the mouse peritoneal macrophages (Chavanpatil et al., 2006).

Size also plays an important role in elimination and degradation of the drug. Small-sized nanoparticles <100 nm can affect the mononuclear phagocytic system, which is an important physiological character to eliminate the drug. Smaller nanoparticles also defy the reticuloendothelial system that targets foreign bodies for degradation. This way, smaller nanoparticles increase the drug-coated nanoparticles' flow in the bloodstream. As the size increases, clearance rate also increases and so the drug is eliminated very fast. In situ studies in rats have demonstrated that cells can readily uptake nanoparticles. Nanoparticles have 15–250-fold greater uptake in the cell than the micro-particles. So, the nano-sized particles are more likely to deliver drug on target with a sustained release (Cooper et al., 2014).

22.7.2 Surface charge/particle charge

Particle charges also determine the efficiency and stability of nanoparticles. Nanoparticles with higher charge will have strong repulsion. Net repulsive forces prevent nanoparticle aggregation and thus help in stability and dispersions of nanoparticles. Nanoparticles with high positive charges are interactive with anionic polyelectrolyte properties of mucus, resulting in enhanced mucoadhesion and retention of NPs within the mucus layer. For example, trimethyl chitosan chloride nanoparticles have a high positive charge and are excellent mucoadhesives (Wei et al., 2009). These trimethyl chitosan nanoparticles are effectively used against mucus infections. Positive nanoparticles with positive charges are effective against bacterial negatively charged cell membrane (Chang et al., 2017). On the other hand, negatively charged nanoparticles will not be taken up by cell membranes due to the repulsive forces (Cooper et al., 2014).

22.7.3 Particle shape

Particles' shapes have a significant effect on the biological properties of nanoparticles. For example, polymer micelles of shorter stature show an increased total blood circulation time following IV injections. Micelles of shorter sphere undergo greater uptake by cells as compared to the larger spheres. It has been demonstrated that particle length also defines the nanoparticles' adhesion to the cells. Nanoparticles with greater lengths are less adhesive to cells as compared to smaller-length particles (Cooper et al., 2014).

22.7.4 Surface modifications and cell targeting

For cellular targeting, most of the time nanoparticles are required to be modified. These modifications are of various types, from small molecules to antibodies, peptides, and sugar residues. These modifications act as ligands that bind to specific receptors present on the surface of cells. When a ligand binds to the receptor on the cell surface, a series of events leads to endocytosis of nanoparticle. Most cells express receptors for transferring and folic acid. For example, folate-formulated polymer-based nanoparticles bind the folate receptors on the tumor cells and initiate the entry. A simple representation of magnetic nanoparticles modifications is given in Fig. 22.5 (Chavanpatil et al., 2006; Cooper et al., 2014).

22.8 Cellular uptake mechanisms

Mammalian cells have a variety of mechanisms to internalize molecules. The process is called endocytosis, which means uptake of foreign material inside the cell. Endocytosis can be divided into phagocytosis (cell eating), pinocytosis (cell drinking), clathrin-dependent endocytosis, and caveolae-mediated endocytosis. Depending upon the nature of nanoparticles (size, surface modification, shape, etc.) and cells, the mechanism of endocytosis varies (Chavanpatil et al., 2006).

22.8.1 Passive diffusion

Passive diffusion is the uptake of materials by the cell without receptor-mediated mechanisms. The process of passive diffusion occurs when lipophilic nanoparticles cross the membrane (Fig. 22.5).

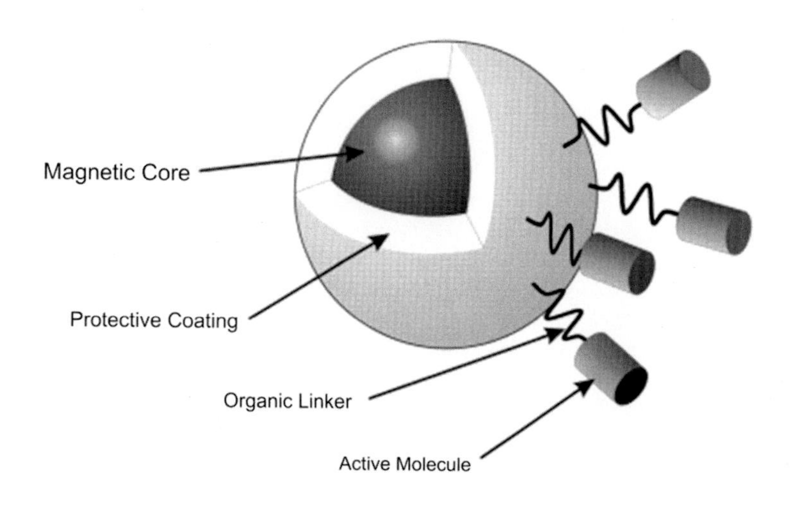

FIGURE 22.5

A typical design of a magnetic nanoparticle and surface modifications.

Diffusion of nanoparticles depends on various physical properties of nanoparticles, i.e., size, surface charge, and shape. NP surface charge can influence the degree to which substances passively diffuse across the lipophilic plasma membrane of various cells. Positive-charged small nanoparticles can cross the cell membrane, leading to membrane rupture and noticeable cytotoxic effects. Membrane disruption can be avoided or reduced by a suitable design of the surface structure and charge density. Cholesterol-containing lipid bilayers increase the rate of passive transport across the membrane (Cooper et al., 2014; Treuel et al., 2013).

22.8.2 Endocytosis

Endocytosis can be divided into two categories: receptor-mediated endocytosis and phagocytosis. Receptor-mediated endocytosis involves engulfment of a variety of ligand-receptor complexes into a coated pit. Clathrin-protein-coated pits are the most common receptor-mediated pathways. Another type of endocytosis is known as caveolae-mediated endocytosis. Like clathrin-mediated endocytosis, caveolae endocytosis is also receptor mediated. The phagocytosis is used when particles are larger, i.e., >0.5 μm. Receptor-mediated endocytosis includes a variety of receptors; most important are folate receptor-mediated, transferrin receptor-mediated, biotin receptor-mediated, and antibody-mediated endocytosis (McBain et al., 2008). The phagocytosis pathways are actin dependent and restricted to professional phagocytes, such as macrophages, dendritic cells, and neutrophils. Macropinocytosis is another endocytosis pathway, which is a nonspecific process to internalize fluids and particles together into the cell (Fig. 22.6) (McBain et al., 2008; Oh et al., 2014).

22.9 Conclusion

Technology has changed the world and our lives. Nanotechnology is one of the prominent technological drift that is revolutionizing every industry of the world from food technology and electronics to

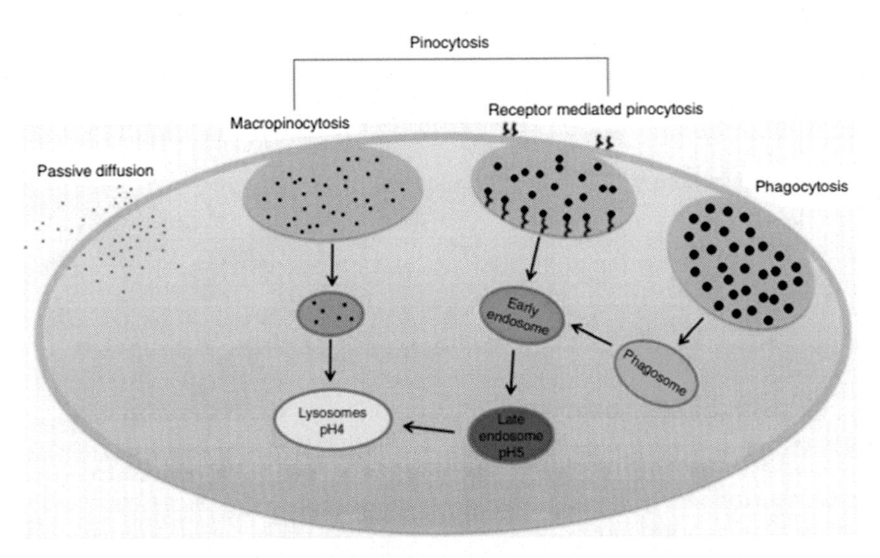

FIGURE 22.6

Simple representation of passive diffusion and endocytosis.

agriculture and medicines. As alternatives to antibiotics to constraint bacterial resistance, nanoparticles and synergistic effects of nanoparticles with drugs are proving to be a potential strategy against multiple drug-resistant pathogens.

It is concluded that AgNPs, AuNPs, CuO NPs, Fe_3O_4 NPs, and carbon-based NPs serve as a potential antimicrobial therapeutic weapon against microbes with multidrug resistivity. AuNPs hinder bacterial growth by inhibiting tRNA attachment to activation site of ribosomal subunit, RecA protein activation, and ROS generation. Likewise, CuO NPs and Fe_3O_4 NPs damage microbial cells through oxidative stress and ROS generation. As well, fullerenes, GO films, and CNTs exhibit antimicrobial effects through photochemical pathways, by interfering with the bacterial cell membrane and damaging the respiratory chain. Magnetic nanoparticles are also excellent carriers of antimicrobial drugs as they can be controlled by external magnetic fields. The most-studied nanoparticles for MDR resistance are chitosan nanoparticles, which are biodegradable, biocompatible, and, most importantly, antimicrobial. Shape, size, surface charge, and other modifications affect the drug targeted delivery in biological systems by controlling these properties, making nanoparticles useful against MDR bacteria.

References

Abrahamse, H., Hamblin, M.R., 2016. New photosensitizers for photodynamic therapy. Biochemical Journal 473 (4), 347–364.

Al-Jumaili, A., Alancherry, S., Bazaka, K., Jacob, M., 2017. Review on the antimicrobial properties of carbon nanostructures. Materials 10 (9), 1066.

Ali, J., Ali, N., Jamil, S.U.U., Waseem, H., Khan, K., Pan, G., 2017. Insight into eco-friendly fabrication of silver nanoparticles by *Pseudomonas aeruginosa* and its potential impacts. Journal of Environmental Chemical Engineering 5 (4), 3266–3272.

Ali, J., Ali, N., Wang, L., Waseem, H., Pan, G., 2019. Revisiting the mechanistic pathways for bacterial mediated synthesis of noble metal nanoparticles. Journal of Microbiological Methods.

Anwar, A., Masri, A., Rao, K., Rajendran, K., Khan, N.A., Shah, M.R., Siddiqui, R., 2019. Antimicrobial activities of green synthesized gums-stabilized nanoparticles loaded with flavonoids. Scientific Reports 9 (1), 3122.

Aras, A., Iqbal, M.J., Naqvi, S., Gercek, Y.C., Boztas, K., Gasparri, M.L., Shatynska-Mytsyk, I., Fayyaz, S., Farooqi, A.A., 2014. Anticancer activity of essential oils: targeting of protein networks in cancer cells. Asian Pacific Journal of Cancer Prevention 15, 8047–8050.

Atiyeh, B.S., Costagliola, M., Hayek, S.N., Dibo, S.A., 2007. Effect of silver on burn wound infection control and healing: review of the literature. Burns 33 (2), 139–148.

Baptista, P.V., McCusker, M.P., Carvalho, A., Ferreira, D.A., Mohan, N.M., Martins, M., Fernandes, A.R., 2018. Nano-strategies to fight multidrug resistant bacteria—"A Battle of the Titans". Frontiers in Microbiology 9.

Behera, S.S., Patra, J.K., Pramanik, K., Panda, N., Thatoi, H., 2012. Characterization and Evaluation of Anti-bacterial Activities of Chemically Synthesized Iron Oxide Nanoparticles.

Castellano, J.J., Shafii, S.M., Ko, F., Donate, G., Wright, T.E., Mannari, R.J., Payne, W.G., Smith, D.J., Robson, M.C., 2007. Comparative evaluation of silver-containing antimicrobial dressings and drugs. International Wound Journal 4 (2), 114–122.

Chang, L., Liu, X.L., Dai Di Fan, Y.Q.M., Zhang, H., Ma, H.P., Liu, Q.Y., Ma, P., Xue, W.M., Luo, Y.E., Fan, H.M., 2016. The efficiency of magnetic hyperthermia and in vivo histocompatibility for human-like collagen protein-coated magnetic nanoparticles. International Journal of Nanomedicine 11, 1175.

Chang, T.-Y., Chen, C.-C., Cheng, K.-M., Chin, C.-Y., Chen, Y.-H., Chen, X.-A., Sun, J.-R., Young, J.-J., Chiueh, T.-S., 2017. Trimethyl chitosan-capped silver nanoparticles with positive surface charge: their catalytic activity and antibacterial spectrum including multidrug-resistant strains of Acinetobacter baumannii. Colloids and Surfaces B: Biointerfaces 155, 61–70.

Chaubey, P., Mishra, B., 2014. Mannose-conjugated chitosan nanoparticles loaded with rifampicin for the treatment of visceral leishmaniasis. Carbohydrate Polymers 101, 1101–1108.

Chavanpatil, M.D., Khdair, A., Panyam, J., 2006. Nanoparticles for cellular drug delivery: mechanisms and factors influencing delivery. Journal of Nanoscience and Nanotechnology 6 (9–10), 2651–2663.

Chouhan, R., Bajpai, A.K., 2010. Release dynamics of ciprofloxacin from swellable nanocarriers of poly (2-hydroxyethyl methacrylate): an in vitro study. Nanomedicine: Nanotechnology, Biology and Medicine 6 (3), 453–462.

Cieplik, F., Deng, D., Crielaard, W., Buchalla, W., Hellwig, E., Al-Ahmad, A., Maisch, T., 2018. Antimicrobial photodynamic therapy—what we know and what we don't. Critical Reviews in Microbiology 44 (5), 571–589.

Cooper, D.L., Conder, C.M., Harirforoosh, S., 2014. Nanoparticles in drug delivery: mechanism of action, formulation and clinical application towards reduction in drug-associated nephrotoxicity. Expert Opinion on Drug Delivery 11 (10), 1661–1680.

Cui, Y., Zhao, Y., Tian, Y., Zhang, W., Lü, X., Jiang, X., 2012. The molecular mechanism of action of bactericidal gold nanoparticles on *Escherichia coli*. Biomaterials 33 (7), 2327–2333.

Darabpour, E., Kashef, N., Mashayekhan, S., 2016. Chitosan nanoparticles enhance the efficiency of methylene blue-mediated antimicrobial photodynamic inactivation of bacterial biofilms: an in vitro study. Photodiagnosis and Photodynamic Therapy 14, 211–217.

De Oliveira, J.F.A., Saito, Â., Bido, A.T., Kobarg, J., Stassen, H.K., Cardoso, M.B., 2017. Defeating bacterial resistance and preventing mammalian cells toxicity through rational design of antibiotic-functionalized nanoparticles. Scientific Reports 7 (1), 1326.

Dhand, C., Dwivedi, N., Loh, X.J., Ying, A.N.J., Verma, N.K., Beuerman, R.W., Lakshminarayanan, R., Ramakrishna, S., 2015. Methods and strategies for the synthesis of diverse nanoparticles and their applications: a comprehensive overview. RSC Advances 5 (127), 105003–105037.

Dizaj, S.M., Mennati, A., Jafari, S., Khezri, K., Adibkia, K., 2015. Antimicrobial activity of carbon-based nanoparticles. Advanced Pharmaceutical Bulletin 5 (1), 19.

Duceppe, N., Tabrizian, M., 2010. Advances in using chitosan-based nanoparticles for in vitro and in vivo drug and gene delivery. Expert Opinion on Drug Delivery 7 (10), 1191–1207.

Dwyer, D.J., Belenky, P.A., Yang, J.H., MacDonald, I.C., Martell, J.D., Takahashi, N., Chan, C.T., Lobritz, M.A., Braff, D., Schwarz, E.G., 2014. Antibiotics induce redox-related physiological alterations as part of their lethality. Proceedings of the National Academy of Sciences of the United States of America 111 (20), E2100–E2109.

Fedlheim, D.L., Foss, C.A., 2001. Metal Nanoparticles: Synthesis, Characterization, and Applications. CRC press.

Gnanadhas, D.P., Thomas, M.B., Thomas, R., Raichur, A.M., Chakravortty, D., 2013. Interaction of silver nanoparticles with serum proteins affects their antimicrobial activity in vivo. Antimicrobial Agents and Chemotherapy 57 (10), 4945–4955.

Guevara-Flores, A., Martínez-González, J., Rendón, J., del Arenal, I., 2017. The architecture of thiol antioxidant systems among invertebrate parasites. Molecules 22 (2), 259.

Hindi, K.M., Ditto, A.J., Panzner, M.J., Medvetz, D.A., Han, D.S., Hovis, C.E., Hilliard, J.K., Taylor, J.B., Yun, Y.H., Cannon, C.L., 2009. The antimicrobial efficacy of sustained release silver–carbene complex-loaded L-tyrosine polyphosphate nanoparticles: characterization, in vitro and in vivo studies. Biomaterials 30 (22), 3771–3779.

Holmgren, A., Masayasu, H., Zhao, R.E., 2002. A drug targeting thioredoxin and thioredoxin reductase. In: Free Radical Biology and Medicine. Pergamon-Elsevier Science Ltd The Boulevard, Langford Lane, Kidlington.

Hoque, J., Adhikary, U., Yadav, V., Samaddar, S., Konai, M.M., Prakash, R.G., Paramanandham, K., Shome, B.R., Sanyal, K., Haldar, J., 2016. Chitosan derivatives active against multidrug-resistant bacteria and pathogenic fungi: in vivo evaluation as topical antimicrobials. Molecular Pharmaceutics 13 (10), 3578–3589.

Huang, K.-S., Shieh, D.-B., Yeh, C.-S., Wu, P.-C., Cheng, F.-Y., 2014. Antimicrobial applications of water-dispersible magnetic nanoparticles in biomedicine. Current Medicinal Chemistry 21 (29), 3312–3322.

Ingle, A.P., Duran, N., Rai, M., 2014. Bioactivity, mechanism of action, and cytotoxicity of copper-based nanoparticles: a review. Applied Microbiology and Biotechnology 98 (3), 1001–1009.

Iqbal, M.J., Ali, S., Rashid, U., Kamran, M., Malik, M.F., Sughra, K., Zeeshan, N., Afroz, A., Saleem, J., Saghir, M., 2018. Biosynthesis of silver nanoparticles from leaf extract of Litchi chinensis and its dynamic biological impact on microbial cells and human cancer cell lines. Cellular and Molecular Biology (Noisy-le-Grand, France) 64 (13), 42–47.

Iravani, S., Korbekandi, H., Mirmohammadi, S.V., Zolfaghari, B., 2014. Synthesis of silver nanoparticles: chemical, physical and biological methods. Research in Pharmaceutical Sciences 9 (6), 385.

Jamil, B., Habib, H., Abbasi, S., Nasir, H., Rahman, A., Rehman, A., Bokhari, H., Imran, M., 2016. Cefazolin loaded chitosan nanoparticles to cure multi drug resistant gram-negative pathogens. Carbohydrate Polymers 136, 682–691.

Kang, S., Herzberg, M., Rodrigues, D.F., Elimelech, M., 2008. Antibacterial effects of carbon nanotubes: size does matter! Langmuir 24 (13), 6409–6413.

Kiesslich, T., Gollmer, A., Maisch, T., Berneburg, M., Plaetzer, K., 2013. A comprehensive tutorial on in vitro characterization of new photosensitizers for photodynamic antitumor therapy and photodynamic inactivation of microorganisms. BioMed Research International.

Kohanski, M.A., Dwyer, D.J., Hayete, B., Lawrence, C.A., Collins, J.J., 2007. A common mechanism of cellular death induced by bactericidal antibiotics. Cell 130 (5), 797–810.

Kudgus, R., 2010. Gold Nanocrystal Therapeutics: Treatment of Multidrug Resistant Pathogens and Disrupting Protein/Protein Interactions.

Lansdown, A.B., 2002. Silver I: its antibacterial properties and mechanism of action. Journal of Wound Care 11 (4), 125–130.

Lee, H., Lee, D.G., 2018. Gold nanoparticles induce a reactive oxygen species-independent apoptotic pathway in *Escherichia coli*. Colloids and Surfaces B: Biointerfaces 167, 1–7.

Li, X., Robinson, S.M., Gupta, A., Saha, K., Jiang, Z., Moyano, D.F., Sahar, A., Riley, M.A., Rotello, V.M., 2014. Functional gold nanoparticles as potent antimicrobial agents against multi-drug-resistant bacteria. ACS Nano 8 (10), 10682–10686.

Li, X., Wang, L., Fan, Y., Feng, Q., Cui, F.-z., 2012. Biocompatibility and toxicity of nanoparticles and nanotubes. Journal of Nanomaterials 2012, 6.

Mahmoudi, M., Serpooshan, V., 2012. Silver-coated engineered magnetic nanoparticles are promising for the success in the fight against antibacterial resistance threat. ACS Nano 6 (3), 2656–2664.

McBain, S.C., Yiu, H.H., Dobson, J., 2008. Magnetic nanoparticles for gene and drug delivery. International Journal of Nanomedicine 3 (2), 169.

McDonnell, G., Russell, A.D., 2001. Antiseptics and disinfectants: activity, action, and resistance. Clinical Microbiology Reviews 14 (1), 227.

Mocan, T., Matea, C.T., Pop, T., Mosteanu, O., Buzoianu, A.D., Suciu, S., Puia, C., Zdrehus, C., Iancu, C., Mocan, L., 2017. Carbon nanotubes as anti-bacterial agents. Cellular and Molecular Life Sciences 74 (19), 3467–3479.

Mohammed, M., Syeda, J., Wasan, K., Wasan, E., 2017. An overview of chitosan nanoparticles and its application in non-parenteral drug delivery. Pharmaceutics 9 (4), 53.

Nagpal, K., Singh, S.K., Mishra, D.N., 2010. Chitosan nanoparticles: a promising system in novel drug delivery. Chemical and Pharmaceutical Bulletin 58 (11), 1423–1430.

Oh, N., Park, J.-H., 2014. Endocytosis and exocytosis of nanoparticles in mammalian cells. International Journal of Nanomedicine 9 (Suppl. 1), 51.

Prabhu, Y., Rao, K.V., Kumari, B.S., Kumar, V.S.S., Pavani, T., 2015. Synthesis of Fe_3O_4 nanoparticles and its antibacterial application. International Nano Letters 5 (2), 85–92.

Rai, M., Deshmukh, S., Ingle, A., Gade, A., 2012. Silver nanoparticles: the powerful nanoweapon against multidrug-resistant bacteria. Journal of Applied Microbiology 112 (5), 841–852.

Rawal, T., Parmar, R., Tyagi, R.K., Butani, S., 2017. Rifampicin loaded chitosan nanoparticle dry powder presents an improved therapeutic approach for alveolar tuberculosis. Colloids and Surfaces B: Biointerfaces 154, 321–330.

Rejinold, N.S., Muthunarayanan, M., Divyarani, V., Sreerekha, P., Chennazhi, K., Nair, S., Tamura, H., Jayakumar, R., 2011. Curcumin-loaded biocompatible thermoresponsive polymeric nanoparticles for cancer drug delivery. Journal of Colloid and Interface Science 360 (1), 39–51.

Rudramurthy, G., Swamy, M., Sinniah, U., Ghasemzadeh, A., 2016. Nanoparticles: alternatives against drug-resistant pathogenic microbes. Molecules 21 (7), 836.

Shahwar, D., Iqbal, M.J., Nisa, M.-u., Todorovska, M., Attar, R., Sabitaliyevich, U.Y., Farooqi, A.A., Ahmad, A., Xu, B., 2019. Natural product mediated regulation of death receptors and intracellular machinery: fresh from the pipeline about TRAIL-mediated signaling and natural TRAIL sensitizers. International Journal of Molecular Sciences 20 (8), 2010.

Tartaj, P., Morales, M., Gonzalez-Carreno, T., Veintemillas-Verdaguer, S., Serna, C., 2005. Advances in magnetic nanoparticles for biotechnology applications. Journal of Magnetism and Magnetic Materials 290, 28–34.

Thangamani, S., Younis, W., Seleem, M.N., 2015. Repurposing ebselen for treatment of multidrug-resistant staphylococcal infections. Scientific Reports 5, 11596.

Treuel, L., Jiang, X., Nienhaus, G.U., 2013. New views on cellular uptake and trafficking of manufactured nanoparticles. Journal of The Royal Society Interface 10 (82), 20120939.

Vecitis, C.D., Zodrow, K.R., Kang, S., Elimelech, M., 2010. Electronic-structure-dependent bacterial cytotoxicity of single-walled carbon nanotubes. ACS Nano 4 (9), 5471–5479.

Vinoj, G., Pati, R., Sonawane, A., Vaseeharan, B., 2015. In vitro cytotoxic effects of gold nanoparticles coated with functional acyl homoserine lactone lactonase protein from Bacillus licheniformis and their antibiofilm activity against proteus species. Antimicrobial Agents and Chemotherapy 59 (2), 763–771.

Wang, L., Hu, C., Shao, L., 2017. The antimicrobial activity of nanoparticles: present situation and prospects for the future. International Journal of Nanomedicine 12, 1227.

Waseem, H., Ali, J., Sarwar, F., Khan, A., Rehman, H.S.U., Choudri, M., Arif, N., Subhan, M., Saleem, A.R., Jamal, A., 2019a. Assessment of knowledge and attitude trends towards antimicrobial resistance (AMR) among the community members, pharmacists/pharmacy owners and physicians in district Sialkot, Pakistan. Antimicrobial Resistance and Infection Control 8 (1), 67.

Waseem, H., Jameel, S., Ali, J., Saleem Ur Rehman, H., Tauseef, I., Farooq, U., Jamal, A., Ali, M., 2019b. Contributions and challenges of high throughput qPCR for determining antimicrobial resistance in the environment: a critical review. Molecules 24 (1), 163.

Waseem, H., Williams, M.R., Stedtfeld, R.D., Hashsham, S.A., 2017. Antimicrobial resistance in the environment. Water Environment Research 89 (10), 921–941.

Wei, D., Sun, W., Qian, W., Ye, Y., Ma, X., 2009. The synthesis of chitosan-based silver nanoparticles and their antibacterial activity. Carbohydrate Research 344 (17), 2375–2382.

Weist, K., Högberg, L.D., 2016. ECDC publishes 2015 surveillance data on antimicrobial resistance and antimicrobial consumption in Europe. Euro Surveillance 21 (46).

Wu, X., Tan, S., Xing, Y., Pu, Q., Wu, M., Zhao, J.X., 2017. Graphene oxide as an efficient antimicrobial nanomaterial for eradicating multi-drug resistant bacteria in vitro and in vivo. Colloids and Surfaces B: Biointerfaces 157, 1–9.

Zhao, H., Ding, R., Zhao, X., Li, Y., Qu, L., Pei, H., Yildirimer, L., Wu, Z., Zhang, W., 2017. Graphene-based nanomaterials for drug and/or gene delivery, bioimaging, and tissue engineering. Drug Discovery Today 22 (9), 1302–1317.

Zheng, K., Setyawati, M.I., Leong, D.T., Xie, J., 2017. Antimicrobial gold nanoclusters. ACS Nano 11 (7), 6904–6910.

Zou, F., Zhou, H., Jeong, D.Y., Kwon, J., Eom, S.U., Park, T.J., Hong, S.W., Lee, J., 2017. Wrinkled surface-mediated antibacterial activity of graphene oxide nanosheets. ACS Applied Materials & Interfaces 9 (2), 1343–1351.

Zou, L., Lu, J., Wang, J., Ren, X., Zhang, L., Gao, Y., Rottenberg, M.E., Holmgren, A., 2017. Synergistic antibacterial effect of silver and ebselen against multidrug-resistant Gram-negative bacterial infections. EMBO Molecular Medicine 9 (8), 1165–1178.

Zou, L., Wang, J., Gao, Y., Ren, X., Rottenberg, M.E., Lu, J., Holmgren, A., 2018. Synergistic antibacterial activity of silver with antibiotics correlating with the upregulation of the ROS production. Scientific Reports 8 (1), 11131.

CHAPTER

Treatment technologies and management options of antibiotics and AMR/ARGs

23

Luqman Riaz[1,7], Muzammil Anjum[3,7], Qingxiang Yang[1,2], Rabia Safeer[3], Anila Sikandar[3], Habib Ullah[4], Asfandyar Shahab[5], Wei Yuan[6], Qianqian Wang[1,2]

[1]*College of Life Sciences, Henan Normal University, Xinxiang, China;* [2]*Henan International Joint Laboratory of Agricultural Microbial Ecology and Technology (Henan Normal University), Xinxiang, China;* [3]*Department of Environmental Sciences, Pir Mehr Ali Shah Arid Agriculture University, Rawalpindi, Pakistan;* [4]*CAS Key Laboratory of Crust Mantle Materials and the Environments, School of Earth and Space Sciences, University of Science and Technology of China, Hefei, China;* [5]*College of Environmental Science and Engineering, Guilin University of Technology, Guilin, China;* [6]*School of Environmental and Municipal Engineering, North China University of Water Resources and Electric Power, Zhengzhou, China;* [7]*School of Materials Science and Engineering, Sun Yat-sen University, Guangzhou, China*

23.1 Introduction

The worldwide population is increasing exponentially, putting increased demand on food and better health facilities. The developments in science and technology have helped to produce a large number of antibiotics that have wide applications not only to cure various diseases in humans and animals but also extensively employed as growth stimulants in livestock, aquaculture to meet dietary demands. In earlier decades the antibiotics consumption pattern for therapeutic and nontherapeutic uses was observed much higher in developed countries, including the United States, France, and Italy but in the past few years, a drastic increase has been observed in China, India, and Pakistan (Klein et al., 2018), and it is expected that this usage will reach at 105,596 tons/year in 2030 (Pan and Chu, 2017). As a consequence, this high usage of antibiotics has generated a greater quantities of waste, which are spreading in the biosphere and posing serious threats to ecology and human health. Antibiotics reach the soil and water environments with varying concentrations through continual input from domestic waste, livestock manure, municipal wastewater treatment plants (MWWTPs), and hospital wastewater. These point sources are not only spreading antibiotics but also serving as hot spots for antibiotic resistant bacteria (ARB) and antibiotic resistance genes (ARGs). The presence of antibiotics, ARB, and ARGs in environment is a major risk to ecology and public health, which has been an important issue for many decades. Meanwhile, the persistence of antibiotics in the environment for a long time at subinhibitory concentrations causes selective pressure on microbial communities, which also results in the development of multidrug-resistant bacteria carrying multiple ARGs and recruitment of this resistance into clinics.

Antibiotics and Antimicrobial Resistance Genes in the Environment. https://doi.org/10.1016/B978-0-12-818882-8.00023-1

Antibiotic resistance has been declared as top risk by the World Health Organization (WHO) since the development of new antibiotics is very slow and the complications are increasing rapidly to treat various common infections (Carlet et al., 2012). Many developing regions of the world are facing this challenge, because the effective therapies lack for many life-threatening infections (Walsh, 2003). In the past, great attention was paid to antibiotic resistance outside the clinical premises, where the focus has been given to the factors that are contributing to spread and accretion of ARGs among resistant pathogens in the natural environment. The diffusion of antibiotic resistance in clinically relevant pathogens has increased the hospitalization and mortality rates in humans. Recent trends claim 23,000 deaths every year in the United States and 25,000 in Europe due to antibiotic resistance or infections caused by resistant bacteria (Sachdeva et al., 2017) and this will reach up to 10 million lives by 2050 globally, which will be higher than deaths due to cancer and diabetes (O'Neill, 2017). According to the WHO, around 250,000 people die each year due to drug resistance to tuberculosis, and this is the reason the WHO has declared antibiotic resistance as a major global threat to human health in the 21st century (Organization, 2017). Extensive efforts have recently been made to address the problem of antibiotic and antibiotic resistance pollution through treatment of waste by biological, physical, and chemical methods. However, in the recent past more attention has been paid by governments and international health organizations in response to increased occurrence of antibiotics and antibiotic resistance genes in the environment (Spellberg et al., 2016; Duarte et al., 2019) through various management options.

23.2 Antibiotics and antimicrobial resistance

Antibiotics are secondary metabolites produced by microorganisms (Köberl et al., 2013) considered as low-molecular and chemically heterogeneous organic compounds possessing the ability to inhibit the growth of other microorganisms (Gottlieb, 1967; Thomashow, 2002). They are active against gram-negative and gram-positive bacteria that are responsible for causing many infections related to gastrointestinal tract, respiratory tract, skin infections, and many sexually transmitted diseases. Until now, several classes of antibiotics are established, which were produced naturally as well as synthetically, helped to minimize the mortality rates due to the infections that were lethal. Based on the structure and mode of action, antibiotics have been divided into aminoglycosides, β-lactams, macrolides, quinolones, sulfonamides, and tetracyclines. The antibiotic used in humans and livestock depends upon the type of antibiotic, animal, season, and country. Antibiotics use is gradually increased in the past years and reached between 100,000 and 200,000 tons per year (Wise, 2002), making them emerging contaminants due to their long-term and synergistic effects, when they are present together in the environment even at lower concentrations (Lombardo-Agüí et al., 2014; Oliveira et al., 2015). Antibiotics when administered orally are excreted 4%–30% in the feces and 50%–80% in the urine as parent compounds in human body (Jjemba, 2006; Verlicchi et al., 2012; Al Aukidy et al., 2014). Similarly, in livestock the excretion of rate of antibiotics as parent compounds are 60% lincosamides, 50%–90% macrolides, and 75%–80% tetracyclines (Van Epps and Blaney, 2016). The maximal doses for poultry production are 29 mg/kg tetracycline and 25 mg/kg beta-lactam, but the antibiotic residues detected in the poultry manure exceed this level, indicating the misuse of antibiotics in poultry feed (McEwen and Fedorka-Cray, 2002). Various studies have reported the occurrence and prevalence of different antibiotics in wastewater (Riaz et al., 2017b, 2018), surface water (Hu et al., 2018), drinking water (Burke et al., 2016), and sediments (Li et al., 2018) due to their incomplete removal in treatment plants. The antibiotics were partially removed through biodegradation and possess high adsorption to sludge in the treatment plants. Also they are detected with varying concentrations in animal manure and enter into agricultural soil system through application of industrial sludge and manure as organic amendments. Among various classes of antibiotics, aminoglycosides and β-lactams cleavage on hydrolysis (Huang et al., 2011) and seldom are detected in the environment; whereas fluoroquinolones, macrolides, sulfonamides, and tetracyclines showed greater persistence in the wastewater and posing threats to nontarget life forms in the ecosystem and human health (Runnalls et al., 2010; Brandt et al., 2015). On the other hand sometime metabolites produced due to antibiotic consumption remain active and result in the transformation to parent compounds in the environment, e.g., N4-acetylsulfapyridine and N4-acetylsulfamethazine have changed to sulfonamide (Bonvin et al., 2012).

In the 21st century, antibiotic resistance is considered a great challenge to human health. Antibiotic resistance is the ability of bacteria to grow and proliferate in the presence of antimicrobials that were made to kill them or inhibit their growth. The long-term exposure to antimicrobials in the ecosystem enhances the selective pressure on bacteria, causing higher mutation level and introduction of resistance genes, resulting in the development of antibiotic resistant bacteria that have more rapid diffusion efficiency. More frequent genetic recombination events due to increased mutations may result in various types of resistance mechanisms involving various types of ARGs in different functions associated with antimicrobial resistance. Generally there are three types of resistance mechanisms present in bacteria, including intrinsic, acquired, and adaptive (Davies, 1994; Alekshun and Levy, 2007; Hughes and Andersson, 2017). Intrinsic resistance is the ability of bacteria to resist antibiotics independent of the antibiotic selective pressure. Acquired resistance is the result of the chromosomal mutation or horizontal gene transfer of plasmids, integrons, transposons, and gene islands through conjugation, transduction, and transformation. Adaptive resistance refers to the response of bacteria to gradual increase in the antibiotic concentration and thus help them to develop resistance profiles. Livestock manure, MWWTPs, and hospital wastewater serve as reservoirs of antibiotic resistance due to high loads of antibiotics. These reservoirs provide combination of abiotic factors (nutrient, pH, temperature, antimicrobial agents, and metals) and biotic factors (microflora, cell density), which not only set favorable growth conditions for multidrug-resistant bacteria but also determine stress induced by antimicrobial agents. In most cases, acquired resistance in bacteria was always resistant to antimicrobial agents at a low level, but some sensitive strains can also gain high ability to become resistant to various antimicrobials. Plasmids serve as vectors for ARGs transfer through horizontal gene transfer (HGT) (Luo et al., 2010) and the frequency of genetic material acquisition through HGT varied approximately from 0% to 22% in 88 prokaryotic bacterial genomes. Previously, ARGs, mobile genetic elements (MGEs), and their transferable potentials have been extensively investigated in various environments, among which livestock manure, hospital wastes or wastewater, municipal wastewater treatment systems, aquaculture systems, etc. are hot spots of ARBs, multiple antibiotic-resistant bacteria (MARB) and ARGs. Recently, different environmental compartments including soil and water environments were considered to be not only major recipients but also declared as sources and reservoirs of ARGs that have clinical importance (Martinez, 2009; Wright, 2010).

23.3 Environmental implications of antibiotics and AMR/ARGs

Antibiotics are considered as the successful therapeutic agents to cure various human and animal diseases. However, their extensive use has resulted in the occurrence and prevalence of antibiotics in environment at varying concentrations ranging from ng/L to mg/L or ng/kg to mg/kg posing serious threats to ecology and public health. The widespread use of different antibiotics has resulted in the detection of their residues in wastewater, surface water, and sediments in the range of 463 ng/L to 31 mg/L (Hu et al., 2018; Li et al., 2018; Riaz et al., 2018). The concentration of antibiotics in sludge, manure, and soil also showed greater variation and was found in range of 5.1 μg/kg to 764 mg/kg (Thiele-Bruhn, 2003; Martínez-Carballo et al., 2007; Lillenberg et al., 2010; Massé et al., 2014; Van Doorslaer et al., 2014). Various antibiotics when present in the environment are degraded slowly, which is affected by type of antibiotics, temperature (Dolliver and Gupta, 2008), type of soil, and moisture content (Stoob et al., 2007). Most importantly the release of antibiotics from livestock manure, hospital wastewater, MWWTPs, and industrial wastewater is a continuous process and the nontarget organisms are constantly exposed to these pollutants (Lindberg et al., 2007). This way the impact on metabolic activity and structure of microbial populations remains even and triggers transcriptional responses in bacteria (Tsui et al., 2004; Davies, 2006; Davies et al., 2006; Linares et al., 2006; Yim et al., 2007; Fajardo and Martínez, 2008). The presence of antibiotics and antibiotic resistance genes in natural ecosystem challenging structure and physiology of microbial populations and altering many biogeochemical cycling processes important for the ecosystem. The ARGs that were detected in clinical settings are now present in the natural environments even without any evidence of antibiotic pollution. Earlier it was reported that antibiotics at low concentrations can serve as signaling molecules and help to shape the microbial community structure (Linares et al., 2006; Yim et al., 2007; Fajardo and Martínez, 2008) but at high concentrations they will shift the original functions (Calabrese, 2004) and play their role as shields to antibiotics (Martínez, 2008). The environmental pollution by antibiotics and ARGs poses risks to algae, *Microcystis aeruginosa*, and cyanobacteria in aquatic environment (Riaz et al., 2017b) leading to accumulation in fish (Llor and Bjerrum, 2014). The antibiotics that enter the soil system cause toxicity to soil fauna and flora (Riaz et al., 2017a, 2018). The biosolids application in soil for organic amendments results in the enrichment of ARGs, which can alter community structure and spread in crops and transfer to humans through the food chain.

23.4 Treatment technologies

Antibiotics are responsible for adverse effects such as acute and chronic toxicity, disruption of photosynthetic organisms in the aquatic environment, and induction of microbial ARGs. Therefore, the treatment of these antibiotics has significant importance in the environment. The treatment of antibiotics and antimicrobial genes could be conducted using various biological, physical, and chemical methods (Fig. 23.1). The reported methods are presented in the following sections.

23.4.1 Biological methods

Many biological methods have been used for the treatment of antibiotics and antimicrobial resistance genes. These methods include aerobic treatment, anaerobic treatment, and constructed wetlands.

(i) Aerobic Treatment

Aerobic treatment is used for treatment of waste containing antibiotic residues and AMR/ARGs under oxygen-rich conditions using microorganisms. Aerobic treatment is known as an effective method that can lower the biological and chemical hazards in organic wastes (Xie et al., 2016; Gou et al., 2018). Antibiotics such as monensin, chlortetracycline, and tylosin concentrations were recorded to be reduced by 54%–99%, using aerobic treatment (Dolliver et al., 2008). During aerobic treatment, most pathogenic organisms and bacteria bearing ARGs are rapidly removed under high-temperature thermophilic aerobic digestion due to self-heating (Wang et al., 2015), however, certain ARGs may persist in the aerobic digestion (Xie et al., 2016; Gou et al., 2018). The advancements in aerobic treatment technology have been boosted in recent years, and some of these methods are explained in the following discussion.

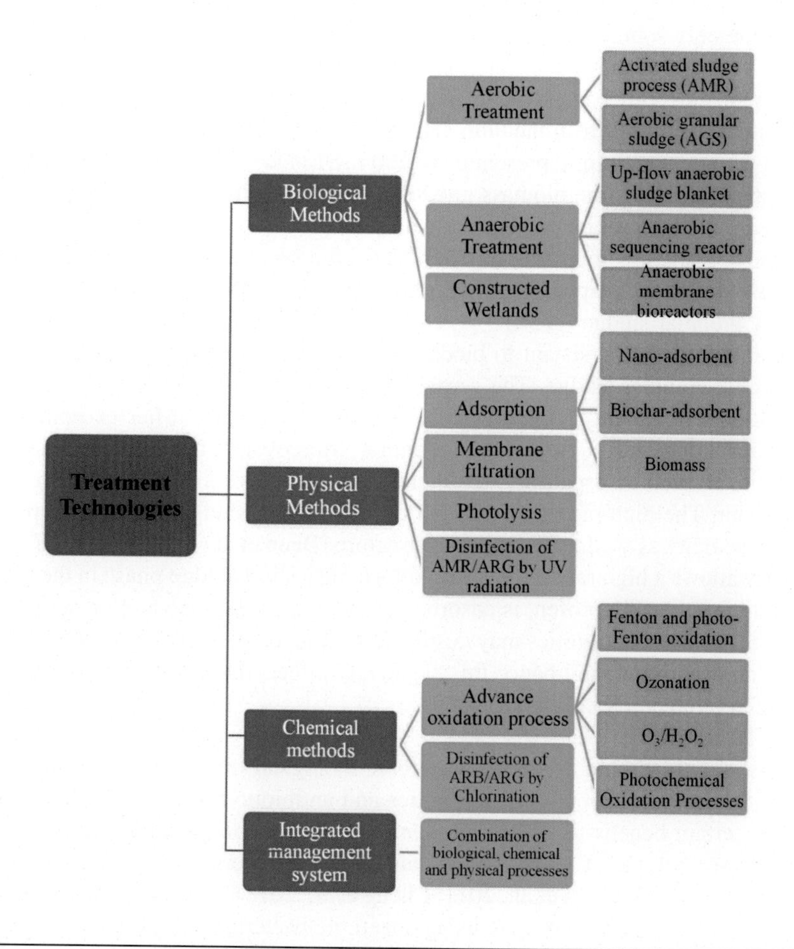

FIGURE 23.1

Applied treatment technologies for antibiotics and AMR/ARGs.

(a) Activated sludge process (AMR)

The activated sludge process is an aerobic method that can be applied for treatment of antibiotics inoculated with activated sludge. It operates at 25°C usually as batch system on draw and fill basis. The bioreactor is refilled with equal quantity of feed drawn from the reactor after a regular interval of days or specific retention time (Drillia et al., 2005). In some cases the enrichment of seed microorganisms is required in the reactor, which is done by applying some nutrient source such as $Na_2HPO_4.2H_2O$, $MgSO_4.7H_2O$, CH_3COONa, $CaCl_2.2H_2O$, KH_2PO_4, and NH_4NO_3, an antibiotic to be treated such as sulfamethoxazole, and some trace elements. Furthermore, the solution of trace elements is prepared using $FeSO_4.7H_2O$, $MnSO_4.H_2O$, $CuCl_2.2H_2O$, $Na_2MoO_4.2H_2O$, H_3BO_3 and concentrated H_2SO_4 and applied in the system. It must be noted that these chemicals are important for growth of degradation bacteria. During operation of the activated sludge process, at the time when antibiotic begins to degrade, the consortium must be supplied with increasing concentrations of antibiotic to be treated. At this stage, the bioreactor's feed must not include CH_3COONa, so that the only source of organic carbon in the bioreactor is an antibiotic. The nitrogen source NH_4NO_3 of the consortium also must be suspended to inhibit the growth of autotrophic nitrifying bacteria. The treatment process is then transferred to a new batch of the bioreactor containing fresh medium and antibiotic concentrations. The suitability of this system is to treat antibiotic present up to 100 mg/L. Due to the low organic carbon supply to the bioreactor's feed, the biomass growth decreases gradually. For the establishment of an enriched mixed microbial population, the whole process must be repeated at least five times after the degradation of an antibiotic. The bacterial species that are commonly present in the activated sludge are responsible for the biodegradation of common antibiotics.

(b) Aerobic granular sludge (AGS)

Some antibiotics are resistant to biodegradation as many pharmaceuticals are adsorbed by the aerobic granular sludge. The aerobic granular sludge has many advantages over conventional treatment technologies for antibiotic treatment. It has exceptional settling properties along with high biomass retention. Moreover, it has excellent biosorption properties as well as capability to simultaneously remove nitrogen and phosphate from wastewater. The high biomass retention of the granular sludge is due to its property of fast settling velocity as well as its compact structure (Beun et al., 1999; Liu and Tay, 2004). This property allows a high rate of organic loading and a short sludge phase in the bioreactor. The antibiotic, such as ibuprofen, is adsorbed at the granular sludge and precipitates out at the end. The adsorbed antibiotics may desorb depending on the conditions in the bioreactor, like metal ion desorption, and hence this negatively affects the antibiotic removal such as ibuprofen removal (Niemi et al., 2009).

(ii) Anaerobic Treatment

In comparison with aerobic digestion, anaerobic biological treatment is a desirable option for treatment of wastewater containing antibiotics and antibiotic residues. The later treatment method has certain benefits over aerobic treatment such as biogas production as a source of bioenergy, lesser volume of waste sludge, and decreased energy costs during operation (Angelidaki et al., 2003; Chen et al., 2011; Cheng et al., 2018). Anaerobic treatment is conducted in a closed system free from oxygen, using anaerobic bacteria. The anaerobic digestion process comprises of four basic steps, including hydrolysis, acidogenesis, acetogenesis, and methanogensis. The organic pollutants are converted into biogas and byproducts in the series of these steps. A brief representation of these anaerobic digestion processes is presented in Fig. 23.2. Various closed-system reactors have been developed, which include up-flow anaerobic sludge blanket, anaerobic sequencing reactor, and anaerobic membrane bioreactors, etc., and these are described in the following sections.

(a) Up-flow anaerobic sludge blanket

Up-flow anaerobic sludge blanket (UASB) reactor is one of the best-known anaerobic reactor types, which has the capability for both low and high temperature and high-rate process for treatment of antibiotics in wastewater. Compared to the other anaerobic methods, UASB has some extraordinary advantages, which include greater efficiency, high organic loadings, short hydraulic retention time (HRT), low energy demand, and easy reactor development. The design of UASB reactor is based on four major components: (A) sludge bed, (b) sludge blanket, (C) gas-solids separator, (D) settlement compartment (Akbarpour Toloti and Mehrdadi, 2011). For operation of UASB, the bioreactor consisting of

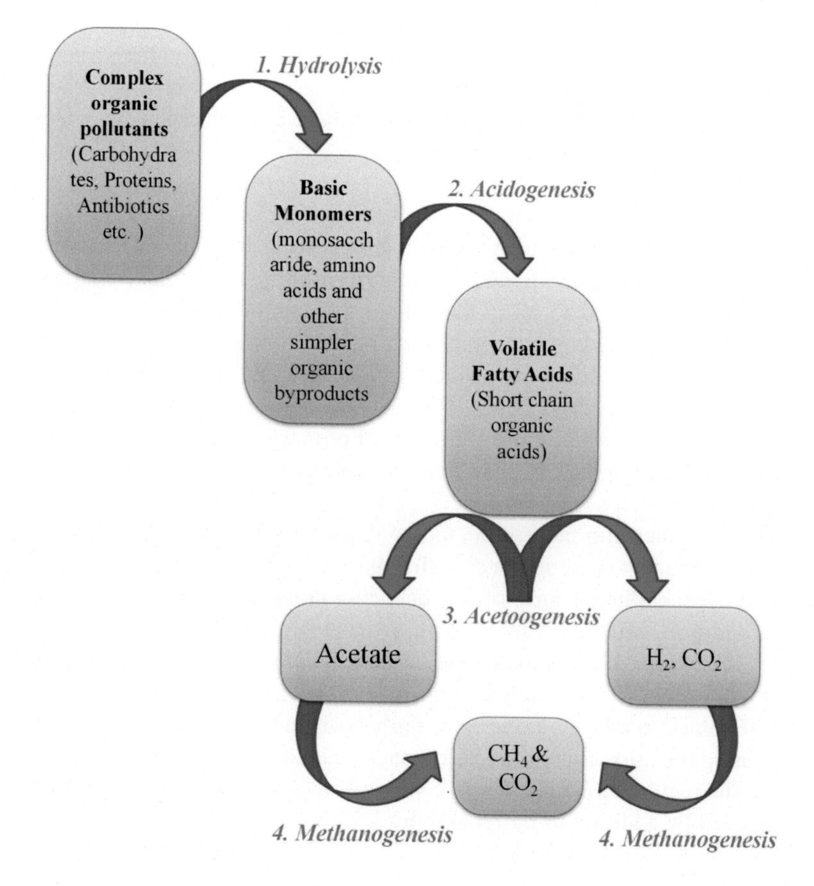

FIGURE 23.2

Anaerobic digestion processes.

upflow anaerobic sludge blanket is supplied by the wastewater containing antibiotics on a daily basis. In advance UASB, the reactor is equipped with two types of sensors; the first is set for sensing the minimum water level, and the second is set for the maximum water level in the reactor. A discharge valve is automatically operated with these two water level sensors, as it automatically opens and closes with water at it maximum and minimum levels, respectively. In some UASB systems, one manual valve is also equipped in the bioreactor for the discharge of sludge on a manual basis. The UASB can be operated with supply of more than 100 liters of wastewater daily with HRT of up to 13 days (Sponza and Demirden, 2007). For desired efficiency of the reactor, the wastewater dilution must be decreased to increase the organic loading of the reactor. The UASB system is also equipped with the heat control system; among these, the most efficient system uses a heat jacket to maintain the temperature of the reactor. Normally, the temperature at 35°C is maintained as a most suitable range for UASB (Coskun et al., 2012).

Besides several advantages, UASB has some issues that hinder the treatment efficiency of antibiotics. For application of UASB, one of the main barriers is its long start-up time, which is usually initially 2–8 months required formation of granules. This duration varies with respect to conditions and type of substrate used. Generally, the stability and performance of UASB depends mainly on sludge granulation, however, the success of the process is not only guaranteed with this, since there are numerous other factors involved (Akbarpour Toloti and Mehrdadi, 2011).

(b) Anaerobic sequencing reactor

The anaerobic sequencing reactors for treatment of antibiotics in wastewater and AMR are recognized as high-rate treatment methods that process by following four cyclic phases: feed, aerobic reaction, settling, and decantation. In the initial step, substrate (wastewater containing antibiotics) is added to the reactor and continuously mixed, thereafter, the degradation is initiated. In sequencing reactors, the volume of substrate wastewater depends on various factors, including the desired organic loading rate, HRT, and expected settling properties (Akil and Jayanthi, 2012). The sequencing reactor system exhibits several benefits such as high biomass concentration, absence of primary and secondary settler, better effluent quality control, high organic removal and gas generation production and flexible control, and high metabolic activity.

In anaerobic sequencing bioreactors the system is inoculated by granular sludge that was previously used in an anaerobic reactor for the treatment of antibiotics from pharmaceutical wastewater. Generally, the anaerobic sequencing bioreactor works on a 24 h cycle, which is of small duration required for feeding while the rest of the time is reaction time. The reactor initially spends some time settling and thereafter takes some time for the withdrawal of liquid phase (Aydin et al., 2015). Afterward, the reactor sets in to achieve a steady state to reach the point determined to add antibiotics on a daily basis for treatment. During the whole process, the hydraulic retention time is set at 2.5 days whereas the solid retention time is set for 30 days. The required pH for anaerobic sequencing reactor is in the range of 6.8–7.2, which is maintained by addition of buffer solution.

(c) Anaerobic membrane bioreactors

Anaerobic membrane bioreactors (AnMBs) combine the advantages of anaerobic digestion process with membrane separation system for the treatment of antibiotics. The high operational stability of AnMBs makes this technology efficient for treating effluent under extreme conditions, including high suspended solids content, high salinity and poor biomass granulation, etc (Dvorak et al., 2016; Radjenović et al., 2009). In membrane bioreactors, the contact time to sludge-substrate increases with the retention of activated sludge and hence this improves the biodegradability of pollutants. In AnMBs, the slow-growing bacteria are retained on the membrane and facilitate the microbial community to develop and target contaminants to be degraded (Gobel et al., 2007). AnMBs also aid in lowering the growing population of pathogenic microorganisms having more antibiotic resistance. In some cases the presence of antibiotics in AnMRs could possibly boost the rate of membrane fouling and abridge the membrane fouling cycle because of the effect of antibiotics on the microbial communities and anaerobic sludge AnMRs (Cheng et al., 2018; Zhu et al., 2018).

(iii) Constructed Wetlands

Constructed wetlands are one of the emerging wastewater treatment technologies for antibiotics. Constructed wetlands contribute to the degradation of pharmaceuticals by using natural processes with the help of plants, sunlight, and microorganisms (Dordio et al., 2010). Where the microbial community is present in constructed wetlands at larger area provide more exposure to antibiotics (Nolvak et al., 2013). The type of bacterial community present in the constructed wetland determines its ability to remove the antibiotics from the pharmaceutical waste. The high concentration of antibiotics might induce changes in the composition of the bacterial community and hence reduce major impacts on the functionality of the constructed wetlands (Scholz and Lee, 2005). The constructed wetlands are energy efficient systems as they require a small amount of energy during the treatment process. The constructed wetlands require high ratio of surface per equivalent inhabitant to acquire quality parameters for wastewater, thus it is better to construct them in small urban areas. Various studies have reported the efficient removal of antibiotics using wetlands, for instance, sulfamethoxazole and sulfapyridine antibiotics were successfully removed using constructed wetlands (Park et al., 2009).

23.4.2 Physical methods

(i) Adsorption

The removal of antibiotics by the adsorption method using adsorbent material is a cost-effective and efficient process. Adsorption is one of the best physical methods of antibiotic removal due to its unique characteristics (Ali and Gupta, 2006; Ali, 2014). Due to the high removal and adsorption capacity for certain organic compounds, the adsorption is widely used to remove organic antibiotics from pharmaceutical wastewater. Current research has shown that multiple mechanisms of action of adsorption process, for instance, π-π interactions, hydrophobic effect, covalent bonding, hydrogen bonds, and electrostatic interactions (Peng et al., 2016). The effectiveness of the adsorption reflects the physicochemical properties of the adsorbent material used, such as porosity, specific surface area, morphology of the material and surface polarity, as well as the adsorbate characteristics such as polarity, size, hydrophobicity, and ionic charge (Huizar-Felix et al., 2019). For adsorption process, various types of nano-adsorbent materials have been successfully applied for removal of antibiotics such as nanomaterial biochar and biomass.

(a) Nano-adsorbent
In recent years nano-adsorbents, including iron particles, graphene-like materials, and carbon nanotubes, have been widely used for several organic compounds including antibiotics. The nano-adsorbents not only have the capacity to adsorb antibiotics but also capacity to adsorb other contaminants from wastewater (Ali, 2012). For testing efficiency of nanomaterial adsorbents, various lab-based experiments have been successfully conducted. Under laboratory-controlled conditions, the adsorption is usually conducted in hermostatic water bath shaker and a batch reactor can be used for desorption with nano-adsorbents. The optimum pH for iron-based nanomaterial is in the range of 4–6, whereas temperature could be in the range of 20–30°C. In recent years environmentally friendly methods for nano-adsorbent synthesis have been established, which are called green methods (Ali et al., 2015, 2016a). In order to further improve the properties of nano-adsorbents, the composite materials are successfully synthesized, for example, iron nanoparticles [Fe(II) to Fe(0)] and carboxymethyl cellulose could be used prepare the composite iron nano-adsorbents. In general, the adsorption of antibiotics using nano-adsorbents is a well-established, fast, and economically feasible approach (Ali et al., 2016b).

(b) Biochar-adsorbent
Biochar is a carbon-rich material developed by high-temperature decomposition (pyrolysis) of various kinds of biomass including wood, leaves, manure, agriculture residues and waste, etc. The biochar is prepared in a limited oxygen supply in complete absence of oxygen (Qadeer et al., 2014). Thermal pyrolysis of biomass produces a biologically and chemically stable pool that contains a high-carbon fraction within the produced biochar (Pratt and Moran, 2010). The use of biochar as adsorbent could achieve high-adsorption efficiency toward antibiotics. Biochar provides a larger surface area and pore volume to antibiotics for adsorption, however, the adsorption capacity of the system depends on the pH of the solution, humic acid, and electrolytes (Zeng et al., 2018). Biochar-adsorption is an economically feasible method used for the removal of organic and inorganic pollutants as well as heavy metals from wastewater (Jiang et al., 2017). Biochar obtained from carbon-rich biomass has unique benefits such as soil amendment, pollution control, and carbon sequestration (Tan et al., 2015). It is more efficient than activated carbon in terms of removing various pollutants (Ahmad et al., 2014). The most unique advantage of biochar use as adsorbent is its low cost with high-adsorption capacity. The quality of the biochar widely depends upon the conditions during the pyrolysis process, which greatly manipulate the biochar-adsorption efficiency for different antibiotics. Temperature also plays an important role in speeding up the biochar's carbonization as more carbon content is produced at higher temperature; high temperature during pyrolysis results in a loss of hydrogen, nitrogen, and sulfur that ultimately breaks the weak bonds of the biochar structure (Demirbas, 2004).

(ii) Membrane filtration
Membrane filtration appears to deliver as a rational option, and has been widely used in the water purification of pharmaceutical industries. Until now, most of the membranes that have been developed have been based on polymeric materials, including polysulfone, polyimide, and cellulose (Liu et al., 2017), however, nano-material-based membranes have also been developed in recent years. The use of nano-filtration and reverse osmosis membranes are the most efficient membrane filtration techniques for the antibiotics' removal from liquid waste (Nghiem et al.,

2005). In membrane filtration reverse osmosis technique is applied for the removal of pollutants from pharmaceutical wastewater. In the reverse osmosis filtration a flat sheet membrane is used for treatment purposes (Li et al., 2004). The temperature of the reverse osmosis filtrate is maintained between 21 and 23°C by using a chiller. For the removal of particulate matters from the waste liquor a coarse ultra-filtration method is used prior to reverse osmosis filtration. The pH of the wastewater containing antibiotics is required to be in range of 4.0–2.0 for membrane filtration, which is maintained by adding acidic buffer by the use of membrane containing reverse osmosis along with ultra-filtration, and the recovery of antibiotics such as oxytetracycline may efficiently be achieved from wastewater. In addition to antibiotics, the removal of trace organic compounds is also possible by the use of activated carbon adsorption and membrane filtration (Koyuncu et al., 2008). It is a little difficult to use filtration membranes in the wastewater treatment plants as there are multiple species that may interact with the membrane surfaces.

(iii) Photolysis

Antibiotics are more highly soluble in water than other pharmaceuticals used by humans and animals. Due to this fact, oxidative or transformative processes for antibiotic treatment such as photolysis are more effective than sorption-based treatment technologies. Moreover, in nitrate photolysis of effluents, the oxidative radicals are produced other than OH radicals when exposed to 200–240 nm wavelength of light (Keen and Linden, 2013). These radicals possibly take part in reactions with antibiotics. The direct photolysis of antibiotics, such as erythromycin, dehydrates the antibiotics at faster rate even at low or medium pressures. This dehydration rate is not fast enough for some antibiotics exposed to direct light, as some antibiotics have low absorbance at wavelengths greater than 215 nm (Dreassi et al., 2000). Antibiotics that have transformation rate of less than 15% exposed to direct photolysis are characterized as nonsusceptible antibiotics. Direct photolysis of antibiotic can occur under sunlight absorption, however, indirect photolytic process involves the reactions with highly reactive species induced under exposure of light, theses species includes, hydroxyl radicals (HO•), singlet oxygen (1O2), and chromophoric dissolved organic matter with the triplet excited state (3CDOM) (Lastre-Acosta et al., 2019).

(iv) Disinfection of AMR/ARGs by UV radiation

Antibiotics and ARGs are of great concern in the environment as they engenders harm to human beings as well as to the environment. The use of UV radiation is one of the highly recognized effective disinfection processes applied in industrial wastewater. While applying UV radiation, during the gradual increase in dosage during the treatment process through ultraviolet disinfection, many of the antibiotic-resistant genes start to decrease. Moreover, the ARGs' inactivation increases with the increase in the intensity of ultraviolet irradiation. The RNA and DNA of the cells absorb the UV light energy l via ultraviolet-transparent structures (Dodd, 2012). So, the continuous exposure to UV results in rapid inactivation of ARGs. For efficient utilization of technology the UV irradiation dosage is required to be optimized to achieve maximum efficiency. Low UV dosage causes a slight change in the number of bacteria; as well, no change occur in the cell permeability. On the other hand, the high dosage of UV irradiation results in the transfer of antibiotic resistant genes (Guo et al., 2015). The use of UV irradiation does not cause any damage to the cell membrane but it directly affects the plasmids with antibiotic resistant genes resulting in the loss of donor or receiver. The UV irradiation is more beneficial than the use of chlorination (Guo et al., 2015) as chlorination enhances the transfer of AGR in wastewater containing NH_3–N.

23.4.3 Chemical methods

In many cases the biological and physical processes are insufficient for complete removal of antibiotics and AMR/ARGs, therefore chemical treatment or oxidation are applied for complete degradation of contaminants and efficient disinfection of wastewater. These mainly include advanced oxidation processes such as ozonation, Fenton, photo-penton, and photocatalysis etc, and chlorination for disinfection.

(i) Advanced oxidation process
Oxidation of organic contaminants is the process in which complex and harmful organic compounds are converted into small and comparatively less nontoxic inorganic molecules using advance oxidation process (Ince and Apikyan, 2000). These complex compounds are usually unable to degrade through biological processes, whereas the advanced oxidation processes (AOPs) produce the hydroxyl radical (OH) in adequately large concentrations so that they act as very strong oxidants in the antibiotic treatment process (Azbar et al., 2005). These hydroxyl radicals are considered as very strong oxidants that are produced in sufficiently high concentration to affect the quality of water (Kurt et al., 2017). AOPs are a novel and proficient technique for the treatment of antibiotics in wastewater. Hydroxyl radicals that are formed as a result of AOPs have the capability to degrade all compounds whose oxidation is impossible through conventional oxidants like chlorine, ozone, and oxygen (Munter, 2001). There are different ways of production of hydroxyl radicals through advance oxidation processes that enhance the versatility of this technique by facilitating compliance with the definite treatment necessities. There are two basic steps in advanced oxidation process: (1) production of hydroxyl radical and (2) oxidation reactions (Mandal et al., 2004). During AOP the dissolved organic pollutants are converted into CO_2 and H_2O.
Several studies have reported on the effectiveness of AOP treatment for removal of antibiotics in wastewater effluents (Adams et al., 2002; Arslan-Alaton and Dogruel, 2004; Naddeo et al., 2009; Elmolla and Chaudhuri, 2011). The effectiveness of oxidative processes for degrading antibiotics will be largely determined by the specific water matrix. However, the effects of water matrix quality on antibiotics removal are less well understood than for other technologies. For example, the presence of natural dissolved organic matter can result in the formation of oxidation by-products that may cause water quality to deteriorate beyond its initial state of contamination. In the same way, the efficiency of selected advanced oxidation process eventually reduced due to the presence of nitrates and carbonates that impede with the demolition of target antibiotics. The various types of oxidants/catalysts used for removal of antibiotics or AMR/ARGs in AOPs are summarized in Box 23.1.
AOPs are usually classified into two groups: nonphotochemical and photochemical. In nonphotochemical processes ozonation, cavitation, Fenton and Photo-Fenton, ozone/hydrogen peroxide and wet air oxidation, etc. are included. Whereas photochemical AOPs include homogeneous and heterogeneous processes such as photocatalysis (Litter, 2005; Kurt et al., 2017). Among these, the technologies applied for removal of antibiotics and AMR/ARGs are explained in the following sections.

(a) Fenton and Photo-Fenton oxidation
Fenton oxidation process is a very effective process for the degradation of antibiotics and disinfection of various ARB/ARGs. In AOP the highly active OH oxidative species are

produced as a result of Fenton oxidative reactions using H_2O_2 and iron salt (Fe(II)) as described below (Pliego et al., 2015).

$$Fe^{2+} + H_2O_2 \rightarrow Fe^{3+} + OH^- + \cdot OH$$

Compared to the Fenton oxidation, Photo-Fenton is a slightly modified process that occurs by the combination of UV radiation, H_2O_2 and Fe(II) or Fe(III). In this system, H_2O_2 as an oxidizing agent and the iron salts act as photocatalysts that produce hydroxyl radicals at greater extent. The Photo-Fenton reaction is described below (Kurt et al., 2017):

$$Fe(OH)^{+2} + \text{UV radiation} \rightarrow \bullet OH + Fe^{+3}$$

A study was conducted by Karaolia et al. (2014) in which they found that the Fenton oxidation successfully degrades various antibiotics like sulfamethoxazole, clarithromycin, etc., and also play a key role in disinfection and inactivation of *Enterococci* and ARB resistant to sulfamethoxazole and clarithromycin, respectively. Solar Fenton method also works efficiently in removal antibiotics and ARB/ARGs present in wastewater. Through solar Fenton

BOX 23.1 TYPES OF OXIDANTS/CATALYSTS USED IN AOPS FOR REMOVAL OF ANTIBIOTICS OR AMR/ARGS

Ozonation:

- O_3

Fenton:

- Fe^{2+}
- $Fe^{2+} + H_2O_2$

Photo-Fenton:

- TiO_2 UV-Fe^{2+}
- Solar-Fe^{2+}
- Solar $Fe^{2+} + H_2O_2$
- $Fe_2+ + H_2O_2$
- VUV/Fe^{2+}
- Fe–TiO_2
- Fe_2O_3–TiO_2 film

Photocatalyst:

- TiO_2
- ZnO
- g–C_3N_4
- TiO_2@g–C_3N_4
- Fe_3O_4/ZnO
- Cu@TiO_2
- N–TiO_2/graphene
- BiOBr/TiO_2
- TiO_2/carbon dots
- Au–CuS–TiO_2
- S/TiO_2
- N/TiO_2

process 5-log reduction in the *Enterococci* resistance to antibiotics was also observed (Karaolia et al., 2014).

(b) Ozonation

Ozone is considered as the most powerful oxidant and it is specifically used for wastewater treatment process (Litter, 2005). Ozonation generates strong oxidizing agents that have potential to react with antibiotics and therefore can help in removal. Many studies have focused on the treatment process of antibiotics from wastewater through ozonation. The general treatment process during ozonation follows the generation of hydroxyl radical:

$$O_3 + HO^- \rightarrow HO_2^- + O_2$$

$$O_3 + HO_2^- \rightarrow HO_2\cdot + O_3^-$$

$$O_3^- + H^+ \rightarrow HO_3$$

$$HO_3 \rightarrow OH + O_2$$

Huber et al. (2005) and Hollender et al. (2009) considered ozonation as a most promising approach to treat antibiotics. With the help of ozone, several antibiotics like b-lactams, sulfonamides, macrolides, quinolones, trimethoprim, and tetracyclines get distorted through direct oxidation during ozonation of wastewater. However, many other antibiotics like G, cephalexin and N4-acetylsulfamethoxazole were greatly change by hydroxyl radicals (Dodd et al., 2006). The main function of ozonation of wastewater is that it deactivates the antibiotic bacterial characteristics by changing or breaking their functional sites (Lange et al., 2006; Dodd et al., 2009), like aniline moieties of sulfonamides (Huber et al., 2005), thioether groups of penicillin, unsaturated bonds of cephalosporin, and the phenol ring of trimethoprim (Dodd et al., 2009). Antibiotics removal from wastewater depends on the concentration of ozone used in the treatment process and by their contact time. Recent studies reported that the 80% removal of sulfonamides, trimethoprim, and macrolides was observed by using ozone in treatment process with a concentration of 2 mg/L Huber et al. (2005) and Hollender et al. (2009). As the ozone concentration was increased (7.1 mg/L) the antibiotics removal rate was also increased upto 95% (Hollender et al., 2009). The treatment of antibiotics through ozonation was proved to be an efficient process for the degradation of b-lactams, macrolides, sulfonamides, trimethoprim, quinolones, tetracyclines, and lincosamides (Michael et al., 2013).

(c) Photochemical oxidation processes

Photochemical oxidation is a well-established process that operates with semiconductor material for removal of organic and microbial pollutants from wastewater such as antibiotics and antibiotic resistant bacteria (Anjum et al., 2017a, 2017b, 2018a, 2018b; Alafif et al., 2018, 2019). A semiconductor material comprises of two energy bands, which include a valence band (low energy) and a conduction band (high energy). This type of photoinduced chemical oxidation is responsible for the generation of OH radicals in solution through semiconducting material that acts as catalyst. The antibiotics can be efficiently degraded into nontoxic simple species under exposure of sunlight, artificial visible light, or UV light, due to the generation of active oxidizing species, i.e., OH, O_2 by photocatalysts (Li et al., 2019).

Among various semiconductor catalyst, TiO_2, ZnO, and strontium titanium trioxide have been applied for commercial implementation. The conduction band and valence bands of a catalyst material are usually distinguished by their energy band gap (Kurt et al., 2017). Each type of catalyst has its own band gap and capability to absorb a specific wavelength of light spectrum, thus called a photocatalyst. A brief description of the mechanism of photocatalytic oxidation of antibiotics and antimicrobial resistant bacteria is presented in Fig. 23.3.

Among various photocatalysts, TiO_2 is the most common and widely used material having strong UV exposure that is shown to degrade various antimicrobial chemicals such as triclocarban (265 nm, 300 W) (Ding et al., 2013), triclosan (365 nm, 15 W) (Yu et al., 2006), 1,4-dioxane (10–400 nm, 100 W) (Alvarez-Corena et al., 2016), and to inhibit growth of microorganisms by disruption of cell membranes and/or by disintegrating stands of DNA (Kim et al., 2013; Liu and Yang, 2003; Sirelkhatim et al., 2015). In comparison to TiO_2, ZnO can be used as an alternative photocatalyst due to its low cost and light absorbance capability in broad UV spectrum, 245–380 nm (Akhmal Saadon et al., 2016; Hwangbo et al., 2019).

(ii) Disinfection of ARB/ARGs by chlorination

Among various disinfection methods, chlorination is the most common disinfection process widely applied worldwide for the treatment of wastewater. Pharmaceutical wastewater contains harmful pathogens and ARB/ARGs that can be removed by chlorination, through oxidation of cellular materials. Moreover, chlorination is considered as the most cost-effective and extensively applied technology for disinfection process in the world (Destiani and Templeton, 2019). Studies have also been conducted reporting use of chlorination process for the removal of antibiotics and ARB/ARGs in wastewater treatments. Chlorination of wastewater requires high concentration of

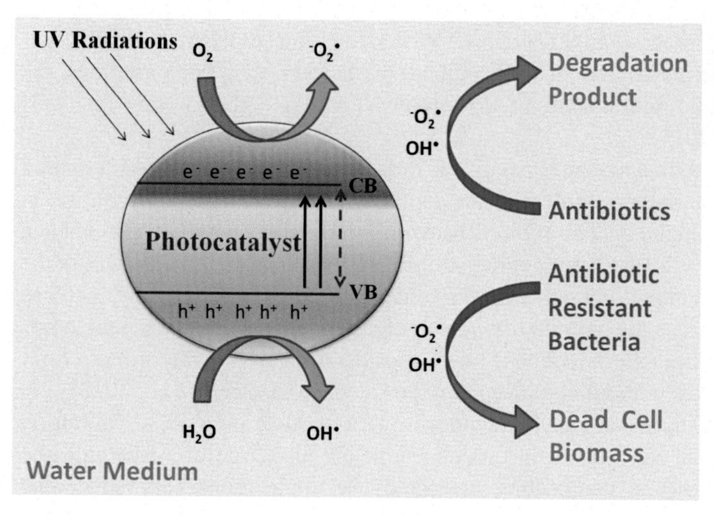

FIGURE 23.3

Mechanism of photocatalytic removal of antibiotics and antibiotic resistant bacteria from wastewater.

free chlorine and long retention time. The chemical reactions that occur during chlorination are as follows:

$$Cl_2 + H_2O \rightarrow HOCl + HCl$$

$$NaOCl + H2O \rightarrow OCl^- + OH^-$$

In one study Li and Zhang (2011) reported up to 91% removal of cephalexin after secondary treatment, and combination improved the removal of cephalexin up to 99%. Different chlorinated species have different levels of potential of removing antibiotics from wastewater. For example, utmost standard oxidation potential was observed in hypochlorite, i.e., (E0 ¼ 1.48 V), whereas, chlorine gas and chlorine dioxide showed less oxidation potential as compared to hypochlorite, i.e., (E0 ¼ 1.36 V) and (E0 ¼ 0.95 V), respectively (Homem and Santos, 2011). Some studies also reported that antibiotic resistance genes couldn't be wiped out as a result of chlorination process, but this process has tendency to inactivate and reduce the number of counts of bacterial ARB. Besides many advantages, there are some drawbacks of using chlorination process for antibiotic removal. First, since chlorine is a very harmful gas, there are numerous potential dangers linked with their usage. Secondly, the produced byproducts as a result of disinfection process may also have some potential risks (Li and Zhang, 2011).

23.4.4 Integrated treatment system

The integrated approach plays a great role in complete treatment of antibiotics and AMR/ARG containing wastewater. The combination of various biological technologies with pre- or post-treatment has special significance to improve water quality. Generally, the main objective of wastewater treatment is to manage biodegradable organic waste, commonly pursued by biological methods such as activated sludge process and integrated with physicochemical process either as primary treatment or tertiary treatment. Physicochemical treatments are not only applied as a prebiological step but sometimes also as a posttreatment process either alone or in combination, especially for pharmaceutical wastewater. Presently, the integrated pharmaceutical wastewater treatment employs the following methods: coagulation and sedimentation, adsorption, flotation, AOPs, photocatalysis, electrolysis, and so on (Li and Li, 2015).

There are numerous integrated management systems in order to treat antibiotic residues and mitigate antibiotic resistance in wastewater treatment plants, agriculture sector, and in aquaculture. In wastewater treatment plants, use of high concentration of bacteria in activated sludge might enhance the growth of pathogens flushed down the drain containing antibiotics resistance genes. Whereas in the agriculture sector numerous amounts of antibiotics are used in order to keep livestock healthy, increase food quality as well as quantity, increase the germination rate and growth of crops, and keep the crops resistant from various diseases. Such practices require a lot of consideration globally in order to limit their use and in the use of alternatives without compromising productivity (Pruden et al., 2013).

In the livestock sector, animal manure contains high concentration of ARB and ARGs. Therefore, many management practices are used for manure in order to limit the discharge of ARGs into the environment. Management practices like composting and land use applications can reduce the intensity of antibiotic resistance genes in soil and a lot of research work in this area is now under consideration. In order to reduce the runoff into nearby streams and lakes to control the flow of different pathogens, buffer zones for the manure application and related products are commonly implemented as a management approach. These management approaches and their implementation are now emerging with great interest. According to Safety and Modernization Act in the United States for harvesting of vegetables, farmers must wait 120 days after the application of manure in soil in order to control antibiotic resistance. This waiting period also depends on the type of vegetables and soil. Most soils are more effectual for antibiotic resistance attenuation. In the case of type of vegetables, underground vegetables have more exposure to soil bacteria whereas leafy ones have larger surface areas for bacterial colonization (Pruden et al., 2013).

In aquaculture most of the antibiotics are directly used for treating various diseases of aquatic animals and these practices are of greater concern than any other practices used in agriculture or treatment plants, because direct use of antibiotics in water in high quantity causes serious impacts. Therefore, their direct use in the water should be avoided in order to control antibiotic resistance in the environment. Antibiotic manufacturing industries are also a main cause of antibiotic resistance prevalence in the environment. Most of the antibiotics are discharged into the environment from the manufacturing industries in the form of waste. According to a study conducted in 2008, it has been reported that the quantity of antibiotics found in an Indian river near to an antibiotic manufacturing site was much greater than the typical dose of antibiotics prescribed to a patient (Fick et al., 2009). Therefore, waste of these manufacturing industries should be managed before their release to avoid antibiotic prevalence in the environment (Pruden et al., 2013).

23.5 Management options to minimize antibiotic and AMR release

Antibiotics serve as effective therapies against various types of infections caused by gram-positive and gram-negative bacteria, but their misuse results in the spread of antibiotics and antibiotic resistance in the environment. The spread of antibiotics and antibiotic resistance is a global challenge and must be addressed at the local or regional levels to minimize the effects caused by these pollutants. For effective management regimes at local and regional levels of the use of antibiotics in clinics and agriculture together, the promotion of sanitation and hygiene is required. Once effective management options are employed at the local level, this will have positive impact at regional or global levels. Here are few suggestions for the effective management of antibiotics to reduce the spread of antibiotic resistance:

- The use and type of critical antibiotics should be restricted for subtherapeutic use and optimized use of antibiotics in the agriculture sector should be promoted.
- The use of alternatives such as biocides and heavy metals with antibiotics in animal feed should be banned since they have enhanced the spread and propagation of antibiotic resistance.
- Treatment of animal waste containing antibiotics and antibiotics resistance genes through composting before application to agricultural soil as organic fertilizer. This will help to degrade antibiotics and minimize copies of resistance genes.
- Effective strategies for sanitation and treatment of waste, particularly in the developing world, where they lack these facilities.

- Strict rules and regulations following implementation of National Environmental Quality Standards.

23.6 Conclusion

Antibiotic residues in environment are emerging concern as dangerous contaminants, which have also promoted the distribution of antibiotic resistance genes and antibiotic resistant bacteria. Due to complexity and extensive variety of these contaminants, the current wastewater treatment plants are insufficient in design and treatment efficiency for the removal of antibiotics and AMR/ARGs, pose serious hazards to health and ecosystems. The treatment of these contaminants can be accomplished by various methods such as biological, chemical, and physical means. However, the integrated approach has a great deal for complete treatment of antibiotics and AMR/ARGs in wastewater. The combination of different technologies could be useful in harvesting the specific potential, thus providing special ability to improve water quality. Generally, the basic aim of wastewater treatment is to treat biodegradable organic waste using various methods such as activated sludge process or anaerobic digestion, which can be combined with advanced chemical and physical methods. Adsorption using nanoadsorbent materials and photochemical oxidation systems have gained great importance in this regard compared to the other methods as mentioned in the previous section. As well, treatment is not the only option; use of prior management practices could also lower the problem of antibiotics contamination. Thus, effective management regimes are of utmost importance and required together with promotion of sanitation and hygiene in clinics, industry, and wherever applicable with respect to use of antibiotics.

References

Adams, C., Wang, Y., Loftin, K., Meyer, M., 2002. Removal of antibiotics from surface and distilled water in conventional water treatment processes. Journal of Environmental Engineering 128, 253–260.

Ahmad, M., Moon, D.H., Vithanage, M., Koutsospyros, A., Lee, S.S., Yang, J.E., Lee, S.E., Jeon, C., Ok, Y.S., 2014. Production and use of biochar from buffalo-weed (*Ambrosia trifida* L.) for trichloroethylene removal from water. Journal of Chemical Technology and Biotechnology 89, 150–157.

Akbarpour Toloti, A., Mehrdadi, N., 2011. Wastewater treatment from antibiotics plant. International Journal of Environmental Research 5, 241–246.

Akhmal Saadon, S., Sathishkumar, P., Mohd Yusoff, A.R., Hakim Wirzal, M.D., Rahmalan, M.T., Nur, H., 2016. Photocatalytic activity and reUASBility of ZnO layer synthesized by electrolysis, hydrogen peroxide and heat treatment. Environmental Technology 37, 1875–1882.

Akil, K., Jayanthi, S., 2012. Anaerobic sequencing batch reactors and its influencing factors: an overview. Journal of Environmental Science and Engineering 54, 317–322.

Al Aukidy, M., Verlicchi, P., Voulvoulis, N., 2014. A framework for the assessment of the environmental risk posed by pharmaceuticals originating from hospital effluents. The Science of the Total Environment 493, 54–64.

Alafif, Z.O., Anjum, M., Ansari, M.O., Kumar, R., Rashid, J., Madkour, M., Barakat, M.A., 2019. Synthesis and characterization of S-doped-rGO/ZnS nanocomposite for the photocatalytic degradation of 2-chlorophenol and disinfection of real dairy wastewater. Journal of Photochemistry and Photobiology A: Chemistry 377, 190–197.

Alafif, Z.O., Anjum, M., Kumar, R., Abdelbasir, S.M., Barakat, M.A., 2018. Synthesis of CuO—GO/TiO_2 visible light photocatalyst for 2-chlorophenol degradation, pretreatment of dairy wastewater and aerobic digestion. Applied Nanoscience 1—13.

Alekshun, M.N., Levy, S.B., 2007. Molecular mechanisms of antibacterial multidrug resistance. Cell 128, 1037—1050.

Ali, I., 2012. New generation adsorbents for water treatment. Chemical Reviews 112, 5073—5091.

Ali, I., 2014. Water treatment by adsorption columns: evaluation at ground level. Separation and Purification Reviews 43, 175—205.

Ali, I., Alothman, Z., Al-Warthan, A., 2016b. Sorption, kinetics and thermodynamics studies of atrazine herbicide removal from water using iron nano-composite material. International Journal of Environmental Science and Technology 13, 733—742.

Ali, I., Al-Othman, Z.A., Alharbi, O.M., 2016a. Uptake of pantoprazole drug residue from water using novel synthesized composite iron nano adsorbent. Journal of Molecular Liquids 218, 465—472.

Ali, I., ALOthman, Z.A., Sanagi, M.M., 2015. Green synthesis of iron nano-impregnated adsorbent for fast removal of fluoride from water. Journal of Molecular Liquids 211, 457—465.

Ali, I., Gupta, V., 2006. Advances in water treatment by adsorption technology. Nature Protocols 1, 2661.

Alvarez-Corena, J.R., Bergendahl, J.A., Hart, F.L., 2016. Advanced oxidation of five contaminants in water by UV/TiO_2: reaction kinetics and byproducts identification. Journal of Environmental Management 181, 544—551.

Angelidaki, I., Ellegaard, L., Ahring, B.K., 2003. Applications of the anaerobic digestion process. In: Biomethanation II. Springer, pp. 1—33.

Anjum, M., Oves, M., Kumar, R., Barakat, M.A., 2017a. Fabrication of ZnO-ZnS@polyaniline nanohybrid for enhanced photocatalytic degradation of 2-chlorophenol and microbial contaminants in wastewater. International Biodeterioration & Biodegradation 119, 66—77.

Anjum, M., Kumar, R., Barakat, M.A., 2017b. Visible light driven photocatalytic degradation of organic pollutants in wastewater and real sludge using ZnO—ZnS/Ag_2O—Ag_2S nanocomposite. Journal of the Taiwan Institute of Chemical Engineers 77, 227—235.

Anjum, M., Kumar, R., Abdelbasir, S.M., Barakat, M.A., 2018a. Carbon nitride/titania nanotubes composite for photocatalytic degradation of organics in water and sludge: pre-treatment of sludge, anaerobic digestion and biogas production. Journal of Environmental Management 223, 495—502.

Anjum, M., Kumar, R., Barakat, M.A., 2018b. Synthesis of Cr_2O_3/C_3N_4 composite for enhancement of visible light photocatalysis and anaerobic digestion of wastewater sludge. Journal of Environmental Management 212, 65—76.

Arslan-Alaton, I., Dogruel, S., 2004. Pre-treatment of penicillin formulation effluent by advanced oxidation processes. Journal of Hazardous Materials 112, 105—113.

Aydin, S., Ince, B., Cetecioglu, Z., Arikan, O., Ozbayram, E.G., Shahi, A., Ince, O., 2015. Combined effect of erythromycin, tetracycline and sulfamethoxazole on performance of anaerobic sequencing batch reactors. Bioresource Technology 186, 207—214.

Azbar, N., Kestioğlu, K., Yonar, T., 2005. Application of advanced oxidation processes (AOPs) to wastewater treatment. In: Burk, A.R. (Ed.), Case Studies: Decolourization of Textile Effluents, Detoxification of Olive Mill Effluent, Treatment of Domestic Wastewater. Water Pollution: New Research. Nova Science Publishers, New York, pp. 99—118.

Beun, J., Hendriks, A., Van Loosdrecht, M., Morgenroth, E., Wilderer, P., Heijnen, J., 1999. Aerobic granulation in a sequencing batch reactor. Water Research 33, 2283—2290.

Bonvin, F., Omlin, J., Rutler, R., Schweizer, W.B., Alaimo, P.J., Strathmann, T.J., McNeill, K., Kohn, T., 2012. Direct photolysis of human metabolites of the antibiotic sulfamethoxazole: evidence for abiotic back-transformation. Environmental Science & Technology 47, 6746—6755.

Brandt, K.K., Amézquita, A., Backhaus, T., Boxall, A., Coors, A., Heberer, T., Lawrence, J.R., Lazorchak, J., Schönfeld, J., Snape, J.R., 2015. Ecotoxicological assessment of antibiotics: a call for improved consideration of microorganisms. Environment International 85, 189–205.

Burke, V., Richter, D., Greskowiak, J., Mehrtens, A., Schulz, L., Massmann, G., 2016. Occurrence of antibiotics in surface and groundwater of a drinking water catchment area in Germany. Water Environment Research 88, 652–659.

Calabrese, E.J., 2004. Hormesis: a revolution in toxicology, risk assessment and medicine: Re-framing the dose–response relationship. EMBO Reports 5, S37–S40.

Carlet, J., Jarlier, V., Harbarth, S., Voss, A., Goossens, H., Pittet, D., 2012. Ready for a world without antibiotics? The pensières antibiotic resistance call to action. BioMed Central.

Chen, Z., Wang, H., Chen, Z., Ren, N., Wang, A., Shi, Y., Li, X., 2011. Performance and model of a full-scale up-flow anaerobic sludge blanket (UASB) to treat the pharmaceutical wastewater containing 6-APA and amoxicillin. Journal of Hazardous Materials 185 (2–3), 905–913.

Cheng, D., Ngo, H.H., Guo, W., Liu, Y., Chang, S.W., Nguyen, D.D., Nghiem, L.D., Zhou, J., Ni, B., 2018. Anaerobic membrane bioreactors for antibiotic wastewater treatment: performance and membrane fouling issues. Bioresource Technology.

Coskun, T., Kabuk, H., Varinca, K., Debik, E., Durak, I., Kavurt, C., 2012. Antibiotic Fermentation Broth Treatment by a pilot upflow anaerobic sludge bed reactor and kinetic modeling. Bioresource Technology 121, 31–35.

Davies, J., 1994. Inactivation of antibiotics and the dissemination of resistance genes. Science 264, 375–382.

Davies, J., 2006. Are antibiotics naturally antibiotics? Journal of Industrial Microbiology and Biotechnology 33, 496–499.

Davies, J., Spiegelman, G.B., Yim, G., 2006. The world of subinhibitory antibiotic concentrations. Current Opinion in Microbiology 9, 445–453.

Demirbas, A., 2004. Effects of temperature and particle size on bio-char yield from pyrolysis of agricultural residues. Journal of Analytical and Applied Pyrolysis 72, 243–248.

Ding, S.L., Wang, X.K., Jiang, W.Q., Meng, X., Zhao, R.S., Wang, C., Wang, X., 2013. Photodegradation of the antimicrobial triclocarban in aqueous systems under ultraviolet radiation. Environmental Science and Pollution Research 20, 3195–3201.

Dodd, M.C., 2012. Potential impacts of disinfection processes on elimination and deactivation of antibiotic resistance genes during water and wastewater treatment. Journal of Environmental Monitoring 14, 1754–1771.

Dodd, M.C., Buffle, M.-O., Von Gunten, U., 2006. Oxidation of antibacterial molecules by aqueous ozone: moiety-specific reaction kinetics and application to ozone-based wastewater treatment. Environmental Science & Technology 40, 1969–1977.

Dodd, M.C., Kohler, H.P.E., Von Gunten, U., 2009. Oxidation of antibacterial compounds by ozone and hydroxyl radical: elimination of biological activity during aqueous ozonation processes. Environmental Science & Technology 43, 2498–2504.

Dolliver, H., Gupta, S., Noll, S., 2008. Antibiotic degradation during manure composting. Journal of Environmental Quality 37, 1245–1253.

Dolliver, H., Gupta, S., 2008. Antibiotic losses in leaching and surface runoff from manure-amended agricultural land. Journal of Environmental Quality 37, 1227–1237.

Dordio, A., Carvalho, A.P., Teixeira, D.M., Dias, C.B., Pinto, A.P., 2010. Removal of pharmaceuticals in microcosm constructed wetlands using *Typha* spp. and LECA. Bioresource Technology 101, 886–892.

Destiani, R. and Templeton, M.R., 2019. Chlorination and ultraviolet disinfection of antibiotic-resistant bacteria and antibiotic resistance genes in drinking water. AIMS Environmental Science, 6, 222–241.

Dreassi, E., Corti, P., Bezzini, F., Furlanetto, S., 2000. High-performance liquid chromatographic assay of erythromycin from biological matrix using electrochemical or ultraviolet detection. Analyst 125, 1077–1081.

Drillia, P., Dokianakis, S., Fountoulakis, M., Kornaros, M., Stamatelatou, K., Lyberatos, G., 2005. On the occasional biodegradation of pharmaceuticals in the activated sludge process: the example of the antibiotic sulfamethoxazole. Journal of Hazardous Materials 122, 259–265.

Duarte, D.J., Oldenkamp, R., Ragas, A.M., 2019. Modelling environmental antibiotic-resistance gene abundance: a meta-analysis. The Science of the Total Environment 659, 335–341.

Dvořák, L., Gómez, M., Dolina, J., Černín, A., 2016. Anaerobic membrane bioreactors—a mini review with emphasis on industrial wastewater treatment: applications, limitations and perspectives. Desalination and Water Treatment 57, 19062–19076.

Elmolla, E.S., Chaudhuri, M., 2011. The feasibility of using combined TiO_2 photocatalysis-SBR process for antibiotic wastewater treatment. Desalination 272, 218–224.

Fajardo, A., Martínez, J.L., 2008. Antibiotics as signals that trigger specific bacterial responses. Current Opinion in Microbiology 11, 161–167.

Fick, J., Söderström, H., Lindberg, R.H., Phan, C., Tysklind, M., Larsson, D., 2009. Contamination of surface, ground, and drinking water from pharmaceutical production. Environmental Toxicology and Chemistry 28, 2522–2527.

Göbel, A., McArdell, C.S., Joss, A., Siegrist, H., Giger, W., 2007. Fate of sulfonamides, macrolides, and trimethoprim in different wastewater treatment technologies. The Science of the Total Environment 372, 361–371.

Gottlieb, D., 1967. Antibiotics and cell metabolism. Hindustan Antibiotics Bulletin 10, 123.

Gou, M., Hu, H.-W., Zhang, Y.-J., Wang, J.-T., Hayden, H., Tang, Y.-Q., He, J.-Z., 2018. Aerobic composting reduces antibiotic resistance genes in cattle manure and the resistome dissemination in agricultural soils. The Science of the Total Environment 612, 1300–1310.

Guo, M.T., Yuan, Q.B., Yang, J., 2015. Distinguishing effects of ultraviolet exposure and chlorination on the horizontal transfer of antibiotic resistance genes in municipal wastewater. Environmental Science & Technology 49, 5771–5778.

Hollender, J., Zimmermann, S.G., Koepke, S., Krauss, M., McArdell, C.S., Ort, C., Singer, H., von Gunten, U., Siegrist, H., 2009. Elimination of organic micropollutants in a municipal wastewater treatment plant upgraded with a full-scale post-ozonation followed by sand filtration. Environmental Science & Technology 43, 7862–7869.

Homem, V., Santos, L., 2011. Degradation and removal methods of antibiotics from aqueous matrices–a review. Journal of Environmental Management 92, 2304–2347.

Hu, Y., Yan, X., Shen, Y., Di, M., Wang, J., 2018. Antibiotics in surface water and sediments from Hanjiang River, Central China: occurrence, behavior and risk assessment. Ecotoxicology and Environmental Safety 157, 150–158.

Huang, C.-H., Renew, J.E., Smeby, K.L., Pinkston, K., Sedlak, D.L., 2011. Assessment of potential antibiotic contaminants in water and preliminary occurrence analysis. Journal of Contemporary Water Research and Education 120, 4.

Huber, M.M., GÖbel, A., Joss, A., Hermann, N., LÖffler, D., McArdell, C.S., Ried, A., Siegrist, H., Ternes, T.A., von Gunten, U., 2005. Oxidation of pharmaceuticals during ozonation of municipal wastewater effluents: a pilot study. Environmental Science & Technology 39, 4290–4299.

Hughes, D., Andersson, D.I., 2017. Environmental and genetic modulation of the phenotypic expression of antibiotic resistance. FEMS Microbiology Reviews 41, 374–391.

Huízar-Félix, A.M., Aguilar-Flores, C., Martínez-de-la Cruz, A., Barandiarán, J.M., Sepúlveda-Guzmán, S., Cruz-Silva, R., 2019. Removal of tetracycline pollutants by adsorption and magnetic separation using reduced graphene oxide decorated with α-Fe_2O_3 nanoparticles. Nanomaterials 9, 313.

Hwangbo, M., Claycomb, E.C., Liu, Y., Alivio, T.E., Banerjee, S., Chu, K.-H., 2019. Effectiveness of zinc oxide-assisted photocatalysis for concerned constituents in reclaimed wastewater: 1,4-Dioxane, trihalomethanes, antibiotics, antibiotic resistant bacteria (ARB), and antibiotic resistance genes (ARGs). The Science of the Total Environment 649, 1189–1197.

Ince, N.H., Apikyan, I.G., 2000. Combination of activated carbon adsorption with light-enhanced chemical oxidation via hydrogen peroxide. Water Research 34, 4169–4176.

Jiang, L., Liu, Y., Liu, S., Zeng, G., Hu, X., Hu, X., Guo, Z., Tan, X., Wang, L., Wu, Z., 2017. Adsorption of estrogen contaminants by graphene nanomaterials under natural organic matter preloading: comparison to carbon nanotube, biochar, and activated carbon. Environmental Science & Technology 51, 6352–6359.

Jjemba, P.K., 2006. Excretion and ecotoxicity of pharmaceutical and personal care products in the environment. Ecotoxicology and Environmental Safety 63, 113–130.

Karaolia, P., Michael, I., García-Fernández, I., Agüera, A., Malato, S., Fernández-Ibáñez, P., Fatta-Kassinos, D., 2014. Reduction of clarithromycin and sulfamethoxazole-resistant Enterococcus by pilot-scale solar-driven Fenton oxidation. The Science of the Total Environment 468, 19–27.

Keen, O.S., Linden, K.G., 2013. Degradation of antibiotic activity during UV/H_2O_2 advanced oxidation and photolysis in wastewater effluent. Environmental Science & Technology 47, 13020–13030.

Kim, S., Ghafoor, K., Lee, J., Feng, M., Hong, J., Lee, D.-U., Park, J., 2013. Bacterial inactivation in water, DNA strand breaking, and membrane damage induced by ultraviolet-assisted titanium dioxide photocatalysis. Water Research 47, 4403–4411.

Klein, E.Y., Van Boeckel, T.P., Martinez, E.M., Pant, S., Gandra, S., Levin, S.A., Goossens, H., Laxminarayan, R., 2018. Global increase and geographic convergence in antibiotic consumption between 2000 and 2015, 201717295. In: Proceedings of the National Academy of Sciences.

Köberl, M., Schmidt, R., Ramadan, E.M., Bauer, R., Berg, G., 2013. The microbiome of medicinal plants: diversity and importance for plant growth, quality and health. Frontiers in Microbiology 4, 400.

Koyuncu, I., Arikan, O.A., Wiesner, M.R., Rice, C., 2008. Removal of hormones and antibiotics by nanofiltration membranes. Journal of Membrane Science 309, 94–101.

Kurt, A., Mert, B.K., Özengin, N., Sivrioğlu, Ö., Yonar, T., 2017. Treatment of antibiotics in wastewater using advanced oxidation processes (AOPs). In: Physico-Chemical Wastewater Treatment and Resource Recovery. IntechOpen.

Lange, F., Cornelissen, S., Kubac, D., Sein, M.M., Von Sonntag, J., Hannich, C.B., Golloch, A., Heipieper, H.J., Möder, M., Von Sonntag, C., 2006. Degradation of macrolide antibiotics by ozone: a mechanistic case study with clarithromycin. Chemosphere 65, 17–23.

Lastre-Acosta, A.M., Barberato, B., Parizi, M.P.S., Teixeira, A.C.S., 2019. Direct and indirect photolysis of the antibiotic enoxacin: kinetics of oxidation by reactive photo-induced species and simulations. Environmental Science and Pollution Research 26, 4337–4347.

Li, B., Zhang, T., 2011. Mass flows and removal of antibiotics in two municipal wastewater treatment plants. Chemosphere 83, 1284–1289.

Li, M.-f., Liu, Y.-g., Zeng, G.-m., Liu, N., Liu, S.-b., 2019. Graphene and graphene-based nanocomposites used for antibiotics removal in water treatment: a review. Chemosphere.

Li, S., Shi, W., Li, H., Xu, N., Zhang, R., Chen, X., Sun, W., Wen, D., He, S., Pan, J., 2018. Antibiotics in water and sediments of rivers and coastal area of Zhuhai City, Pearl River estuary, south China. The Science of the Total Environment 636, 1009–1019.

Li, S.-z., Li, X.-y., Wang, D.-z., 2004. Membrane (RO-UF) filtration for antibiotic wastewater treatment and recovery of antibiotics. Separation and Purification Technology 34, 109–114.

Li, X., Li, G., 2015. A review: pharmaceutical wastewater treatment technology and research in China. In: 2015 Asia-Pacific Energy Equipment Engineering Research Conference. Atlantis Press.

Lillenberg, M., Yurchenko, S., Kipper, K., Herodes, K., Pihl, V., Lõhmus, R., Ivask, M., Kuu, A., Kutti, S., Litvin, S., 2010. Presence of fluoroquinolones and sulfonamides in urban sewage sludge and their degradation as a result of composting. International journal of Environmental Science and Technology 7, 307–312.

Linares, J.F., Gustafsson, I., Baquero, F., Martinez, J., 2006. Antibiotics as intermicrobial signaling agents instead of weapons. Proceedings of the National Academy of Sciences 103, 19484–19489.

Lindberg, R.H., Björklund, K., Rendahl, P., Johansson, M.I., Tysklind, M., Andersson, B.A., 2007. Environmental risk assessment of antibiotics in the Swedish environment with emphasis on sewage treatment plants. Water Research 41, 613–619.

Litter, M.I., 2005. Introduction to photochemical advanced oxidation processes for water treatment. In: Environmental Photochemistry Part II. Springer, pp. 325–366.

Liu, H.-L., Yang, T.C.-K., 2003. Photocatalytic inactivation of *Escherichia coli* and Lactobacillus helveticus by ZnO and TiO_2 activated with ultraviolet light. Process Biochemistry 39, 475–481.

Liu, M.-k., Liu, Y.-y., Bao, D.-d., Zhu, G., Yang, G.-H., Geng, J.-F., Li, H.-T., 2017. Effective removal of tetracycline antibiotics from water using hybrid carbon membranes. Scientific Reports 7, 43717.

Liu, Y., Tay, J.-H., 2004. State of the art of biogranulation technology for wastewater treatment. Biotechnology Advances 22, 533–563.

Llor, C., Bjerrum, L., 2014. Antimicrobial resistance: risk associated with antibiotic overuse and initiatives to reduce the problem. Therapeutic advances in drug safety 5, 229–241.

Lombardo-Agüí, M., Cruces-Blanco, C., García-Campaña, A.M., Gámiz-Gracia, L., 2014. Multiresidue analysis of quinolones in water by ultra-high performance liquid chromatography with tandem mass spectrometry using a simple and effective sample treatment. Journal of Separation Science 37, 2145–2152.

Luo, Y., Mao, D., Rysz, M., Zhou, Q., Zhang, H., Xu, L., Alvarez, P.J.J., 2010. Trends in antibiotic resistance genes occurrence in the Haihe River, China. Environmental Science & Technology 44, 7220–7225.

Mandal, A., Ojha, K., De Asim, K., Bhattacharjee, S., 2004. Removal of catechol from aqueous solution by advanced photo-oxidation process. Chemical Engineering Journal 102, 203–208.

Martínez, J.L., 2008. Antibiotics and antibiotic resistance genes in natural environments. Science 321, 365–367.

Martinez, J.L., 2009. The role of natural environments in the evolution of resistance traits in pathogenic bacteria. In: Proceedings of the Royal Society B: Biological Sciences, vol. 276, pp. 2521–2530.

Martínez-Carballo, E., González-Barreiro, C., Scharf, S., Gans, O., 2007. Environmental monitoring study of selected veterinary antibiotics in animal manure and soils in Austria. Environmental Pollution 148, 570–579.

Massé, D., Saady, N., Gilbert, Y., 2014. Potential of biological processes to eliminate antibiotics in livestock manure: an overview. Animals 4, 146–163.

McEwen, S.A., Fedorka-Cray, P.J., 2002. Antimicrobial use and resistance in animals. Clinical Infectious Diseases 34, S93–S106.

Michael, I., Rizzo, L., McArdell, C., Manaia, C., Merlin, C., Schwartz, T., Dagot, C., Fatta-Kassinos, D., 2013. Urban wastewater treatment plants as hotspots for the release of antibiotics in the environment: a review. Water Research 47, 957–995.

Munter, R., 2001. Advanced oxidation processes – current status and prospective. In: Proceedings of the Estonion Academy of Scienses. Chemistry, vol. 50, pp. 59–80.

Naddeo, V., Meriç, S., Kassinos, D., Belgiorno, V., Guida, M., 2009. Fate of pharmaceuticals in contaminated urban wastewater effluent under ultrasonic irradiation. Water Research 43, 4019–4027.

Nghiem, L.D., Schäfer, A.I., Elimelech, M., 2005. Pharmaceutical retention mechanisms by nanofiltration membranes. Environmental Science & Technology 39, 7698–7705.

Niemi, R.M., Heiskanen, I., Heine, R., Rapala, J., 2009. Previously uncultured β-Proteobacteria dominate in biologically active granular activated carbon (BAC) filters. Water Research 43, 5075–5086.

Nõlvak, H., Truu, M., Tiirik, K., Oopkaup, K., Sildvee, T., Kaasik, A., Mander, Ü., Truu, J., 2013. Dynamics of antibiotic resistance genes and their relationships with system treatment efficiency in a horizontal subsurface flow constructed wetland. The Science of the Total Environment 461, 636–644.

O'Neill, J., 2017. Securing New Drugs for Future Generations: The Pipeline of Antibiotics. 2015.

Oliveira, T.S., Murphy, M., Mendola, N., Wong, V., Carlson, D., Waring, L., 2015. Characterization of Pharmaceuticals and Personal Care products in hospital effluent and wastewater influent/effluent by direct-injection LC-MS-MS. The Science of the Total Environment 518, 459–478.

Organization, W.H., 2017. Antibacterial Agents in Clinical Development: An Analysis of the Antibacterial Clinical Development Pipeline, Including Tuberculosis. World Health Organization.

Pan, M., Chu, L., 2017. Fate of antibiotics in soil and their uptake by edible crops. Science of The Total Environment 599, 500–512.

Park, N., Vanderford, B.J., Snyder, S.A., Sarp, S., Kim, S.D., Cho, J., 2009. Effective controls of micropollutants included in wastewater effluent using constructed wetlands under anoxic condition. Ecological Engineering 35, 418–423.

Peng, B., Chen, L., Que, C., Yang, K., Deng, F., Deng, X., Shi, G., Xu, G., Wu, M., 2016. Adsorption of antibiotics on graphene and biochar in aqueous solutions induced by π-π interactions. Scientific Reports 6, 31920.

Pliego, G., Zazo, J.A., Garcia-Muñoz, P., Munoz, M., Casas, J.A., Rodriguez, J.J., 2015. Trends in the intensification of the Fenton process for wastewater treatment: an overview. Critical Reviews in Environmental Science and Technology 45, 2611–2692.

Pratt, K., Moran, D., 2010. Evaluating the cost-effectiveness of global biochar mitigation potential. Biomass and Bioenergy 34, 1149–1158.

Pruden, A., Larsson, D.J., Amézquita, A., Collignon, P., Brandt, K.K., Graham, D.W., Lazorchak, J.M., Suzuki, S., Silley, P., Snape, J.R., 2013. Management options for reducing the release of antibiotics and antibiotic resistance genes to the environment. Environmental Health Perspectives 121, 878–885.

Qadeer, S., Batool, A., Rashid, A., Khalid, A., Samad, N., Ghufran, M.A., 2014. Effectiveness of biochar in soil conditioning under simulated ecological conditions. Soil & Environment 33.

Radjenović, J., Petrović, M., Barceló, D., 2009. Fate and distribution of pharmaceuticals in wastewater and sewage sludge of the conventional activated sludge (CAS) and advanced membrane bioreactor (MBR) treatment. Water Research 43, 831–841.

Riaz, L., Mahmood, T., Coyne, M.S., Khalid, A., Rashid, A., Hayat, M.T., Gulzar, A., Amjad, M., 2017a. Physiological and antioxidant response of wheat (*Triticum aestivum*) seedlings to fluoroquinolone antibiotics. Chemosphere 177, 250–257.

Riaz, L., Mahmood, T., Kamal, A., Shafqat, M., Rashid, A., 2017b. Industrial release of Fluoroquinolones (FQs) in the wastewater bodies with their associated ecological risk in Pakistan. Environmental Toxicology and Pharmacology.

Riaz, L., Mahmood, T., Khalid, A., Rashid, A., Siddique, M.B.A., Kamal, A., Coyne, M.S., 2018. Fluoroquinolones (FQs) in the environment: a review on their abundance, sorption and toxicity in soil. Chemosphere 191, 704–720.

Runnalls, T.J., Margiotta-Casaluci, L., Kugathas, S., Sumpter, J.P., 2010. Pharmaceuticals in the aquatic environment: steroids and anti-steroids as high priorities for research. Human and Ecological Risk Assessment 16, 1318–1338.

Sachdeva, S., Palur, R.V., Sudhakar, K.U., Rathinavelan, T., 2017. *E. Coli* group 1 capsular polysaccharide exportation nanomachinary as a plausible antivirulence target in the perspective of emerging antimicrobial resistance. Frontiers in Microbiology 8, 70.

Scholz, M., Lee, B.H., 2005. Constructed wetlands: a review. International Journal of Environmental Studies 62 (4), 421–447.

Sirelkhatim, A., Mahmud, S., Seeni, A., Kaus, N.H.M., Ann, L.C., Bakhori, S.K.M., Hasan, H., Mohamad, D., 2015. Review on zinc oxide nanoparticles: antibacterial activity and toxicity mechanism. Nano-Micro Letters 7, 219–242.

Spellberg, B., Srinivasan, A., Chambers, H.F., 2016. New societal approaches to empowering antibiotic stewardship. JAMA 315, 1229–1230.

Sponza, D.T., Demirden, P., 2007. Treatability of sulfamerazine in sequential upflow anaerobic sludge blanket reactor (UASB)/completely stirred tank reactor (CSTR) processes. Separation and Purification Technology 56, 108–117.

Stoob, K., Singer, H.P., Mueller, S.R., Schwarzenbach, R.P., Stamm, C.H., 2007. Dissipation and transport of veterinary sulfonamide antibiotics after manure application to grassland in a small catchment. Environmental Science & Technology 41, 7349–7355.

Tan, X., Liu, Y., Zeng, G., Wang, X., Hu, X., Gu, Y., Yang, Z., 2015. Application of biochar for the removal of pollutants from aqueous solutions. Chemosphere 125, 70–85.

Thiele-Bruhn, S., 2003. Pharmaceutical antibiotic compounds in soils—a review. Journal of Plant Nutrition and Soil Science 166, 145–167.

Thomashow, L.S., 2002. Antibiotic production by soil and rhizosphere microbes in situ. In: Manual of Environmental Microbiology.

Tsui, W.H., Yim, G., Wang, H.H., McClure, J.E., Surette, M.G., Davies, J., 2004. Dual effects of MLS antibiotics: transcriptional modulation and interactions on the ribosome. Chemistry & Biology 11, 1307–1316.

Van Doorslaer, X., Dewulf, J., Van Langenhove, H., Demeestere, K., 2014. Fluoroquinolone antibiotics: an emerging class of environmental micropollutants. The Science of the Total Environment 500, 250–269.

Van Epps, A., Blaney, L., 2016. Antibiotic residues in animal waste: occurrence and degradation in conventional agricultural waste management practices. Current Pollution Reports 2, 135–155.

Verlicchi, P., Al Aukidy, M., Galletti, A., Petrovic, M., Barceló, D., 2012. Hospital effluent: investigation of the concentrations and distribution of pharmaceuticals and environmental risk assessment. The Science of the Total Environment 430, 109–118.

Walsh, C., 2003. Opinion—anti-infectives: where will new antibiotics come from? Nature Reviews Microbiology 1, 65.

Wang, J., Ben, W., Zhang, Y., Yang, M., Qiang, Z., 2015. Effects of thermophilic composting on oxytetracycline, sulfamethazine, and their corresponding resistance genes in swine manure. Environmental Science Processes and Impacts 17, 1654.

Wise, R., 2002. Antimicrobial resistance: priorities for action. Journal of Antimicrobial Chemotherapy 49, 585–586.

Wright, G.D., 2010. Antibiotic resistance in the environment: a link to the clinic? Current Opinion in Microbiology 13, 589–594.

Xie, W.-Y., Yang, X.-P., Li, Q., Wu, L.-H., Shen, Q.-R., Zhao, F.-J., 2016. Changes in antibiotic concentrations and antibiotic resistome during commercial composting of animal manures. Environmental Pollution 219, 182–190.

Yim, G., Huimi Wang, H., Davies Frs, J., 2007. Antibiotics as signalling molecules. Philosophical Transactions of the Royal Society B: Biological Sciences 362, 1195–1200.

Yu, J.C., Kwong, T.Y., Luo, Q., Cai, Z., 2006. Photocatalytic oxidation of triclosan. Chemosphere 65, 390–399.

Zeng, Z.w., Tian, S.r., Liu, Y.g., Tan, X.f., Zeng, G.m., Jiang, L.h., Yin, Z.h., Liu, N., Liu, S.b., Li, J., 2018. Comparative study of rice husk biochars for aqueous antibiotics removal. Journal of Chemical Technology and Biotechnology 93, 1075–1084.

Zhu, Y., Wang, Y., Zhou, S., Jiang, X., Ma, X., Liu, C., 2018. Robust performance of a membrane bioreactor for removing antibiotic resistance genes exposed to antibiotics: role of membrane foulants. Water Research 130, 139–150.

Index

'*Note*: Page numbers followed by "f" indicate figures and "t" indicates tables.'

B

C

F

G

H

I

K

L

M

N

O

P

Q

R

S

Printed in the United States
By Bookmasters